Endogenous and Exogenous Regulation and Control of Physiological Systems

Biomedical Engineering Series

Edited by Michael R. Neuman

Published Titles

Electromagnetic Analysis and Design in Magnetic Resonance Imaging, Jianming Jin

Endogenous and Exogenous Regulation and Control of Physiological Systems, Robert B. Northrop

Artificial Neural Networks in Cancer Diagnosis, Prognosis, and Treatment, Raouf N.G. Naguib and Gajanan V. Sherbet

Medical Image Registration, Joseph V. Hajnal, Derek Hill, and David J. Hawkes

Introduction to Dynamic Modeling of Neuro-Sensory Systems, Robert B. Northrop

Noninvasive Instrumentation and Measurement in Medical Diagnosis, Robert B. Northrop

Forthcoming Titles

Handbook of Neuroprosthetic Methods, Warren E. Finn and Peter G. LoPresti

The **BIOMEDICAL ENGINEERING** Series
Series Editor: Michael Neuman

Endogenous and Exogenous Regulation and Control of Physiological Systems

Robert B. Northrop

CHAPMAN & HALL/CRC

A CRC Press Company
Boca Raton   London   New York   Washington, D.C.

QH
508
.N67
2000

Library of Congress Cataloging-in-Publication Data

Northrop, Robert B.
　　Endogenous and exogenous regulation and control of physiological systems / Robert B. Northrop.
　　　　p. cm.
　　Includes bibliographical references and index.
　　ISBN 0-8493-9694-8 (alk. paper)
　　　1. Biological control systems. 2. Pharmacokinetics. 3. Biomedical engineering. 4. Homeostasis. I. Title.
　　QH508.N67   1999
　　571.7—dc21   99-40282
　　　CIP

This book contains information obtained from authentic and highly regarded sources. Reprinted material is quoted with permission, and sources are indicated. A wide variety of references are listed. Reasonable efforts have been made to publish reliable data and information, but the author and the publisher cannot assume responsibility for the validity of all materials or for the consequences of their use.

Neither this book nor any part may be reproduced or transmitted in any form or by any means, electronic or mechanical, including photocopying, microfilming, and recording, or by any information storage or retrieval system, without prior permission in writing from the publisher.

The consent of CRC Press LLC does not extend to copying for general distribution, for promotion, for creating new works, or for resale. Specific permission must be obtained in writing from CRC Press LLC for such copying.

Direct all inquiries to CRC Press LLC, 2000 N.W. Corporate Blvd., Boca Raton, Florida 33431.

Trademark Notice: Product or corporate names may be trademarks or registered trademarks, and are used only for identification and explanation, without intent to infringe.

Visit the CRC Press Web site at www.crcpress.com

© 2000 by Chapman & Hall/CRC

No claim to original U.S. Government works
International Standard Book Number 0-8493-9694-8
Library of Congress Card Number 99-40282
Printed in the United States of America   2 3 4 5 6 7 8 9 0
Printed on acid-free paper

B1961509

Preface

PURPOSE

This text is intended to be used in a one- or two-semester classroom course for senior and graduate biomedical engineers who are interested in learning about the endogenous and exogenous regulation and control of certain physiological systems. It also will be of interest to graduate students and workers in pharmacology interested in closed-loop drug delivery systems. It has been written based on the author's experience in teaching EE 271/300, *Physiological Control Systems,* for over 25 years in the Electrical & Systems Engineering Department at the University of Connecticut, Department of Electrical & Systems Engineering. The contents of this course have evolved with technology and our knowledge of human physiology, molecular biology, and medicine. One of the greatest factors impacting our understanding of regulatory physiology has been the revolution in our knowledge of cell biology and biochemistry brought about by the need to conquer diseases such as AIDS and cancer. Another factor influencing the design of exogenous control systems for physiological systems has been the technological development of microcomputers, digital signal-processing algorithms, and adaptive and self-tuning control strategies.

Because the field of physiological control systems is very broad, the author has chosen to focus attention on certain "wet" physiological systems (i.e., those that use endocrine hormones as parametric control substances). Also described in detail are certain physiological systems in which exogenous control has been effective clinically (e.g., plasma glucose regulation, mean arterial pressure regulation, and pain analgesia). Plant dynamics are derived and various controller designs are considered. Nonlinear control is stressed. (Not considered in this text are the respiratory, temperature, and short-term blood pressure regulators; nor are various neuromuscular systems considered, including oculomotor, eye/hand, and postural control.)

This text includes the newest technologies in its 10 chapters; however, it does not neglect the traditional approaches to nonlinear control system design. An analytical and quantitative approach is stressed in describing the basic "components" of physiological regulators and control systems (PRCs). The inherent nonlinearity in PRCs is stressed, and the classical means of analyzing and designing nonlinear control systems are presented. The relation of system failure to disease is considered.

READER BACKGROUND

Readers are assumed to be students studying biomedical engineering at the senior or graduate level. They should have taken core electrical engineering curriculum courses, including introductory courses on *signals and systems* and *systems analysis* including linear control systems. Because biomedical engineering is an interdisciplinary area, the reader is assumed also to have had introductory courses on human physiology, biochemistry, and cell biology.

Specific *engineering* skills required of the reader include mastery of the elements of introductory circuit theory. The reader should also have familiarity with both time- and frequency-domain methods of describing linear dynamic systems characterized by ordinary, linear, differential, or difference equations including Laplace, Fourier and z-transforms, state methods, transfer functions, and Bode plots. The basic concept of linear system stability and tests for it should be known. The root locus technique is very important in both design and analysis of closed-loop linear systems. The reader should also be familiar with the computer simulation of dynamic systems with applications such as Matlab® or Simnon™.

Specific skills in *physiology* should include a knowledge of how the major organ systems of the body work, a knowledge of the major hormones that regulate biochemical processes in the body, a knowledge of how the nervous system contributes to homeostasis, etc. (I have long used Guyton's *Textbook of Medical Physiology* in teaching my classroom course on physiological control systems.) In the area of cell biology and biochemistry, the reader should have an ability to describe the basic structures of cells and the biochemistry of nucleic acids, protein synthesis, cell membranes, receptors, etc.

SCOPE OF THE TEXT

The text is organized into 10 chapters. Chapter 1, "Introduction to Physiological Regulators and Control Systems," provides an overview of the properties of physiological systems. Common properties of physiological systems are given in which their nonlinearity and the concept of parametric control are described. Models for the regulation of the pressure of the aqueous humor in the eyes are used as examples of parametric control.

Chapter 2, "Physical and Chemical Factors Governing the Behavior of Physiological Regulators and Control Systems," describes the physical chemical, and biophysical laws governing the behavior of physiological systems. Diffusion, mass-action, and fluidic systems with their electrical analogs are treated.

In Chapter 3, "Introduction to Linear SISO Control Systems and Systems with Delays," classical linear controller design is reviewed, including proportional-plus derivative, proportional-plus integral, and proportional-plus-integral-plus derivative architectures. Root locus is presented as a pole placement design tool. Simple criteria for the stability of linear and nonlinear systems are given, including the Nyquist criterion, describing functions, and the Popov method.

The problems associated with controlling linear systems with delays (transport lags) are treated; the model reference Smith compensator is presented as one means of stabilizing high-gain closed-loop systems with significant delays in their loop gains.

Chapter 4, "Introduction to Compartmental Modeling and Pharmacokinetic Systems," describes mathematical models for how internalized drugs are distributed and metabolized in the body. Before an exogenous drug or hormone can be effective in modifying a physiological state, the substance must reach the target organ or tissue. Compartmental analysis is a means of describing the dynamics of drug distribution and is necessary to include in the design of any closed-loop drug administration system.

Chapter 5, "Special Types of Closed-Loop Drug Input Controllers," first considers the analysis and design of nonlinear, limit-cycling, on/off drug infusion controllers using the describing function method. Also examined is the suitability of using nonlinear decoupling controllers for drug administration. They are found unsuitable for compartmental pharmacokinetic applications because infusion rates can only be nonnegative. This chapter also introduces the integral pulse-frequency modulation (IPFM) bolus drug injector and discusses its suitability in certain types of exogenous physiological control applications.

Finally, Chapter 5 uses classical phase-plane analysis to treat nonlinear parameter-switching controllers in which plant output and system error (and their derivatives) determine controller gains to achieve robust control.

In Chapter 6, "Hormonal Regulation of Sodium, Potassium, Calcium, and Magnesium Ions," the natural endogenous regulation of those ions is reviewed. The kidneys are described in detail as key elements in the parametric (hormonal) regulation of Na^+, K^+, Ca^{++}, and Mg^{++} ions in the plasma.

In Chapter 7, "Regulation of Blood Glucose," first the normal endogenous regulation of plasma glucose concentration is examined. The roles of insulin-sensitive and noninsulin-sensitive cells, the liver, the kidneys, and the hormones insulin and glucagon are described. Several mathematical models are presented to describe the nonlinear dynamics of normal glucoregulation. Type 1 diabetes mellitus is described, and the design of various, type 0, "artificial beta cells" is reviewed. Type 1 controllers for exogenous glucoregulation are also presented.

In Chapter 8, "Control of Mean Arterial Pressure by Sodium Nitroprusside Injection," the pharmacology and pharmacokinetics of the vasodilator sodium nitroprusside (SNP) are described. Various controllers which have been used successfully for SNP infusion are presented, including variable-parameter nonlinear designs, optimal controllers, self-tuning model-reference controllers, controllers using Smith compensation, and the author's controller design which uses a Smith delay compensator coupled to a proportional-plus-integral element that drives an IPFM SNP bolus injector. Problems associated with the on-line identification of the mean arterial pressure/SNP plant are discussed.

Chapter 9, "Control of Postoperative Pain by Self-Administered Opioids," describes the design of a discrete intervention controller that regulates the injection of the opioid analgesic fentanyl in response to patient demand (button pushing). The pharmacology and pharmacokinetics of fentanyl are described, and dynamic models for patient response to pain are introduced. The results of Reasbeck and Jacobs' stochastic controller for postoperative pain on patients are described, as is the IPFM pain controller of Liu and Northrop. Results of simulation studies are also described.

Chapter 10, "The Human Immune System Seen from a Biomedical Engineering Viewpoint," tackles the very complex problem of making nonlinear dynamic models to describe the operation of the human immune system. This chapter is unique in that, to the author's knowledge, no text on physiological control systems or modeling has included this challenging topic. The goal of creating complex immune system models is to be able to investigate novel, untried immunotherapies for cancers and treatments for certain autoimmune diseases and AIDS. The chapter begins with a description of the components of the immune system, including macrophages, T-cells, cytotoxic T-cells, natural killer cells, mast cells, B-cells, antibodies, complement, and the myriad immunocytokines which effect immune system growth and performance. Simulations are given to illustrate model validations and the putative control of cancer and HIV proliferation.

SOFTWARE

Example simulations given in the text and certain problems at the end of the chapters are given as programs using the Simnon simulation language for nonlinear systems. Simnon is recommended by the author because the model simulated is entered directly as a set of nonlinear, first-order state equations. Such equations arise directly in physiological systems modeling. Other simulation languages can be used, but they are generally more costly and not as well suited to handle first-order, nonlinear ordinary differential equations. (See the appendix for a discussion of simulation languages for physiological systems.)

BIBLIOGRAPHY AND REFERENCES

The references cited encompass a wide time span, from the 1950s to the present. There are many entries of review articles and specialized texts that will help the reader to pursue detailed interests, in-depth. There are also citations of certain World Wide Web sites that have provided a rich, contemporary source of information on rapidly changing topics in physiological control systems, such as leptin, AIDS, and the immune system. Web references are often ephemeral, which limits their long-term usefulness unless they are printed out.

ACKNOWLEDGMENTS

I thank Dr. Joseph Palladino of Trinity College, Hartford, CT, for his helpful comments on the manuscript and my editor, Bob Stern, at CRC Press for his guidance in the development of my manuscript. I am greatly indebted to my wife, Adelaide, for her moral support, criticisms, encouragement, and her patience.

DEDICATION

I dedicate this work to my students.

Robert B. Northrop

The Author

Robert B. Northrop majored in electrical engineering at MIT, graduating with a bachelor's degree in 1956. At the University of Connecticut Graduate School, he received a master's degree in control engineering in 1958. As the result of long-standing interest in physiology, he entered a Ph.D. program at UCONN in physiology, doing research on the neuromuscular physiology of catch muscles. He received his Ph.D. in 1964.

In 1963, he rejoined the UCONN Electrical Engineering Department as a lecturer and was hired as an Assistant Professor in 1964. In collaboration with Dr. Edward G. Boettiger, he secured a 5-year training grant in 1965 from NIGMS (NIH), and started one of the first interdisciplinary Biomedical Engineering graduate training programs in New England

Throughout his career, Dr. Northrop's areas of research have been broad and interdisciplinary and have been centered around biomedical engineering. He has done sponsored research on the neurophysiology of insect vision and theoretical models for visual neural signal processing. He also did sponsored research on electrofishing and developed, in collaboration with Northeast Utilities, effective, working systems for fish guidance and control in hydroelectric plant waterways using underwater electric fields.

Still another area of sponsored research has been in the design and simulation of nonlinear, adaptive, digital controllers to regulate *in vivo* drug concentrations or physiological parameters, such as pain, blood pressure, or blood glucose in diabetics. An outgrowth of this research led to his development of mathematical models for the dynamics of the human immune system that have been used to investigate theoretical therapies for autoimmune diseases, cancer, and HIV infection.

Biomedical instrumentation also has been an active research area; a grant supported studies on the use of the ocular pulse to detect obstructions in the carotid arteries. Minute pulsations of the cornea from arterial circulation in the eyeball were sensed using a no-touch ultrasound technique. Ocular pulse waveforms were shown to be related to cerebral blood flow in rabbits and humans.

Most recently, he has been addressing the problem of noninvasive blood glucose measurement for diabetics. Starting with a Phase I SBIR grant, Dr. Northrop has been developing a means of estimating blood glucose by reflecting a beam of polarized light off the front surface of the lens of the eye, and measuring the very small optical rotation resulting from glucose in the aqueous humor, which in turn is proportional to blood glucose. As an offshoot of techniques developed in micropolarimetry, he developed a sample chamber for glucose measurement in biotechnology applications.

Another approach being developed will use percutaneous, long-wave IR light in a nondispersive spectrometer to noninvasively measure blood glucose.

Dr. Northrop has been on the Electrical and Systems Engineering faculty at UCONN until his retirement in June, 1997. Throughout this time, he was program director of the Biomedical Engineering Graduate Program. As Emeritus Professor, he now teaches graduate courses in Biomedical Engineering, writes texts, tinkers, sails, and travels. His current project is a text on modeling neurosensory systems. He lives in Chaplin, CT with his wife and two cats.

Table of Contents

Chapter 1
Introduction to Physiological Regulators and Control Systems 1
   1.0   Introduction 1
   1.1   Physiology and Physiological Regulation Defined 2
   1.2   The Purposes of Modeling 3
   1.3   Some General Properties of PRCs 4
   1.4   Some General Properties of Linear, Time-Varying, and
         Nonlinear Systems 8
         1.4.1   Introduction 8
         1.4.2   Properties of Time-Invariant Linear Systems 8
         1.4.3   Time-Variable Linear Systems 10
         1.4.4   Properties of Nonlinear Systems 11
   1.5   Parametric Control 15
   1.6   Chapter Summary 25
   Problems 27

Chapter 2
Physical and Chemical Factors Governing the Behavior of Physiological
Regulators and Control Systems 31
   2.0   Introduction 31
   2.1   Behavior of Fluidic Systems Under Laminar Flow Conditions 31
         2.1.1   Fluid Resistance 32
         2.1.2   Fluid Capacitance 32
         2.1.3   Fluid Inertance 34
         2.1.4   Transfer Functions of Fluidic Systems 35
   2.2   Diffusion Dynamics and Fick's First Law 38
   2.3   Mass-Action Kinetics in Chemical Systems 41
         2.3.1   Examples of Mass-Action Kinetics 41
   2.4   Chapter Summary 45
   Problems 46

Chapter 3
Introduction to SISO Control Systems and Systems with Delays 55
   3.0   Introduction 55
   3.1   Conventional Controllers for Linear Plants with Real Poles 56
   3.2   The Use of Root Locus in Linear Control System Design 60
   3.3   Some Tests for Stability in Linear and Certain Classes of
         Nonlinear Control Systems 73
         3.3.1   Introduction 73

	3.3.2	The Routh–Hurwitz Stability Test 74

      3.3.2 The Routh–Hurwitz Stability Test 74
      3.3.3 The Nyquist Stability Criterion 77
      3.3.4 Describing Functions and Stability 90
      3.3.5 Popov's Stability Criterion 102
  3.4 Linear Closed-Loop Systems with Delay: Stability Problems 107
  3.5 Chapter Summary 121
  Problems 121

Chapter 4
Introduction to Compartmental Modeling and Pharmacokinetic Systems 127
  4.0 Introduction 127
  4.1 Basic Compartmental Analysis.. 128
  4.2 The Use of Washout Curves and Bolus Inputs to Identify the
        Parameters of Simple CPK Systems 140
  4.3 Overview of Drug Input Controllers 145
  4.4 Chapter Summary 146
  Problems 146

Chapter 5
Special Types of Closed-Loop Drug Input Controllers 153
  5.0 Introduction 153
  5.1 Design of On/Off Limit-Cycling Controllers for CPK/P Systems 154
  5.2 Nonlinear Decoupling Controllers for Two-Input/Two-Output
        CPK Systems.. 162
  5.3 Integral Pulse-Frequency Modulation (Bolus) Controllers 170
  5.4 Introduction to Nonlinear Parameter-Switching Controllers: Analysis in
        the Phase Plane 184
  5.5 Chapter Summary 209
  Problems 210

Chapter 6
Hormonal Regulation of Sodium, Potassium, Calcium, and Magnesium Ions ... 217
  6.0 Introduction 217
  6.1 Implicit Summing Points in Hormonal Regulation................................ 219
  6.2 Introduction to the Physiology of the Kidneys 222
  6.3 The Regulation of Plasma Sodium Ion Concentration and
        Osmolarity 227
  6.4 Regulation of Plasma Potassium Ions by the Aldosterone System 233
  6.5 Calcium Ion Regulation 237
  6.6 Magnesium Ion Regulation 244
  6.7 Chapter Summary 245
  Problems 245

Chapter 7
Regulation of Blood Glucose 247
- 7.0 Introduction 247
- 7.1 Molecules Important in Normoglycemic Regulation 248
- 7.2 The Normal Blood Glucose Regulation System 249
- 7.3 A Model for Normal Blood Glucose Regulation 254
- 7.4 The Cobelli and Mari Model for Glucoregulation 266
- 7.5 Diabetes Mellitus 272
- 7.6 Exogenous Glucoregulation: The Artificial Beta Cell 273
  - 7.6.1 Introduction 273
  - 7.6.2 Early Type 0 Artificial Beta Cells 275
  - 7.6.3 Type 1 Controllers for Artificial Beta Cells 282
- 7.7 Chapter Summary 289
- Problems 290

Chapter 8
Control of Mean Arterial Pressure by Sodium Nitroprusside Injection 295
- 8.0 Introduction 295
- 8.1 The Pharmacology and Pharmacokinetics of SNP 297
- 8.2 MAP/SNP Controllers Using Proportional-Plus-Integral Processing and Smith Delay Compensation 300
- 8.3 Adaptive and Self-Tuning Controllers for MAP 304
- 8.4 Chapter Summary 319
- Problems 321

Chapter 9
Control of Postoperative Pain by Self-Administered Opioids 325
- 9.0 Introduction 325
- 9.1 The Neurophysiological and Pharmacological Basis for Pain 326
  - 9.1.1 Sources of Pain 326
  - 9.1.2 The Neuroanatomy of Pain 327
  - 9.1.3 The Pharmacology of Pain 330
- 9.2 The PCA System of Reasbeck and Jacobs 333
  - 9.2.1 The Pharmacology of IV Fentanyl 333
  - 9.2.2 The Button-Pressing Model of Reasbeck and Jacobs 336
  - 9.2.3 The Drug Injection Controller 338
- 9.3 The PCA Model of Liu and Northrop 341
  - 9.3.1 The Patient Pain Model of Liu and Northrop 341
  - 9.3.2 Model Validation by Simulation 344
  - 9.3.3 IPFM/Smith Delay Compensator Controller Design 344
  - 9.3.4 Results of Simulations 346
- 9.4 Chapter Summary 347
- Problems 349

Chapter 10
The Human Immune System Seen from a Biomedical Engineering
 Viewpoint 353
 10.0 Introduction 353
 10.1 Immune System Cells 355
  10.1.1 Macrophages 356
  10.1.2 T-Lymphocytes: Helper T-Cells, Cytotoxic T-Cells,
   Suppressor T-Cells 357
  10.1.3 B-Cells and Antibodies 360
  10.1.4 Natural Killer Cells 363
  10.1.5 Mast Cells, Platelets, and Complement 365
 10.2 Antigen Presentation 368
 10.3 Immunocytokines: Their Sources, Effects, and Targets 370
  10.3.1 The Interleukins 372
  10.3.2 The Interferons 376
  10.3.3 Tumor Necrosis Factors 376
  10.3.4 Prostaglandins and Other Eicosanoids 377
  10.3.5 Section Summary 380
 10.4 Apoptosis in the Immune System 381
 10.5 The Art of Mathematically Modeling Scenarios in the Immune
  System 382
  10.5.1 Introduction 382
  10.5.2 The Immune System vs. Cancer 385
  10.5.3 Models of HIV Infection and AIDS 394
 10.6 Chapter Summary 424
 Problems 427

Bibliography and References 435

Appendix: Discussion of Simulation Languages for Physiological/
Pharmacokinetic and Chemical Kinetic Systems 445

Index 449

1 Introduction to Physiological Regulators and Control Systems

1.0 INTRODUCTION

Three major issues are addressed in this text: (1) how to accurately model and describe the dynamics of certain physiological regulators and control systems (PRCs) in order to predict their behavior when they are subject to unusual conditions caused by diseases, drugs, or physical damage; (2) how to use a validated model of a compartmental pharmacokinetic system to design a closed-loop drug administration system to maintain a desired concentration of a drug in a target organ; and (3) how to use a validated mathematical model of a compartmental pharmacokinetic system and a physiological system to design an effective closed-loop drug administration system in which the drug is used to manipulate a physiological parameter.

To accurately describe a physiological system in mathematical terms (as by a set of nonlinear ordinary differential equations), one must appreciate the enormous complexity of living systems. This complexity extends to the molecular level, where some expertise in biochemistry and cell and molecular biology is needed, as well as a knowledge of the physics of molecular diffusion and physical–chemical kinetics. In short, an interdisciplinary approach is necessary.

The horns of the dilemma created in trying to model a complex physiological system are *complexity vs. reductionism.* An accurate description of the behavior of a complex nonlinear system lies in its detailed parametric interactions. The inclusion of many states or parameters in a mathematical model is not as much of a challenge for today's computers as it is for today's bioengineers, physiologists, and biochemists. Most mathematical models of physiological systems require rate constants; many of them are unknown, as they have never been measured or, worse, are not now measurable. We are therefore forced to simplify or reduce the order and complexity of the model, thereby possibly missing interesting system behavior that is the result of nonlinear complexity.

Physiological systems are generally self-regulatory closed-loop systems. External physical or chemical factors or the presence of disease can destroy the "natural balance" of a physiological regulator, leading to a physiological parameter going out of its natural bounds. Such deviation can lead to malfunction in other PRCs or organs dependent on the out-of-bounds parameter (e.g., blood glucose in diabetes mellitus). By measuring certain critical physiological parameters, it is possible to design an external controller that will administer a drug that will allow a physiological parameter to be brought back into a "normal" range in a safe manner. As will be demonstrated, the design of closed-loop drug administration systems is made more challenging by the fact that the dynamics of how a drug is metabolized by the body and how it affects its target organ(s) change with time.

1.1 PHYSIOLOGY AND PHYSIOLOGICAL REGULATION DEFINED

Arthur C. Guyton[59] defines *physiology* as the "…study of function in living matter, attempting to explain the physical and chemical factors that are responsible for the origin, development and progression of life." Guyton goes on to remark that each type of life (e.g., viruses, bacteria, plants, animals, insects, fish, humans, etc.) has a subspeciality of physiology associated with it. In addition, there is mathematical, systems, regulatory, and medical physiology, to cite some other major specializations.

Whether examining a bacterium or a human, it is evident that *function* can be broken down into a series of biochemical events which generally take place intracellularly; however, some reactions occur on cell surfaces, while others take place in the extracellular volume. The sum of these biochemical events is responsible for *homeostasis*, which is the maintenance of the steady-state conditions in the internal environment. Individual cells exhibit homeostasis, as does an intact organism which is built of cells. As examples of parameters under homeostatic maintenance in humans, consider body temperature, calcium ion concentration in the blood, blood glucose concentration, and mean arterial blood pressure.

Homeostatic maintenance of a physiological parameter is the result of a number of *negative feedback loops,* in which nature forms a *regulator* to stabilize the parameter against external and internal influences that would force it into a range that would harm the organism or decrease its probability of propagating its species. The importance of physiological regulation cannot be overestimated. Just about every clinically measurable physiological parameter is under some sort of internal feedback regulation. Regulation is necessary because the living organism is continually being perturbed by physical and chemical changes in its (external) environment. Also, for example, genetically and hormonally activated *internal events,* such as the biosynthesis of something as basic as an eggshell in birds, create a tremendous sink for calcium ions, which must be regulated to compensate for the loss.

In this text, the distinction will be made between *PRCs* and the *external control of physiological systems.* A physiological regulator acts to maintain a constant level of the regulated parameter (blood calcium, for example) in the face of fluctuating dietary input and the biosynthesis of bone or eggshell. Its action is involuntary; that is, we cannot consciously change the output. A regulator has a virtual, fixed, internal dc *set point* that ideally is the nominal parameter value. Most physiological homeostatic feedback systems are regulators. Regulators are a subset of feedback control systems.

An example of a physiological regulator that can be overridden is the system that controls our rate of breathing and hence blood hemoglobin oxygen saturation. We can override its regulatory function by conscious thought. For example, we can hold our breath when swimming underwater. Another regulator that can be overridden is the ocular tracking system. In addition, *biofeedback conditioning* makes it possible to consciously modulate other physiological regulator parameters, such as heart rate and skin temperature.

The output parameters of many PRCs can be controlled by the externally controlled input of drugs or hormones. Such exogenous control is treated in Chapters 6 through 10 in this text. The purpose of such external control is to correct for internal maladjustment of the regulator caused by disease, injury, etc. and to reestablish a normal balance of physiological parameters in the body.

1.2 THE PURPOSES OF MODELING

A dynamic mathematical model of a physiological system is a set of mathematical relations between quantitative measures of system behavior (states). The relations can be linear or nonlinear and generally involve first-order ordinary differential equations. The equations are based on assumptions and observations about the physical and chemical properties of the system.

Finkelstein and Carson[41] define three uses of models: *descriptive, explanatory,* and *predictive*. Descriptive and explanatory models help us to understand both qualitatively and quantitatively the relationships between states in complex PRCs. They are useful for teaching medical students and biomedical engineers systems physiology. The predictive use of detailed models is stressed in this text.

We want to be able to *predict* the behavior of a PRC under novel conditions, including where the system's gains and rate constants are altered by disease, externally applied drugs or hormones, or physical injury. Such prediction gives us insight into potential methods of treating diseases and how to design artificial organs and closed-loop drug administration systems.

An accurate mathematical model of a PRC can be difficult to realize. Often, gains, rate constants, receptor densities, and the details of biochemical pathways can only be estimated. The estimates can be justified by simulating system behavior under "normal" conditions and adjusting estimate values until known *in vivo* behavior is seen, thus "validating" the model. Also, intermediate states in a PRC are often not available for measurement, or data on their behavior are not available in the literature. This is especially true in massively complex systems such as the human immune system.

The worst-case condition for modeling a PRC is when we have only the input and output variables with which to work. Modeling then becomes the synthesis of a "black-box" empirical model, the order of whose linear dynamics may be in question. The position of obvious nonlinearities in the box is also often left to conjecture. Such empirical models do little to aid in the understanding of how the physiology works and have limited use for predicting behavior because they are usually based on one class of inputs.

Fortunately, many PRCs have been studied in detail, and so the dynamics of the component subsystems are known. An example of such a system is *glucoregulation* (the regulation of plasma glucose concentration). The pancreatic α and β cells, the liver and its glucose/glycogen biochemistry, insulin- and noninsulin-dependent cells, the kidneys, the gastrointestinal tract, etc. have all been characterized, so that comprehensive, detailed, nonlinear dynamic models of glucoregulation can be constructed. Such a comprehensive, detailed model where there is a one-to-one correspondence between mathematical features and physiological features is called a *structural* or *isomorphic model*.[41] (See, for example, the detailed glucoregulation models of Cobelli et al.,[25-27] Cramp and Carson,[33] and Finkelstein and Carson,[41] as well as the author's models in Chapter 7.)

Other detailed models may rely to some extent on theoretical relations as well as empirical, measured relations between the system's states. Such an empirical/theoretical model can be called a *"gray-box" model*.[41]

All models require validation. That is, the model output(s) must agree with its observed living counterparts, given natural input(s). Once validated, one has more confidence in using the model for prediction.

In summary, a detailed, validated, mathematical model of a physiological system can be used to teach systems physiology to biomedical engineers and medical students, to predict system behavior under altered physiological conditions (thereby saving research animals), to study the effectiveness of drug therapy, and to design closed-loop drug injectors.

1.3 SOME GENERAL PROPERTIES OF PRCS

In this section, some general properties of PRCs are described. These are bulk properties (i.e., they apply to entire systems) Some properties may seem rather obvious; others are not.

By definition, PRCs are closed-loop systems that employs one or more causal feedback loops. Feedback is generally negative. Loop gains tend to be low.

"Wet" PRCs generally have realpole loop gains. (An exception is found in the horizontal tracking control system for the eyes. Here, a central-nervous system-controller time constant is sufficiently long so that we have what is in essence an integrator [pole at the origin], creating a *type 1 closed-loop system*.[107,130] Also, in the eye movement control system, the plant [the eyeball] generally has complex-conjugate poles because of its moment of inertia coupled with the damped elasticity of the extraocular muscles.) Other neuromuscular control systems may also have complex-conjugate poles because of the mass or moment of inertia of a moving object such as a limb.

As a consequence of real-pole loop gains, PRCs are generally type 0 systems that have some steady-state error in response to a constant set point input.

PRCs are generally multiple-input/multiple-output systems. They generally have cross-coupling and interactions with related PRCs. Cross-coupling can arise from shared nervous system pathways, shared effector organs (such as the kidneys), or a hormone affecting more than one class of target cells.

PRCs generally use parametric control. Output variables, especially in "wet" systems, are regulated by the alteration of diffusion and chemical reaction rate constants and loss rates by hormones. Hormones also affect the active "pumping" of ions across cell membranes. Neurotransmitters alter specific ion permeabilities or conductances.

PRCs often have transport lags (dead time) in their loop gains. Such lags can arise from the time it takes a hormone to activate a multistep biosynthetic pathway, as well as the finite time it takes nerve impulses to propagate down an axon and reach the target organ (muscle, gland, etc.). Transport lags create stability problems in exogenous, high-gain, closed-loop control systems, mandating special controller designs. Endogenous PRCs seldom have stability problems because their loop gains are generally low.

PRCs are generally stable or, in certain cases, may exhibit bounded, limit cycle oscillations. Such oscillations are seen under certain conditions in the lens accommodation control system[130] and in the canine and human fasting glucoregulatory systems.[74,147]

All PRCs are nonlinear. This is not surprising because there are no negative concentrations or hormone release rates. Nonlinearities also enter PRCs from mass-action kinetics, saturation of a finite number of cell membrane receptors with a hormone, and intrinsic biochemical regulatory mechanisms that regulate biosynthesis.

PRCs are massively parallel systems. Every organ is composed of hundreds of thousands of cells that have similar functions, and each cell or functional group of cells may receive inputs in parallel (e.g., from nerve fibers serving a common purpose or a hormone that affects all the cells simultaneously). Such redundant, parallel architecture ensures robust behavior under conditions of injury or infection.

Some "wet" PRCs are push–pull in the sense that two regulatory hormones that have opposing effects are coupled into the system to regulate a common parameter. An example of this physiological control strategy is in the regulation of blood glucose. The pancreatic hormone insulin causes glucose in the blood to lower, by it being taken up by the liver for storage as glycogen and by it diffusing into insulin-sensitive cells. The pancreatic hormone glucagon causes blood glucose to rise, mostly by causing glycogen stored in the liver to be broken down and released into the blood as glucose. Glucagon also stimulates gluconeogenesis,[59] which is the synthesis of glucose from metabolic substrates such as fats and amino acids by liver cells.

Physiological regulators often have no uniquely identifiable summing points where error is generated between the set point and controlled variable. (Again, an exception is in eye movement control, where the angular position of a retinal image relative to the fovea can be considered error.) PRCs involving hormones have what

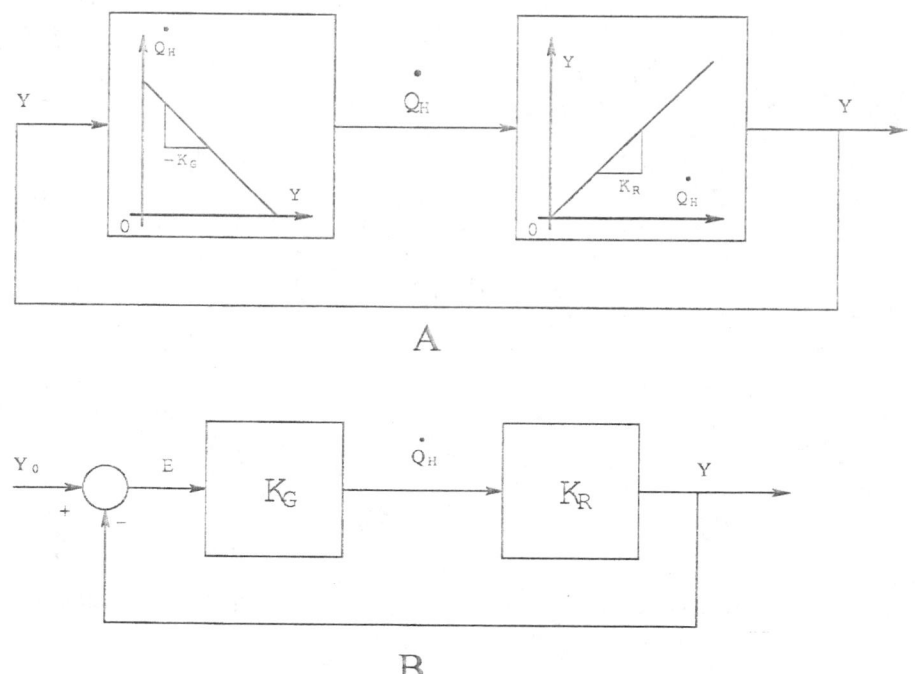

FIGURE 1.1 (A) Block diagram of a simple linear hormonal regulator with an implicit summing point. (B) The hormonal regulator of Figure 1.1A reduced to show effective gains, the summing point, and the virtual set point, Y_o.

is considered to be *implicit summing points*.[70] Figure 1.1A illustrates a hypothetical linear hormonal regulatory feedback system operating in the steady state. Block 1 contains an endocrine gland whose hormonal secretion rate is a linear, decreasing function of the concentration of the controlled variable. The hormone secretion rate, in the steady state, affects the controlled variable linearly. Mathematically, we have:

$$\dot{Q}_H = \dot{Q}_{Ho} - K_G Y = K_G Y_o - K_G Y = K_G \left(Y_o - Y \right) \quad (1.1)$$

$$Y = K_R \dot{Q}_H \quad (1.2)$$

Thus an implicit summing point is created where Y_o is the effective set point. Figure 1.1B illustrates the conventional feedback system block diagram with the implicit summing point. By inspection, the steady-state level of the controlled variable is

$$Y = Y_o \frac{K_G K_R}{1 + K_G K_R} \quad (1.3)$$

Introduction to Physiological Regulators and Control Systems

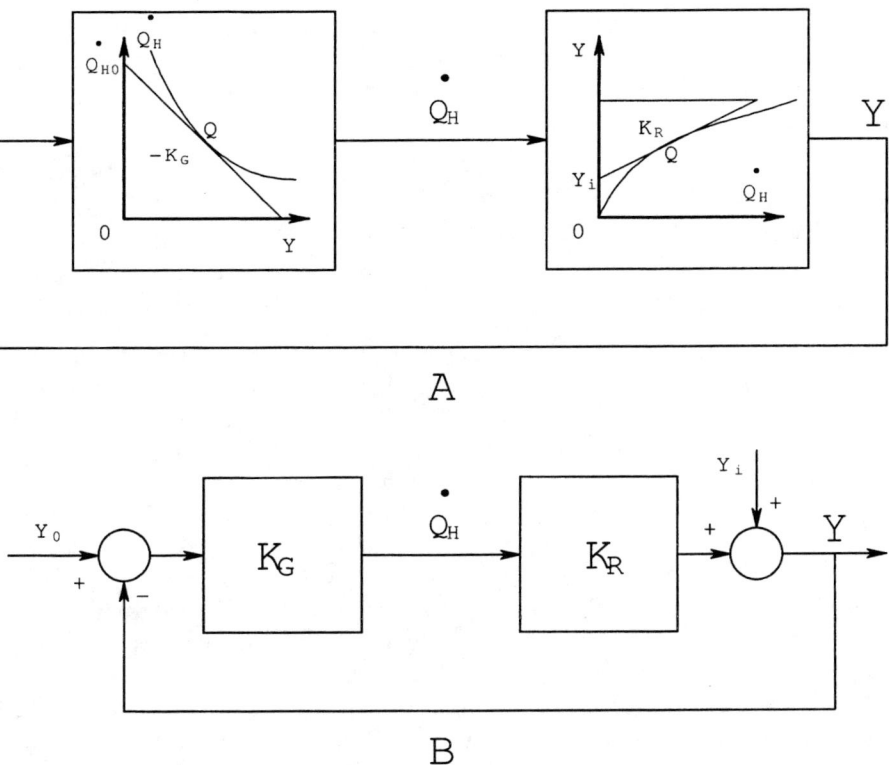

FIGURE 1.2 (A) Block diagram of a simple nonlinear hormonal regulator with an implicit summing point. A tangent is drawn to the nonlinear curves at the nominal operating point of the regulator. The slopes of the tangents are the effective linearized gains, and their y-intercepts determine virtual dc inputs to the linearized system. (B) Linearized block diagram showing implicit summing points and virtual inputs.

Jones[70] shows that the same strategy works to find the implicit summing point when the gland and regulated variable relations are nonlinear, as shown in Figure 1.2A. Straight-line tangent approximations to $\dot{Q}_H = f(Y)$ and $Y = g(\dot{Q}_H)$ are drawn where the two curves intersect, and the procedure shown above is applied to the two resulting linear equations (see Figure 1.2B). The linearized gain is

$$Y = (Y_o - Y_i) \frac{K_G K_R}{1 + K_G K_R} \tag{1.4}$$

Linearization around an operating point is an established technique in engineering analysis. (As an example, consider the small-signal h-parameters used to describe the circuit behavior of a bipolar junction transistor.)

Twenty years ago, the fact that all PRCs are nonlinear was a major setback for physiologists and engineers who wished to model them and predict their behavior.

Now, with the aid of modern engineering software and PCs, it is possible to model complex nonlinear PRCs and gain an increased understanding about certain diseases and possible therapies.

1.4 SOME GENERAL PROPERTIES OF LINEAR, TIME-VARYING, AND NONLINEAR SYSTEMS

1.4.1 INTRODUCTION

We are driven by the need to simplify complex systems through inequality assumptions and linearization techniques. However, when dealing with complex, nonlinear, high-order physiological systems, such reductionism, while making a solution easier to implement, they often rob the investigator of the details of the system's complex behavior. Such behavior may only show up in the whole complex system for a certain range of states or initial conditions and be hidden elsewhere. Hence, caution must be exercised when simplifying the model of a complex physiological system, lest behavioral detail be lost. On the other hand, there is a practical limit to how much mathematical detail should be included. Programming time and computational time are costs to be considered in simulating a high-order nonlinear system. Also, there is a point in the complexity of modeling when several system rate constants are unknown and must be estimated. Such uncertainty creates fuzzy or gray-box models. At this point, solutions become more speculation than simulated system behavior, and model validation becomes more difficult.

As we have seen in Section 1.3, *all* PRCs are generally nonlinear. However, over a limited range of operating space, most nonlinear systems can be linearized, and thus all the powerful tools of linear systems analysis can be brought to bear. As we shall see, if a system cannot be linearized, there are few tools useful for analysis and design of nonlinear control systems. Very often, nonlinear design is done by trial and error using computer simulations. There is a plethora of textbooks on linear feedback system analysis and design, but very few that consider the problems of nonlinear control.[107,135,140]

In the following sections, the properties of *linear systems* will be listed. By logical exclusion, all systems that *do not* obey the properties of linear systems must be nonlinear. Nonlinear systems also have certain unexpected properties which are also described below.

1.4.2 PROPERTIES OF TIME-INVARIANT LINEAR SYSTEMS

As taught in introductory engineering systems analysis, the key characteristic of all linear systems is that they obey the property of *superposition*. As a consequence of superposition, a linear system's output can be given by the *convolution integral*, in which the system's input is convolved with the system's weighting function (also known as impulse response). Further, the (linear) convolution integral can be Laplace or Fourier transformed to find the system's transfer function or frequency response. Nonlinear systems do not enjoy this property, although one sees models in which a linear transfer function is followed (or preceded) by a static nonlinearity. Of course, such a series system is nonlinear. Linear systems (LS) obey *all* of the following properties; nonlinear systems do not obey one or more of the following properties: $x(t)$ is the input, $y(t)$ is the output, and $h(t)$ is the LS impulse response or weighting function.

$$x_1 \to \text{LS} \to y_1 = x_1 \otimes h \qquad \text{real convolution} \qquad (1.5)$$

If

$$\xrightarrow{a_2 x_2} \text{LS} \xrightarrow{\begin{array}{c} y_2 = x_2 \otimes h \\ y_2' = a_2(x \otimes h) \end{array}} \qquad \text{scaling} \qquad (1.6)$$

then

$$\xrightarrow{a_1 x_1 + a_2 x_2} \text{LS} \xrightarrow{y = a_1 y_1 + a_2 y_2} \qquad \text{superposition} \qquad (1.7)$$

$$\xrightarrow{\begin{array}{c} x_1(t_1 - t_1) \\ + x_2(t_2 - t_2) \end{array}} \text{LS} \xrightarrow{\begin{array}{c} y = y_1(t - t_1) \\ + y(t - t_2) \end{array}} \qquad \text{shift variance} \qquad (1.8)$$

Linear single-loop control systems can be classified by their ability to follow transient inputs, including steps, ramps, and parabolas. In general, the *loop gain* of a *negative feedback system* can be written:

$$A_L(s) = \frac{-K\beta(s+b_1)(s+b_2)\dots(s+b_M)}{s^k(s+a_1)(s+a_2)\dots(s+a_{N-k})} = G(s)H(s) \qquad (1.9)$$

Note that there are k poles at the origin ($k = 0, 1, 2, \dots$). Assuming a single-loop architecture, the system error is easily seen to be

$$E(s) = \frac{R(s)}{1 - A_L(s)} = \frac{R(s)}{1 + \dfrac{K\beta(s+b_1)(s+b_2)\dots(s+b_M)}{s^k(s+a_1)(s+a_2)\dots(s+a_{N-k})}}$$

$$E(s) = R(s) \frac{s^k(s+a_1)(s+a_2)\dots(s+a_{N-k})}{s^k(s+a_1)(s+a_2)\dots(s+a_{N-k}) + K\beta(s+b_1)(s+b_2)\dots(s+b_M)}$$

$$(1.10)$$

To find the steady-state error given $R(s) = 1/s$, $1/s^2$, and $2/s^3$ [$r(t) = U(t)$, $t\,U(t)$, and $t^2\,U(t)$, respectively], we use the *Laplace final value theorem*:

$$e(t)\Big|_{\lim t\to\infty} = sR(s)\frac{s^k(s+a_1)(s+a_2)\dots(s+a_{N-k})}{s^k(s+a_1)(s+a_2)\dots(s+a_{N-k}) + K\beta(s+b_1)(s+b_2)\dots(s+b_M)}\Big|_{\lim s\to 0} \quad (1.11)$$

Now if $k = 0$ (type 0 system) and $r(t)$ is a unit step, there is finite steady-state error:

$$e(\infty) = s(1/s)\frac{(a_1 a_2 \dots a_N)}{(a_1 a_2 \dots a_N) + K\beta(b_1 b_2 \dots b_M)}\Big|_{\lim s\to 0} \quad (1.12)$$

If the input is a ramp or parabolic, it is easy to see that the steady-state error goes to infinity for the type 0 system. Now let $k = 1$ (*one pole at the origin in the loop gain is a type 1 system*) and consider a *step input*:

$$e(\infty) = s(1/s)\frac{s^1(a_1 a_2 \dots a_N)}{[s^1(a_1 a_2 \dots a_N) + K\beta(b_1 b_2 \dots b_M)]}\Big|_{\lim s\to 0} \to 0 \quad (1.13)$$

That is, there is zero steady-state error to a step input for a type 1 system. A *ramp input* produces a finite steady-state error in a type 1 control system:

$$e(\infty) = s(1/s^2)\frac{s^1(a_1 a_2 \dots a_N)}{[s^1(a_1 a_2 \dots a_N) + K\beta(b_1 b_2 \dots b_M)]}\Big|_{\lim s\to 0} \to \frac{(a_1 a_2 \dots a_N)}{K\beta(b_1 b_2 \dots b_M)} \quad (1.14)$$

A *parabolic input* to a type 1 system gives infinite error, etc.:

$$e(\infty) = s(2/s^3)\frac{s^1(a_1 a_2 \dots a_N)}{[s^1(a_1 a_2 \dots a_N) + K\beta(b_1 b_2 \dots b_M)]}\Big|_{\lim s\to 0} \to \infty \quad (1.15)$$

PRCs are generally type 0 with real poles and zeros in their loop gains. Two notable exceptions, mentioned above, are the lateral eye movement control system and the eye–hand tracking system, where there is evidence for a central nervous system "integrator" in the loop,[130] giving a type 1 system with improved dynamic performance.

1.4.3 Time-Variable Linear Systems

Time-varying linear systems are ones in which the system can be summarized by the ordinary differential equations (ODEs)

$$\dot{\mathbf{x}} = \mathbf{A}(t)\mathbf{x} + \mathbf{B}(t)\mathbf{u} \quad (1.16)$$

Introduction to Physiological Regulators and Control Systems

Note that one or more elements of the **A** and **B** matrices vary with time. That is, one or more rate constants change with time. Solutions in state space are possible if $\mathbf{A}(t)$ and $\mathbf{B}(t)$ are piecewise continuous functions of t in an interval, $t_o \leq t \leq t_1$. Ogata[107] shows that a general solution for the states is given by:

$$\mathbf{x}(t) = \Phi(t, t_o) \mathbf{x}(t_o) + \int_{t_o}^{t} \Phi(t, \tau) \mathbf{B}(\tau) \mathbf{u}(\tau) d\tau \tag{1.17}$$

where $\Phi(t, t_o)$ is the unique matrix satisfying the equations

$$\dot{\Phi}(t, t_o) = \mathbf{A}(t) \Phi(t, t_o), \quad \Phi(t_o, t_o) = \mathbf{I} \tag{1.18}$$

Needless to say, effective solutions of high-order, time-variable linear systems require computer simulations.

1.4.4 Properties of Nonlinear Systems

If a sinusoidal input is given to a dynamic linear system, then the output is a sinusoid of the same frequency but with (in general) different phase and amplitude. When a nonlinear system is given a sinusoidal input, the steady-state output will generally be periodic, but not purely sinusoidal. The periodic output can be written as a Fourier series, showing the existence of the fundamental frequency and higher order harmonics. The amplitude distribution of the harmonics will, in general, depend on the amplitude of the sinusoidal input signal. A nonlinear system can also generate *intermodulation distortion* terms at its output, given an input which is the sum of two or more sinusoids of different frequencies. The intermodulation process can be illustrated by assuming a simple power-law static nonlinearity:

$$y = a_0 + a_1 x^1 + a_2 x^2 + a_3 x^3 \tag{1.19}$$

If we let $x = b_1 \sin(\omega_1 t) + b_2 \sin(\omega_2 t)$, then the output can be written:

$$\begin{aligned} y = a_0 &+ a_1 \left[b_1 \sin(\omega_1 t) + b_2 \sin(\omega_2 t) \right] \\ &+ a_2 \left[b_1^2 \sin^2(\omega_1 t) + 2 b_1 b_2 \sin(\omega_1 t) \sin(\omega_2 t) + b_2^2 \sin^2(\omega_2 t) \right] \\ &+ a_3 \left[\begin{array}{l} b_1^3 \sin^3(\omega_1 t) + b_2^3 \sin^3(\omega_2 t) + 3 b_1^2 b_2 \sin^2(\omega_1 t) \sin(\omega_2 t) \\ + 3 b_1 b_2^2 \sin(\omega_1 t) \sin^2(\omega_2 t) \end{array} \right] \end{aligned} \tag{1.20}$$

From trigonometric identities, we see that $y(t)$ contains not only terms at the input frequencies but also dc terms, plus terms with frequencies of $(\omega_1 - \omega_2)$, $(\omega_1 + \omega_2)$,

$2\omega_1$, $2\omega_2$, $3\omega_1$, $3\omega_2$, $(2\omega_1 - \omega_2)$, $(2\omega_1 + \omega_2)$, $(2\omega_2 - \omega_1)$, and $(2\omega_2 + \omega_1)$. (The situation becomes even more complex if an x^4 term is present.)

Certain kinds of nonlinear systems are seen to generate subharmonics at their outputs. Tomović[135] gives an example of a subharmonic dynamic nonlinear system:

$$\ddot{y} = -y - 0.2\,y^3 + A\cos(6t) \tag{1.21}$$

This system is easily simulated with Simnon™; the state equations are

$$\dot{x}_1 = -x_2 - 0.2\,x_2^3 + A\cos(6t) \tag{1.22A}$$

$$\dot{x}_2 = x_1,\; y = x_2 \tag{1.22B}$$

A Simnon™ program to simulate Equations 1.22A and B is

```
continuous system tomovic1
STATE x1 x2
DER dx1 dx2
TIME t
"
dx1 = A*cos(6*t) - x2 - 0.2*x2^3
dx2 = x1
in = 0.3*cos(6*t)
"
A:1
zero:0
"
END
```

The simulation shows that $y(t)$ contains terms $B\cos(6t) + C\cos(2t)$ (see Figure 1.3). The **C** term is the subharmonic term.

The behavior of nonlinear systems is dependent on the initial conditions and the input amplitude and frequency. Certain nonlinear systems can exhibit output limit cycles, which are bounded, periodic oscillations of the output. Nonlinear systems can also exhibit input amplitude-dependent damping, where the system's output in response to an input step shows less and less damping as the input amplitude is increased. In some cases, the damping can go to zero or a negative value, causing an unbounded output. Initial conditions can also determine whether a nonlinear system's response is stable, oscillatory, or unbounded for a given input.

The nonlinearity in a nonlinear system can often be modeled by a functional nonlinearity in series with otherwise linear dynamics. The functional nonlinearity can be half-wave rectification, otherwise known as nonnegativity, common to all physiological concentrations, or it can be an odd saturating nonlinearity such as $y = A\tanh(bx) = A(1 - e^{-2bx})/(1 + e^{-2bx})$. Many physiological systems exhibit rate saturation. For example, the steady-state secretion rate of a gland in response to a hormone concentration, $[H]$, can be written as the (hyperbolic) *Hill function*:

Introduction to Physiological Regulators and Control Systems

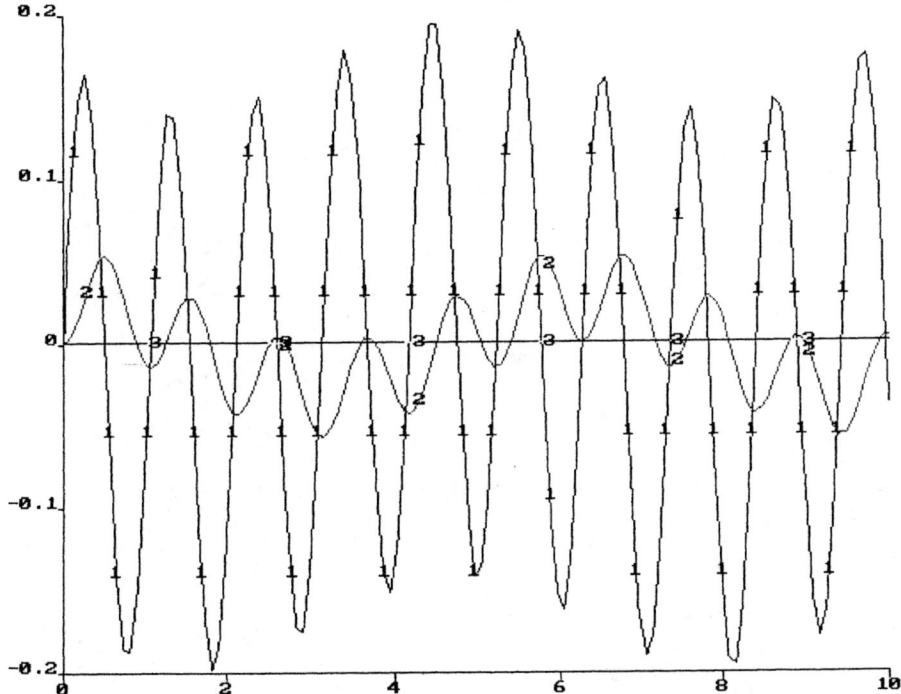

FIGURE 1.3 Steady-state behavior of a nonlinear system "tomovic1" that exhibits subharmonic oscillations. Results of Simnon simulation. Trace 1 is state $x1$ (cf. Equation 1.22A), and trace 2 is state $x2 = y$ (cf. Equations 1.21 and 1.22B).

$$\dot{Q}_G = \frac{[H]\dot{Q}_{GMAX}K}{(1+[H]K)}, \quad \text{for } [H] > 0 \tag{1.23}$$

To be nonlinear, a nonlinear system need not have a specific functional nonlinearity. The nonlinear behavior can be due to a set of nonlinear ODEs. This is generally the case in modeling physiological systems involving chemical kinetics. Such state equations can contain terms that are products of states and states raised to integral powers. These ODEs can be "stiff," which means that certain terms can $\rightarrow 0$ while other terms are very large. There are often problems in the numerical solutions of sets of stiff ODEs; special integration routines must be used to prevent numerical over- and underflows.[62]

Under sinusoidal excitation, certain algebraic nonlinear systems can exhibit the phenomena of *multivalued responses* and *jump resonances*. These interesting nonlinear characteristics may be demonstrated on a simple, mechanical, second-order nonlinear system consisting of a mass in series with a linear dashpot and a nonlinear spring, as shown in Figure 1.4. The ODE governing the mass position, x, can be written as:

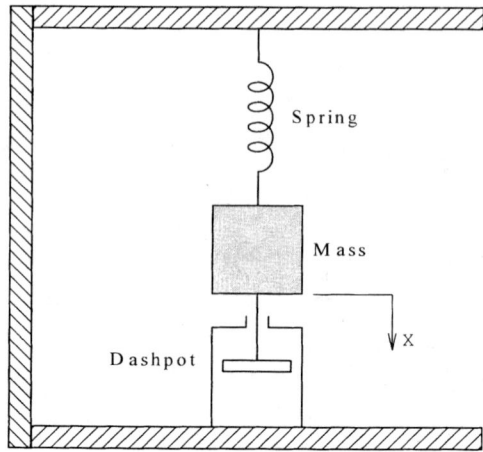

FIGURE 1.4 Nonlinear mechanical system consisting of a series spring, mass, and dashpot. The spring is made nonlinear to examine jump resonance phenomena.

$$M\ddot{x} + D\dot{x} + kx + k'x^3 = P\cos(\omega t) \qquad (1.24)$$

This equation is known as Duffing's equation. It can be rewritten as:[135]

$$\ddot{x} + k_2 \dot{x} + \omega_o^2 x + k_1 x^3 = A\cos(\omega t) \qquad (1.25)$$

Equation 1.25 can be put in state form and solved using Simnon:

$$\dot{x}_1 = -k_2 x_1 - k_1 x_2^3 - \omega_o^2 x_2 + A\cos(\omega t) \qquad (1.26A)$$

$$\dot{x}_2 = x_1 \qquad (1.26B)$$

ω is swept by setting $\omega = (\omega_s + \Delta\omega t)$. The amplitude jump phenomenon with increasing frequency is shown in Figure 1.5. The straight line is a plot of $10\omega(t)$. Note that in the *stiff spring system*, $k_1 > 0$, and the amplitude of x abruptly decreases at about $t = 110$ sec where $\omega \cong 1.3$ r/sec. The sharpness of the jump in amplitude and the frequency at which it occurs depend on the driving cosine's amplitude, A. $A = 15$ in this case. In Figure 1.6, the frequency of the source is swept down from 2.5 r/sec to 1.875 r/sec Now the amplitude of x increases to a peak at $t = 100$ ($\omega = 2.25$) and then linearly decreases with frequency.

Figure 1.7 shows a more pronounced amplitude jump behavior with slowly decreasing frequency. In this case, the system has a "soft" spring ($k_1 < 0$). The jump in this case occurs when ω reaches 1.45 r/sec. A transitional behavior on increasing frequency for the soft spring system is illustrated in Figure 1.8. At $\omega \geq 0.96$ r/sec, the peak amplitude of x begins to decrease linearly in the soft spring system.

Introduction to Physiological Regulators and Control Systems

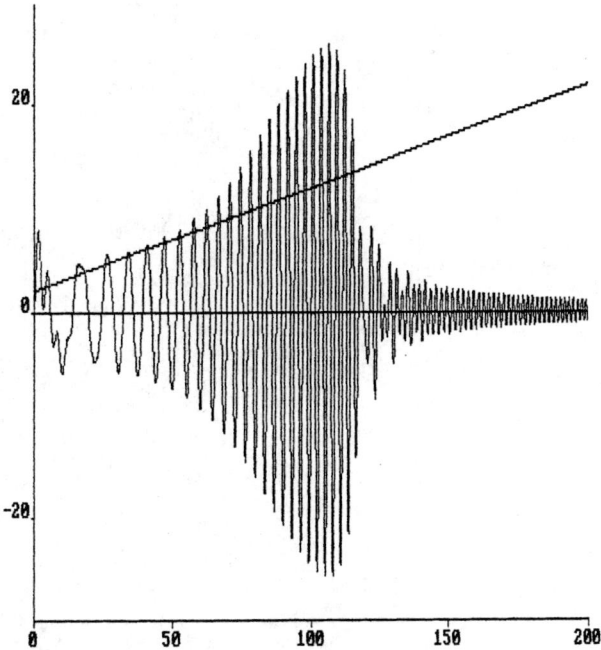

FIGURE 1.5 Response of the stiff-spring Duffing system (cf. Equation 1.25) to a linearly increasing driving frequency. Straight line is frequency (Hz). Note the sudden decrease in the amplitude of the mass displacement, $x(t)$.

In conclusion, it is important to stress that nature provides nonlinear, noisy, and time-variable physiological systems. Modeling such systems used to require linearization and severe reduction. Now, computer simulations allow us to tackle reasonable complexity (system order), nonlinearity, noise, and time-varying coefficients directly, preserving the interesting and often unexpected behavior.

1.5 PARAMETRIC CONTROL

An important property of many PRCs is that they use *parametric control*. In a parametric controller, the difference between the set point or desired output and a function of the output is used to manipulate one or more system gains, rate constants, or parameters, in order to force the output toward the desired value. By contrast, in a conventional control system or regulator, the control (input to the plant), u, is varied as a function of the difference between the plant's input and output. To illustrate parametric control, three examples will be presented: (1) a theoretical cellular system; (2) a fluidic, flow-regulating valve; and (3) the regulation of intraocular pressure.

FIGURE 1.6 Response of Duffing system to linear decreasing frequency.

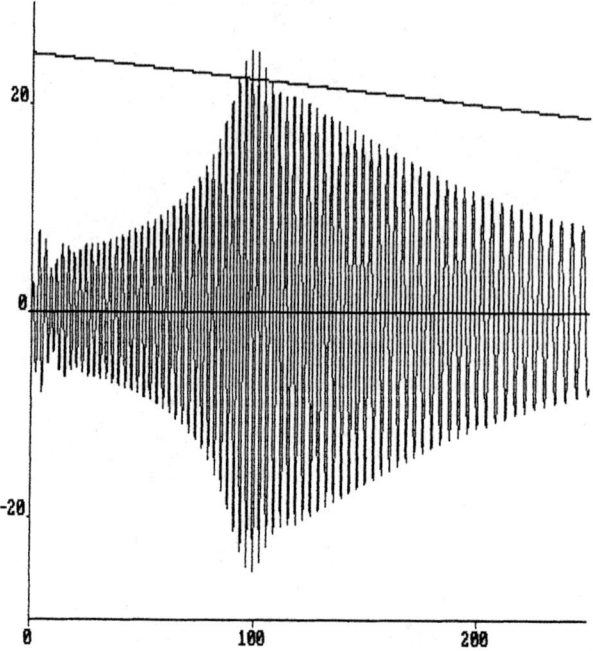

FIGURE 1.7 Nonlinear displacement response of a Duffing system with a "soft" spring to decreasing frequency.

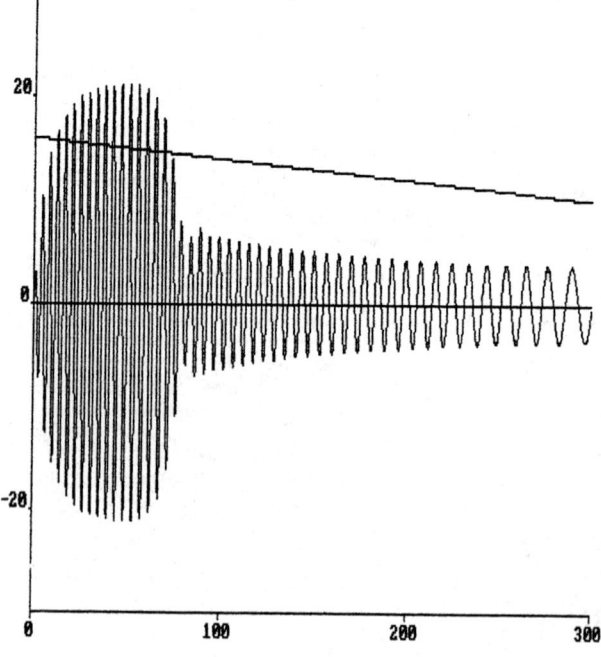

Introduction to Physiological Regulators and Control Systems

FIGURE 1.8 Response of the soft-spring Duffing system to linearly increasing frequency.

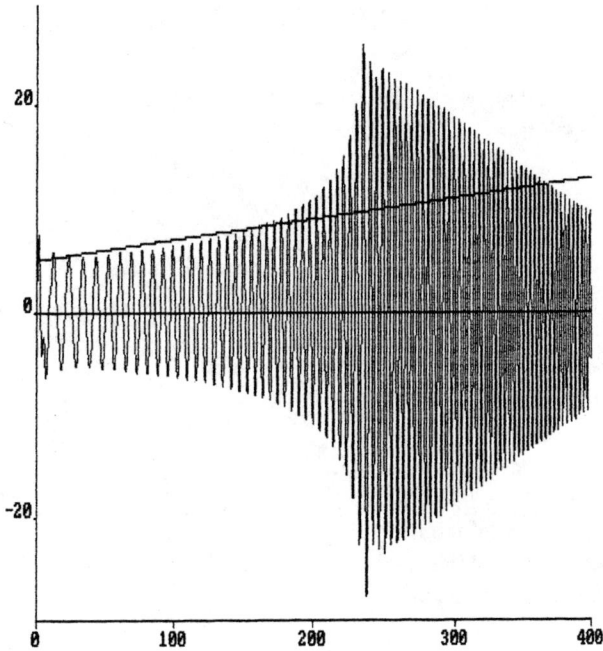

In the *first example*, a *theoretical cellular diffusion system*, we assume the regulated parameter is the concentration, x_2, of a certain substance, X, inside a cell. The substance is present outside the cell in concentration $x_1 > x_2$. Concentrations are in micrograms per liter. The substance diffuses passively into the cell according to Fick's first law and is metabolized inside the cell at a rate proportional to its internal concentration. These processes are illustrated in Figure 1.9 and are described by a linear first-order ODE: K_D is the *diffusion rate constant*, K_L is the loss rate constant, and V is the cell's volume.

FIGURE 1.9 Schematic cell used to illustrate parametric control. Substance X diffuses into the cell, where it is metabolized according to first-order kinetics.

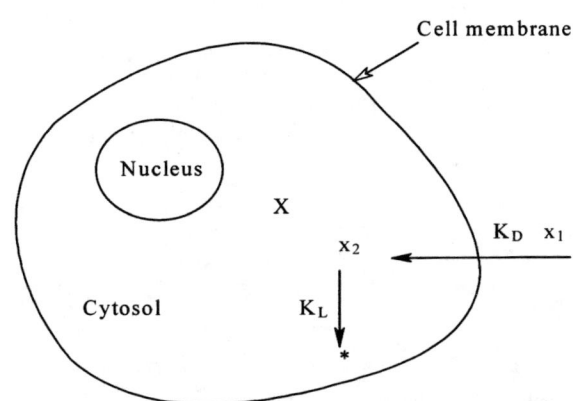

$$\dot{x}_2 = (K_D/V)(x_1 - x_2) - (K_L/V)x_2 \quad \mu g/(\ell \times \min) \tag{1.27}$$

This ODE can be rearranged and Laplace transformed to yield the (linear) transfer function:

$$\frac{X_2}{X_1}(s) = \frac{K_D/V}{s + (K_D + K_L/V)} \tag{1.28}$$

From the transfer function, the steady-state internal concentration can be written:

$$x_{2SS} = \frac{x_1}{1 + K_L/K_D} \tag{1.29}$$

Figure 1.10 shows a plot of x_{2SS} vs. K_D. It is clear that manipulation of K_D will control the value of x_{2SS}.

To examine how parametric control works, we assume a biochemical feedback mechanism exists which causes the diffusion constant, K_D, to decrease linearly as x_2 increases. That is,

$$\begin{aligned} K_D &= K_{D0} - \rho x_2, & 0 \le x_2 \le K_{D0}/\rho \\ &= 0, & x_2 > K_{D0}/\rho \end{aligned} \tag{1.30}$$

This relation is plotted in Figure 1.11. Note that K_D is linear over $0 \le x_2 \le K_{D0}/\rho$. Thus, in the steady state, where x_2 lies in the linear range for K_D, we can write:

$$x_{2SS} = \frac{x_1}{1 + K_L/(K_{D0} - \rho x_{2SS})} \tag{1.31}$$

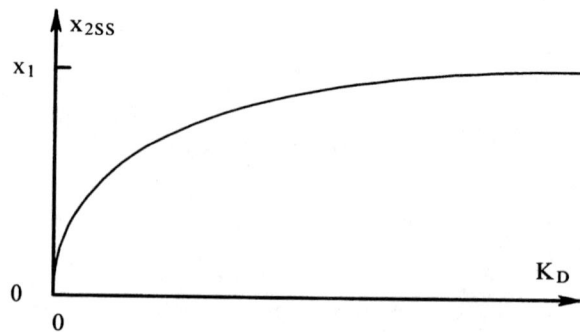

FIGURE 1.10 Plot of the steady-state concentration of X inside the cell (x_2) vs. the diffusion constant, K_D.

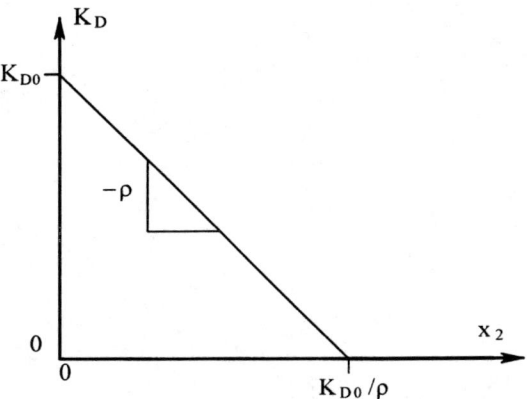

FIGURE 1.11 Assumed parametric control law where the internal concentration of X, x_2, determines K_D according to the law $K_D = K_{D0} - \rho\, x_2$ for $0 \leq x_2 \leq K_{D0}/\rho$, else 0.

Equation 1.31 can be written in standard quadratic form for x_{2SS}:

$$x_{2SS}^2 - x_{2SS}\left[(K_{D0} + K_L)/\rho + x_1\right] + x_1 K_{D0}/\rho = 0 \tag{1.32}$$

This quadratic equation can be solved in the usual manner:

$$x_{2SS} = \left[(K_{D0} + K_L)/\rho + x_1\right]/2 \\ \pm \sqrt{\left[(K_{D0} + K_L)/\rho + x_1\right]^2 - 4x_1(K_{D0}/\rho)}\,/2 \tag{1.33}$$

If we factor out the $[(K_{D0} + K_L)/\rho + x_1]$ term from the square root and consider the − root, we can write:

$$x_{2SS} \cong \frac{x_1(K_{D0})/(K_{D0} + K_L)}{1 + \rho(x_1)/(K_{D0} + K_L)} \tag{1.34}$$

If the gain, ρ, is large, Equation 1.34 reduces to:

$$x_{2SS} \approx \frac{x_1(K_{D0})}{\rho(x_1)} = \frac{K_{D0}}{\rho} \tag{1.35}$$

which is independent of x_1. In other words, the sensitivity, $S_{x2} = dx_{2SS}/dx_1 \to 0$.

In the *second example*, we consider a practical, nonphysiological, industrial example of a parametric regulator, in this case a flow regulator valve, illustrated in

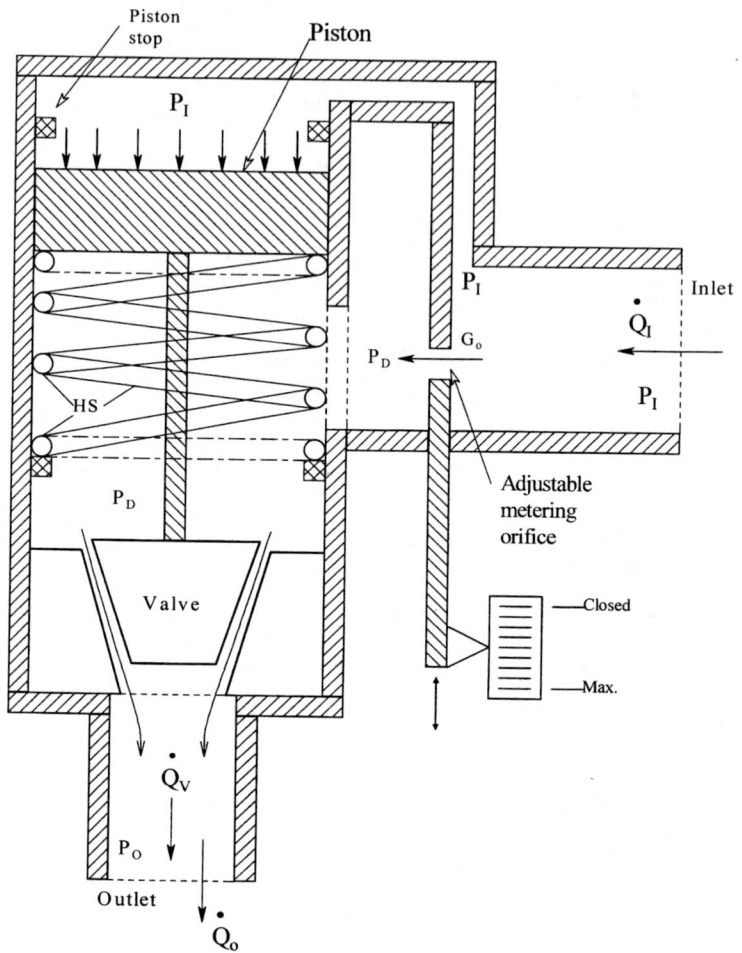

FIGURE 1.12 Cross-sectional schematic of a volume flow regulator valve using parametric control. Ideally, the output flow is independent of pressure.

a sectional schematic in Figure 1.12. The parametric flow regulator works as follows: If the input pressure P_I is low, $(P_I < P_S)$, the force of the helical spring forces the piston up, opening the metering valve wide so that it has negligible hydraulic resistance, R_o, compared to the metering orifice. Thus $P_D \approx P_O = 0$, and $\dot{Q}_I \approx \dot{Q}_V \approx \dot{Q}_o = P_I G_o$. When $P_I X$ (piston area) exceeds the force of the helical spring, the piston is forced down, closing the variable valve so that its resistance increases so that it is no longer negligible. In the steady state, when the valve is regulating, there is flow continuity and we can show that the output flow rate $\dot{Q}_o$ is independent of P_I, providing certain restrictions are met. Flow continuity means:

$$\dot{Q}_I = \dot{Q}_V = \dot{Q}_o \qquad (1.36)$$

Introduction to Physiological Regulators and Control Systems

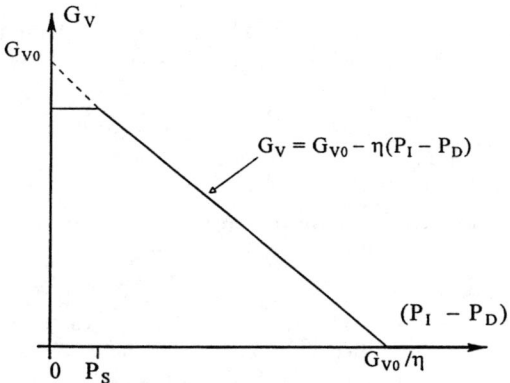

FIGURE 1.13 The parametric control paradigm: valve hydraulic conductance as a function of the pressure difference across the metering orifice.

$$\dot{Q}_I = (P_I - P_D)G_o \tag{1.37}$$

where G_o is the hydraulic conductance set for the metering orifice. Also, the valve law for variable valve with conductance G_V is illustrated in Figure 1.13. The valve's conductance is given by:

$$G_V = G_{V0} - \eta(P_I - P_D) \tag{1.38}$$

The chamber pressure is:

$$P_D = P_I - \dot{Q}_o R_o \tag{1.39}$$

Hence:

$$\begin{aligned}\dot{Q}_V = P_D G_V &= \left(P_I - \dot{Q}_o R_o\right)\left[G_{V0} - \eta\left(P_I - P_D\right)\right]\\ &= \left(P_I - \dot{Q}_V R_o\right)\left[G_{V0} - \eta\left(P_I - \{P_I - \dot{Q}_V R_o\}\right)\right]\end{aligned} \tag{1.40}$$

This equation is put into standard quadratic form:

$$\dot{Q}_V^2 - \dot{Q}_V\left[1 + R_o\left(\eta P_I + G_{V0}\right)\right]/R_o^2\eta + P_I G_{V0}/R_o^2\eta = 0 \tag{1.41}$$

To facilitate solution, we assume that $\eta P_I \gg G_{V0}$, $\eta P_I R_o \gg 1$, and $\sqrt{1-\varepsilon} \cong 1 - \varepsilon/2$. If we take the − root, we obtain the output flow rate as:

$$\dot{Q}_V \approx G_{V0}/R_o\eta \tag{1.42}$$

Note that $\dot{Q}_V$ is independent of the input pressure, P_I. Hence, in the range where the assumptions above are valid, the output flow rate is set by the metering orifice's resistance parameter, R_o.

As a *third example* of a parametric regulator, we examine the physiological system that regulates the pressure of the aqueous humor (AH) in the eye. The AH is found around the lens and in the anterior chamber of the eye. It serves as a nutrient solution for the lens and the inside of the cornea, and its pressure helps to maintain the proper shape of the eyeball.

A cross-sectional view of a human eyeball is shown in Figure 1.14. AH is continuously formed in a normal eye at a rate of about $Q_{AH} = 2$ mm³/min by the cells of the ciliary process, a tissue that lies behind the lens and has an exposed area of about 6 cm².[59] AH is formed by the active "pumping" of Na⁺ ions from inside ciliary process cells to the perilenticular region inside of the eyeball. There is also active transport of ascorbate and certain amino acids into the eyeball. Water, chloride, glucose, and bicarbonate ions follow the ions pumped into the eyeball due to osmotic pressure and diffusion down concentration gradients. AH contains mainly

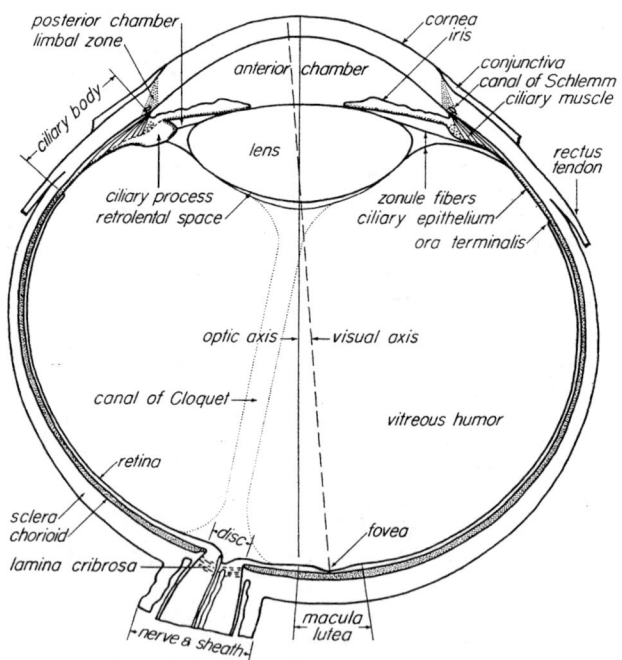

FIGURE 1.14 Cross-section through a human eyeball showing the cilliary process where aqueous humor is produced and the path by which it leaves the posterior chamber between the lens and iris. Aqueous humor leaves the anterior chamber through the canal of Schlemm. See text for details. (From Walls, G.L., *The Vertebrate Eye and Its Adaptive Radiation*, Haffner Publishing, New York, 1942. Reprinted by permission of the Cranbrook Institute of Science.)

low-molecular-weight substances, including Na⁺, K⁺, HCO_3^-, citrate, ascorbate, urea, glucose, etc.

Clearly, in the steady state, the AH must exit the eye at the same volume flow rate at which it enters. Outflow of AH is through the *canal of Schlemm*, into the episcleral veins, then into the main venous circulation, etc. The eyeball is slightly elastic, with most of its compliance coming from the thin, clear cornea. Normal intraocular pressure (IOP) is about 16 mmHg. If there is an increase in the outflow resistance, the normal IOP rises, and if the IOP exceeds its normal high range (about 30 mmHg), the condition known as glaucoma exists. In extreme situations, the IOP can exceed 60 to 80 mmHg. Such acute glaucoma sharply reduces normal arterial blood flow to the retina, causing poor oxygenation and impaired nutrition of retinal neurons and glial cells. If prolonged, glaucoma can lead to the death of retinal neurons, including the loss of retinal ganglion cells which comprise the optic nerve. Such neuron loss is irreversible, and it causes loss of visual acuity and even total blindness.

Because of the importance of vision to survival, it is reasonable to hypothesize that we have evolved regulatory mechanisms that maintain the IOP within normal bounds. Thus glaucoma can be viewed as a failure of this IOP regulator. Figure 1.15 illustrates a simple electric circuit analog to the AH system. Q_{AH} is the volume flow rate of AH into the eyeball. It is analogous to a dc current source. C_{PC} is the hydraulic capacitance of the posterior chamber of the eye (behind the lens), and C_{AC} is the hydraulic capacitance of the anterior chamber of the eye, including the cornea. Hydraulic capacitance, C_H, is defined as (see Section 2.1. for a detailed description of hydraulic analog parameters):

$$C_H \equiv \Delta V / \Delta P \tag{1.43}$$

where ΔV is the increase in volume of a chamber resulting from an increase in its internal fluid pressure, ΔP. The hydraulic resistance R_{IL} is due to reduced space between the front of the lens and the iris; normally, R_{IL} is negligibly small. R_{TS} is the hydraulic resistance of the trabecular network and the canal of Schlemm; R_{TS} is not negligible. (Hydraulic resistance is defined analogously to Ohm's law: $R_H = \Delta P / \dot{Q}$. ΔP is analogous to voltage drop, and $\dot{Q}$ is analogous to current [cf. Section

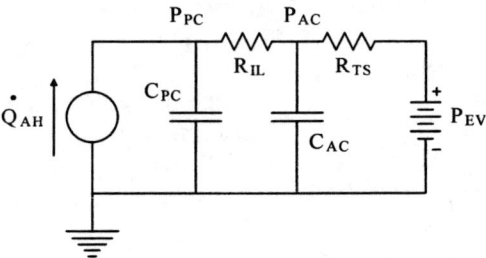

FIGURE 1.15 Electrical analog circuit describing the hydraulic dynamics of aqueous humor production and loss.

2.1.1].) P_{EV} is the pressure in the episcleral veins. Clearly, for the AH to exit the eye, $P_{AC} > P_{EV}$. P_{AC} is the AH pressure in the anterior chamber.

We can write equations for the steady-state pressures, P_{PC} and P_{AC}, using Kirchoff's voltage (pressure) law and Ohm's law:

$$P_{PC} = \dot{Q}_{AH}\left(R_{IL} + R_{TS}\right) + P_{EV} \qquad (1.44A)$$

$$P_{AC} = \dot{Q}_{AH} R_{TS} + P_{EV} \qquad (1.44B)$$

We now postulate a parametric mechanism whereby P_{AC} is regulated: We assume that increased IOP stretches the trabeculae and the canal of Schlemm, increasing their radii hence *decreasing* R_{TS}. We will see that decreased R_{TS} lowers both P_{PC} and P_{AC} (see equations above). Let us assume a linear relation between the hydraulic resistance of the trabeculae and the canal of Schlemm and the pressure of AH in the anterior chamber:

$$R_{TS} = R_{TS0} - \rho P_{AC} \qquad (1.45)$$

Substituting Equation 1.45 into Equation 1.44B, we can write:

$$P_{AC} = \frac{P_{EV} + \dot{Q}_{AH} R_{TS0}}{1 + \rho \dot{Q}_{AH}} \qquad (1.46)$$

Now Equations 1.44B and 1.45 are substituted into Equation 1.44A, and we solve for P_{PC}:

$$P_{PC} = \frac{P_{EV} + \dot{Q}_{AH} R_{TS0}}{1 + \rho \dot{Q}_{AH}} + \dot{Q}_{AH} R_{IL} \qquad (1.47)$$

Thus the sensitivities dP_{AC}/dP_{EV}, $dP_{AC}/d\dot{Q}_{AH}$, dP_{PC}/dP_{EV}, and $dP_{PC}/d\dot{Q}_{AH}$ are reduced by this mode of feedback. The second term in Equation 1.47 is normally small, so the fact that $dP_{PC}/d\dot{Q}_{AH}$ and dP_{PC}/dR_{IL} are unchanged is not significant.

A second postulated mechanism of parametric IOP regulation is the reduction of AH production rate due to high P_{PC}. In this scenario, we assume that as the P_{PC} rises, the arterioles carrying blood to the ciliary process are gradually compressed, reducing blood flow to the secretory cells. The reduced blood flow results in reduced $\dot{Q}_{AH}$ according to the simplified rule:

$$\begin{aligned}\dot{Q}_{AH} &= \dot{Q}_{AH0} - \gamma P_{PC}, \quad \text{where } P_{PC} > 0 \\ &= 0, \qquad \text{for } P_{PC} > \dot{Q}_{AH0}/\gamma \end{aligned} \qquad (1.48)$$

Introduction to Physiological Regulators and Control Systems

When Equation 1.48 for $\dot{Q}_{AH}$ is substituted into Equations 1.44A and B, we get:

$$P_{PC} = \frac{P_{EV} + \dot{Q}_{AH0}(R_{IL} + R_{TS})}{1 + \gamma(R_{IL} + R_{TS})} \qquad (1.49)$$

$$P_{AC} = \frac{P_{EV}(1 + \gamma R_{IL}) + \dot{Q}_{AH0} R_{TS}}{1 + \gamma(R_{IL} + R_{TS})} \qquad (1.50)$$

In this case, sensitivities against changes in P_{EV} and $\dot{Q}_{AH0}$ are reduced for both P_{AC} and P_{PC}. It should be stressed that both mechanisms of parametric IOP regulation are conjectural and are presented here to illustrate how parametric regulation could work in this system.

In summary, the three examples in this section illustrate in detail how parametric negative feedback mechanisms in many cases result in reduced sensitivity of the regulated parameter to changes in system inputs or system parameters. We shall see that many biochemical processes are parametrically regulated by the concentrations of certain enzymes and hormones which speed (or slow) reaction rates. The concentrations of these enzymes are often affected by the concentrations of intermediary reaction products or reaction end products.

In Chapter 7, another important physiological system that uses parametric control will be examined in detail: the blood glucose regulatory system. The pancreatic hormone insulin is seen to increase the rate at which glucose diffuses into insulin-sensitive cells by effectively increasing the transmembrane diffusion rate constant for glucose. The pancreatic hormone glucagon acts parametrically to increase blood glucose concentration by stimulating the production of key enzymes in liver cells that catalyze the breakdown of glycogen into glucose and increase its release rate into the bloodstream. Many other examples of parametric control and regulation will also be described in this text.

1.6 CHAPTER SUMMARY

In this introductory, *physiology, physiological regulation,* and *homeostasis* were defined and examples given. The study of physiology was shown to be applicable to all living organisms, plant and animal, and to extend from the organ system level to subcellular biochemical mechanisms. Homeostasis is the result of physiological regulation. Steady-state internal conditions are maintained through various negative-feedback regulatory pathways or loops to protect the organism, organ, or cell from external fluctuations in its physical or biochemical environment.

Mathematical modeling of physiological systems was discussed. The purpose of modeling is to be able to predict the complex, dynamic behavior of a physiological system subject to parametric disturbances and to externally induced changes in various states. Physiological models generally consist of sets of nonlinear first-order ODEs. Nonlinear behavior can be described by algebraic relations or by functional blocks. Nonlinear behavior can arise from the nonnegativity of chemical concentrations and from biochemical mass-action kinetic equations. If nonlinearity can be ignored, as in small-signal behavior around an operating point, then powerful linear systems techniques can be exploited in the identification and analysis of physiological regulatory systems.

Section 1.3 described some general properties of PRCs. These include the fact that PRCs are parallel, redundant systems with real-pole loop gains. They are almost all type 0 systems (no loop gain pole at the origin of the s-plane) with low loop gains. They are also stable systems, with few examples of some with bounded limit cycles in the controlled variable under certain conditions. Natural selection has favored stability for obvious reasons. Parametric control was seen to be the dominant means of effecting regulation or control in physiological systems where hormones are found.

In *parametric control,* the controlled variable is manipulated by the control (u) altering one or more system parameters, such as gains, diffusion constants, or reaction rate constants. Typically, hormones effect parametric control. Examples of parametric control were given. Note that all parametric control systems are nonlinear.

General properties of linear vs. nonlinear systems were described in Section 1.4. In summary, stationary linear systems obey superposition. They can be described by convolution, Laplace and Fourier transfer functions, and sets of state equations. Nonlinear systems, on the other hand, generally cannot be described by convenient linear systems techniques, except in the case of certain system architectures which can be separated into a linear portion and a functional nonlinearity in the loop gain. Such systems are analyzable by *describing functions* under certain conditions; otherwise one must resort to simulations. Nonlinear systems can be unpredictable; they can exhibit conditionally stable behavior that changes to unstable behavior when the value of a critical state crosses a threshold. Initial conditions can also determine whether a linear system is well behaved or is unstable. Often, stable behavioral modes are hidden in nonlinear systems until a state crosses a threshold. A classic example of this behavior is the simulated generation of a nerve action potential by the well-known Hodgkin–Huxley equations. Other examples of bizarre behavior in complex nonlinear physiological systems can be found in the immune system. For example, the amount and rate by which a new antigen is introduced to the immune system can result in desensitization to subsequent doses of the antigen (e.g., as in allergy shots) or can lead to a life-threatening anaphylactic reaction. Such bistable behavior is characteristic of complex nonlinear systems.

In the next chapter, the details of the key physical and chemical phenomena that occur in physiological systems are examined.

PROBLEMS

1.1 For the following hormones or autacoids, give the source(s), molecular structure if known, typical concentration in the blood, half-life, target organ(s), and effects:
  a. Leptin
  b. Renin
  c. Insulin
  d. Cortisol
  e. Epinephrine
  f. Oxytocin
  g. Calcitonin
  h. Thyroid hormone
  i. Adrenocorticotropin (ACTH)
  j. Aldosterone

1.2 Consider the theoretical regulated diffusion system of the first example in Section 1.5. Assume that the diffusion constant is approximated by $K_d \cong \gamma/x_1$, $x_1 > 0$.
  A. Derive an algebraic expression for the steady-state x_2, given a constant input $x_1 > 0$.
  B. What condition on γ is required to make x_{2ss} = a constant?

1.3 Equation 1.23 is a Hill function describing soft saturation of a hormone secretion rate. Give at least three other mathematical functions that describe a soft saturation with increasing input parameter X value. Make $\dot{Q}_G \to \dot{Q}_{GMAX}$ as $X \to \infty$ and $\dot{Q}_G(0) = \dot{Q}_{G0}$.

1.4 One form of the nonlinear van der Pol equation is

$$\ddot{x} + a(x^2 - b)\dot{x} + x = 0$$

Note that this system must be written as two first-order ODEs for simulation. Simulate this nonlinear ODE for $a = 1$ and $b = 1$. Try $x(0) = 1$. View the limit cycles in the time domain and in the phase plane ($\dot{x}$, x). Investigate the effect of parameter and initial condition (IC) values on the system behavior. Are there parameter values for which limit cycles do not occur?

1.5 The nonlinear ODEs below are a form of the Lotka–Volterra equations which can be used to model predator–prey interactions in an oversimplified ecosystem. x and y are the population densities of the two competing species.

$$\dot{x} = x(a - by)$$
$$\dot{y} = y(cx - d)$$

Simulate the system and observe its behavior for different ICs and parameter values. Begin with $x(0) = 1$, $y(0) = 1$, $a = 2.7$, $b = 0.7$, $c = 1$, $d = 3$.

1.6 The Lorenz equation can be given as a system of three ODEs:

$$\dot{x} = a(y - x)$$
$$\dot{y} = bz - y - xz$$
$$\dot{z} = -cx + xy$$

Let $a = 10$, $b = 28$, $c = 8/3$, $x(0) = -8$, $y(0) = -8$, and $z(0) = 24$. Simulate this system using Simnon. Observe (z, x). Vary the parameters and observe system behavior. This is a deterministic system that possesses unexpected (chaotic) behavior.

1.7 The venerable Hodgkin–Huxley (*J. Physiol.*, 117: 500–544, 1952) nonlinear ODE set models the generation of nerve axon action potentials. These equations are a good example of a nonlinear system that exhibits an "all-or-nothing" threshold behavior. A Simnon model of the Hodgkin–Huxley equations is given below:

```
continuous system HODHUX2      " 1/27/97
STATE v m n h                  " v in mV. v is depolarization if < 0.
DER dv dm dn dh                " Vm is actual transmembrane potential.
"                                Vmo = resting potential = - 70 mV.
TIME t                         " t in ms.
"
dv = Im/Cm - Ik/Cm - Ina/Cm - Il/Cm    " HH membrane patch ODE.
Ik = gk*(n^4)*(v - Vk)
Ina = gna*(m^3)*h*(v - Vna)
Il = gl*(v - Vl)
Inet = Ik + Ina + Il
"
dn = - n*(an + bn) + an        " K+ activation parameter.
dm = - m*(am + bm) + am        " Na+ activation parameter.
dh = - h*(ah + bh) + ah        " Na+ inactivation parameter
"
an = 0.01*(v + 10)/(exp(0.1*v + 1) - 1)
bn = 0.125*exp(v/80)
am = 0.1*(v + 25)/(exp(0.1*v + 2.5) - 1)
bm = 4*exp(v/18)
ah = 0.07*exp(v/20)
bh = 1/(exp(0.1*v + 3) + 1)
"
Vm = -(v + 70)     " Transmembrane voltage measured from inside out
                     in mV.
"
" CONSTANTS:
zero:0
Vk:12          " mV
Vna:-115       " mV
VL:-10.613     " mV
"
gl:0.3         " mS/cm^2
gk:36          " "
```

```
gna:120              "   "
Cm:.333              "   mF/cm^2
f:0.10               "   Hz
pi:3.14159
"
"   INPUTS:
"
"Im1 = IF t > to THEN K*(t - to) ELSE 0     " Delayed current ramp
                                              input.
"Im2 = IF t > t2 THEN -K*(t - t2) ELSE 0
t2:10
to:2                                         " Im in micro A > 0
                                               inward.
K:-0.3
Im1 = IF t > to THEN Imo ELSE 0              " Delayed current pulse
                                               input.
Im2 = IF t > (to + delt) THEN -Imo ELSE 0
Imo:-10
Imac:1
delt:5
"Im1 = IF t > to THEN Imo ELSE 0             " Delayed current step input.
"Im2 = Imac*sin(2*pi*f*t)                    " Sinusoidal current input.
Im = Im1 + Im2          " Im < 0 is + charges inward which depolarize
                          membrane.
"
"   INITIAL CONDS.
m:0.0527             "   These ICs are important!
n:0.3171
h:0.5734
v:0
"
END
```

    A. Set the program parameters to obtain a 5-msec *negative* pulse of current beginning at $t_o = 2$ msec. Set the RKF45 integrator precision to 1.E-6. Make a series of plots of V_m, I_m, and I_{net} vs. t over 10 msec. Find the critical I_m value below which an "action potential" will not occur.

    B. Now repeat part A for positive pulses of I_m. Why does an action potential occur at the end of the hyperpolarizing pulse?

1.8 Use the Hodgkin–Huxley model of Problem 1.7 to investigate the behavior of the model as a *current-to-frequency converter*. Make the I_m pulses 98 msec in length and vary I_{mo} in steps from −1. Simulation time = 100 ms. Use Euler integration with $\Delta T = 0.001$. Plot the pulse frequency vs. I_{mo}. Find the I_{mo} threshold below which sustained, periodic pulses are not produced. What happens to the *amplitude* of $V_m(t)$ as I_{mo} is made more negative? Comment on the linearity of the Hodgkin–Huxley model as a current-to-frequency converter.

1.9 The circuit uses parametric control to charge a capacitor to $v_c = V_D$. As v_c approaches V_D, the conductance $G(v_c)$ decreases linearly to 0 at V_D. That is, $G(v_c) = G_o - kv_c$ for $0 \leq v_c \leq V_D$, and 0 for $v_c > V_D$. $k = G_o/V_D$. Simulate the

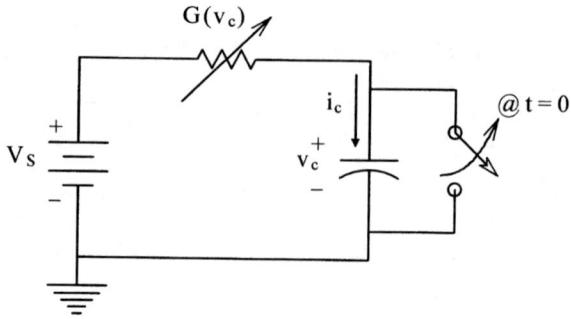

PROBLEM 1.9

behavior of the circuit as the capacitor charges. Let $V_S = 10$ V, $V_D = 1$ V, $G_o = 10^{-3}$ S, and $C = 10^{-3}$ F. $v_c(0) = 0$. Plot $v_c(t)$ V, $i_c(t)$ (mA), and tau$(t) = $ C/G sec. Vertical axis $= 0, 10$; time axis $= 0, 1.5$ sec.

1.10 See the circuit of Problem 1.9: In this case, G is affected parametrically by v_c according to the law $G(v_c) = G_o \exp(-v_c/2V_D)$ S, $0 \le v_c \le V_D$. When $v_c > V_D$, $G = 0$. Simulate and plot $v_c(t)$, $i_c(t)$, and tau(t). Use the parameter values of Problem 1.9.

2 Physical and Chemical Factors Governing the Behavior of Physiological Regulators and Control Systems

2.0 INTRODUCTION

The bulk (macro) dynamic behavior and nonlinear properties of physiological regulators and control systems can be related to the physical and chemical events taking place within them on a microscale. Because physiological systems are generally "wet," the physics governing fluid flow and molecular diffusion are often relevant in formulating quantitative dynamic models. In addition, the biochemical reactions underlying all physiological processes can often be described by the laws of chemical mass action. Mass-action descriptions generally involve nonlinear ordinary differential equations (ODEs). Fluidics and diffusion often require the use of linear ODEs in their descriptions. Partial differential equations need to be invoked to model detailed fluidic and diffusion processes. However, in this introductory text, an effort will be made to try to avoid them wherever possible through the use of approximations.

In the following sections of this chapter, the dynamics of fluidic systems under laminar flow conditions, diffusion phenomena in one dimension, and chemical mass-action kinetics used in describing the dynamics of biochemical reactions are examined.

2.1 BEHAVIOR OF FLUIDIC SYSTEMS UNDER LAMINAR FLOW CONDITIONS

Several physiological systems involve fluid dynamics in their descriptions. These include the respiratory system, the regulatory systems for intraocular pressure and cerebrospinal fluid pressure, and systems involving the circulatory system (e.g., blood pressure regulation, short term and long term). In the simplified descriptions of fluidic systems below, we assumed that laminar flow conditions are present (i.e., there is no turbulence). Laminar flow is generally present when the *Reynolds number* is less than 1000. The Reynolds number of a fluid flowing in a cylindrical tube is given by:

$$R_e = \frac{\rho \bar{v} r}{\eta} \quad (2.1)$$

where η is the viscosity of the fluid in Poise (cgs units), ρ is the density of the fluid (g/cm^3), r is the radius of the tube, and $\bar{v}$ is the average laminar flow velocity (cm/sec).

Viscosity is defined in terms of the time T it takes a liquid to flow a volume V (into a beaker, for example) through a cylindrical tube of length L and radius r, given a pressure drop ΔP over the length of the tube. The viscosity is given by:[90]

$$\eta = \frac{\pi \Delta P \, r^4 \, T}{8 V L} \quad \text{(when cgs units used)} \quad (2.2)$$

The dimensions of η are M L^{-1} T^{-1}.

2.1.1 Fluid Resistance

Fluid resistance is defined as $R_F \equiv \Delta P/\dot{Q}$. $\dot{Q}$ is volume flow (cm^3/sec) through the fluid-resistive element, and ΔP is the pressure difference between the ends of the tube (dyn/cm^2). From the definition of viscosity above, we can write the R_F of a cylindrical tube in which there is laminar flow:

$$R_F = \frac{8 L \eta}{\pi r^4} \quad (2.3)$$

Note that fluid resistance is analogous to electrical resistance, in which ΔP is analogous to the potential difference across the resistor, and the volume flow rate, $\dot{Q}$, is analogous to electrical current. Fluid resistance has the basic dimensions of M L^{-4} T^{-1}.

2.1.2 Fluid Capacitance

Fluid capacitance is defined by $C_F = \Delta V/\Delta P$. C_F is analogous to electrical capacitance, where ΔV is analogous to the change in charge on the capacitor divided by the change in pressure (voltage). Fluid capacitance has the dimensions of M^{-1} L^4 T^2. Several examples of fluid capacitance are given next.

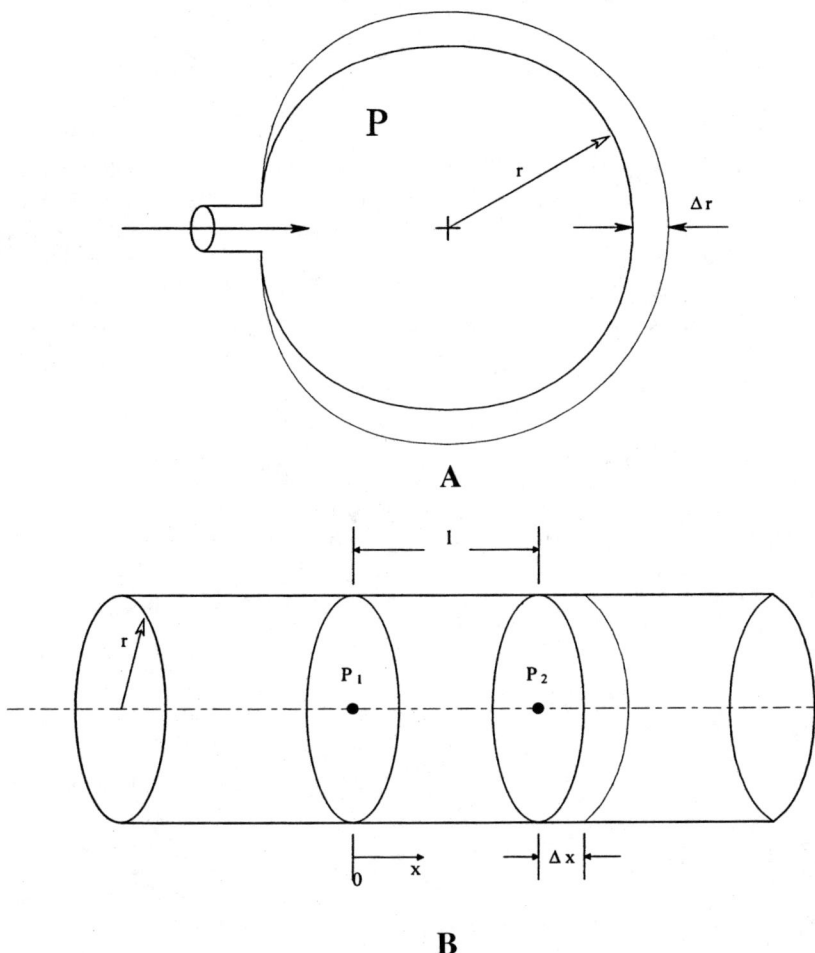

FIGURE 2.1 (A) An elastic spherical vessel. An increase in pressure, ΔP, causes an increase in radius, Δr. (B) Stiff-walled cylindrical vessel illustrating inertance.

1. Consider the C_F of a *spherical elastic vessel* (see Figure 2.1A). (In an anatomical context, this could be a bladder or an eyeball.) For a sphere, the volume $V = (4/3)\pi r^3$. If we assume the pressure causes the radius to vary as $r = f(P) = \sigma P^{(1/3)}$, then $C_F = dV/dP = 4\pi\sigma^3/3 =$ constant. If $r = \sigma P$, then $C_F = 4\pi\sigma^3 P^2$; that is, C_F varies with the square of the pressure.
2. Another important anatomical case is the *elastic cylindrical vessel*. Such a structure has not only capacitance (per unit length) but also fluid resistance, thus forming a distributed-parameter RC transmission line. The volume/length of the elastic cylinder is $v = \pi r^2$. If we assume that

$r = \sqrt{r_o + \sigma P}$, then the capacitance/length is simply, $C_F/L = \pi \sigma$ (a constant). If $r = r_o + \sigma P^{1/2}$, then $C_F/L = \pi \sigma (r_o/P^{1/2} + \sigma) \to \pi \sigma^2$ as $P \to \infty$.

3. A third example of fluid capacitance can be seen for the case of an *open-topped cylindrical tank* that has a constant area as a function of height. This is not a physiological or anatomical analog, but is given to illustrate manipulations of fluidic parameters. The input pipe is at the bottom of the tank. The pressure at the bottom of the tank is simply $P = \rho Y$, where ρ is the density of the fluid and Y is the height of the fluid in the tank. Thus the volume of fluid in the tank is $V = AY$. A is the cross-sectional area of the tank. C_F in this simple case is easily shown to be A/ρ.

2.1.3 FLUID INERTANCE

Fluid inertance, M_F, is analogous to electrical inductance. In fluidic terms, $\Delta P = M_F \ddot{Q}$. Here, $\ddot{Q}$ is the fluid volume acceleration; it is analogous to the rate of change of current, di/dt. In electrical terms, $v = L\, di/dt$. An expression for M_F for a cylindrical tube can be derived from Newton's second law: Consider the fluid-filled tube shown in Figure 2.1B. We have:

$$F = ma = A(P_1 - P_2) = A\, \Delta P = (A L \rho/g) a \tag{2.4}$$

The velocity of the fluid can be written:

$$\dot{Q} = A \frac{\Delta x}{\Delta t} = A \frac{dx}{dt} = A \dot{x} \tag{2.5}$$

The acceleration of the fluid is simply $\ddot{x} = a = \ddot{Q}/A$. Finally, we have:

$$\Delta P = (L\rho/g)(\ddot{Q}/A) = [L\rho/gA]\ddot{Q} = [M_F]\ddot{Q} \tag{2.6}$$

Clearly, $M_F = L\rho/gA$. Inertance has the curious dimensions of $ML^{-5} T^{-2}$.

If a long, thin tube has elastic walls, it is possible that under certain conditions it behaves like a distributed-parameter RLC transmission line. That is, each unit length has a resistance/length, a capacitance/length, and an inertance/length. For a "lossless" electrical transmission line, the *characteristic impedance* is defined as:

$$Z_o \equiv \sqrt{L/C} \quad \text{ohms} \tag{2.7}$$

In the case of a fluidic line,

$$Z_{oF} = \sqrt{\frac{\rho/gA}{\pi \sigma}} \quad \text{"fluidic ohms"}$$

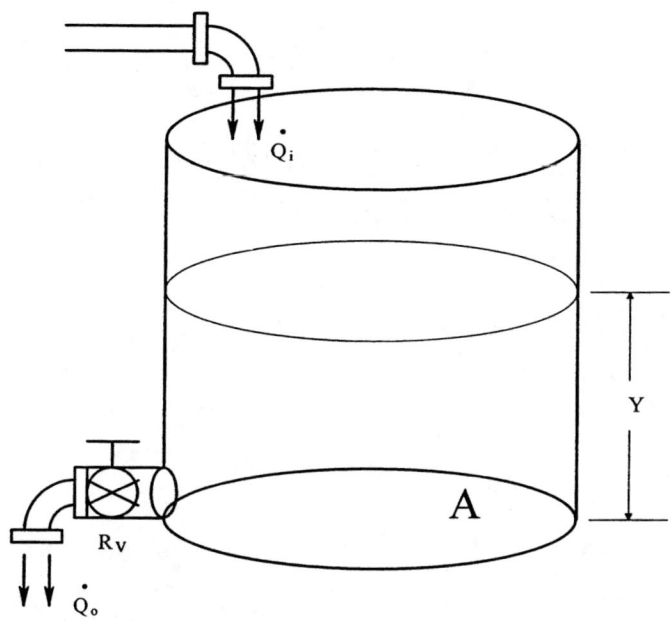

FIGURE 2.2 The liquid level in the open tank is Y.

2.1.4 TRANSFER FUNCTIONS OF FLUIDIC SYSTEMS

We now examine several simple examples of the formulation of linear fluidic systems dynamics. The guiding principle is flow continuity, analogous to Kirchoff's current law.

In the *first example,* illustrated in Figure 2.2, a cylindrical tank is filled at a rate $\dot{Q}_i$. The liquid can leave the tank through a drain valve at the bottom of the tank. The valve has resistance R_V. From flow continuity, we can write that the liquid volume in the tank is

$$Q_T = A Y = \int \left(\dot{Q}_i - \dot{Q}_o \right) dt \qquad (2.8)$$

Differentiating Equation 2.8, we can write:

$$A \dot{Y} = \dot{Q}_i - \rho Y / R_v \qquad (2.9)$$

which reduces to the state equation:

$$\dot{Y} = \dot{Q}_i / A - \rho Y / A R_v \qquad (2.10)$$

This ODE can be Laplace transformed and written as a linear transfer function:

$$\frac{Y(s)}{Q_i(s)} = \frac{1/A}{(s+\rho/AR_v)} \qquad (2.11)$$

This transfer function can be used to find $y(t)$, the filling characteristic of the tank, given $\dot{Q}_i(t)$. Let us assume $\dot{Q}_i(t) = \dot{Q}_{io} U(t)$ (step input). Now the Laplace transform of the liquid height is

$$Y(s) = \frac{\dot{Q}_{io}}{s} \frac{1/A}{(s+\rho/AR_v)} \qquad (2.12)$$

From a table of inverse Laplace transforms, it is easy to show that:

$$y(t) = \dot{Q}_{io} R_v/\rho \left(1 - e^{-t\rho/AR_v}\right) \qquad (2.13)$$

which indicates that the liquid level, y, rises exponentially from zero to a steady-state value of $\dot{Q}_{io} R_V/\rho$ with a time constant $\tau = R_F C_F = AR_V/\rho$ sec.

In the *second example*, we consider the hydraulic system described in Figure 2.3A. A pressure source P_1 (analogous to a voltage source) is considered to be the input to the system. A tube with a resistance R_1 and inertance M is connected to a short tube with resistance R_2 which is terminated in a constant, fluidic capacitance to ground, C_H. A third tube with resistance R_3 and negligible inertance and capacitance makes a "T" at the junction and is attached to a pressure sensor. Because there is no flow in the third tube, there is no pressure drop along it, so the pressure sensor responds to P_2 at the T. Using the electrical analogy shown in Figure 2.3B, we can write a loop equation for the circuit and solve for P_2. Using the voltage-divider relation:

$$P_2 = P_1 \frac{R_2 + 1/sC}{R_2 + 1/sC + sM + R_1} = P_1 \frac{sCR_2 + 1}{s^2 CM + sC(R_1 + R_2) + 1} \qquad (2.14)$$

To use inverse Laplace transforms, the transfer function above must be put in Laplace format:

$$P_2(s) = P_1(s) \frac{(R_2/M)(s + 1/CR_2)}{s^2 + s(R_1 + R_2)/M + 1/CM} \qquad (2.15)$$

The denominator of Equation 2.15 is of the standard quadratic form: $s^2 + s\,2\zeta\omega_n + \omega_n^2$. Obviously, the undamped resonant frequency is $\omega_n = \sqrt{1/CM}$ r/sec, and the damping factor is

Physical and Chemical Factors

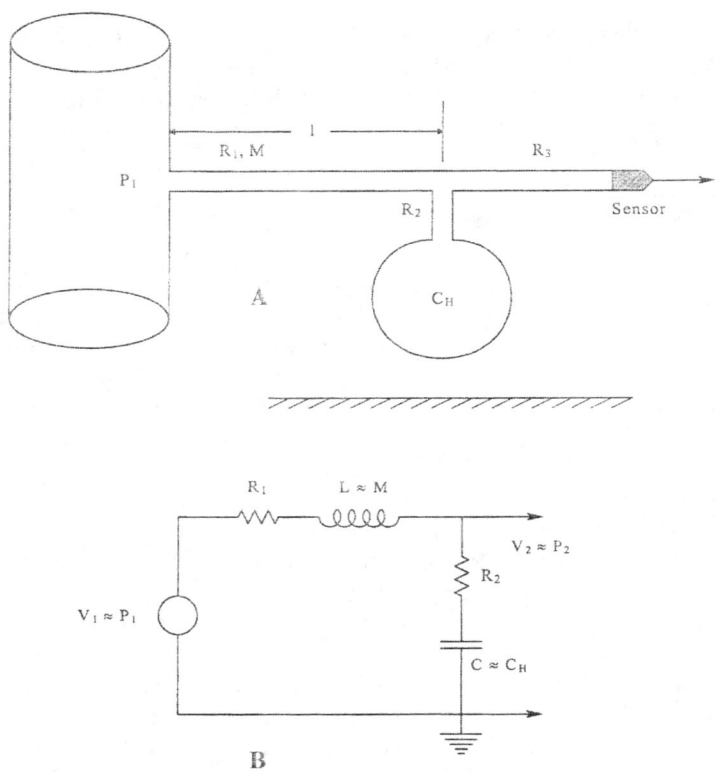

FIGURE 2.3 (A) A hydraulic pressure measurement system. P_2 is the pressure at the "T" and at the sensor. (B) Electrical circuit analog of the hydraulic pressure measurement system.

$$\zeta = \frac{R_1 + R_2}{2}\sqrt{C/M} \tag{2.16}$$

The response of this system to an impulse of $p_1(t) = P_1\,\delta(t)$ can easily be shown to be, using inverse Laplace transform tables,

$$p_2(t) = (P_1\,R_2/M)\frac{1}{\omega_d}\left[\sqrt{(1/CR_2 - \zeta\omega_n)^2 + \omega_d^2}\right]\exp(-\zeta\omega_n t)\sin(\omega_d t + \phi) \tag{2.17}$$

where $\omega_d = \omega_n\sqrt{1-\zeta^2}$ and

$$\phi = \tan^{-1}\left(\frac{\omega_d}{1/CR_2 - \zeta\omega_n}\right)$$

The waveform of Equation 2.17 is an exponentially damped oscillatory transient (assuming $\zeta < 0.5$) due to the inertance/capacitance energy storage in the hydraulic system. Note that air bubbles in a blood pressure measurement catheter introduce capacitance and can lead to underdamped second-order transient response of the measurement system. It is also important to use nonelastic tubing to minimize distributed capacitance in the system.

Dynamic models of the circulatory system can also be constructed in a piecewise-linear method, using R_F, C_F, and M elements or distributed-parameter transmission line analogs. Such modeling is, because of the branching nature of the blood vessels, extremely complicated.

2.2 DIFFUSION DYNAMICS AND FICK'S FIRST LAW

On a microscopic scale, all physiological systems contain cells, as well as molecules and ions suspended or dissolved in physiological fluids. The molecules and ions are in constant random motion as the thermodynamic result of being above 0 K. In general, the higher the temperature, the more rapid and chaotic their movement, and the higher the pressure the particles can exert on the walls of their containers by elastic collisions. The molecules and ions, through their random movements, can also collide with like or different molecules and ions. In such collisions, chemical reactions can occur, or a transfer of momentum can occur in which electrostatic and internuclear forces play a role.

In physiological systems, Fick's laws for *diffusion* describe the average or bulk movement of ions or molecules in response to concentration gradients set up by chemical reactions producing source/sink conditions for the ions or molecules in a volume. Physiological diffusion generally occurs through cell membranes. The ions or molecules generally pass through the membrane at discrete sites, through protein receptors or "pores," which may have specificity for a particular species of ion or molecule. If the protein pores combine either physically or chemically with the ion or molecule diffusing through the membrane, the process is called *facilitated* or *carrier-mediated diffusion*.[59] If we plot the rate of diffusion (ng/min/µm^2) vs. concentration difference across the membrane, we see that the rate saturates above a critical concentration difference. This saturation may be due to (1) the finite number of diffusion sites/areas on the membrane and/or (2) the molecular processes whereby an ion or molecule binds to the receptor molecule(s), causing a configurational change which transports the ion or molecule across the membrane, occurs slowly (i.e., the transport process is rate limiting).[59]

In some cases, a second messenger molecule or ion, if present, can modulate the permeability of the pore, producing *ligand-gated diffusion*. For example, the hormone insulin increases the diffusion for glucose molecules at glucose pore sites. In the presence of insulin, the bulk permeability for glucose (in insulin-sensitive cells) rises, increasing the rate of diffusion of glucose into those cells. The glucose diffuses

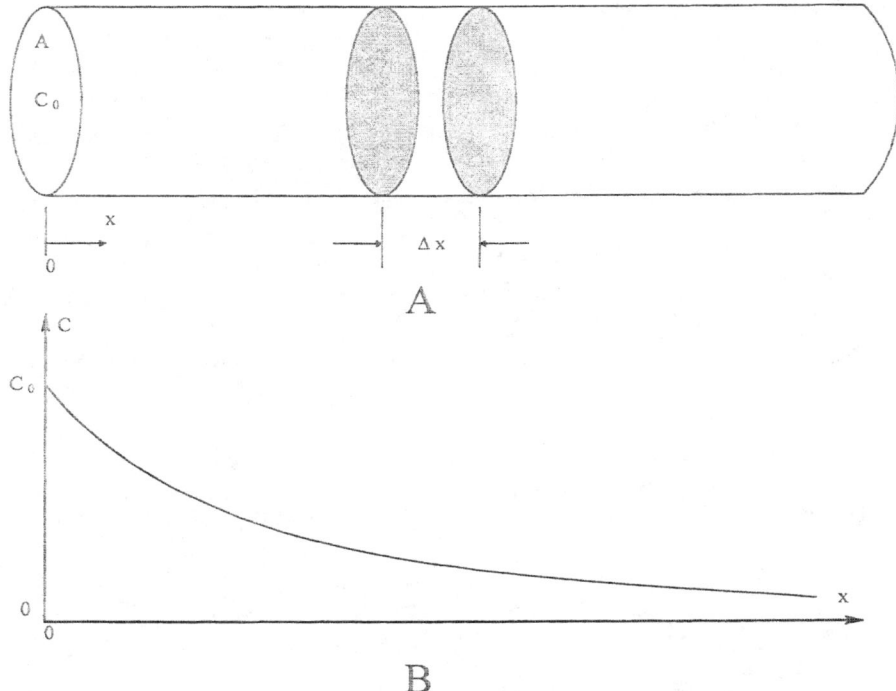

FIGURE 2.4 (A) Diffusion in one dimension along a tube. (B) Distribution of concentration along the tube at some time t_1.

from a higher extracellular concentration to a lower intracellular concentration. Other types of pores are opened by a change in the transmembrane potential difference, producing *voltage-gated diffusion*. Voltage-gated diffusion figures largely in the generation of nerve impulses, or in their inhibition, and in the triggering of muscle contraction.

In deriving a quantitative description of simple diffusion, we consider the one-dimensional model shown in Figure 2.4. The tube has cross-sectional area A. The concentration of a molecule S at x_1 is C_1; at x_2 it is C_2. Let $C_1 > C_2$. We assume that the probability of a molecule of S jumping in either $-x$ or $+x$ direction is equal. Furthermore, the mass/time of S going from plane 1 to plane 2 is proportional to C_1 @ x_1. Likewise, the mass/time going from plane 2 to 1 is proportional to C_2 @ x_2. The mass rates are also proportional to A. Thus we can write:

$$\frac{\Delta m}{\Delta t} = \beta A (C_1 - C_2) = -\beta A \Delta C \quad (2.18)$$

β is approximated by $k/(x_2 - x_1) = k/\Delta x$. $\Delta C \equiv (C_2 - C_1) < 0$.

Substituting this relation into Equation 2.18 and letting Δs approach 0, we can write:

$$\frac{\partial m}{\partial t} = -k A \frac{\partial C}{\partial x} \tag{2.19}$$

In the above one-dimensional diffusion equation, m is the mass diffusing, k is Fick's constant with dimensions $L^2 \, T^{-1}$, and $\partial C/\partial x$ is the concentration gradient in the $-x$ direction. When a thin membrane separates C_1 and C_2, the diffusion equation can be simplified to:

$$dm/dt = -k A \Delta C = k A(C_1 - C_2) \quad \mu g/sec \tag{2.20}$$

Because cells individually have small intracellular volumes compared with the total extracellular volume, the prolonged influx of m will raise C_2, lowering the concentration difference across the membrane and hence reducing the diffusion rate, dm/dt. If nothing happens to m once it diffuses into the cell, eventually $C_2 = C_1$ in the steady state, and $dm/dt = 0$.

The diffusion equation can be put in a general three-dimensional form:[1]

$$\frac{\partial c}{\partial t} = k \left[\frac{\partial^2 c}{\partial x^2} + \frac{\partial^2 c}{\partial y^2} + \frac{\partial^2 c}{\partial z^2} \right] \tag{2.21}$$

Here, c is the volume distribution of concentration at time t, $c = c(x,y,z)$. In vector notation, Equation 2.21 becomes

$$\frac{\partial c}{\partial t} = k \nabla^2 c \tag{2.22}$$

The mass current density is defined as:

$$\mathbf{J} = \left(\frac{1}{A} \frac{\partial m}{\partial t} \right) \mathbf{n} = k \nabla c \tag{2.23}$$

where $\mathbf{n}$ is a unit vector normal to the surface A. If the molecule diffusing is being manufactured in a region in x,y,z space at some rate per unit volume, we can write the *inhomogeneous diffusion equation*:

$$\frac{\partial c}{\partial t} = k \nabla^2 c + q(x, y, z) \tag{2.24}$$

Note that some diffusion equations are written in terms of mass/time and others in terms of concentration/time; concentration is simply mass/volume. In most descriptions of diffusion in physiological modeling, we try to use the simple approach as shown in Equation 2.20.

2.3 MASS-ACTION KINETICS IN CHEMICAL SYSTEMS

Biochemical reactions, such as the regulated synthesis of hormones and proteins, underlie nearly all physiological regulatory and control processes. All chemical reactions proceed at rates governed by the concentrations of the reactants, the temperature of the reaction, and the number of molecules of reactants required to produce a molecule of product. One underlying assumption in formulating mass-action descriptions of chemical kinetics is that the reactants are uniformly distributed (in a compartment) and are free to move around and collide with a probability that is proportional to their concentration(s). Another assumption is that mass-action formulations are based on large numbers of molecules interacting; mass action is therefore a bulk formulation of chemical behavior.[90]

As we shall see, mass-action systems generally involve solutions of nonlinear ODEs. Thus, dynamic solutions are best done by computer simulation. Some simple examples will be given here that illustrate the procedures to be used and also some typical biochemical reaction architectures.

2.3.1 Examples of Mass-Action Kinetics

As a *first example*, let us consider a bimolecular reaction in which one molecule of A combines irreversibly with one molecule of B to form one molecule of product, C. That is, in chemical notation, $A + B \xrightarrow{k_1} C$. In the first kinetic formulation, we start at $t = 0$ with a moles of A, b moles of B, and no C. $x =$ moles of C made at time t. By mass action, we can write the ODE:

$$\dot{x} = k_1(a-x)(b-x) = k_1\left[ab - x(a+b) + x^2\right] \qquad (2.25)$$

k_1 is the reaction rate constant, generally an increasing function of Kelvin temperature.

An alternative mass-action ODE uses the running concentration (RC) of $A = y = (a - x)$, the RC of $B = z = (b - x)$, and the RC of $C = x$. Initial conditions are used. Thus the rate of appearance of C is

$$\dot{x} = k_1 yz \qquad (2.26)$$

and the rate of disappearance of A is

$$\dot{y} = k_1 yz \qquad (2.27)$$

A *second example* of mass-action kinetics is given for the *reversible* oxidation of nitrogen oxide. The chemistry is

$$2NO + O_2 \underset{k_2}{\overset{k_1}{\rightleftharpoons}} 2NO_2 \qquad (2.28)$$

Let k_1 = the forward rate constant, k_2 = the reverse rate constant, a = the initial concentration of NO, b = the initial concentration of O_2, and x = the amount of O_2 reacted at t. The forward reaction rate is

$$\dot{x}_f = k_1(a - 2x)^2(b - x)$$

The reverse reaction rate is

$$\dot{x}_r = k_2(2x)^2$$

The net reaction rate is

$$\dot{x}_{net} = \dot{x}_f - \dot{x}_r = k_1(a - 2x)^2(b - x) - k_2(2x)^2$$

(Note that when two molecules react, the concentration is squared; when three molecules react, it is cubed.)

For a *third example,* we consider the physiologically important formation and decomposition of carbonic acid:

$$H_2CO_3 \underset{k_R}{\overset{k_F}{\rightleftharpoons}} H_2O + CO_2 \qquad (2.29)$$

Let x = the concentration of carbonic acid, w = the concentration of water, and g = the concentration of dissolved CO_2. Thus we can write for the rate of appearance of carbonic acid:

$$\dot{x} = -k_F\, x + k_R(w\, g) \qquad (2.30)$$

The rate of appearance of CO_2 is simply:

$$\dot{g} = +k_F\, x - k_R(w\, g) = -\dot{x} \qquad (2.31)$$

If water is present in excess, then the w factor can be eliminated from the ODEs above and its effect incorporated in k_R.

For a *fourth example,* we examine a typical biochemical "two-step" reaction, in which two reactants combine reversibly to form a complex, and then the complex is rapidly converted to the end product with the release of the E reactant unchanged. The product P decays at rate $k_4 P$. This reaction form is known as the *Michaelis–Menten* architecture.[55,95]

$$X + E \underset{k_2}{\overset{k_1}{\rightleftharpoons}} E*X \xrightarrow{k_3} E + P \xrightarrow{k_4} * \qquad (2.32)$$

where x_1 = the running concentration of substrate X, u = the running concentration of free enzyme E, x_2 = the running concentration of the complex, x_3 = the running concentration of product P, and k_1, k_2, k_3 = reaction rate constants. The reaction is assumed to take place in a closed vessel, so enzyme is conserved. That is, $u_o = u + x_2$, or $u = u_o - x_2$. The nonlinear mass-action ODEs can be written:

$$\dot{x}_1 = -k_1(u_o - x_2)x_1 + k_2 x_2 \tag{2.33}$$

$$\dot{x}_2 = k_1(u_o - x_2)x_1 - (k_2 + k_3)x_2 \tag{2.34}$$

$$\dot{x}_3 = k_3 x_2 - k_4 x_3 \tag{2.35}$$

From enzyme conservation, we see that $\dot{x}_2 = -\dot{u}$. If we assume steady-state conditions, $\dot{x}_1 = \dot{x}_2 = \dot{x}_3 = 0$, and we have:

$$x_{2SS} = \frac{k_1 x_1 u_o}{(k_2 + k_3) + k_1 x_1} = \frac{x_1 u_o}{x_1 + K_M} \tag{2.36}$$

$K_M \equiv (k_2 + k_3)/k_1$ is the well-known (to biochemists) *Michaelis constant*.

If we set $x_1 \equiv x_{1o}$ (constant) and $u \equiv u_o$ (constant) and again assume steady-state conditions where all derivative terms $\to 0$, the equilibrium concentration of P is easily shown to be

$$x_{3SS} = \frac{k_3 x_2}{k_4} = \frac{k_3}{k_4} \frac{k_1 u_o x_{1o}}{(k_2 + k_3)} = \frac{k_3 u_o x_{1o}}{k_4 K_M} \tag{2.37}$$

It is clear that dynamic solutions of the three Michaelis–Menton equations are best done by computer simulation.

As a *fifth example*, consider a compartment surrounded by a diffusion membrane (perhaps this is a cell). In the compartment is an enzyme in excess concentration that catalyzes the reversible transformation of A into B. Furthermore, A must diffuse into the compartment, and B diffuses out. The concentration of A outside the membrane is a_o; inside the compartment it is a. Likewise, the concentration of B outside the compartment is b_o, and it is b inside. Refer to Figure 2.5. The rate of increase of A and B inside the compartment is thus governed by both mass action and diffusion. The ODEs are

$$\dot{a} = K_{da}(a_o - a) + k_{-1} b - k_1 a \quad \text{kg/sec} \tag{2.38A}$$

$$\dot{b} = -K_{db}(b - b_o) + k_1 a - k_{-1} b \quad \text{kg/sec} \tag{2.38B}$$

K_{da} and K_{db} are diffusion rate constants. The k_1 and k_{-1} terms relate to simple mass action. Note that in this special case, the system is linear and can be solved for its states, a and b, using conventional linear algebraic methods and Laplace transforms.

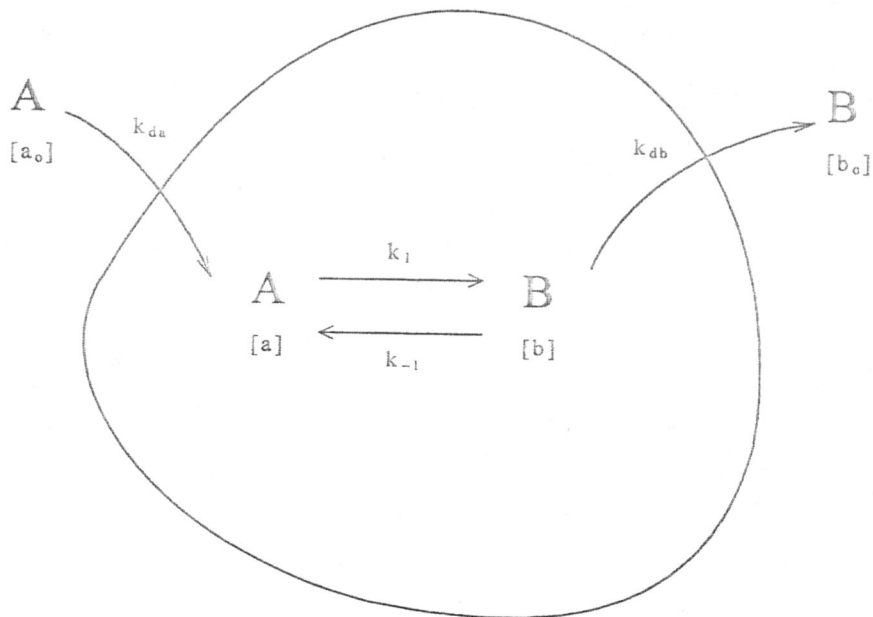

FIGURE 2.5 Schematic of events in the example 5: A diffuses into the "cell" and is transformed into B. B then diffuses out across the membrane.

Diffusion and mass-action dynamics are often necessarily used together to describe intracellular biochemical events, such as the synthesis of a hormone in response to a signal from a control hormone. The control hormone must enter the cell, and the synthesized hormone must diffuse out through the cell's membrane to enter the bloodstream in order to reach its intended remote target cells. In addition, reaction substrates, such as glucose and oxygen, must diffuse into the cell. The mass-action dynamics governing the hormone's synthesis are generally nonlinear and high order; no simple Laplace solution is possible, as in the fourth example above.

In the *sixth example*, we consider the hypothetical situation where there is a finite number, N, of cell membrane receptors per unit volume. At any instant, there is a density F of free receptors which can bind with molecules of an input hormone, H_1, to form a complex. Once bound, the complex initiates the biosynthesis and release of a second messenger hormone, H_2. The complex is broken down enzymatically to release an inactivated input hormone molecule, $\overline{H_1}$, and a free membrane receptor. Intuitively, we see that this system will saturate for high concentration of H_1. That is, there can be no further increase in the rate at which H_2 is produced because nearly all of the free receptors are complexed at any time. The processes can be represented by:

$$H_1 + F \underset{k_2}{\overset{k_1}{\rightleftharpoons}} H_1 * F \xrightarrow{k_3} \overline{H_1} + F \qquad (2.39\text{A})$$

Physical and Chemical Factors

$$H_1 * F \xrightarrow{k_4} H_2 \xrightarrow{k_5} * \qquad (2.39B)$$

The total number of receptors $N = f + b$. f is the number of free receptors; b is the number of receptors complexed with H_1. The system mass-action equations are

$$\dot{b} = -(k_2 + k_3)b + k_1 h_1 f = -(k_2 + k_3)b + k_1 h_1 (N - b) \qquad (2.40A)$$

$$\dot{h}_2 = -k_5 h_2 + k_4 b \qquad (2.40B)$$

In the steady state, we see from Equation 2.40A that:

$$b_{SS} = N h_1 / (h_1 + K_M) \to N \qquad (2.41)$$

for $h_1 \gg K_M$. K_M is the Michaelis constant for the system. Also:

$$h_{2SS} = k_4 b_{SS} / k_5 \qquad (2.42)$$

Thus the concentration of H_2 exhibits saturation as a function of H_1 as well.

Remember that mass-action kinetics depend on the assumption that large numbers of molecules of a given species are uniformly distributed in a volume where the reaction is taking place. Many biochemical reactions take place on membrane surfaces where concentration gradients of reactants and products can exist. Such gradients invoke diffusion dynamics in addition to mass-action ODEs. This latter situation is generally neglected in modeling because of the inhomogeneity of tissues surrounding the cell membranes in question.

2.4 CHAPTER SUMMARY

In this chapter, some of the physical and chemical factors that can contribute to the dynamics of physiological systems were examined. Fluidic systems, considered for the simplifying case of laminar flow, were shown to have direct analogy to linear RLC electrical circuits. Pressure is analogous to voltage, volume flow is analogous to electric current, and the fluidic parameters of resistance, capacitance, and inertance are analogous to electrical resistance, capacitance, and inductance. Thus linear circuit analysis techniques (e.g., Kirchoff's laws, node equations, loop equations, Thevenin's theorem, and Norton's theorem) are directly applicable to the analysis of any fluidic system operating in its linear range. Mathematical descriptions of the flows and pressures in the peripheral circulatory system, the aqueous humor of the eye, the cerebrospinal fluid, and the filtrate in the kidneys can be formulated in terms of lumped- and distributed-parameter fluidic RLC models.

Descriptions of diffusion of biomolecules based on Fick's law were seen to be important in modeling transcellular mass-flow dynamics, such as the hormonally controlled uptake and release of glucose by liver cells (net hepatic glucose balance).

Classical three-dimensional diffusion descriptions are seldom used in physiological modeling because tissues are generally not uniform and are anisotropic.

The dynamics of biochemical metabolism are described by chemical mass-action kinetics. Reaction rates in biochemical systems were shown to be modeled by systems of first-order nonlinear ODEs. Parametric control of reaction rates may be through hormones or intermediate and end product concentrations. Modulation of substrate availability is another means of governing biochemical reaction rates. Biosynthesis mass-action and diffusion ODEs are often part of a coherent system description. Necessarily, there must be sources and sinks of biomolecules. Enzymes, hormones, and neurotransmitters are broken down or metabolized, and their component molecules and radicals are biochemically recycled. Sometimes the breakdown products are reassembled locally, as in the case of the neurotransmitter acetylcholine, and other molecules such as proteins are broken down into their individual amino acids, which serve as metabolic currency for many types of cells and biochemical reactions.

When viewed on the level of cell physiology, biochemical homeostasis is very complex. If cells are viewed as biochemical factories, then external hormones act as signals to produce or to stop production. Many reactions take place at specific sites on membranes or in organelles such as mitochondria or cell nuclei. Other reactions take place in the cytoplasmic volume and thus better fit the framework of the mass-action assumptions. In this text, we shall examine bulk properties of physiological regulators and control systems, largely relieving ourselves of the necessity to model detailed, intracellular, biochemical events.

PROBLEMS

2.1 Reactants A and B combine reversibly to form compound P inside a cell. P diffuses out of the cell to "zero" concentration. Assume that reactant B has

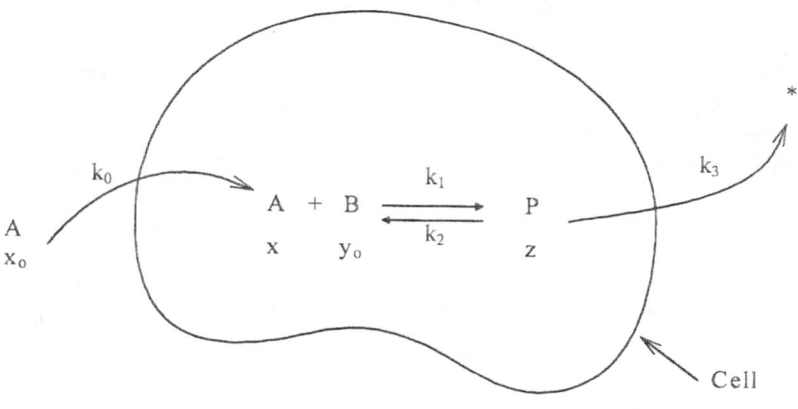

PROBLEM 2.1

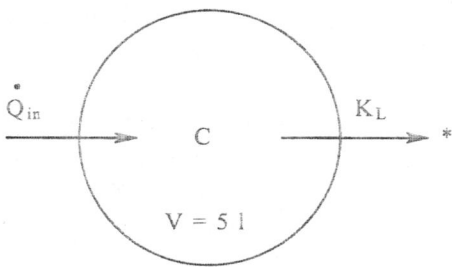

PROBLEM 2.3

a constant concentration, y_o, inside the cell. A diffuses into the cell from outside, where its concentration is x_o.
  A. Write the state equations for x and z, the running concentrations of A and P, respectively. Is the system linear?
  B. Give an expression for the steady-state z, given $x_o > 0$.

2.2 Refer to Problem 2.1. Now assume that reactant B is made inside the cell at rate $\dot{y}_o$ (i.e., B no longer has a fixed concentration). A biochemical feedback system has been discovered which acts according to the rule $\dot{y} = \dot{y}_{oo} - \beta z$, where $\dot{y}$, $\beta > 0$, and $z \geq 0$.
  A. Write the state equations for the system.
  B. Give the expression for the net outward diffusion rate of P in the steady state.

2.3 A drug is infused into a single pharmacokinetic compartment at rate $\dot{Q}_{in}$ (μg/min). The volume of the compartment is 5 l. In the steady state, the concentration in the compartment is 10 μg/l. If the infusion stops, the concentration c in the compartment falls exponentially. After 2.7 hr, $c = 5$ μg/l.
  A. Make a signal flow graph for the linear system. Give its transfer function, $C(s)/\dot{Q}_{in}(s)$.
  B. Find the steady-state input infusion rate, $\dot{Q}_{in}$.
  C. Give the system's loss rate constant, K_L, and its time constant, τ, in minutes.

2.4 A hormone H controls the diffusion of a molecule M into a cell. Let h = the extracellular hormone concentration (ng/ml), m_e = the extracellular molecule concentration (ng/ml), and m_i = the intracellular molecule concentration (ng/ml). The diffusion rate constant is given by $K_d = K_{do} + a H^2$. K_L is the loss rate constant for m_i.
  A. Write the state equation for m_i.
  B. Draw a block diagram describing the system's dynamics.
  C. Find an expression for the steady-state m_i vs. h. Assume $K_L \gg K_{do}$.
  D. Plot and dimension the system's natural frequency, ω_o, vs. h.

2.5 A proposed eight-state model for the storage dynamics for thyroid hormone (TH) and 3-iodotyrosine (3IT) is shown in the figure. Note that there are two compartment volumes (extracellular fluid and intracellular fluid) as well as

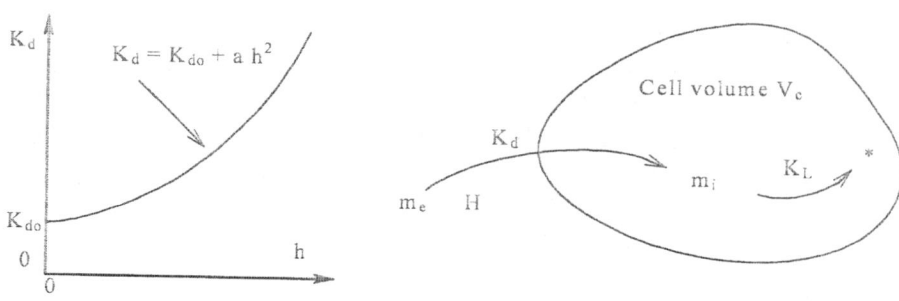

PROBLEM 2.4

storage equilibrium reactions for TH and 3IT in both volumes. The states are concentrations. In addition to the storage equilibria, TH is deiodinized to form 3IT in both volumes. Assume that the concentration of thyroid-hormone-binding globulin is constant everywhere at $x_0 = 1$ ng/ml.

A. Write the remaining seven state equations for the system. Assume that the diffusion rates are proportional to concentration differences. The first state equation is

$$\dot{x}_1 = -x_1(k_1 x_0 + k_5 + k_7) + x_2 k_2 + x_5 k_7 + u_1$$

B. Assume $k_1 = \ldots = k_{13} = 1$, $k_{14} = 10$. Let $u_1(t) = 10\delta(t)$. Plot the states vs. time to scale. Simulate the system with Matlab® or Simnon™.

2.6 In this problem, you will derive an expression for the *fluid capacitance per unit length* of a thin-walled, elastic, cylindrical vessel with internal pressure P_i. The internal pressure exerts a differential radial force on the vessel walls, $P_i \, r \, d\theta \, dL = dF$. The radial force can be resolved into a tangential hoop force

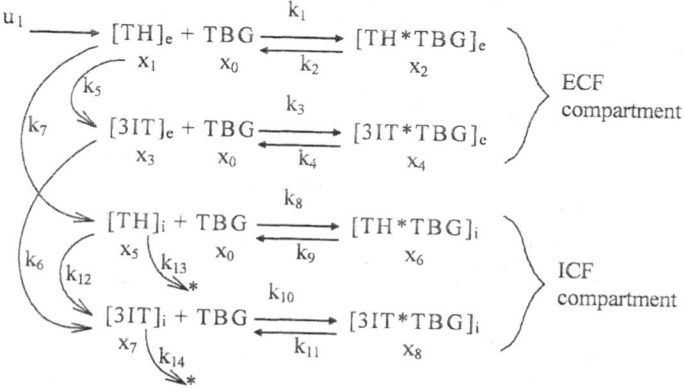

PROBLEM 2.5

per length L of the cylinder, $T = P_i\, r$ N/m. Thus the circumference of the cylinder, c, is increased by the pressure, which in turn increases the volume of the cylinder. The hydraulic capacitance is $C_H = dV/dP_i \cong \Delta V/\Delta P_i$. Therefore, $c = 2\pi r = c_o + K[T]$. K is the elastic constant (m²/N), r is the cylinder radius at pressure P_i, and r_o is the radius when $P_i = 0$, $c_o = 2\pi r_o$.
  A. Derive an expression for the radius of the cylinder as a function of P_i.
  B. Find the capacitance/length for $P_i \to 0$. Plot and dimension C_H/L vs. P_i.

2.7 A Michaelis-type chemical reaction in which parametric control is present is shown in the figure. In this case, the end product concentration, p, affects the rate constant K_1 according to the law $K_1 = \lambda/p$.

$$\begin{array}{ccccccc}
& k_1 & & k_3 & & k_4 & \\
R + E & \rightleftarrows & R{*}E & \longrightarrow & E + P & \longrightarrow & * \\
r \quad e & k_2 & c & & e \quad p & &
\end{array}$$

  A. Write the ODEs governing the system's dynamics. Note that enzyme E is conserved, so that $e + c = e_o$.
  B. In the steady state, given constant r, the derivative terms de/dt and $dp/dt \to 0$. Solve for the steady-state p, given that $r = r_o$ and $K_1 = \lambda/p$. Show that p_{SS} is independent of r.

2.8 An inherent biosynthetic regulation of end product P arises from P competing for the same enzyme binding site as the substrate S. The reactions are shown in the figure. From chemical mass action, we have the equations

$$e_o = c + q + e$$

where e_o is total enzyme (conserved), e is free enzyme, q is enzyme complexed with Q, and c = enzyme complexed with S.

$$\begin{array}{ccccc}
& k_1 & & k_3 & \nearrow^{k_4 \;\; *} \\
S + E & \rightleftarrows & S{*}E & \longrightarrow & P + E \\
& k_2 & & & \\
s \quad e & & c & & p \quad e
\end{array}$$

$$P + E \underset{k_6}{\overset{k_5}{\rightleftarrows}} Q$$
$$p \quad e \qquad\qquad q$$

PROBLEM 2.8

$$\dot{c} = k_1 es - (k_2 + k_3)c$$
$$\dot{p} = k_3 c - k_4 p - k_5 pe + k_6 q$$
$$\dot{q} = k_5 pe - k_6 q$$

In the steady state, the derivatives → 0.
A. Using the steady state assumption, where $s = s_1$ = constant, derive expressions for $p_{SS} = f(e_{SS}, s_1)$ and $e_{SS} = g(p_{SS}, s_1)$.
B. Solve the quadratic equation derived from the results of part A for $p_{SS} = h(s_1)$.
C. Show what happens when we assume that $k_1 s_1 \gg (k_2 + k_3)$.

2.9 In 1961, Spangler and Snell [*Nature*, 191(4787): 457–458] proposed a chemical–kinetic model as a possible mechanism to describe "biological clocks." Their model was based on the concept of cross-competitive inhibition. The four chemical reactions shown below are described by eight state equations.

$$(A) \xrightarrow{J_a} A + E \underset{k_{1r}}{\overset{k_1}{\rightleftharpoons}} B \xrightarrow{k_2} E + P \xrightarrow{k_3} *$$

$$nP + E' \underset{k_{4pr}}{\overset{k_{4p}}{\rightleftharpoons}} I'$$

$$nP' + E \underset{k_{4r}}{\overset{k_4}{\rightleftharpoons}} I$$

$$(A') \xrightarrow{J_{ap}} A' + E' \underset{k_{1pr}}{\overset{k_{1p}}{\rightleftharpoons}} B' \xrightarrow{k_{2p}} E' + P' \xrightarrow{k_{3p}} *$$

Substrates A and A' diffuse into the reaction volume at constant rates J_a and J_{ap}, respectively. A and A' combine with enzyme/catalysts E and E', respectively, and are transformed reversibly into complexes B and B'. B and B' next decompose to release active enzymes and products P and P'. P and P' next diffuse out of the compartment to effective zero concentration with rates k_3 and k_{3p}, respectively. These two main reactions are entirely independent. However, the system becomes cross-coupled and therefore dependent and possibly unstable because of the reversible combination of the products with the complementary free enzyme. The entire system is described by eight state equations (mass-action ODEs): a = running concentration (rc) of A, b = rc of B, p = rc of P, i = rc of I, ap = rc of A', bp = rc of B', pp = rc of P', and ip = rc of I'. There are no ODEs for e and ep. Assume the enzymes are conserved in the compartment (i.e., enzymes are either free or tied up as complexes; enzymes are not destroyed or created). Thus $e = e_0 + b + i$ and $ep = e_{p0} + bp + ip$, where e_0 is the initial (total) concentration of enzyme E and e_{p0} is the initial (total) concentration of E'. Note that n molecules of P

in the compartment combine with one molecule of E' to reversibly form one molecule of complex I', etc. Thus when I' breaks down, n molecules of P are released with one molecule of E', etc.

A. Using the notation above and mass-action kinetics, write the eight ODEs and two algebraic equations describing the coupled Spangler and Snell system.

B. To illustrate the awesome nonlinearity of the system, draw a detailed signal flow graph describing the system. (You will need to use multipliers.) Note that Mason's rule cannot be used on such nonlinear systems.

C. Simulate the system's behavior at "turn-on." Note that the initial conditions of states b and $bp = 4$. Other states have zero initial conditions. The initial amounts of free enzymes e_0 and $e_{p0} = 5$. Try the parameters listed below. Use a (vertical) amplitude scale of 0 to 30 and a time scale of 0 to 3000 (the system is *slow*). Verify that the system has a steady-state limit cycle by plotting dp/dt vs. $p(t)$. Plot the system parameters vs. time to see if you can alter the steady-state oscillation period and/or amplitudes. How critical are parameters to establish sustained steady-state oscillations?

```
CONTINUOUS SYSTEM snell2  "8/24/98 Cross-coupled enzymatic oscillator
STATE  a  b  p  i  ap  bp  pp  ip
DER    da db dp di dap dbp dpp dip
TIME t
"
da  = - k1*ra*re + k1r*rb + Ja           " Ja is input rate of A.
"
db  = k1*ra*re - (k1r + k2)*rb
"
dp  = k2*rb - k3*rp - k4p*rp^n*rep + k4pr*rip*n
"
dip = k4p*rp^n*rep - k4pr*rip
"
re  = e0 - ri - rb             " Initial, constant amount of E = e0.
"
rep = ep0 - rip - rbp          " Initial, constant amount of E' = ep0.
"
dap = - k1p*rap*rep + k1pr*rbp + Jap
"
dbp = k1p*rap*rep - k1pr*rbp - k2p*rbp
"
dpp = -k3p*rpp + k2p*rbp - k4*rpp^n*re + k4r*ri*n
"
di  = k4*re*rpp^n - k4r*ri
"
ra  = IF a  > 0 THEN a  ELSE 0
rb  = IF b  > 0 THEN b  ELSE 0
rp  = IF p  > 0 THEN p  ELSE 0
ri  = IF i  > 0 THEN i  ELSE 0
rap = IF ap > 0 THEN ap ELSE 0
rbp = IF bp > 0 THEN bp ELSE 0
```

```
rpp = IF pp > 0 THEN pp ELSE 0
rip = IF ip > 0 THEN ip ELSE 0
"
Ja:3
Jap:1
e0:5
ep0:5
k1:0.1
k1r:0.05
k1p:0.1
k1pr:0.05
k2:1
k2p:1
k3:0.3
k3p:0.3
k4:0.5
k4p:0.5
k4r:0.5
k4pr:0.5
n:2
"
b:4
bp:4
END
```

—

[snell2] " SNELL10.T parameters; use with SNELL.T pgm. Gives self-sustaining "oscs in P & P'. 8/23/98

b:4., bp:4., k1:0.1, k1r:0.05, Ja:3., Jap:1., k2:1., k3:0.3, k4p:0.5, n:2., k4pr:0.5, e0:5., ep0:5., k1p:0.1, k1pr:0.05, k2p:1., k3p:0.3, k4:0.5, k4r:0.5

2.10 In an interesting paper by Chance, Pye, and Higgins, (*IEEE Spectrum*, pp. 79–86, August 1967), the authors give the architecture for another putative biochemical–chemical–kinetic oscillator. The enzyme is conserved. Thus $E_o(0) = E_aB(0) + E_a(0) + E_I(0)$. E_a is the uncomplexed reactive form of the enzyme, E_aB is complexed, and E_i is the inactive form. Chance et al. assume the chemical reactions follow the architecture:

$$A \xrightarrow{J_o} B$$

$$B + E_a \underset{k_2}{\overset{k_1}{\rightleftharpoons}} E_aB \xrightarrow{k_3} C + E_a$$

$$C \xrightarrow{k_6} D$$

$$E_a \underset{ck_6}{\overset{k_5}{\rightleftharpoons}} E_i$$

A. Write the mass-action ODEs for the running concentrations of *b, c, eab, ea,* and *ei*.
B. Draw the nonlinear signal flow graph for the system.
C. Simulate the system using various [*k*] and initial conditions, and see whether you can make the system exhibit *sustained* oscillations. Begin with

```
J0:.04, k1:1, k2:1, k3:1, k4:.5, k5:1, k6:1.  ICs: b:0,
eab:0, c:0, ea:2, ei:0
```

Note: Similar to Problem 2.9, the simulation time scale will need to be long (0 to 1000) to view any oscillatory system behavior. If sustained oscillations (limit cycles) cannot be obtained, what prevents them?

3 Introduction to Linear SISO Control Systems and Systems with Delays

3.0 INTRODUCTION

In this chapter, the basic analysis and design tools used with single-input/single-output (SISO) linear control systems are reviewed. Attention is focused on the control of real-pole plants that commonly result from modeling linear compartmental pharmacokinetic (CPK) systems or physiological systems. Regulation of a certain state can be *endogenous* (by the organism) or *exogenous* (as by the controlled infusion of drugs). This chapter concentrates on exogenous regulation and the various "classical" means to improve system performance. The basics of compensation of closed-loop system performance by inserting certain filters in the feedback path are reviewed in Section 3.1. Proportional-plus-integral-plus-derivative (PID), lead, lag, and lead/lag filters are described. The use of proportional-plus-integral (PI) and PID controllers is also introduced. PI controllers insert a real zero plus a pole at the origin in the loop gain. PID controllers insert a pair of zeros (they can be complex-conjugate) and a pole at the origin in the loop gain. Thus PI or PID controllers can convert a type 0 regulator acting on a real-pole plant into a type 1 system having zero steady-state error. PID systems must be designed carefully to avoid long settling times caused by the introduced zeros.

Section 3.2 describes in detail the use of the root locus technique to analyze and design closed-loop regulators by closed-loop pole placement. Several examples of the use of root locus are given. The Matlab® root locus package is shown to be very useful in quantitative pole-placement design.

In Section 3.3, some "classic" tests for the stability of linear closed-loop systems are reviewed. The algebraic Routh–Hurwitz test is described and examples are given. The graphical Nyquist stability criterion is derived and application examples are shown. The application of the Nyquist stability test to nonlinear systems in the form of describing function analysis is given in Section 3.3.4. Describing functions are shown to be amplitude- and sometimes frequency-dependent equivalent gains which replace the nonlinear element. The Popov stability criterion for nonlinear systems is described, and its relationship to the Nyquist method and describing functions is given.

The stability problems of linear systems with transport lags in their loops are examined in Section 3.4. Stabilizing controllers for these systems are introduced, with special attention to the Smith delay compensator, a model reference control strategy that can cancel the unstabilizing effects of the transport lag in the loop gain.

3.1 CONVENTIONAL CONTROLLERS FOR LINEAR PLANTS WITH REAL POLES

In designing external feedback control systems for compartmental pharmacokinetic/physiological (CPK/P) systems, one generally wants the controlled variable to assume its steady-state, regulated level in minimum time, *without* overshoot (or undershoot). A CPK/P control system must also be robust. That is, it must maintain the set point in the presence of noise entering the loop and tolerate plant parameter changes with time. Nonstationary plant parameters are characteristic of physiological systems; plant parameters can change as the result of the drug being injected, due to intrinsic, diurnal, biological rhythms and from interaction with other physiological systems.

The plants, as we have observed, generally have real poles, can be frankly nonlinear, can have transport lags, and can have dead zones caused by thresholds. In controlling a CPK/P plant, intermediate states are not always available (although initially someone had to measure them in order to characterize the system). For example, in the glucose regulatory system, the mass of glycogen in the liver may figure in the plant dynamics, but we cannot noninvasively measure liver glycogen in a living animal. Even in the face of these potential difficulties, conventional linear-control-systems design techniques often can be applied to CPK/P systems. Some of these techniques are outlined below. Chapter 5 introduces nonlinear and parameter-switching controllers, which are somewhat off the beaten path for mainstream biomedical control engineers. Their simplicity and effectiveness, however, offer motivation for their introduction in this text.

One generally wants the externally controlled CPK/P plant to have zero steady-state error. This implies either that the set point be multiplied by a gain correction factor in a type 0 system or that the control system be compensated so that it is type 1 (i.e., has one loop gain pole at the origin). A PI or PID controller is generally used to make a type 1 feedback system from a real-pole plant.

Also, in CPK/P systems, the control input to the plant, u, is nonnegative. u is generally the rate of infusion of a drug or hormone. The magnitude of u is determined in the continuous case by the drive speed of an infusion pump. The linear dynamics that determine u are those of the system's controller. In the simplest case, $u = eK_c = (SP - y)K_c$. If the plant has all real poles (no pole at the origin), the loop gain will be type 0, and the closed-loop system will have a steady-state error for a constant SP.[108] The position of the closed-loop system's poles can easily be found either algebraically or by a root locus plot. (Root locus is covered in detail in Section 3.2.)

We are generally interested in a closed-loop CPK/P system's transient response to impulse (bolus), step, or ramp inputs. Sinusoidal inputs (superimposed on a dc level) are rarely encountered. Many excellent computer simulation packages can be used to visualize the time-domain performance of a closed-loop CPK/P system. Simnon™ and MATLAB®/Simulink® are easy-to-use, PC/Windows-compatible simulation packages. The author is partial to Simnon, which handles systems described by high-order, nonlinear, stiff ordinary differential equations with ease. Simnon data files can be exported to Matlab to make use of Matlab's excellent graphics capabilities. Only Matlab, however, allows analysis and design of linear systems in state space.

Given a certain real-pole plant, one can design an external controller that will make the closed-loop system meet certain design specifications on system transient response (i.e., rise time, settling time, percent overshoot, etc.). Part of the design of an effective controller is the implementation of an appropriate *compensation filter* for the system. In simple terms, compensation is the insertion of additional poles and zeros as factors into the loop gain of the closed-loop system, so that, at some specified dc gain, the closed-loop system will have its poles at positions that give it the desired transient response. Root locus plots enable us to see all the possible closed-loop pole positions as the gain is varied. In general, the addition of poles alone to the loop gain function has a destabilizing effect on the closed-loop system. One or more zeros added as factors to the loop gain function have the effect of pulling the system's root locus branches to the left, tending to stabilize the closed-loop system at high gains. Compensation is more complex than simply adding zeros to the system's loop gain, however.

One of the simplest compensation schemes is to give the system *proportional-plus-derivative* (PD) feedback. PD feedback inserts a real zero in the loop gain function, which generally has a stabilizing effect on the closed-loop system. Thus the control is $u = SP - (K_c y + K_d \dot{y})$. Note that the PD compensation operation is in the feedback path. That is, it operates on the plant output, y, and not the system error, e. In practice, $\dot{y}$ is generally not directly available, and it has to be estimated by differentiating y. Various analog and digital means of estimating $\dot{y}$ exist. $\dot{y}$ is estimated from dc up to some maximum frequency where the differentiator becomes low-pass in order to limit excess noise.[136]

Lead and *lag compensation* adds a real pole *and* a real zero to the loop gain function. In *lead* compensation, the zero is at a lower frequency than the pole; in *lag* compensation, the pole is at a lower frequency than the zero. The Bode plot of a typical lead filter is shown in Figure 3.1A; the sinusoidal frequency response of a lag filter is shown in Figure 3.1B.

Lag/lead compensation is done by cascading a lag filter with a lead filter. The frequency response of a lag/lead filter is shown in Figure 3.2. The lag/lead filter is seen to add two real poles and two real zeros to the system's loop gain. Thus lag, lead, and lag/lead filters add poles and zeros to the loop gain function in equal

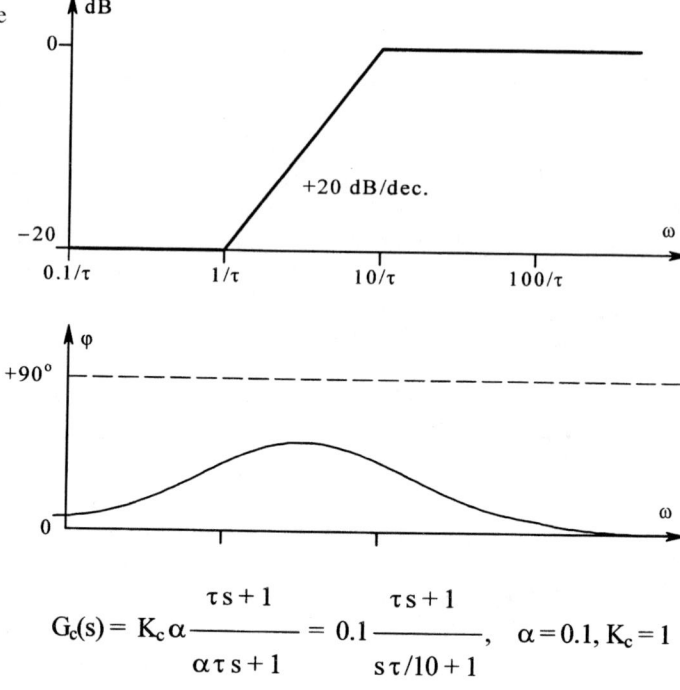

FIGURE 3.1A Bode plot asymptotes and phase vs. frequency for a simple lead filter. The pole is at 10× the zero's frequency.

$$G_c(s) = K_c \alpha \frac{\tau s + 1}{\alpha \tau s + 1} = 0.1 \frac{\tau s + 1}{s\tau/10 + 1}, \quad \alpha = 0.1, K_c = 1$$

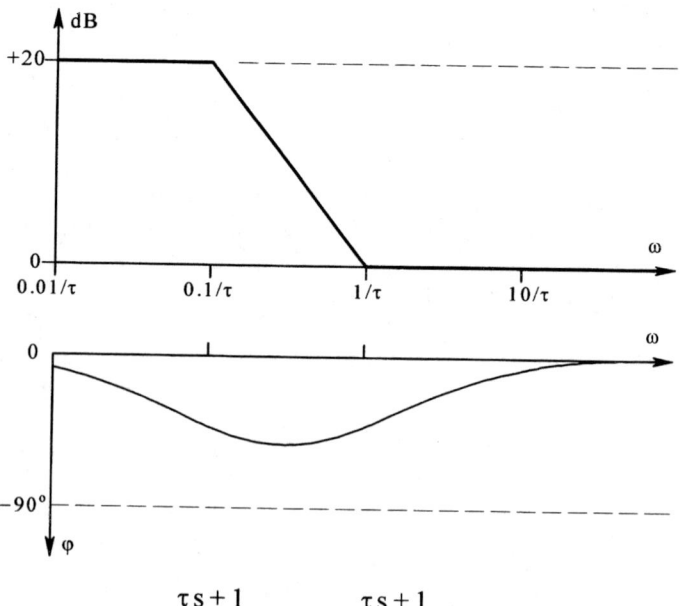

FIGURE 3.1B Bode plot asymptotes and phase vs. frequency for a simple lag filter. The zero is at 10× the pole's frequency.

$$G_c(s) = K_c \beta \frac{\tau s + 1}{\beta \tau s + 1} = 10 \frac{\tau s + 1}{10 s\tau + 1}, \quad \beta = 10, K_c = 1$$

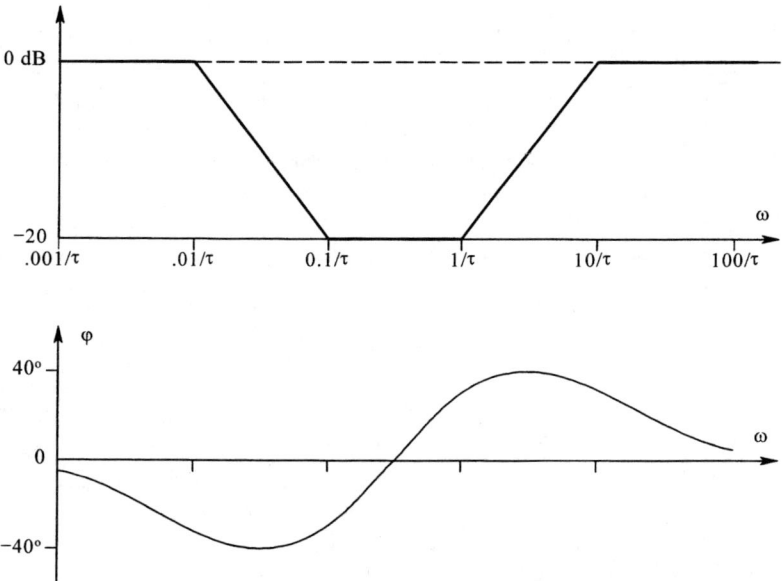

FIGURE 3.2 Bode plot asymptotes and phase vs. frequency for a lag/lead filter. The first real pole is at $\omega_o/100$; there are zeros at $\omega_o/10$ and at ω_o and a final real pole at 10 ω_o. ($\omega_o = 1/\tau$.)

numbers. The number of asymptotes in the root locus plot does not change, but the presence of the compensatory poles and zeros causes the locus branches to bend into more stable regions of the s-plane at certain gains, giving the closed-loop system better damping and bandwidth. Ogata[108] includes an excellent section with examples of compensation with lag, lead, and lag/lead filters.

As mentioned above, it is often desirable to make a closed-loop CPK/P system type 1 so there will be zero steady-state error. Either the *PI or the PID controller* can be used to make the loop gain type 1. A PI controller generally acts on system error; thus:

$$U(s) = E(s)\left[K_c + K_i/s\right] = E(s)\frac{K_c(s + K_i/K_c)}{s} = E(s)\frac{K_c(s+a)}{s} \qquad (3.1)$$

A PID controller inserts two zeros which can be real or complex-conjugate:

$$U(s) = E(s)\left[K_c + K_d s + K_i/s\right] = E(s)\frac{K_d\left(s^2 + [K_c/k_d]s + K_i/K_d\right)}{s} \qquad (3.2)$$

If the zeros are complex-conjugate, they have $\omega_n = \sqrt{(K_i/k_d)}$ and a damping factor equal to $\zeta = [K_c/K_d]/(2\omega_n)$. Note that both the PI and PID compensators are controllers in that they act on system error. Both put a loop gain pole at the origin of the s-plane. Because the PI or PID controllers are in the closed-loop system's forward path, they insert one or two zeros in the closed-loop system's transfer function. Under certain conditions, the zero(s) can cause overshoot to a step, even though other signs of quadratic system underdamped behavior are missing.[108]

Much has been written on the "tuning" of control systems to obtain the desired performance specifications by adjusting compensation parameters (gains or, equivalently, compensator pole and zero positions).[76,101,108] Effective tools to use in the design of conventional feedback control systems are root locus plots and simulations.

Many, many texts have been written on the analysis and design of linear-feedback control systems in the past 50 years (since World War II), beginning with those based on the design of naval fire control systems in World War II at MIT, such as the texts by James et al.[69] in 1947, Chestnut and Mayer[20] in 1951, Truxal[138] in 1955, Dorf[38] in 1967, etc. Some more recent texts include those by Kuo[76] in 1982, Franklin et al.[49] in 1986, Miron[101] in 1989, and Ogata[108] in 1990, to mention some of the author's favorites.

There are also more specialized texts that deal with the on-line identification of systems (plants). On-line identification of plants is necessary to create self-tuning control systems that are robust against nonstationary plant parameters. See the texts by Graupe,[57] Ljung and Söderström,[85] and Ljung.[84] Note that Matlab has a "Systems Identification Toolbox." Self-tuning control is another specialty area with relevance to the design of robust CPK/P control systems; see the texts by Gawthrop[53] and Åström and Wittenmark.[8] The "Nonlinear Control Design Toolbox" from Matlab contains aids for automatically tuning the performance of models of nonlinear control systems. Other authors have also given multivariable (cross-coupled) feedback systems special consideration; see, for example, texts by Owens[109] and Maciejowski.[86] The fact that most physiological control systems are multiple input/multiple output underscores the need for consideration of multivariable feedback systems (see Section 5.3 on nonlinear decoupling).

3.2 THE USE OF ROOT LOCUS IN LINEAR CONTROL SYSTEM DESIGN

Given the knowledge of the positions of the poles and zeros of the loop gain of a linear SISO, single-loop feedback system, the root locus technique allows us to predict precisely where the closed-loop poles of the system will be as a function of system gain. Thus root locus can be used to predict the conditions for instability, as well as to design for a desired closed-loop transient response. Many texts on linear control systems treat the generation and interpretation of root locus plots in detail.[76,108,138] Root locus has also been used in the design of electronic feedback amplifiers, including sinusoidal oscillators.[104] It should be stressed that CPK/P systems are nonlinear, and therefore root locus techniques strictly cannot be applied.

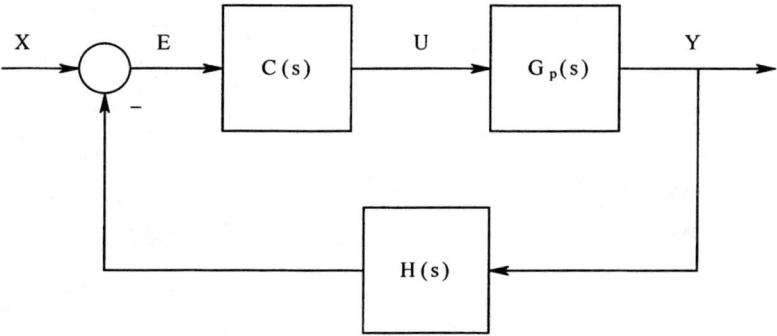

FIGURE 3.3 A simple single-loop, SISO linear control system.

However, if a CPK/P system is operating linearly (all states positive) and the controller nonlinearity is linearized, root locus can be applied and some insight can be gained about the system's closed-loop behavior.

It is often tedious to construct root locus diagrams by hand on paper, except in certain simple cases described below. Detailed, quantitative root locus plots can be generated using a Matlab subroutine, "RLOCUS." Several of the examples in this section are from Matlab plots.

The concept behind the root locus diagram is simple: Figure 3.3 shows a simple one-loop, linear SISO feedback system. The closed-loop gain is

$$\frac{Y}{X}(s) = \frac{C(s)G_p(s)}{1 - A_L(s)} = \frac{C(s)G_p(s)}{1 + C(s)G_p(s)H(s)} = F(s) \qquad (3.3)$$

$C(s)$ is the controller transfer function acting on $E(s)$, $G_p(s)$ is the plant transfer function [input $U(s)$, output $Y(s)$], and $H(s)$ is the feedback path transfer function. The system loop gain is

$$A_L(s) = -C(s)G_p(s)H(s) \qquad (3.4)$$

The root locus technique allows us to plot in the s-plane those complex s values that make the return difference $\rightarrow 0$, thus making the closed-loop transfer function, $|F(s)| \rightarrow \infty$. The s values that make $|F(s)| \rightarrow \infty$ are, of course, by definition, the poles of $\mathbf{F(s)}$. These poles move in the s-plane as a function of a gain parameter in a predictable, continuous manner as the gain is changed. Since finding s values that set $\mathbf{A_L(s)} = 1\angle 0°$ is the same as setting $|F(s)| = \infty$, the root locus plotting rules are based on finding s values that cause the angle $-A_L(s) = -180°$ and $|-A_L(s)| = 1$. The root locus plotting rules are based on satisfying either the angle or magnitude condition on $-A_L(s)$. We use the vector format of $-A_L(s)$ to derive the plotting rules.

An example of how this is done is given below. First, we assume a feedback system with the loop gain $A_L(s)$ given below. Note that there are three equivalent ways of writing $A_L(s)$:

$$A_L(s) = \frac{-K\beta(s\tau_1 + 1)}{(s\tau_2 + 1)(s\tau_3 + 1)} \qquad \text{(time-constant form)} \qquad (3.5A)$$

$$-A_L(s) = \frac{-K\beta\tau_1}{\tau_2\tau_3} \frac{(s + 1/\tau_1)}{(s + 1/\tau_2)(s + 1/\tau_3)} \qquad \text{(Laplace form)} \qquad (3.5B)$$

$$-A_L(s) = \frac{-K\beta\tau_1}{\tau_2\tau_3} \frac{(s - s_1)}{(s - s_2)(s - s_3)} \qquad \text{(vector form)} \qquad (3.5C)$$

where $s_1 = -1/\tau_1$, $s_2 = -1/\tau_2$, and $s_3 = -1/\tau_3$.
For the *magnitude criterion*, s must satisfy:

$$\frac{|s - s_1|}{|s - s_2||s - s_3|} = \frac{\tau_2\tau_3}{K\beta\tau_1} \qquad (3.6)$$

For the *angle criterion*, s must satisfy $\theta_1 - (\theta_2 + \theta_3) = -180°$.

Nine basic root locus plotting rules are derived from the angle and magnitude conditions above and are used for pencil-and-paper construction of root locus diagrams:

1. *Number of branches*: There is one branch for each pole of $A_L(s)$.
2. *Starting points*: Locus branches start at the poles of $A_L(s)$ for $K\beta = 0$.
3. *End points*: The branches end at the finite zeros of $A_L(s)$ for $K\beta \to \infty$. Some zeros of $A_L(s)$ can be at $|s| = \infty$.
4. *Behavior of the loci on the real axis* (from the angle criterion): For a negative feedback system, on-real axis locus branches exist to the left of an odd number of on-axis poles and zeros of $A_L(s)$. If the feedback for some reason is positive, then on-axis locus branches are found to the right of a total odd number of poles and zeros of $A_L(s)$. See the examples below.
5. *Symmetry*: Root locus plots are symmetrical around the real axis in the s-plane.
6. *Magnitude of gain at a point on a valid locus branch*: From the magnitude criterion, at a vector point s on a valid locus branch we have:

$$K\beta = \frac{|s - s_2||s - s_3|\tau_2\tau_3}{|s - s_1|\tau_1} \qquad (3.7)$$

7. *Points where locus branches leave or join the real axis*: The breakaway or reentry point is algebraically complicated to find; also, there are

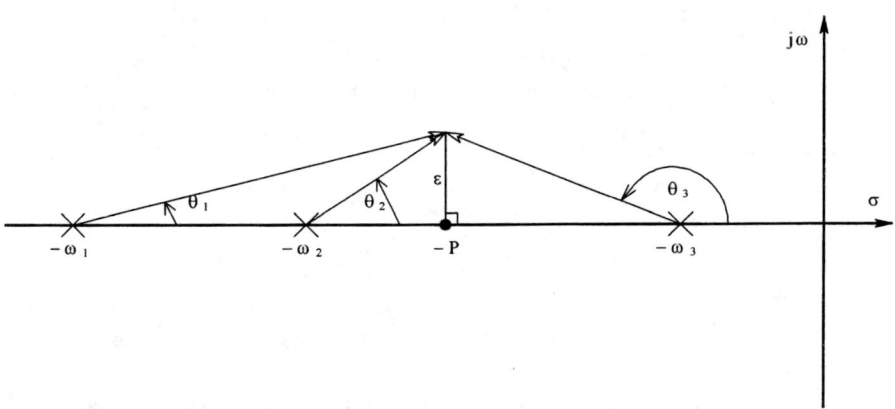

FIGURE 3.4 Diagram showing the geometry in the *s*-plane relevant to finding the breakaway point *P* of the on-axis root locus branches between ω_2 and ω_3.

methods based on both the angle and magnitude criteria. For example, in a negative-feedback-system loop gain function with three real poles, two locus branches leave the two poles closest to the origin and travel toward each other along the real axis as $K\beta$ is raised. At some critical $K\beta$, the branches break away from the real axis, one at +90° and the other at −90°. If we examine the angle criterion at the breakaway point, as shown in Figure 3.4, it is clear that:

$$-(\theta_1 + \theta_2 + \theta_3) = 180° \qquad (3.8)$$

From the geometry on the figure, the angles can be written as arctangents:

$$\tan^{-1}\left[\varepsilon/(\omega_1 - P)\right] + \tan^{-1}\left[\varepsilon/(\omega_2 - P)\right] \\ + \left\{\pi - \tan^{-1}\left[\varepsilon/(P - \omega_3)\right]\right\} = \pi \qquad (3.9)$$

For small arguments, $\tan^{-1}(x) \cong x$, so:

$$\left[\varepsilon/(\omega_1 - P)\right] + \left[\varepsilon/(\omega_2 - P)\right] - \left[\varepsilon/(P - \omega_3)\right] = 0 \qquad (3.10)$$

This equation can be written as a quadratic in *P*:

$$3P^2 - 2(\omega_1 + \omega_2 + \omega_3)P + (\omega_2\omega_3 + \omega_1\omega_2 + \omega_1\omega_3) = 0 \qquad (3.11)$$

The desired *P*-root, of course, lies between $-\omega_2$ and $-\omega_3$.

8. *Breakaway or reentry angles of branches with the real axis*: The loci are separated by angles of 180°/*n*, where *n* is the number of branches

TABLE 3.1
Asymptote Angles with the Real Axis in the s-Plane

Negative Feedback		Positive Feedback	
N − M	φ_k	N − M	φ_k
1	180°	1	0°
2	90°, 270°	2	0°, 180°
3	60°, 180°, 300°	3	0°, ±120°
4	45°, 135°, 225°, 315°	4	0°, 90°, 180°, 270°

intersecting the real axis. In most cases, $n = 2$, so the branches approach the axis perpendicular to it.

9. *Asymptotic behavior of the branches for $\beta \to \infty$:*
   a. The number of asymptotes along which branches approach zeros at $|s| = \infty$ is $N_A = N - M$, where N is the number of finite poles and M is the number of finite zeros of $A_L(s)$.
   b. The angles of the asymptotes with the real axis are φ_k, where $k = 1 \ldots N - M$. φ_k is given in Table 3.1.
   c. The intersections of the asymptotes with the real axis are a value I_A along the real axis. It is given by:

$$I_A = \frac{\Sigma(\text{real parts of finite poles}) - \Sigma(\text{real parts of finite zeros})}{N - M} \quad (3.12)$$

An example of finding the asymptotes and I_A is shown in Figure 3.5. This negative-feedback system has a loop gain with four poles at $s = 0$, $s = -3$, and a complex-conjugate (CC) pair at $s = -1 \pm j2$. There are no finite zeros. Thus $I_A = [(-3 - 1 - 1) - (0)]/4 = -5/4$. The angles are, from the table, ±45° and ±135°. The root locus plot was done with Matlab.

To get a feeling for plotting root locus diagrams with the rules above, it is necessary to examine several representative examples. In *example 1*, we examine the commonly encountered circle root locus. The negative feedback system's loop gain is

$$A_L(s) = \frac{-K_p \beta(s + a)}{(s + b)(s + c)} \quad (3.13)$$

where $a > b > c > 0$. Angelo[6] gave an elegant proof that this system's root locus does indeed contain a circle centered on the zero of $A_L(s)$. The circle's radius R is shown to be the *geometrical mean distance from the zero to the poles*. That is,

$$R = \sqrt{(a - b)(a - c)} \quad (3.14)$$

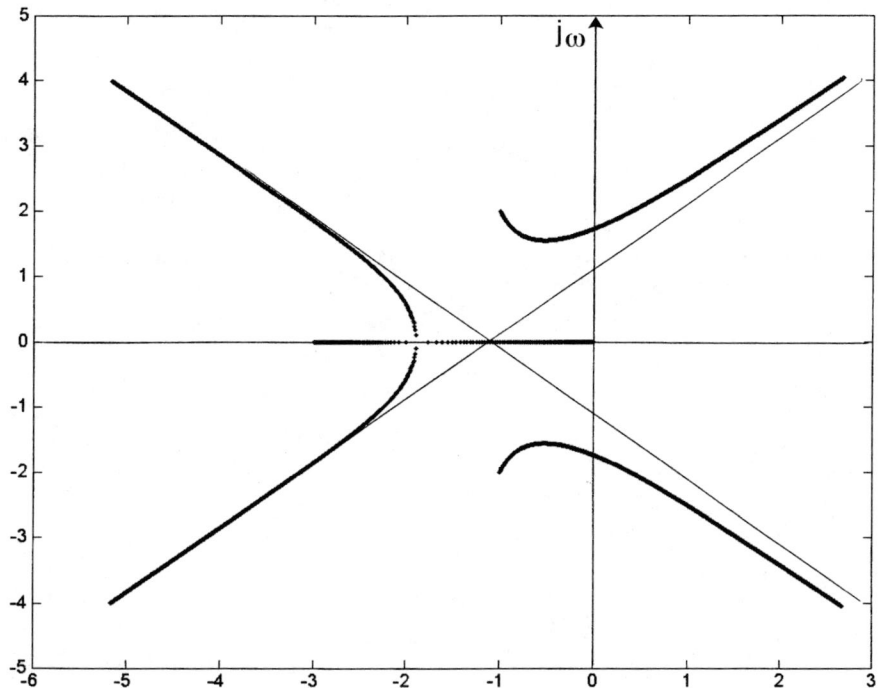

FIGURE 3.5 Root locus diagram of a negative-feedback system with loop gain: $A_L(s) = -K/[s(s+3)(s^2+2s+5)]$. Plot was done using Matlab's *rlocus* function. K ranged from 0 to 10^3 in increments of 0.1. Note that the system becomes unstable for $K > K_{crit}$.

The breakaway and reentry points are easily found from a knowledge that the circle has a radius R and is centered at $s = -a$. The circle root locus is shown in Figure 3.6A. If the loop gain's poles are CC, the root locus is also an interrupted circle, shown in Figure 3.6B. The poles are at $s = -\alpha \pm j\gamma$, and the zero is at $s = -\sigma$. The $A_L(s)$ is

$$A_L(s) = \frac{-K_p \beta (s+\sigma)}{s^2 + s(2\alpha) + (\alpha^2 + \gamma^2)} \quad (3.15)$$

The circle's radius is now found from the Pythagorean theorem:

$$R = \sqrt{(\sigma - \alpha)^2 + \gamma^2} \quad (3.16)$$

Note that as the gain, $K_p\beta$, is increased, the closed-loop system poles become more and more damped, until they both become real, one approaching the zero and the other going to $-\infty$.

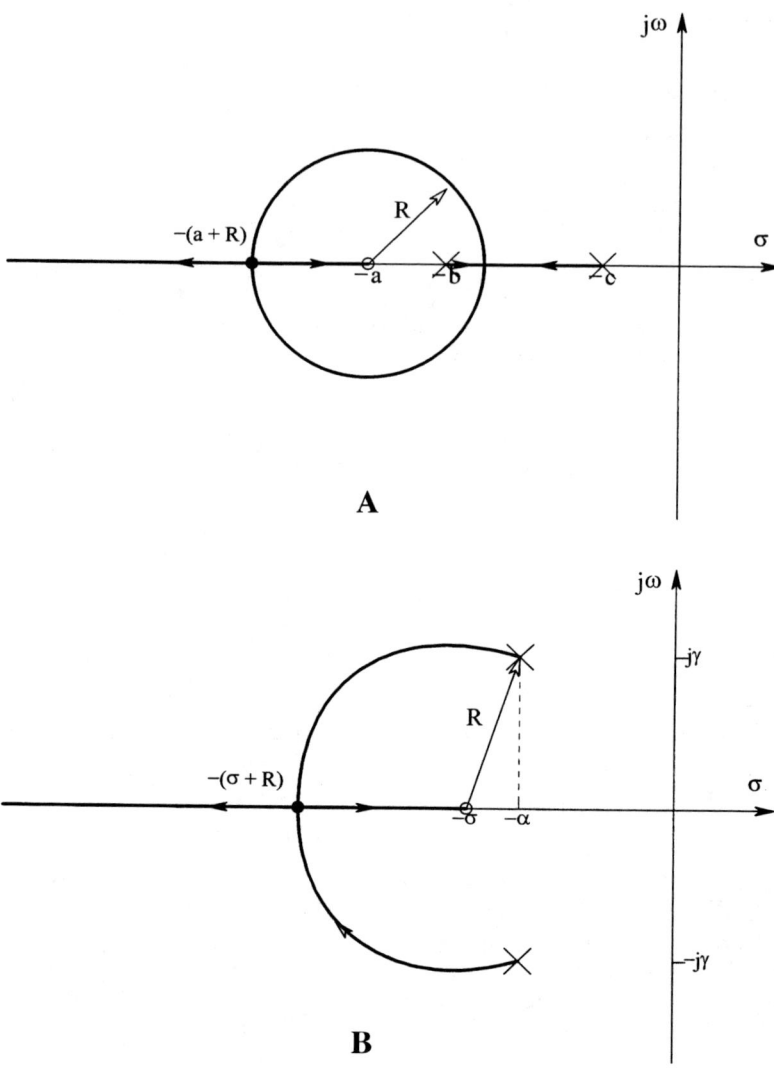

FIGURE 3.6 (A) The circle root locus for the negative-feedback system of example 1. (B) Interrupted circle root locus for a negative-feedback system with complex-conjugate poles.

The damping of a complex-conjugate pole pair in the s-plane can be determined quantitatively by drawing a line from the origin to the upper-half plane pole. The damping factor, ζ, associated with the CC poles can be shown to be the cosine of the angle the line from the origin to the CC pole makes with the negative real axis. That is, $\zeta = \cos(\phi)$. If the poles lie close to the $j\omega$ axis, $\phi \to 90°$, and $\zeta \to 0$. The length of the line from the origin to one of the CC poles is the undamped natural

Introduction to Linear SISO Control Systems and Systems with Delays

frequency, ω_n, of that CC pole pair. Recall that the CC pole pair is the result of factoring the quadratic term, $[s^2 + s(2\zeta\omega_n) + \omega_n^2]$. By way of example, the damping of the open-loop, CC pole pair in Figure 3.6B is $\zeta = \cos[\tan^{-1}(\gamma/\alpha)] = \alpha/\omega_n$.

In *example 2*, we examine the root locus of a simple type 1 negative-feedback system with a transport lag. The system's loop gain is

$$A_L(s) = \frac{-K_p e^{-s\delta}}{(s+a)} \tag{3.17}$$

Let $a = 1$ and $\delta = 1$. We write $-\mathbf{A_L}(s)$ in vector form, noting that, in general, $\mathbf{s} = \sigma + j\omega$.

$$-\mathbf{A_L}(s) = \frac{K_p e^{-\delta\sigma} e^{-j\omega\delta}}{\mathbf{s} - \mathbf{s}_1} \tag{3.18}$$

By the *magnitude criterion*:

$$|-\mathbf{A_L}(s)| = \frac{K_p e^{-\delta\sigma}}{|\mathbf{s}-\mathbf{s}_1|} = 1, \quad \text{or} \quad K_p = |\mathbf{s}-\mathbf{s}_1| e^{+\delta\sigma} \tag{3.19}$$

The *angle criterion* gives:

$$\angle[-\mathbf{A_L}(s)] = -\omega\delta R - \theta_1 = -180°(2k+1), \quad k = 0, 1, 2, \ldots \tag{3.20}$$

The loop gain has one real pole at $s = -1$ and an infinite number of poles at $s = -\infty \pm j2\pi k$, where $k = 0, 1, 2, \ldots$. The root locus plot for this simple delay system is shown in Figure 3.7.

First, we find the ω_o value where the first closed-loop, CC pole pair crosses the $j\omega$ axis. The ω_o value is found by solving the angle criterion by trial and error. That is,

$$\omega_o \delta R + \tan^{-1}(\omega_o) = 180° \tag{3.21}$$

ω_o is found to be 2.029 r/sec. To find the K_p value required to put the closed-loop poles at $\pm j\omega_o$, we substitute ω_o into the magnitude criterion. Thus:

$$K_p = \sqrt{1+\omega_o^2}\, e^{+\sigma\delta} = 2.262 \quad (\text{NB: } \sigma = 0 \text{ for } \mathbf{s} = jw_o) \tag{3.22}$$

The gain at which the primary closed-loop pole pair becomes CC is found for $s = -2$:

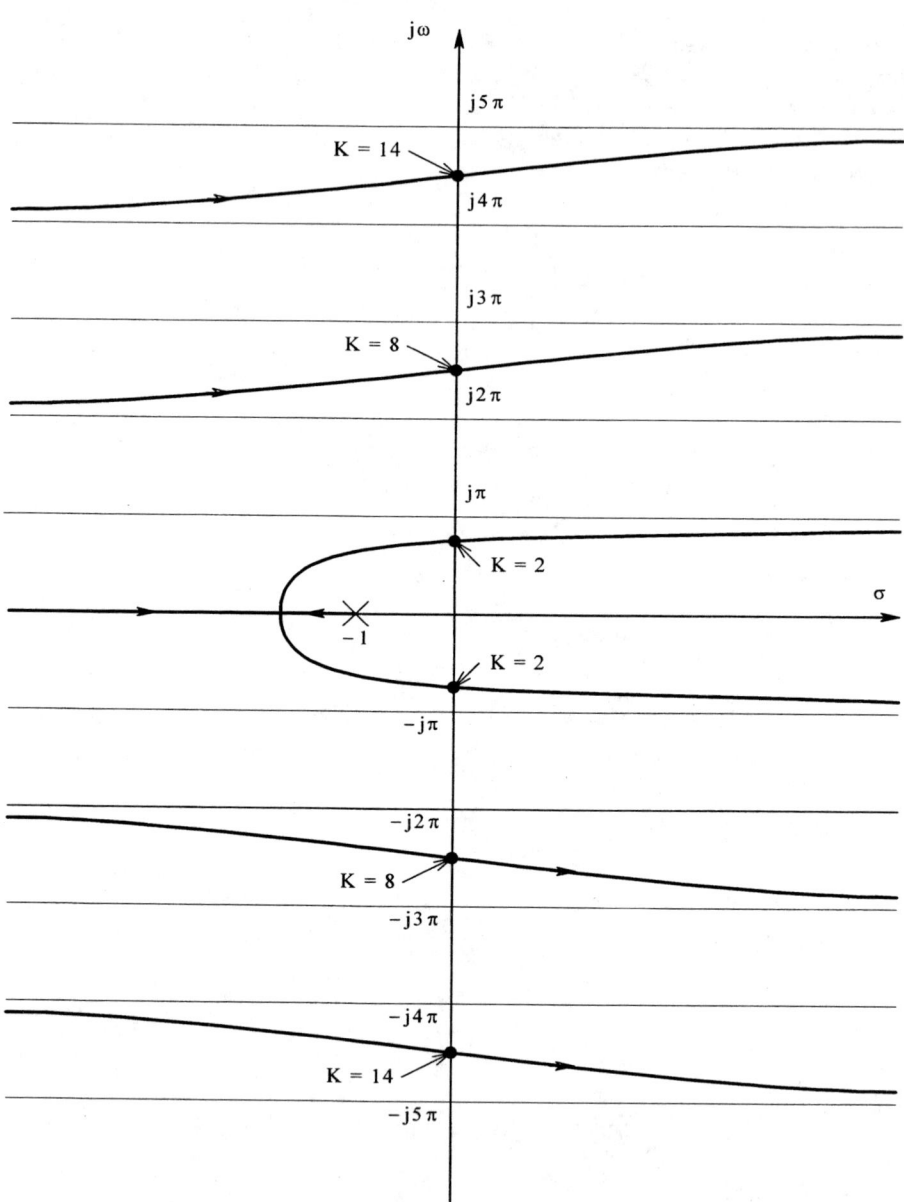

FIGURE 3.7 Root locus diagram for a simple negative-feedback system with $A_L(s) = -Ke^{-s\delta}/(s + 1)$. Note harmonic locus branches originating at poles at $\mathbf{s} = e^{-\delta(\sigma + jn2\pi)}$, $\sigma \to -\infty$, $n = 0, \pm 1, \pm 2, \dots$.

Introduction to Linear SISO Control Systems and Systems with Delays

$$K_p = 1e^{-2(1)} = 0.1353 \tag{3.23}$$

Because the higher order closed-loop pole pairs cross the $j\omega$ axis at progressively higher K_p values, it is the first-order ($k = 0$) pole pair that is dominant in determining system instability. In other words, once the first-order, closed-loop, CC pole pair is in the right-half s plane, the other locus branches are of academic interest only; the system is already unstable.

For a *third example,* let us consider a CPK plant with two real poles and direct, negative feedback. The system loop gain is

$$A_L(s) = \frac{-K_c K_p}{(s+1)(s+5)} \tag{3.24}$$

This is a type 0 system, and its root locus is shown in Figure 3.8. We note that the system is never unstable, but that there is a practical limit to the size of $K_c K_p$ in

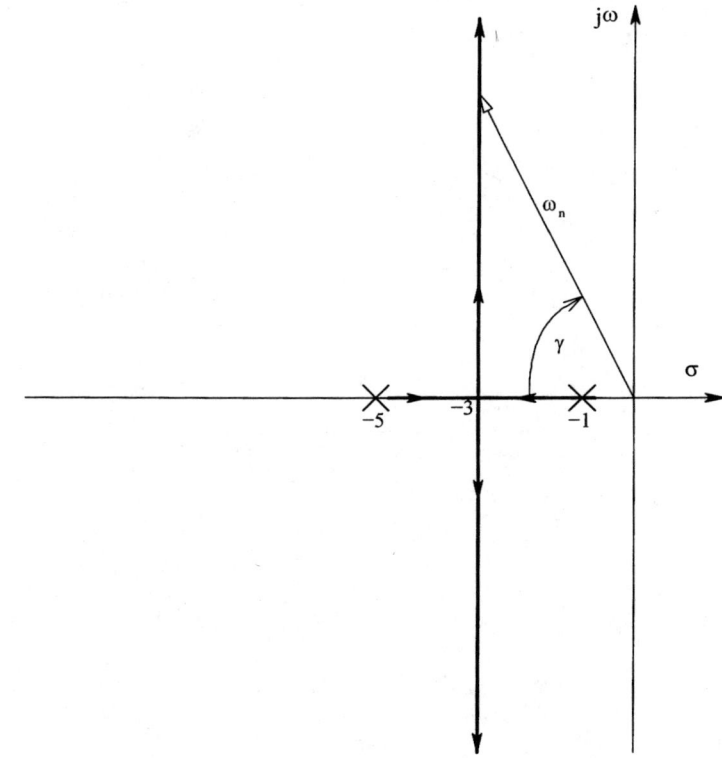

FIGURE 3.8 Root locus of a negative-feedback system with two real poles. See text for $A_L(s)$.

terms of acceptable closed-loop settling time and minimum damping, ζ. To make a useful drug infuser, we wish to give the system unity gain (for zero steady-state error) and be able to raise the gain to reduce settling time while retaining a practical range of closed-loop-system pole damping, say $0.5 \le \zeta_{CL} \le 0.707$.

In *example 4*, we introduce a PD feedback for the system of example 3. We also condition the input (set point) by a factor of $(5 + a\,K_p\,K_c)/K_p$ in order to give the closed-loop system unity dc gain. The PD feedback introduces a zero as $s = -a$ into the loop gain function. If $a > 5$, then the closed-loop system has the circle root locus of example 1. The transfer function of the PD compensator is $C(s) = K_c(s + a)$. The system loop gain is now:

$$A_L(s) = \frac{-K_p\,K_c\,(s+a)}{(s+1)(s+5)} \qquad (3.25)$$

We choose a so the zero lies on the real axis at $s = -15$. Figure 3.9 illustrates the system's root locus diagram. Note that there are two closed-loop system poles; they start out real, then become CC, then real again as the gain $K_p\,K_d$ increases. One then

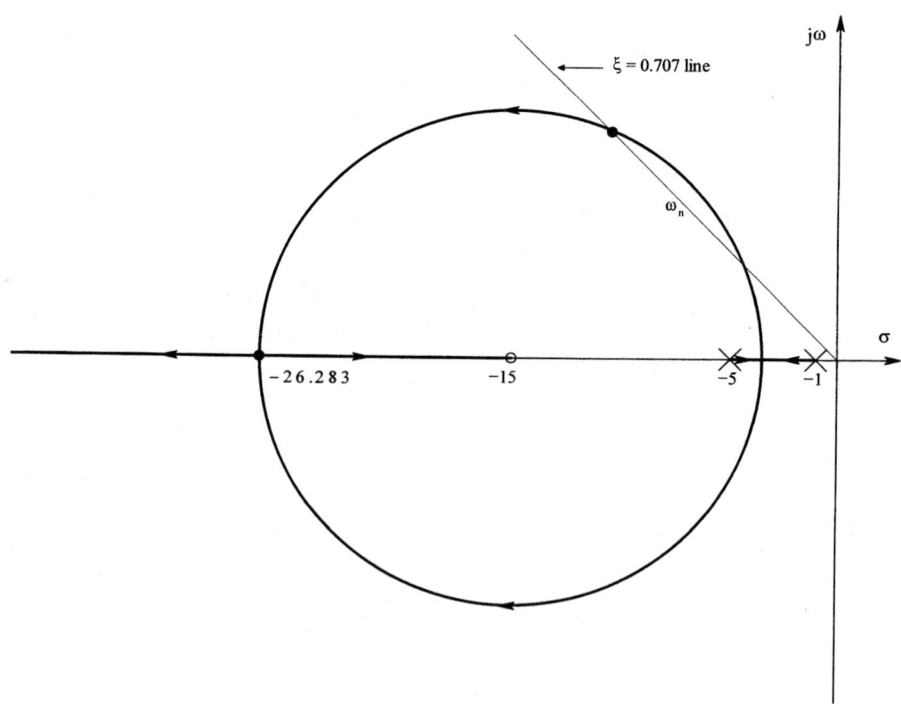

FIGURE 3.9 Circle root locus of a system that has two different ω_n values at $\xi = 0.707$.

Introduction to Linear SISO Control Systems and Systems with Delays 71

approaches the zero and the other goes to $-\infty$. The circle radius is $R = \sqrt{(15-5)(15-1)} = 11.832$. Thus the locus branches leave the real axis at $s = -(15 - 11.832) = -3.168$ and rejoin it at $s = -(15 + 11.832) = -26.832$.

To obtain a good, quick step response for the system with little overshoot, we want the CC poles to have a damping factor $\zeta = 0.707$. Recall that ζ is equal to the cosine of the angle a line drawn from the origin to the upper CC pole makes with the negative real axis. In this case, it is a line at 45°, as shown in Figure 3.9. Where the 45° line intersects the CC root locus branch at its most distant intersection is the desired closed-loop, CC pole position. Once the pole position is known, the root locus magnitude criterion can be used to find the required gain, $K_p K_d$. The length of the line from the origin to the desired closed-loop, CC pole position is the closed-loop system's undamped natural frequency, ω_{nCL}. Graphical measurement on the plot yields $\omega_{nCL} = 15.5$ r/hr. From the root locus magnitude criterion, we can find the corresponding gain:

$$\left|A_L(\omega_{nCL})\right| = 1 = \frac{K_p K_c R}{A B} \longrightarrow K_p K_c = A B/R \qquad (3.26)$$

$$= \sqrt{\left[(11-5)^2 + 11^2\right]} \sqrt{\left[(11-1)^2 + 11^2\right]} / 11.832 = 15.743$$

We can also simulate the system's step response using Simnon. The program is as follows:

```
CONTINUOUS SYSTEM pdex32      "6/20/97 Example 4 in Sec. 3.2.
STATE y1 y2                   " PD control system
DER dy1 dy2
TIME t
"
e = SP - y2
u = SP*(5 + a*Kp*Kc)/Kp - Kc*(a*y2 + dy2)
dy1 = -y1 + u
dy2 = -5*y2 + Kp*y1
"
SP = IF t > 0 THEN SPo ELSE 0
"
SPo:1
Kc:3
Kp:5
a:15
"
END
-
```

Note that the SP*(5 + a*Kp*Kc)/Kp term in the program normalizes the steady-state output gain to 1 for any K_c. Figure 3.10 illustrates the step response of the PD-compensated system of example 4. Output plots of y_2 for K_c values of 1, 3.15, 10, and 50 are shown. The system has the desired closed-loop damping $\zeta = 0.707$ for

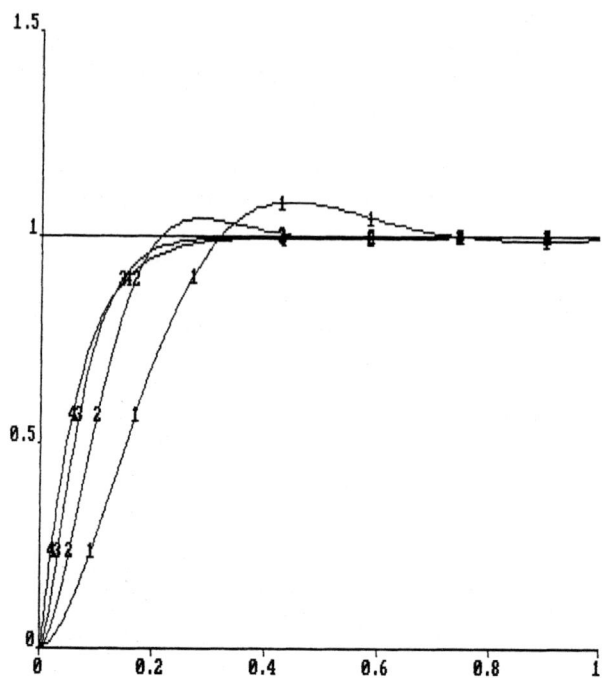

FIGURE 3.10 Unit step response of the PD-compensated system of example 4. Vertical axis = y2, horizontal axis = time (sec). $K_p = 5$. Trace 1, $K_c = 1$; trace 2, $K_c = 3.15$ so $\xi = 0.707$; trace 3, $K_c = 10$; trace 4, $K_c = 50$.

$K_c = 3.15$, $K_p = 5$.

As a final exercise with the PD-compensated system, we use Matlab's *rlocus* subroutine to plot the system's root locus diagram. The dots represent the closed-loop system's pole positions for values of $K_p K_c$ ranging from 0 to 50 in increments of 0.10. Note dot spacing on the real axis and along the circle. Large spacing indicates that the closed-loop pole positions are changing rapidly with gain. Although the closed-loop, CC poles lie on a circular path; the path appears elliptical in Figure 3.11 because of scale distortion (the real axis spans 0 to –35 while the $j\omega$ axis spans –15 to + 15).

Many other interesting examples of root locus plots can be found in the control systems texts by Kuo[76] and Ogata.[108] As demonstrated here, circle root locus plots are easy to construct graphically by hand. More complex root locus plots should be done by computer. In closing, it is important to stress that root locus plots show the closed-loop system's *poles*. Closed-loop zeros can be found algebraically. They are generally fixed (not functions of gain) and do affect system transient response. Merely locating the closed-loop poles in what appears to be a good position does not necessarily guarantee a step response without an objectionable overshoot (the settling time may be short, but overshoot is bad in CPK systems). Such behavior is seen in high-order systems with PID controllers.[108]

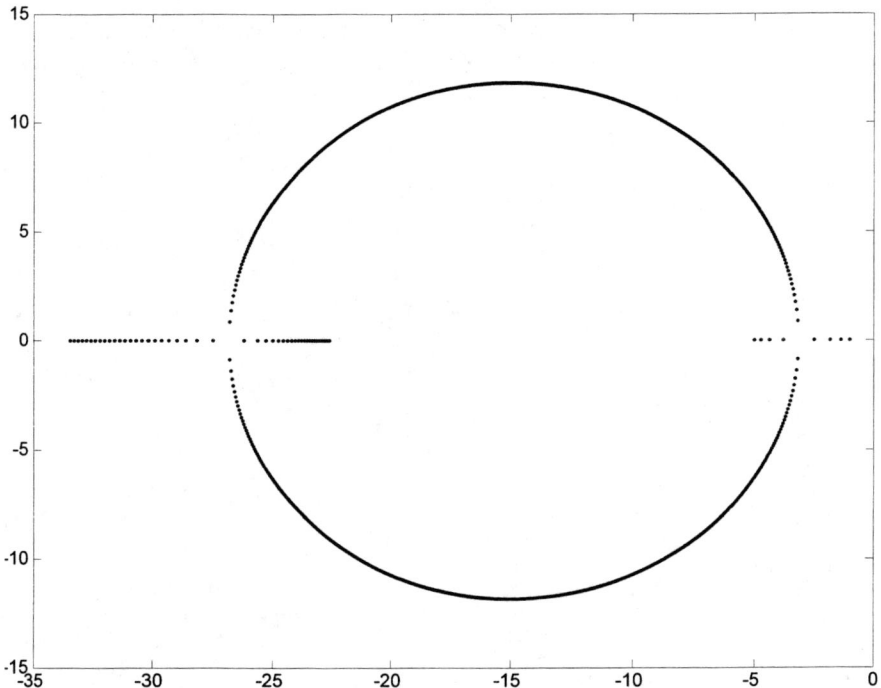

FIGURE 3.11 Root locus plot of the system of example 4.

3.3 SOME TESTS FOR STABILITY IN LINEAR AND CERTAIN CLASSES OF NONLINEAR CONTROL SYSTEMS

3.3.1 INTRODUCTION

Stability tests for linear systems generally examine the system *loop gain* to predict under what conditions one or more poles of the closed-loop system will lie in the right-half s-plane. Closed-loop poles in the right-half s-plane mean that the linear system will respond to any input, however small, with an output that grows in an unbounded manner. From a practical viewpoint, a system does not have to be unstable to be unusable. Closed-loop, CC pole pairs that have damping factors less than 0.35 generally have long settling times and unpleasant ringing in response to transient inputs. This suggests that the root locus technique is very useful in that it predicts the conditions under which a high-order system ceases to be practically useful, before frank instability occurs.

Nonlinear and discontinuous control systems present a special problem in that they can exhibit limit cycle oscillations (in some cases desired), as well as unbounded outputs. Very often, one must resort to simulation to verify appropriate limit cycle behavior or to test for instability. It is often easier to simulate the dynamic behavior of a nonlinear closed-loop system than it is to apply difficult-to-implement stability tests such as Lyupanov's methods.

3.3.2 THE ROUTH–HURWITZ STABILITY TEST

This is a relatively simple algebraic test applied to the coefficients of the expanded denominator polynomial of the closed-loop linear system transfer function. It tells us whether any of the closed-loop system's poles lie in the right-half s-plane, having positive real parts. Poles in the right-half s-plane make the system unstable. The Routh–Hurwitz (RH) test is only applicable to linear stationary systems that have no transport lags in their loop gains. Recall that a simple single-loop feedback system's closed-loop gain can be written:

$$H(s) = \frac{Y(s)}{X(s)} = \frac{G(s)}{[1 - A_L(s)]} \quad (3.27)$$

Recall that $[1 - A_L(s)]$ is the *return difference* and that the zeros of the return difference are the poles of the closed-loop system. Assume a system's loop gain is a rational polynomial of the form:

$$-A_L(s) = \frac{b_m s^m + b_{m-1} s^{m-1} + \ldots + b_1 s + b_0}{a_n s^n + a_{n-1} s^{n-1} + \ldots + a_1 s + a_0} = \frac{P(s)}{Q(s)}, \quad n > m \quad (3.28)$$

Thus the return difference can be written:

$$\begin{aligned}
[1 - A_L(s)] &= \frac{[P(s) + Q(s)]}{Q(s)} \\
&= \frac{a_n s^n + a_{n-1} s^{n-1} + \ldots + a_1 s + a_0 + b_m s^m + b_{m-1} s^{m-1} + \ldots + b_1 s + b_0}{a_0 s^n + a_1 s^{n-1} + \ldots + a_{n-1} s + a_n} \\
&= \frac{c_n s^n + c_{n-1} s^{n-1} + \ldots + c_1 s + c_0}{Q(s)} \quad (3.29) \\
&= \frac{c_n (s - s_1)(s - s_2) \ldots (s - s_{n-1})(s - s_n)}{Q(s)}
\end{aligned}$$

In the last term above, the nth-order polynomial, $[P(s) + Q(s)]$, is written in factored form. The RH test tells us if any of its roots, $s_1, s_2, \ldots s_n$, have positive real parts and therefore are unstable.

Implementation of the RH test is relatively simple. Here we follow the protocol given by Dorf.[38] A Routh array is constructed using the c_k notation above for the numerator of the expanded return difference:

$$
\begin{array}{c|cccc}
s^n & c_n & c_{n-2} & c_{n-4} & \cdots \\
s^{n-1} & c_{n-1} & c_{n-3} & c_{n-5} & \cdots \\
s^{n-2} & d_{n-1} & d_{n-3} & d_{n-5} & \cdots \\
s^{n-3} & e_{n-1} & e_{n-3} & e_{n-5} & \cdots \\
\vdots & \vdots & \cdot & \cdot & \\
s^0 & h_{n-1} & & &
\end{array}
\tag{3.30}
$$

The other coefficients for the array are found:

$$
d_{n-1} = \frac{-1}{c_{n-1}} \begin{vmatrix} c_n & c_{n-2} \\ c_{n-1} & c_{n-3} \end{vmatrix}
\tag{3.31A}
$$

$$
d_{n-3} = \frac{-1}{c_{n-1}} \begin{vmatrix} c_n & c_{n-4} \\ c_{n-1} & c_{n-5} \end{vmatrix}
\tag{3.31B}
$$

$$
e_{n-1} = \frac{-1}{d_{n-1}} \begin{vmatrix} c_n & c_{n-3} \\ d_{n-1} & d_{n-3} \end{vmatrix} \quad \text{etc.}
\tag{3.31C}
$$

A necessary and sufficient condition that the closed-loop system have no poles in the right-half s-plane is that there be no sign changes in the first column of the array. Several examples of the RH test are shown below, adapted from Dorf:[38]

1. Let $[P + Q] = c_2 s^2 + c_1 s + c_0$ (quadratic system). The Routh array is

$$
\begin{array}{c|cc}
s_2 & c_2 & c_o \\
s_1 & c_1 & 0 \\
s_0 & d_1 &
\end{array}
\tag{3.32}
$$

where

$$d_1 = (-1/c_1) \begin{vmatrix} c_2 & c_0 \\ c_1 & 0 \end{vmatrix} = +c_0 \tag{3.33}$$

Thus a stable quadratic system will have all positive coefficients, c_k.

2. Consider the cubic system $[P + Q] = c_3 s^3 + c_2 s^2 + c_1 s + c_0$. The RH array is

$$\begin{array}{c|cc} s^3 & c_3 & c_1 \\ s^2 & c_2 & c_0 \\ s^1 & d_1 & 0 \\ s^0 & e_1 & 0 \end{array} \tag{3.34}$$

where $d_1 = (c_2 c_1 - c_0 c_3)/c_2$ and $e_1 = d_1 c_0/d_1 = c_0$. Therefore, for the cubic system to be stable, all c_k must be > 0, and $c_2 c_1 > c_0 c_3$. When $c_2 c_1 = c_0 c_3$, one pair of poles lies on the $j\omega$ axis in the s-plane.

3. In the fourth-order system, $[P + Q] = s^4 + s^3 + s^2 + s + K$. We want to determine the gain K that results in borderline instability. The RH array is

$$\begin{array}{c|ccc} s^4 & 1 & 1 & K \\ s^3 & 1 & 1 & 0 \\ s^2 & \varepsilon & K & 0 \\ s^1 & c_1 & 0 & 0 \\ s^0 & K & 0 & 0 \end{array} \tag{3.35}$$

If only one element of the first column of the array is zero, it may be replaced with a small positive number, ε, which is allowed to approach zero after completing the array. Now, $c_1 = (\varepsilon - K)/\varepsilon \cong -K/\varepsilon$. Therefore for any $K > 0$, the system is unstable. Also, since the last term in the first column is K, a negative K will also give an unstable system. Therefore the system is unstable for all values of K.,

4. Now let the characteristic equation be $[P + Q] = s^3 + 2s^2 + 4s^1 + K$, where K is the adjustable loop gain. The RH array is

$$\begin{array}{c|cc} s^3 & 1 & 4 \\ s^2 & 2 & K \\ s^1 & (8-K)/2 & 0 \\ s^0 & K & 0 \end{array} \tag{3.36}$$

Thus, for a stable system, we require that $0 \le K \le 8$. When $K = 8$, we have two conjugate poles on the $j\omega$ axis, which is considered to be a borderline instability case. Other interesting examples of the RH test are described by Dorf.[38]

3.3.3 THE NYQUIST STABILITY CRITERION

In this section, the application and limitations of the venerable Nyquist test for the stability of linear single-loop feedback systems are described. While more useful information about the closed-loop behavior of a feedback system may be obtained from root locus considerations when doing design, the Nyquist test, which is based on the steady-state sinusoidal frequency response of the system's loop gain function, is well suited for design based on experimental frequency response data. The Nyquist test is also directly related to the use of describing functions in the analysis and design of certain classes of nonlinear control systems.[107] To introduce the Nyquist test for closed-loop system stability, attention is directed to the conventional SISO linear feedback system shown in Figure 3.12. In this system, no minus sign assumption is made at the summing point. The closed-loop system transfer function is simply:

$$\frac{Y}{X}(s) = \frac{G(s)}{[1 - A_L(s)]} \qquad (3.37)$$

In a negative-feedback system, the loop gain $A_L(s) = -G(s)H(s)$, so we have finally:

$$\frac{Y}{X}(s) = \frac{G(s)}{[1 + G(s)H(s)]} \qquad (3.38)$$

The denominator of the closed-loop transfer function is called the *return difference*, $F(s) = [1 + G(s)H(s)]$. If $F(s)$ is a rational polynomial, then clearly *its zeros are the poles of the closed-loop system function.* The Nyquist test effectively examines the zeros of $F(s)$ to see if any lie in the right-half s-plane. As we know, poles of the closed-loop system in the right-half s-plane produce unstable behavior. In practice, it is more convenient to work with $A_L(s) = 1 - F(s)$ and see if there are (vector) **s** values that make

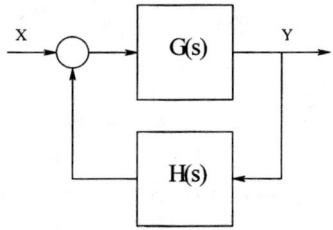

FIGURE 3.12 A general SISO, linear negative-feedback system.

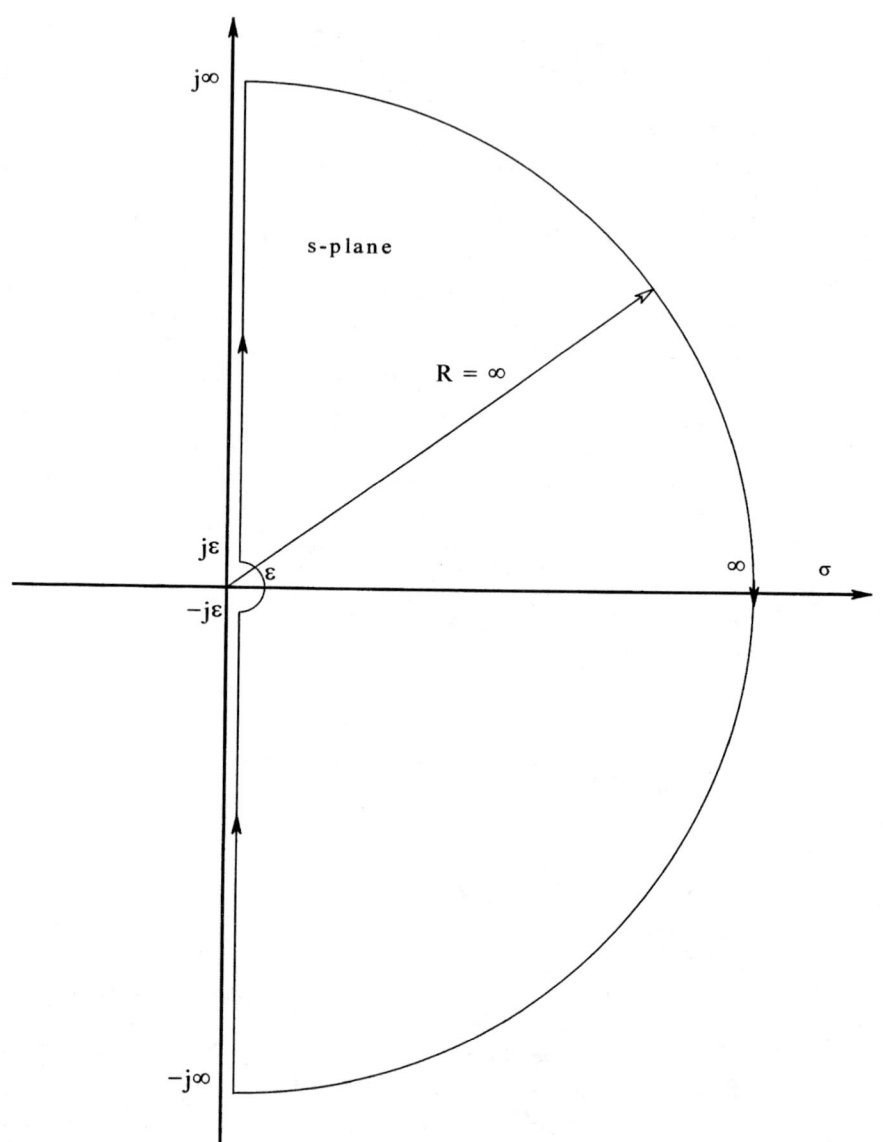

FIGURE 3.13 Clockwise contour enclosing the entire right-half s-plane used in the Nyquist test.

$A_L(s) \to 1 \angle 0°$. The Nyquist test uses a process known as conformal mapping to examine the poles and zeros of $A_L(s)$ and hence the zeros of $F(s)$. In the conformal mapping process used in the Nyquist test, the vector **s** lies on the contour shown in Figure 3.13. This contour encloses the entire right-half s-plane. The infinitesimal semicircle to the left of the origin is to avoid poles of $A_L(s)$ at the origin. Because the test is a vector test, $A_L(s)$ is written in vector difference form. For example:

Introduction to Linear SISO Control Systems and Systems with Delays

$$A_L(s) = \frac{K(s+a)}{(s+b)(s+c)} \longrightarrow A_L(s) = \frac{-K(s-s_1)}{(s-s_2)(s-s_3)}$$

Factored Laplace form   Vector difference form

(3.39)

$$= \frac{K|\mathbf{A}|}{|\mathbf{B}||\mathbf{C}|} \angle \theta_a - \theta_b - \theta_c - \pi$$

Vector polar form

The convention used in this text places a (net) minus sign in the numerator of the loop gain when the feedback system uses negative feedback. The vectors $s_1 = -a$, $s_2 = -b$, and $s_3 = -c$ are negative real numbers. In practice, s has values lying on the contour shown in Figure 3.13. The vector differences $(s - s_1) = \mathbf{A}$, $(s - s_2) = \mathbf{B}$, and $(s - s_3) = \mathbf{C}$ are shown for $s = j\omega_1$ in Figure 3.14. For each s value on the contour C_1, there is a corresponding vector value, $A_L(s)$. A fundamental theorem in conformal mapping says that if the vector s assumes values on the closed contour, C_1, then the $A_L(s)$ vector will also generate a closed contour, the nature of which depends on its poles and zeros.

Before continuing with the treatment of the vector loop gain, let us go back to the vector return difference, $\mathbf{F}(s) = 1 - A_L(s)$ and examine the Nyquist test done on the $F(s)$ of an *unstable* system. Assume:

$$F(s) = \frac{K(s-1)}{(s+2)(s+1)} \rightarrow \mathbf{F}(s) = \frac{K(s-s_1)}{(s-s_2)(s-s_3)} \qquad (3.40)$$

Recall that a right-half s-plane zero of $F(s)$ is a right-half (unstable) pole of the closed-loop system. s assumes values on the contour C_1' in Figure 3.15. Note that C_1' does not need the infinitesimal semicircle around the origin of the s-plane because $F(s)$ has no poles or zeros at the origin. Note that as s goes from $s = j0+$ to $s = +j\infty$, the angle of $\mathbf{F}(s)$ goes from $+180°$ to $-90°$ and $|\mathbf{F}(s)|$ goes from $K/2$ to 0. When $|s| = \infty$, $|\mathbf{F}(s)| = 0$ and its angle goes from $-90°$ to $+90°$ at $s = -j\infty$. The $\mathbf{F}(s)$ vector contour in the $\mathbf{F}(s)$ plane is shown in Figure 3.16. Note that the vector contour for $s = -j\omega$ values is the mirror image of the contour for $s = +j\omega$ values. We also see in this case that the complete $\mathbf{F}(s)$ contour (for all s values on C_1' traversed in a *clockwise* direction in the s-plane) makes one net clockwise encirclement of the origin in the $\mathbf{F}(s)$ plane. This encirclement is the result of the contour C_1' having enclosed the right-half s-plane zero of $F(s)$. The encirclement is the basis for the Nyquist test relation for the return difference:

$$Z = N_{CW} + P \qquad (3.41)$$

Here, Z is the number of right-half s-plane zeros of $F(s)$ *or* right-half s-plane poles of the closed-loop system's transfer function. Obviously, it is desirable for Z to be

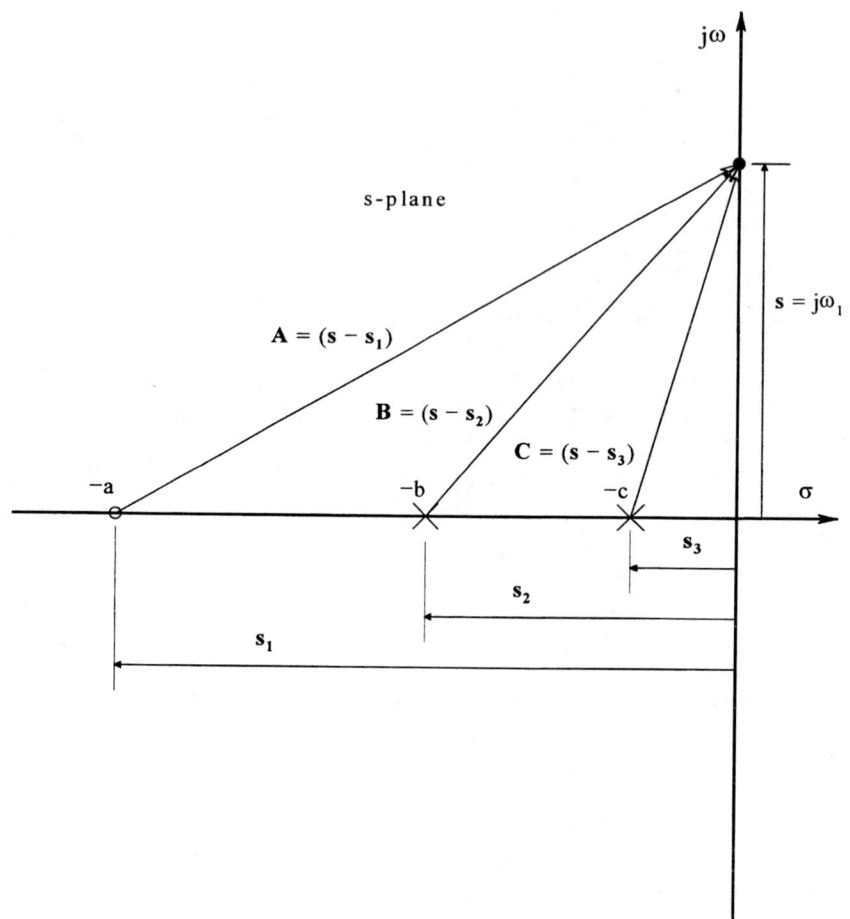

FIGURE 3.14 Vector differences in the s-plane.

zero. N_{CW} is the observed total number of clockwise encirclements of the origin by the $F(s)$ contour. P is the known number of right-half s-plane poles of $F(s)$ (usually zero).

Now we return to the consideration of the more useful loop-gain transfer function. Recall that when $F(s) = 0$, $\mathbf{A_L}(s) = +1$. For the *first example*, we take:

$$A_L(s) = \frac{-K(s-2)}{(s+5)^2(s+2)} \rightarrow \mathbf{A_L}(s) = \frac{-K(s-s_1)}{(s-s_2)^2(s-s_3)} \quad (3.42)$$

Figure 3.17 shows the s-plane with the poles and zeros of $A_L(s)$, the contour C_1', and the vector differences used in calculating $\mathbf{A_L}(s)$ as s traverses C_1' clockwise. Table 3.2 gives values of $|\mathbf{A_L}(s)|$ and $\angle \mathbf{A_L}(s)$ for s values on C_1'. The polar plot of $\mathbf{A_L}(s)$

Introduction to Linear SISO Control Systems and Systems with Delays

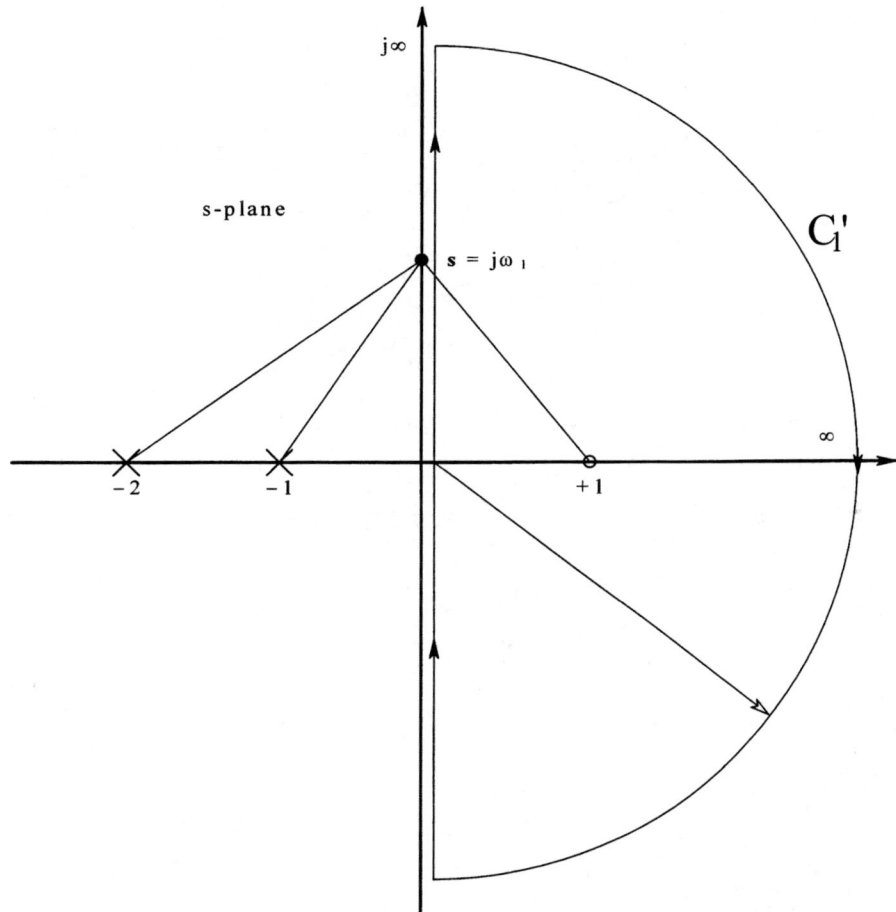

FIGURE 3.15 The clockwise contour C_1' encloses the return difference $F(s)$ zero at $+1$ in the right-half s-plane.

is shown in Figure 3.18. Since we are using $\mathbf{A_L(s)}$ instead of $\mathbf{F(s)}$, the point $\mathbf{A_L(s)} = +1$ is critical for encirclements, rather than the origin. The positive real point of intersection occurs for $\mathbf{s} = j0$. It is easily seen that if $K > 25$, there will be one net clockwise encirclement of the $+1$ point. We can modify the Nyquist conformal equation to:

$$P_{CL} = N_{CW} + P \tag{3.43}$$

P_{CL} is the number of closed-loop-system poles in the right-half s-plane. N_{CW} is the total number of clockwise encirclements of $+1$ in the $\mathbf{A_L(s)}$ plane as $\mathbf{s}$ traverses the contour C_1' clockwise. P is the number of poles of $A_L(s)$ known to be in the right-half s-plane. Thus, for the system in the first example, $P = 0$, and $N_{CW} = 1$ only if

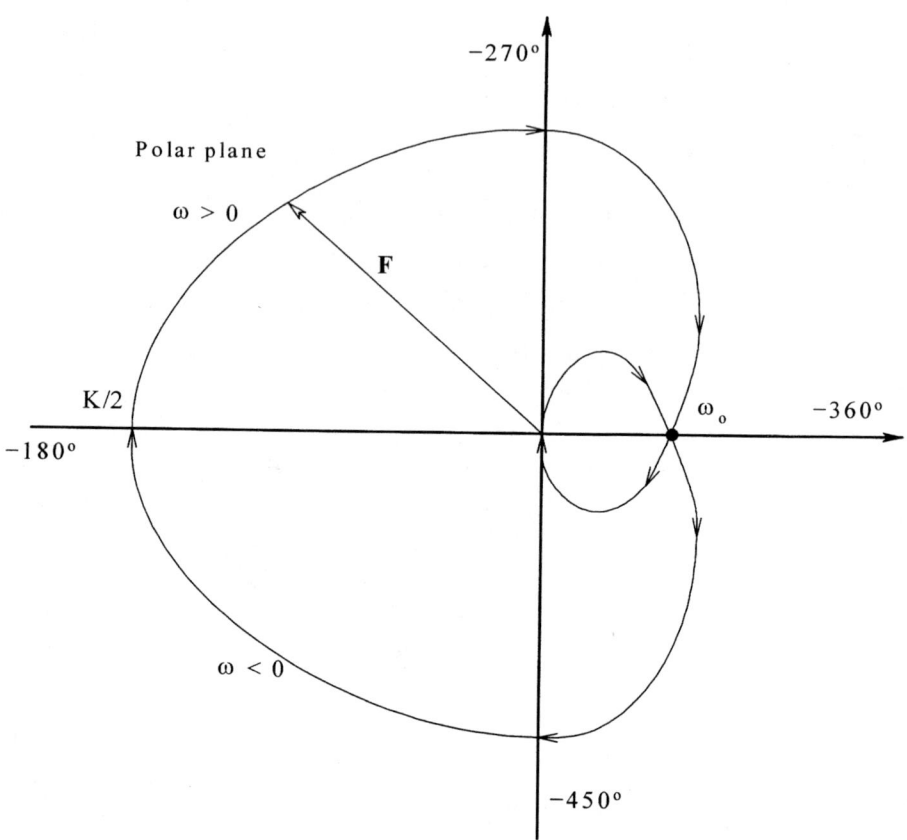

FIGURE 3.16 Polar plane plot of $\mathbf{F}(s)$, where the vector **s** assumes its values on the contour C_1' as it traverses it. There is one net encirclement of the origin.

TABLE 3.2
Values of $A_L(s)$ as s Traverses Contour C_1' Clockwise in the s-Plane

s	$|A_L(s)|$	$\angle A_L(s)$
$\pm j0$	$+K/25$	$0°$
$j2$	$K/29$	$-133.6°$
$j5$	$K/50$	$-226.4°$
$j\infty$	0	$-360°$
$+\infty$	0	$-180°$
$-j\infty$	0	$+360°$
$-j2$	$K/29$	$+133.6°$
$-j5$	$K/50$	$+226.4°$
$-j\infty$	0	$+360°$

Introduction to Linear SISO Control Systems and Systems with Delays

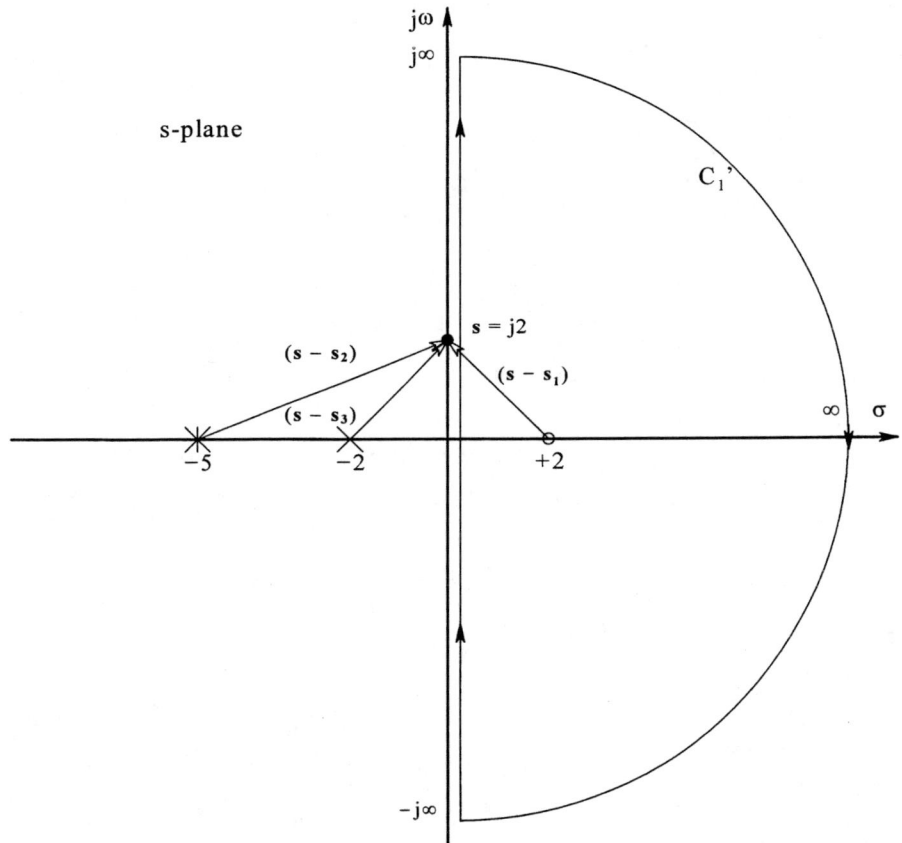

FIGURE 3.17 s-plane of the loop gain, $A_L(s)$, with clockwise Nyquist contour, C_1'.

$K > 25$. The closed-loop system is seen to be unstable with one pole in the right-half s-plane when $K > 25$.

In the *second example*, we give the system of example 1 positive feedback. Thus:

$$A_L(s) = \frac{+K(s-2)}{(s+5)^2(s+2)} \qquad (3.44)$$

Figure 3.19 shows the polar plot of the positive feedback system's $A_L(s)$ as s traverses the contour C_1'. Now we see that if K exceeds a critical value, there will be *two* clockwise encirclements of +1; hence two closed-loop-system poles will be in the right-half s-plane. Under these conditions, the instability will be a sinusoidal oscillation with an exponentially growing amplitude. To find the critical value of K for instability, it is convenient to first find the $s = j\omega_o$ value at which the system

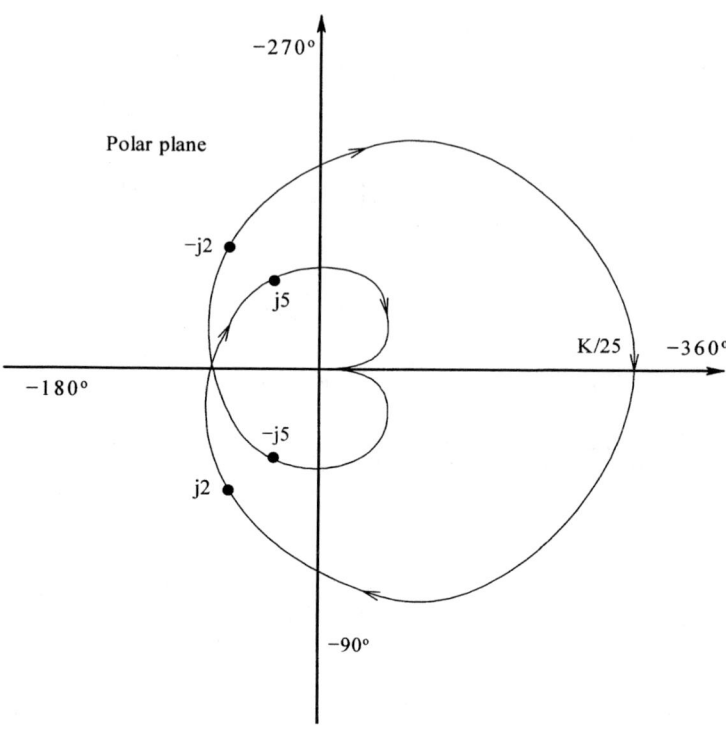

FIGURE 3.18 Polar-plane plot of the vector, $A_L(s)$. Note that the vector contour for $\omega > 0$ lies below the 0° axis, and the locus of $A_L(s)$ for $\omega < 0$ lies symmetrically above it. The +1 point is now critical for stability. One clockwise encirclement of +1 by the complete polar $A_L(s)$ locus will occur for $K > 25$, denoting instability.

will oscillate by examining the phase of $A_L(j\omega_o)$ where A_L crosses the positive real axis:

$$-2\tan^{-1}(\omega_o/5) - \tan^{-1}(\omega_o/2) + \left[180° - \tan^{-1}(\omega_o/2)\right] = 0°$$

$$-2\tan^{-1}(\omega_o/5) - 2\tan^{-1}(\omega_o/2) = -180° \qquad (3.45)$$

$$\tan^{-1}(\omega_o/5) + \tan^{-1}(\omega_o/2) = 90°$$

Trial-and-error solution of the last equation above yields $\omega_o = 3.162$ r/sec. Substitution of this ω_o value into $|A_L(j\omega_o)| = +1$, and solving for K we get $K = 35.00$. Therefore, if $K > 35$, the positive-feedback system is unstable in that it has two net clockwise encirclements of +1 and thus a CC pole pair in the right-half s-plane. Furthermore, the frequency of oscillation at the threshold of instability is $\omega_o = 3.162$ r/sec.

Introduction to Linear SISO Control Systems and Systems with Delays

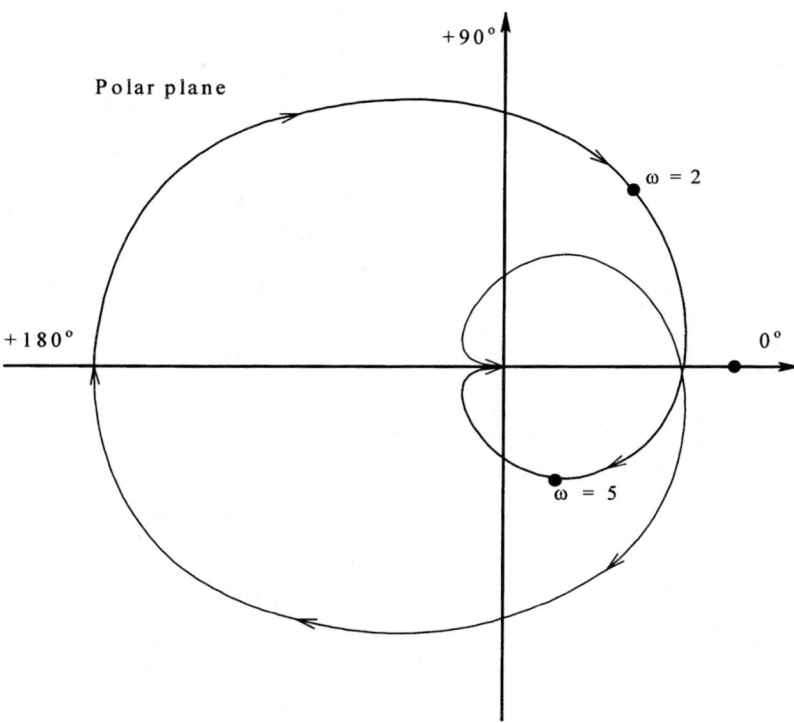

FIGURE 3.19 Polar-plane plot of the $A_L(s)$ locus when the system is given positive feedback. See text for discussion.

As a *third example* of the Nyquist stability criterion, we examine the Nyquist plot of a third-order negative feedback system with loop gain:

$$A_L(s) = \frac{-K\beta}{s(s+5)(s+2)} \tag{3.46}$$

This system has a pole at the origin; therefore, in examining $A_L(s)$, s must follow the contour C_1 that has the infinitesimal semicircle of radius ε avoiding the origin. We will start at $s = j\varepsilon$ and go to $s = j\infty$. At $s = j\varepsilon$, $|A_L(s)| \to \infty$, and $\angle A_L = -270°$. As $s \to j\infty$, $|A_L(s)| \to 0$ and $\angle A_L \to -450°$. Now when s traverses the ∞ radius semicircle on C_1, $|A_L(s)| = 0$ and the phase goes through $0°$ then to $+90°$, all with $|A_L(s)| = 0$. Now as s goes from $-j\infty$ to $-j\varepsilon$, $|A_L(s)|$ grows larger, and its phase goes from $+90°$ to $-90°$ (or $+270°$). As s traverses the small semicircle part of C_1, $|A_L(s)| \to \infty$, and the phase goes from $-90°$ to $-180°$ at $s = \varepsilon$ and then to $-270°$ at $s = j\varepsilon$. The complete contour, $A_L(s)$, is shown in Figure 3.20. Note that if $K\beta$ is large enough, there are two clockwise encirclements of $+1$ by the $A_L(s)$ locus, signifying that there are two closed-loop-system poles in the right-half s-plane and

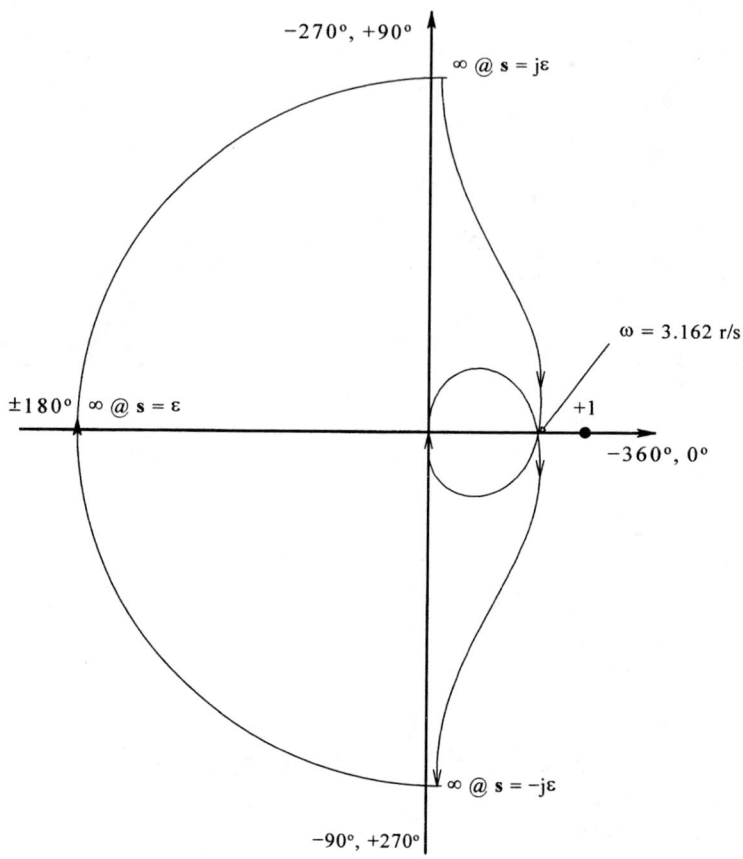

FIGURE 3.20 Polar-plane plot of the vector, $A_L(s)$, for the negative-feedback system of example 3. This is a type 1 system with one pole at the origin. If the gain is high enough, the closed-loop system acquires a pair of CC poles in the right-half s-plane.

hence oscillatory instability. Because $A_L(j\omega_o)$ is real, we can set the imaginary terms in the denominator of $A_L(j\omega_o) = 0$ and solve for ω_o. Thus:

$$j\omega_o(j\omega_o + 5)(j\omega_o + 2) = j\omega_o^3 - 7\omega_o^2 + 10j\omega_o = \text{Real} \tag{3.47}$$

Thus, $(-j\omega_o^3 + 10j\omega_o) = 0$, or $\omega_o = 3.162$ r/sec, and $K\beta > 7\omega_o^2 = 70$ for instability.

As a *fourth and final example* of application of the Nyquist test, consider the simple real-pole plant with a transport lag δ in its feedback path, shown in Figure 3.21. The loop gain function is

$$A_L(s) = \frac{-K_1 K_2/\tau_1 \, e^{-s\delta}}{(s + 1/\tau_1)} \tag{3.48}$$

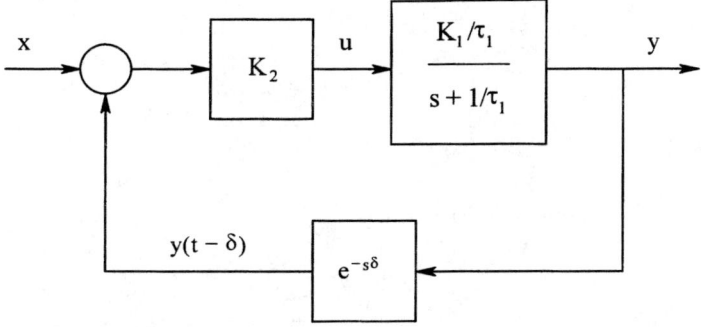

FIGURE 3.21 Block diagram of the delay system of example 4.

The simple contour, C_1', in the s-plane is used to define s values used in evaluating the $\mathbf{A_L}(s)$ contour. Figure 3.22 illustrates the $\mathbf{A_L}(s)$ contour. It begins at $\mathbf{A_L}(j0) = -K_1 K_2$ and spirals clockwise into the origin as $s = j\omega \rightarrow j\infty$. When s traverses the ∞ semicircle in the s-plane, the magnitude of $\mathbf{A_L}(s) = 0$. Then, as s goes from $-j\infty$ to $-j0$, $\mathbf{A_L}(s)$ traces the mirror image of the vector locus for positive $j\omega$. Clearly,

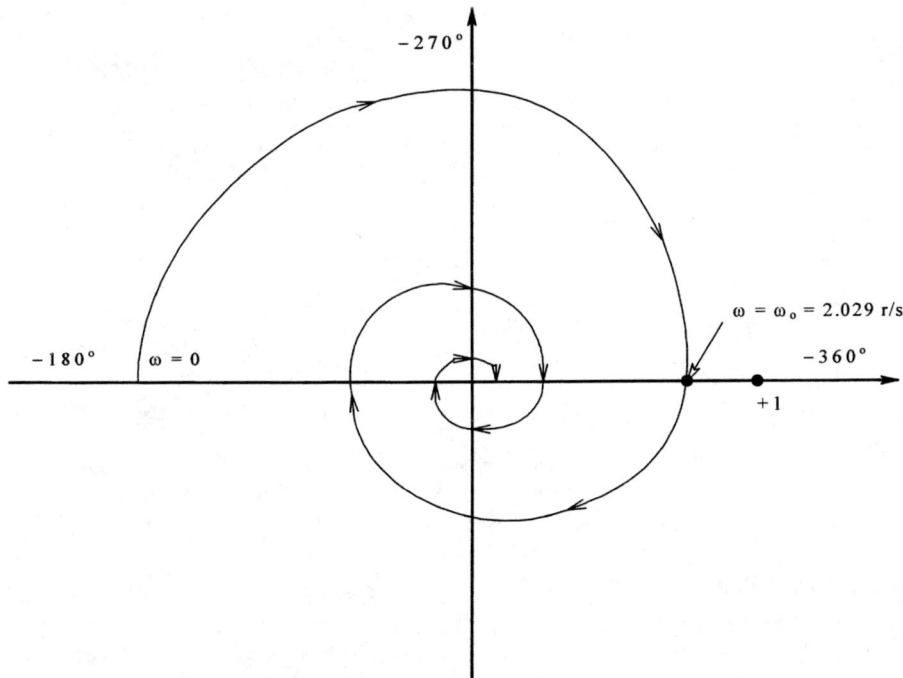

FIGURE 3.22 Polar-plane plot of the vector, $\mathbf{A_L}(s)$, for $0 \leq \omega \leq \infty$, for the delay system of example 4. See text for discussion.

if the gain $K_1 K_2$ is large enough, there can be 2, 4, 6, ... clockwise encirclements of the +1 point, implying 2, 4, 6 ... closed-loop poles in the right-half s-plane. If there were no transport lag ($\delta = 0$), the phase would approach $-270°$, and there could be *no* encirclement of +1 for any positive gain. Hence the system would be stable. However, with $\delta > 0$, the phase of $\mathbf{A_L}(s)$ becomes

$$\phi = -180° - \tan^{-1}(\omega\tau) - \omega\delta R \tag{3.49}$$

$R = 57.3°$/rad. Let $\delta = 1$ sec and $\tau = 1$ sec. The phase equation can be solved for the ω_o at which $\phi = -360°$. This turns out to be $\omega_o = 2.029$ r/sec. By setting $|\mathbf{A_L}(j\omega_o)| = 1$, we find that the system is unstable (two poles in the right-half s-plane) if $K_1 K_2 > 2.262$. At some higher gain, there will be *four* clockwise encirclements of +1, implying that there are *two* pairs of CC poles of the closed-loop system in the right-half s-plane. One might argue that this is quadruple instability. The critical gain, however, is that which produces one CC pole pair in the right-half s-plane (i.e., 2.262).

It is of interest to simulate the system in the fourth example with Simnon and compare the results with those from the Nyquist test. The Simnon program is

```
CONTINUOUS SYSTEM exfour
STATE y              " y is plant output
DER dy
TIME t
"
dy = -y/tau + K1*u/tau      " First-order plant dynamics.
w = K2*DELAY(y, delta)      " w is delayed plant output.
u = r - w                   " u is input to plant
r = IF t < dt THEN 1/dt ELSE 0  " Unit impulse input.
"
dt:.01
delta:1
tau:1
K1:1
K2:1
"
END
```

There are two ways of observing the behavior of this system as $K_1 K_2$ is varied. One is to plot y vs. t and see if the system's impulse response dies out. The second is to plot $dy(y)$ and see if a stable limit cycle is reached for some critical value of $K_1 K_2$. In Figure 3.23A, we see a stable, albeit poorly damped, closed-loop-system impulse response with $K_1 K_2 = 1$. In Figure 3.23B, the gain is increased to 1.5; the system is more oscillatory, showing that a CC pole pair in the left-half s-plane is approaching the $j\omega$ axis more closely. Figure 3.24 illustrates a phase plane plot of the closed-loop system's unit impulse response. $\dot{y}$ is plotted on the vertical axis vs. y on the horizontal axis for the same conditions as in the time-domain plot (Figure 3.23B). Note that the stable trajectory spirals in to the 0.0 point.

Introduction to Linear SISO Control Systems and Systems with Delays

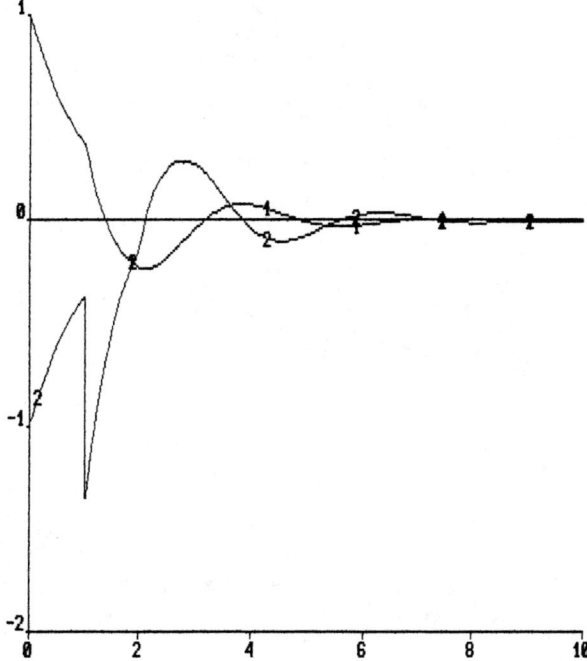

FIGURE 3.23A Impulse response of the delay system of example 4 for $K_1 K_2 = 1$. Trace 1 = y, trace 2 = dy/dt.

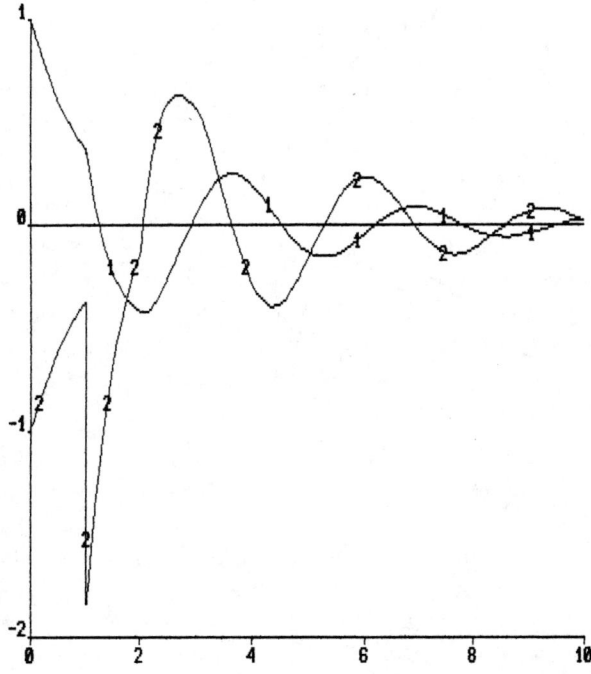

FIGURE 3.23B Same as Figure 3.23A except $K_1 K_2 = 1.5$, making the system more oscillatory.

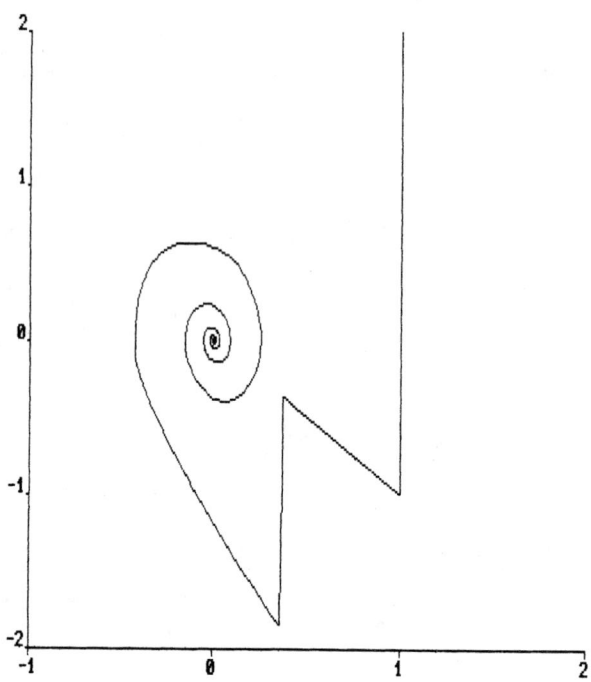

FIGURE 3.24 A phase-plane plot of dy/dt vs. y for the conditions used in Figure 3.23B.

From the Nyquist criterion applied to the system, we found that $K_1 K_2 = 2.262$ was the critical gain, above which the system is unstable. Accordingly, we set $K_1 = 1$, $K_2 = 2.262$ in a Simnon simulation in which the default Runge–Kutta–Fehlberg order 4/5 integrator was used with error limit of 1.E–6. The trajectories in the phase plane slowly shrank, indicating damping. K_2 was slowly increased until 2.2679, when stable trajectories with no observable growth or shrinking were observed (see Figure 3.25). At this critical value, closed-loop-system poles were on the $j\omega$ axis.

We then used Euler integration in the simulation with an interval of 0.0005. The phase portrait (Figure 3.26) showed very slight *outward growth* of the trajectories with time, indicating poles very slightly in the right-half s-plane for $K_2 = 2.262$. The stability of the simulated system is, not surprisingly, slightly dependent on the integration algorithm used. There is good agreement, however, when Euler integration is used with a fine Δt.

3.3.4 Describing Functions and Stability

Describing functions (DFs) are used in conjunction with the Nyquist stability criterion to predict the behavior of closed-loop nonlinear systems, (i.e., determine whether they will be stable, oscillate in a finite-amplitude limit cycle, or will be frankly unstable). The DF method is based on a knowledge of the sinusoidal frequency response of the linear portion of the system's loop gain. In the design of some

Introduction to Linear SISO Control Systems and Systems with Delays 91

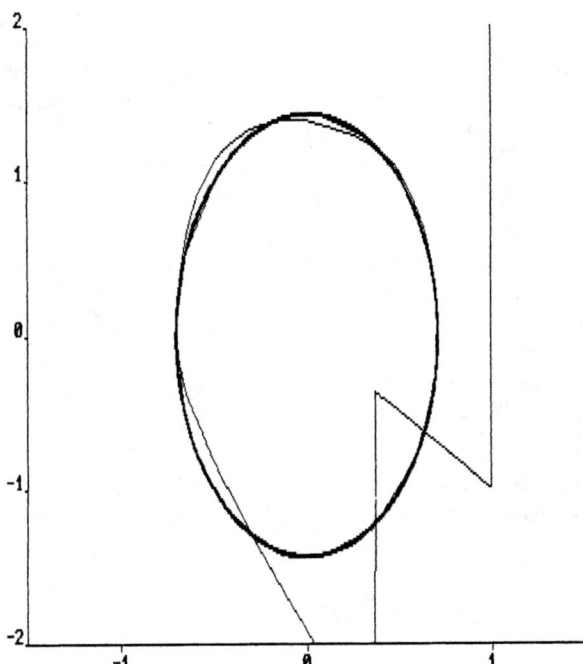

FIGURE 3.25 Phase-plane plot of the delay system's impulse response with the critical gain $K_1 K_2$ = 2.2679. System oscillations produce the ellipses. Closed-loop-system poles are very close to the $j\omega$ axis ($\xi \to 0$). See text for discussion.

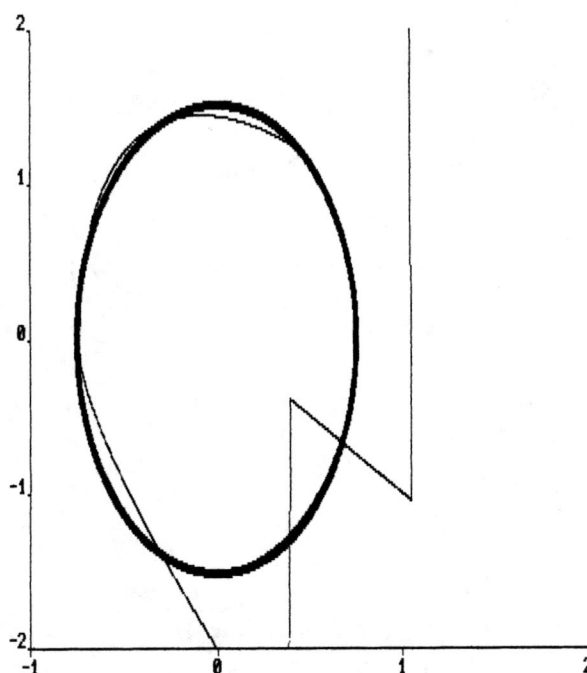

FIGURE 3.26 Another phase-plane plot of the delay system's impulse response under conditions of near instability. Here the oscillation amplitude grows in time. See text for discussion.

control systems with nonlinearities, we do not want the system to oscillate; other nonlinear control systems are designed to oscillate in bounded limit cycles. Examples of the design of limit-cycling control systems are given in Section 6.2.

The DF is the equivalent vector gain (or attenuation) of an *odd nonlinearity* in the feedback loop, usually associated with the controller. That is, for the purpose of computational ease, we replace the nonlinearity with an equivalent vector gain. A DF gain is generally a function of the amplitude of the input to the nonlinearity. The use of DFs is justified and valid under certain conditions described below.[107,135]

The nonlinearity which we replace with a DF has input $e(t)$ and output $u(t)$ and must be an odd nonlinearity, $u(e) = -u(-e)$. Often, a nonlinearity can be made odd for DF analysis by adding constant values to the input and/or the output. In calculating the DF for an odd nonlinearity, *we assume that the input is a sine wave with zero mean*. That is, $e(t) = E \sin(\omega t)$. In general, the output of the nonlinearity will be nonsinusoidal but have the same period as the input, $T = 2\pi/\omega$. Thus, $u(t) = u(t + T)$ and may be represented by a *Fourier series*. There are several formats for writing the Fourier series of a periodic function. One is

$$u(t) = A_o + \sum_{n=1}^{\infty} \left[A_n \cos(n\omega t) + B_n \sin(n\omega t) \right]$$

$$= A_o + \sum_{n=1}^{\infty} C_n \sin(n\omega t + \phi_n)$$

(3.50)

where $C_n = \sqrt{A_n^2 + B_n^2}$, $\phi_n = \tan^{-1}(A_n/B_n)$ and $\omega = 2\pi/T$. A_o is the dc (average) output, here taken to be zero. If the input to the nonlinearity is an odd function of time, and $u = f(e)$ is odd, then $u(t)$ will be odd, and the even (A_n) terms in the Fourier series can be shown to be zero. The harmonic amplitude coefficients are given by:

$$A_o = (1/T) \int_0^T u(t)\, dt = (1/2\pi) \int_0^{2\pi} u(\omega t)\, d\omega t \qquad (3.51\text{A})$$

$$A_n = (2/T) \int_0^T u(t) \cos(n\omega_o t)\, dt = (1/\pi) \int_0^{2\pi} u(\omega t) \cos(n\omega t)\, d\omega t \quad n \geq 1 \qquad (3.51\text{B})$$

$$B_n = (2/T) \int_0^T u(t) \sin(n\omega_o t)\, dt = (1/\pi) \int_0^{2\pi} u(\omega t) \sin(n\omega t)\, d\omega t \quad n \geq 1 \qquad (3.51\text{C})$$

The DF is defined as $N(E) \equiv (C_1/E) \angle \phi_1$. The DF has a nonzero angle if the output of the odd nonlinearity is phase shifted with respect to the input sinusoid, $e(t)$.

Most single-valued nonlinearities give rise to zero-phase DFs, which are positive real functions. One exception is the hysteresis function, encountered in certain temperature controllers and in mechanical systems with gear backlash. (The hysterisis nonlinearity is called a memory nonlinearity.)

We now examine the calculation of the DFs of certain commonly found nonlinearities. In the *first example*, we assume an algebraic odd nonlinear function:

$$u(t) = D\, e^3(t) \tag{3.52}$$

For the sinusoidal input, $u(t) = DE^3 \sin^3(\omega t)$. By trigonometric identity, this becomes

$$u(t) = DE^3(1/4)\left[\, 3\sin(\omega t) - \sin(3\omega t)\,\right] \tag{3.53}$$

Thus, by the definition, $N(E) = DE^2\,(3/4)$. In this case, $N(E)$ is a positive real *increasing* function of E.

In the *second example*, we examine the hard saturation nonlinearity, $u(t) = D\,\mathrm{sgn}(e)$. Here, $u = +D$ for $e(t) \geq 0$, and $u = -D$ for $e(t) < 0$. Thus, an input sine wave of any finite E generates a square wave with peak height D and the same frequency and phase. It is well known that a square wave has a Fourier series containing odd harmonics; it is given by:

$$\begin{aligned} u(t) &= \frac{4D}{\pi} \sum_{n=1}^{\infty} \frac{1}{2n-1} \sin\left[(2n-1)\omega t\right] \\ &= \frac{4D}{\pi}\left[\sin(\omega t) + (1/3)\sin(3\omega t) + (1/5)\sin(5\omega t)\dots\right] \end{aligned} \tag{3.54}$$

So, the DF for the saturation nonlinearity is

$$N(E) = \frac{4D}{\pi E}\ \angle 0°$$

Unlike that for the cubic nonlinearity, this DF is a *decreasing* function of the input amplitude.

The DF for the linear controller with output saturation nonlinearity, shown in Figure 3.27, is considered in the *third example*. In the linear region, the nonlinearity has gain $k = U_m/E_m$. Thus, if $E < E_m$, $u(\omega t) = kE\sin(\omega t)$, and $N(E) = k$. When $E > E_m$, the tops and bottoms of the output sine wave are clipped to $\pm U_m$. This distorted sine wave has a Fourier series in which the (even) A_n coefficients are zero because $u(t)$ is odd in ωt. Thus, to find the DF for this nonlinearity, we only have to evaluate B_1 for $E > E_m$. $u(\omega t)$ saturates in the regions $\beta \leq \omega t \leq (\pi - \beta)$ and $(\pi + \beta) \leq \omega t \leq (2\pi - \beta)$. From simple trigonometry, $\beta = \sin^{-1}(E_m/E)$, $E > E_m$. We calculate the first harmonic peak amplitude, B_1:

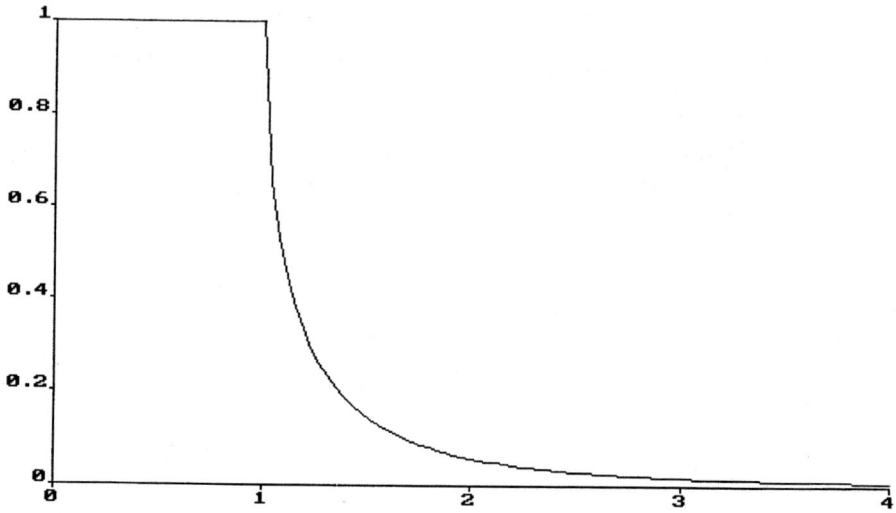

FIGURE 3.27 DF for an odd saturating nonlinearity. The linear gain is 1 and it saturates for $E \geq 1$.

$$B_1 = (1/\pi) \begin{bmatrix} \int_0^\beta kE \sin(\omega t) \sin(\omega t)\, d\omega t + \int_\beta^{\pi-\beta} U_m \sin(\omega t)\, d\omega t \\ \\ + \int_{\pi-\beta}^{\pi+\beta} kE \sin(\omega t) \sin(\omega t)\, d\omega t - \int_{\pi+\beta}^{2\pi-\beta} U_m \sin(\omega t)\, d\omega t \\ \\ + \int_{2\pi-\beta}^{2\pi} kE \sin(\omega t) \sin(\omega t)\, d\omega t \end{bmatrix} \quad (3.55)$$

Calculation of the five definite integrals above is tedious and will not be reproduced here. Division of B_1 by E yields the DF for the saturation nonlinearity:

$$N(E) = \frac{2k}{\pi}\left[\sin^{-1}(E_m/E) - (E_m/E)\sqrt{1-(E_m/E)^2}\right], \quad \text{for } E > E_m \quad (3.56)$$

Table 3.3 illustrates some common nonlinearities and their DFs. Note that some have nonzero phase and some are double-valued in E.

In the application of DFs to the analysis and design of nonlinear feedback systems, we assume that the input to the odd nonlinearity at the *design center* is a zero-mean sine wave. For example, this input can be system error, defined as

TABLE 3.3
Some Common Nonlinearities and Their DFs

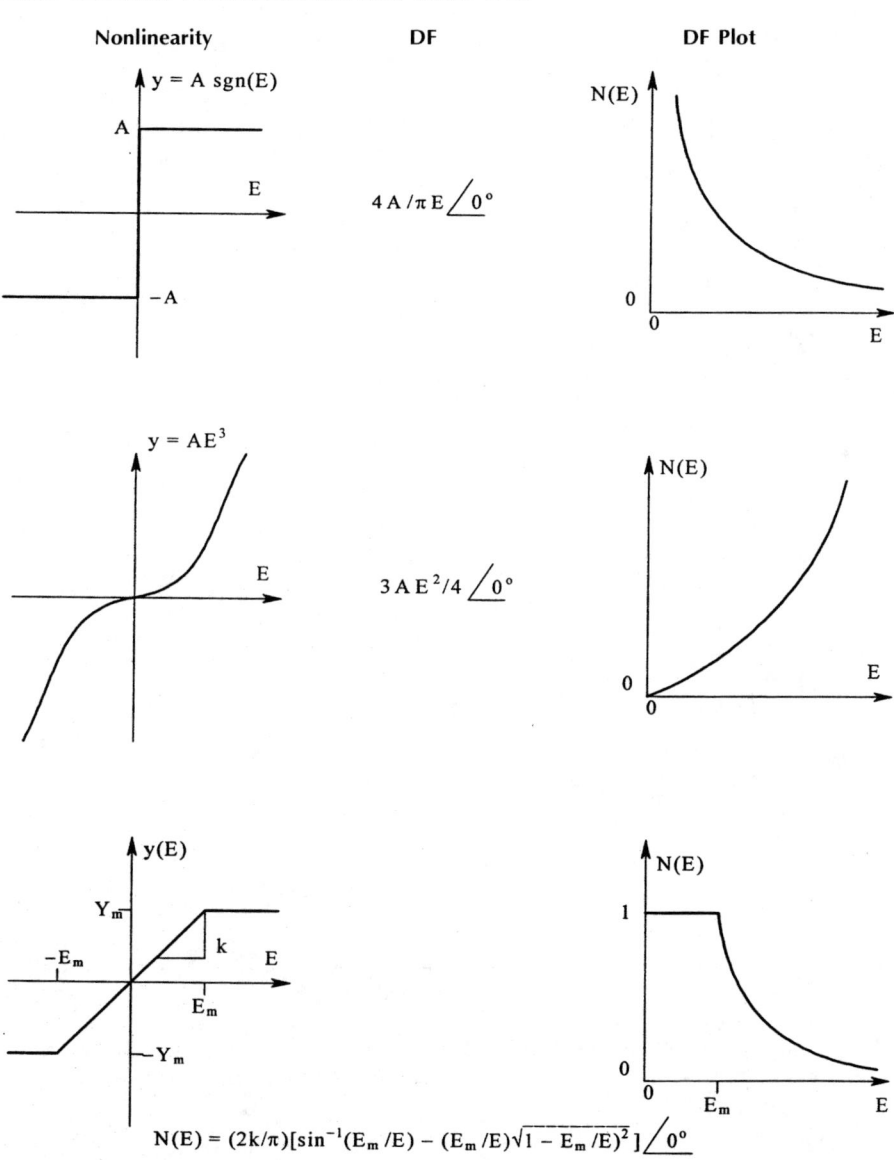

$N(E) = (2k/\pi)[\sin^{-1}(E_m/E) - (E_m/E)\sqrt{1 - E_m/E)^2}] \underline{/0°}$

$e = r - y$. The set point r at the design center is R_o. The plant output, $y(t)$, is assumed to have a dc level, $Y_o = R_o$, and a sinusoidal ripple component, $y(t) = -e(t)$. The plant input, however, is $u(t)$, the output of the nonlinearity. The higher order harmonics from $u(t)$ are assumed to be attenuated by the plant to a negligible level so that the

TABLE 3.3
Some Common Nonlinearities and Their DFs (continued)

Nonlinearity	DF	DF Plot

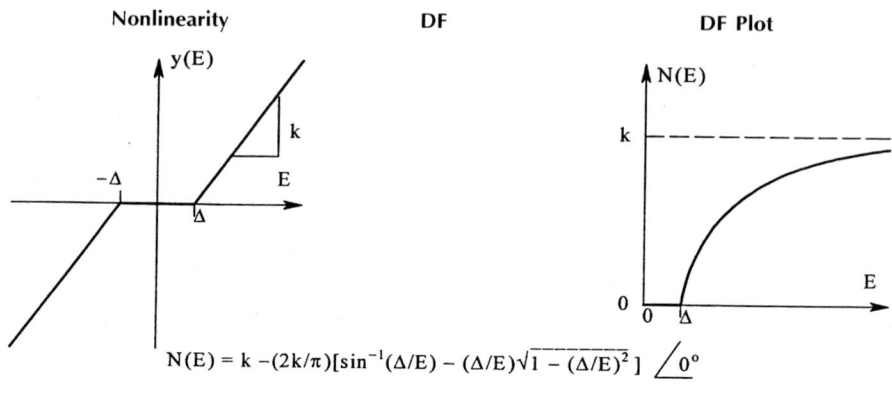

$N(E) = k - (2k/\pi)[\sin^{-1}(\Delta/E) - (\Delta/E)\sqrt{1 - (\Delta/E)^2}] \; \angle 0°$

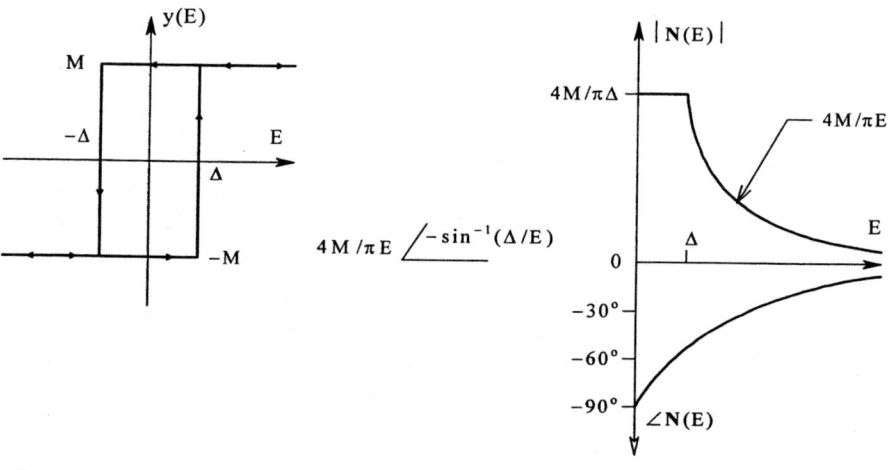

$4M/\pi E \; \angle -\sin^{-1}(\Delta/E)$

sinusoidal output component of the plant is just due to the fundamental frequency in $u(t)$. This attenuation implies that the plant be low-pass in nature when DFs are used.

The *fourth example* shows the use of DF analysis to examine the stability of a drug infusion system. The system is shown in Figure 3.28A. The pharmacokinetic plant has two real poles, and system error is acted on by a PI compensator. The PI compensator makes the system a type 1 system that will have zero steady-state error for a dc set point input. The infusion pump is nonlinear in that it cannot generate negative u (infusion rate), and it has a maximum delivery rate of U_m mg/min. We are interested in using the DF method to see what the stability limits of the system

Introduction to Linear SISO Control Systems and Systems with Delays

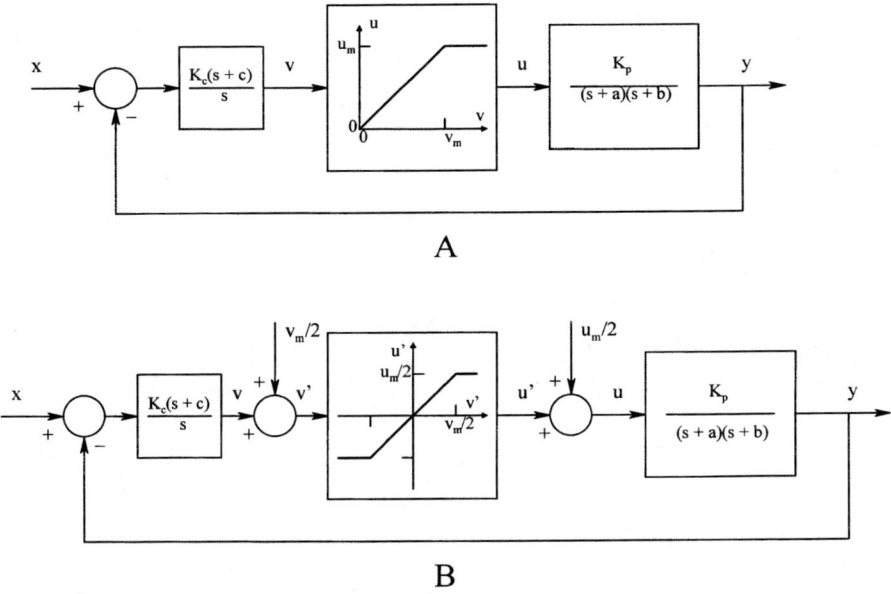

FIGURE 3.28 (A) Block diagram of a drug infusion controller. The nonlinearity describes the action of the infusion pump. Note that the nonlinearity is not an odd function of v. (B) The drug input controller is modified so that the nonlinearity is an odd function of v, permitting DF analysis to be used.

are and what the amplitude of oscillations will be if the system becomes unstable.

In order to use the DF method, we must convert the infusion pump nonlinearity to an odd nonlinearity. This is accomplished, as shown in Figure 3.28B, by subtracting $V_m/2$ from v and adding $U_m/2$ to u', the output of the new odd nonlinearity. Stability of the system is evaluated by using the Nyquist criterion in conjunction with the DF. We examine the closed-loop system's return difference. Are there conditions on system parameters and complex s that will make the return difference → 0? Mathematically, we set:

$$F(s) = \left[1 - N(V')A_L(s)\right] = 0 \tag{3.57}$$

Equivalently,

$$A_L(s) = N^{-1}(V') \tag{3.58}$$

for instability. We plot $A_L(j\omega)$ on polar graph paper with the reciprocal DF, $N^{-1}(V')$. Intersection of $A_L(j\omega)$ and $N^{-1}(V')$ implies that the closed-loop system is unstable. In the case of a saturating nonlinearity, the oscillations are bounded (i.e., a limit cycle). The DF of the odd nonlinearity is

$$\mathbf{N}(V') = (2K_v/\pi) \left[\sin^{-1}(V_m/2V') - (V_m/2V')\sqrt{1-(V_m/2V')^2} \right] \quad (3.59)$$

where $K_v = U_m/V_m$. $\mathbf{N}^{-1}(V')$ is a real increasing function of V', starting at $1/K_v$. The linear loop gain can be written in vector form:

$$\mathbf{A_L}(s) = \frac{-K_p K_c(s-s_1)}{s(s-s_2)(s-s_3)} \quad (3.60)$$

Let $s_1 = -10$, $s_2 = -4$, $s_3 = -1$. The polar plots of $\mathbf{A_L}(j\omega)$ and $\mathbf{N}^{-1}(V')$ are shown in Figure 3.29. Note that $\mathbf{A_L}(j\omega)$ crosses the $-360°$ axis at a frequency of ω_o. If $K_p K_c$

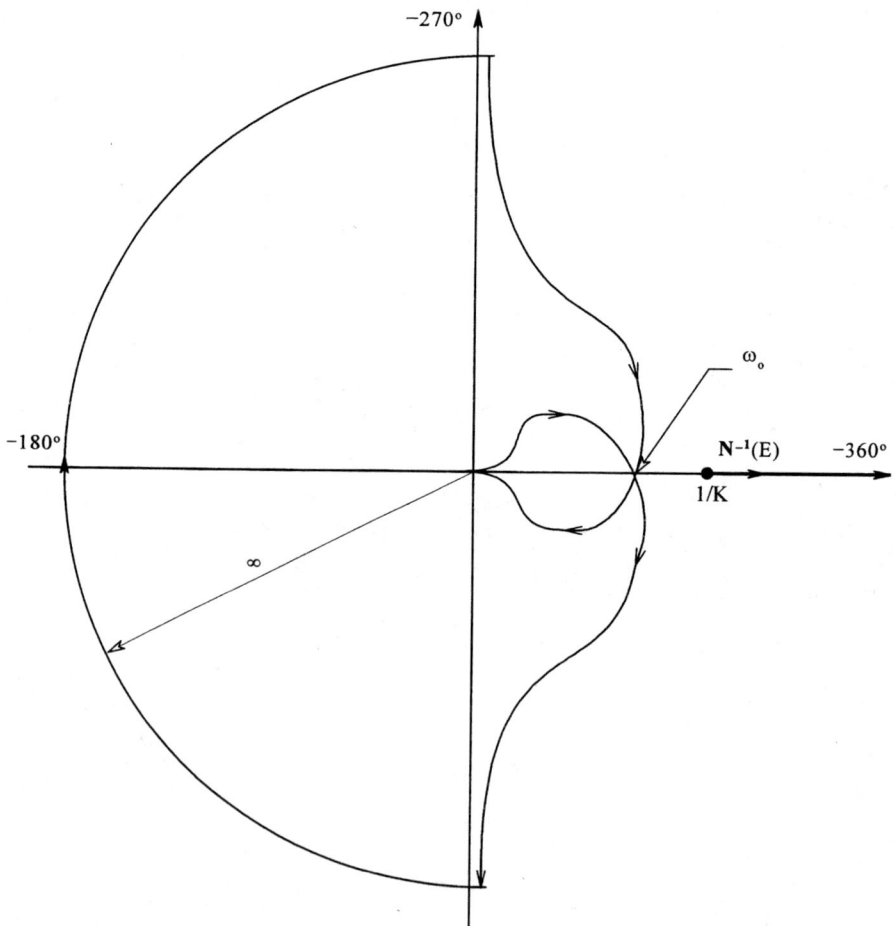

FIGURE 3.29 Polar plot of the linear loop gain and $N^{-1}(V')$ of the saturating drug infusion controller. Note that the system is stable until $K_p K_c$ reaches a critical value.

is large enough, $\mathbf{A_L}(j\omega)$ will intersect $\mathbf{N^{-1}}(V')$, implying instability. Such an intersection gives rise to stable limit-cycle oscillations at radian frequency, ω_o. That the oscillations have a stable amplitude can be seen by noting that if the oscillations grow in amplitude, V' increases, and the V'_o point at the intersection moves to the right on the $\mathbf{N^{-1}}(V')$ locus, implying that V'_o is a stable point of intersection and amplitude. The frequency ω_o where $\angle \mathbf{A_L}(j\omega_o) = -360°$ can be found by solving the phase equation for $\mathbf{A_L}(j\omega)$.

$$-360° = -180° - 90° - \tan^{-1}(\omega_o) - \tan^{-1}(\omega_o/4) + \tan^{-1}(\omega_o/10)$$
$$\rightarrow +90° = \tan^{-1}(\omega_o) + \tan^{-1}(\omega_o/4) - \tan^{-1}(\omega_o/10)$$
(3.61)

Trial-and-error solution finds $\omega_o = 2.828$ r/min.
For linear behavior of the system (i.e., no oscillations), $|\mathbf{A_L}(j\omega_o)| < 1/K_v$.

$$|\mathbf{A_L}(j\omega_o)| = \frac{K_p K_c \sqrt{\omega_0^2 + 100}}{\omega_o \sqrt{(\omega_0^2 + 1)} \sqrt{(\omega_0^2 + 16)}} < 1/K_v \rightarrow$$

$$K_p K_c K_v < \frac{\sqrt{\omega_o(\omega_0^2 + 1)}\sqrt{(\omega_0^2 + 16)}}{K_p K_c \sqrt{(\omega_0^2 + 100)}}$$
(3.62)

Substituting the value for ω_o, we find that for stability, $K_p K_c K_v < 4.00$. If $K_p K_c K_v > 4.00$, the system will exhibit a bounded amplitude limit cycle of frequency ω_o and amplitude V'_o determined by solution of:

$$|\mathbf{A_L}(j\omega_o)| = \frac{K_p K_c \sqrt{\omega_0^2 + 100}}{\omega_o \sqrt{(\omega_0^2 + 1)} \sqrt{(\omega_0^2 + 16)}}$$
(3.63A)

$$\mathbf{N^{-1}}(V'_o) = \frac{1}{\left(2K_v/\pi \left[\sin^{-1}(V_m/2V'_o) - (V_m/2V'_o)\sqrt{1 - (V_m/2V'_o)^2}\right]\right)}$$
(3.63B)

given $K_p K_c K_v$ and ω_o above.

In the *second example* of the application of the DF method, we consider a nonlinear controller defined by the nonlinear algebraic equation

$$u(t) = K_c \left[e + e^2 \dot{e} \right]$$
(3.64)

Let $e(t) = E \sin(\omega t)$, so $\dot{e}(t) = E\omega \cos(\omega t)$. Thus:

$$u(t) = K_c \{E \sin(\omega t) + E^2 \sin^2(\omega t) E\omega \cos(\omega t)\} \quad (3.65)$$

Using trigonometric identities, we find:

$$u(t) = K_c \{E \sin(\omega t) + (E^3 \omega/4) \cos(\omega t) - (E^3 \omega/4) \cos(3\omega t)\} \quad (3.66)$$

$u(t)$ is seen to have two fundamental frequency terms and a third harmonic term. The DF is found by inspection to be a function of E and ω:

$$N(E,\omega) = C_1/E = K_c \sqrt{\left[1 + (E^4 \omega^2/16)\right]} \angle \tan^{-1}(E^2 \omega/4) \quad (3.67)$$

$N^{-1}(E, \omega)$ is

$$N^{-1}(E,\omega) = \frac{1}{K_c \sqrt{\left[1 + (E^4 \omega^2/16)\right]}} \angle -\tan^{-1}(E^2 \omega/4) \quad (3.68)$$

In the polar plane, $N^{-1}(E, \omega)$ plots as a family of curves, starting at $1/K_c$ on the $0°$ axis and going to zero at $-90°$. A typical $N^{-1}(E, \omega)$ locus is shown qualitatively in Figure 3.30 for one frequency. The locus of the linear frequency response, $A_L(j\omega)$, for a two-real-pole plant is also shown in the figure. Note that it can never intersect the $N^{-1}(E, \omega)$ loci, and thus the system is always stable. A second $A_L(j\omega)$ locus for a two-pole plant with a transport lag is also plotted in the figure. This $A_L(j\omega)$ curve intersects the $N^{-1}(E, \omega)$ loci, and thus instability is predicted by the

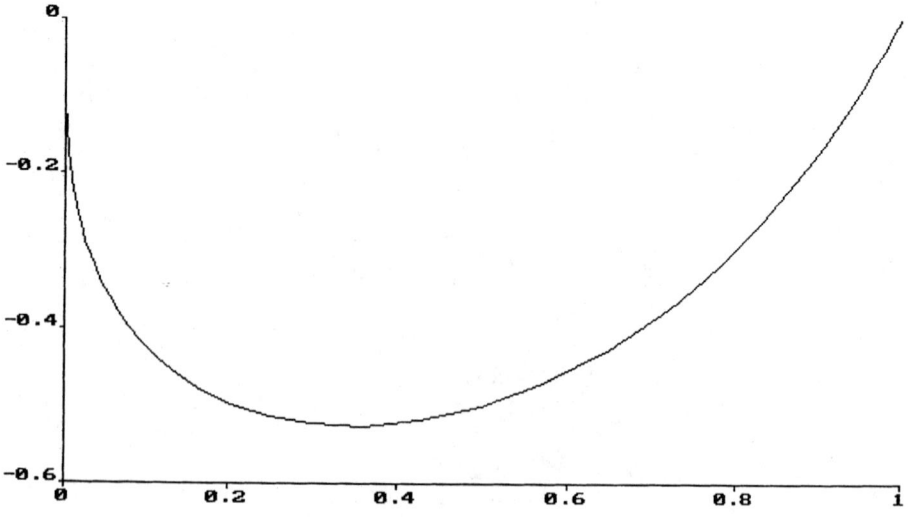

FIGURE 3.30 A polar plot of the vector $N^{-1}(E, \omega)$ for $K_c = 1$ and $\omega = 1$ for increasing E.

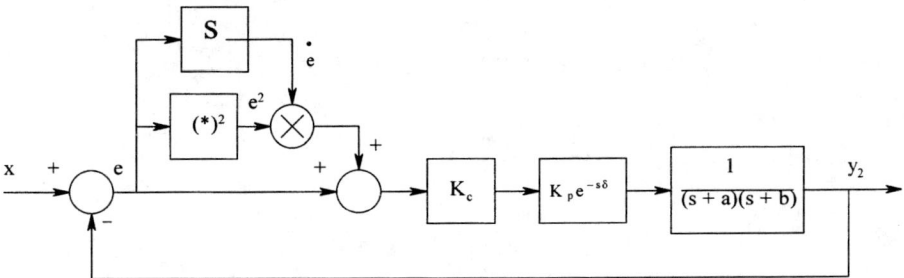

FIGURE 3.31 Block diagram of the nonlinear controller showing delay.

DF/Nyquist criteria, but only if E exceeds the critical value where the loci intersect. If oscillations start, then they will grow, because increasing E places the operating point on the $\mathbf{N}^{-1}(E, \omega)$ locus *inside* the $\mathbf{A}_L(j\omega)$ curve. Further increase in E places the operating point on the $\mathbf{N}^{-1}(E, \omega)$ locus further inside the $\mathbf{A}_L(j\omega)$ curve, signifying unbounded growth of oscillations. This amplitude-dependent instability can be demonstrated by a Simnon simulation. The system is shown in block diagram form in Figure 3.31. The Simnon program is listed below:

```
CONTINUOUS SYSTEM dfpknl1
STATE y1 y2
DER dy1 dy2
TIME t
"
dy1 = -a*y1 + w
dy2 = -b*y2 + y1
e = SP - y2
de = -dy2
u = Kc*(e + e*e*de)
w = Kp*DELAY(u, delta)
"
a:1
b:5
SP:1
Kp:5
Kc:1
delta:.2
"
END
```

Figure 3.32 shows the result of the simulation; system parameters are held fixed and the size of the step input is varied to cause an increase in the value of the initial error. The plant output is plotted along with the set point. For SP ≥ 2.4, the system output oscillates and grows in magnitude so that there is register overflow in the computer, and simulation is stopped. As SP is lowered, the oscillatory response becomes less pronounced, until at SP = 1.5 the output appears well damped. Note that the oscillation frequency is about the same, regardless of SP. This is to be

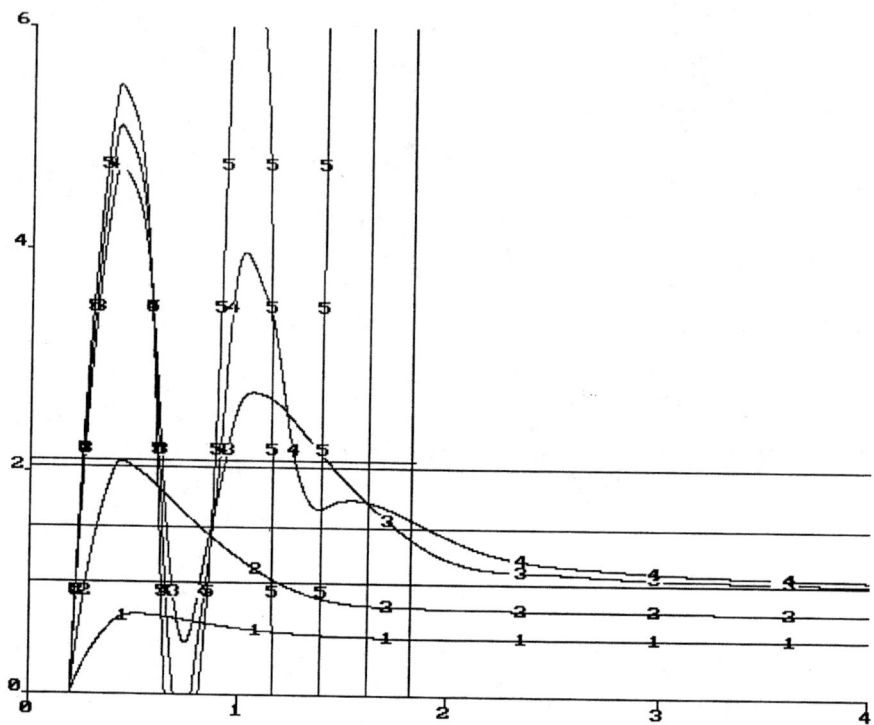

FIGURE 3.32 Step response of the nonlinear controller with delay for various step (set point) sizes. Trace 1, SP = 1; trace 2, SP = 1.5; trace 3, SP = 2.0; trace 4, SP = 2.05; trace 5, SP = 2.1. The system is unstable for SP ≥ 2.1.

expected because there is only one unique intersection of $\mathbf{A_L}(j\omega)$ with the $\mathbf{N}^{-1}(E, \omega_o)$ locus at $\omega = \omega_o$. In a conditionally stable, nonlinear system such as this, instability is implied when the $\mathbf{A_L}(j\omega)$ vector locus lies to the right of and encloses the critical point on the $\mathbf{N}^{-1}(E, \omega)$ vector locus. The critical E_o value is that at the intersection point of $\mathbf{A_L}(j\omega_o)$ with $\mathbf{N}^{-1}(E_o, \omega_o)$. If the system operating conditions are such that a value of $E > E_o$ is generated, as with the step input used in the simulation, the system will be unstable.

In Section 6.2 of this text, there are further applications of the DF method in the design of on/off (switching) limit-cycling controllers for drug infusion to CPK plants. In both of the examples above, oscillation was to be avoided. As we shall see, for the on/off controllers, a bounded limit cycle is part of the design strategy and is desired.

3.3.5 Popov's Stability Criterion

Popov's criterion predicts the absolute stability of single-loop, SISO feedback systems that have a nonlinear element followed by a linear low-pass loop gain

Introduction to Linear SISO Control Systems and Systems with Delays

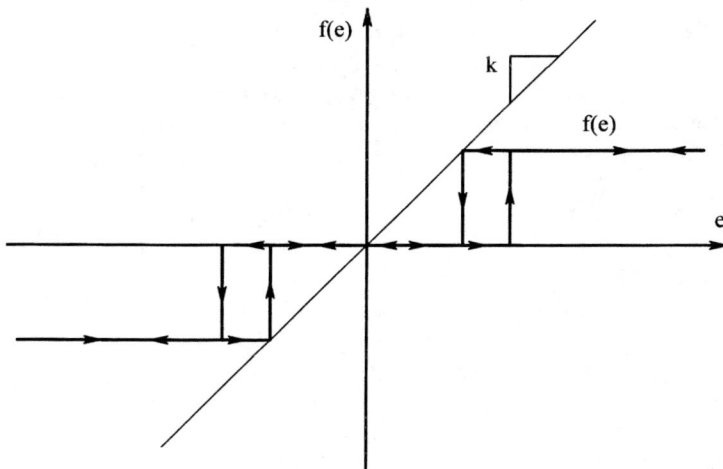

FIGURE 3.33 The Popov sector is the wedge-shaped area between the line $y = ke$ and the e-axis. The nonlinearity is a hysteresis with a dead zone.

function.[135] In fact, the Popov stability criterion is applicable to any system that can be analyzed by DFs. In Popov analysis, we assume that the nonlinearity satisfies the following conditions: $u = f(e)$ is a single-valued, continuous real function and $u(0) = 0$. A *Popov sector* is defined as shown in Figure 3.33. The straight line with slope k passes through the origin and defines a wedge-shaped sector bounded by the line $y = ke$ and the e axis, entirely inside of which the nonlinearity lies. Stated mathematically:

$$0 \leq \frac{f(e)}{e} \leq k \tag{3.69}$$

If $k \to \infty$, then the relation above reduces to:

$$0 \leq \frac{f(e)}{e}, \quad \text{or,} \ ef(e) \geq 0 \tag{3.70}$$

The above relations are true for a type 0 linear loop gain function. It can be shown that for a type 1 system:

$$\varepsilon \leq \frac{f(e)}{e} \leq k, \quad (e \neq 0) \tag{3.71}$$

where ε is an arbitrarily small positive number.

Let us write the system's negative, linear loop gain in rectangular vector form:

$$-\mathbf{A}_L(j\omega) = X(\omega) + jY(\omega) \tag{3.72}$$

We use $-\mathbf{A}_L(j\omega)$ because the Popov method assumes that the error summer subtracts the output from the input. The protocol used throughout this text is to associate the minus sign from negative feedback with the loop gain function. Thus $-\mathbf{A}_L(j\omega)$ has no minus sign.

To exercise the Popov criterion, we plot the modified *Popov vector frequency response* on polar coordinates:

$$\mathbf{P}_L(j\omega) = X(\omega) + j\omega Y(\omega) \tag{3.73}$$

Thus the imaginary term is compressed for $\omega < 1$ and stretched for $\omega > 1$. Figure 3.34 shows polar plots of $-\mathbf{A}_L(j\omega)$ and $\mathbf{P}_L(j\omega)$ for $-\mathbf{A}_L(j\omega) = K_p/[(j\omega + a)(j\omega + b)]$, where $K_p = 4$, $a = 1$, and $b = 4$. Note that $\mathbf{A}_L(j\omega)$ is a rational polynomial with M = highest s power in the numerator and N = highest s power in the denominator, so that:

$$-\mathbf{A}_L(j\omega) = \frac{b_o(j\omega)^M + b_1(j\omega)^{M-1} + \cdots + b_M}{(j\omega)^N + (j\omega)^{N-1} + \cdots + a_N} \tag{3.74}$$

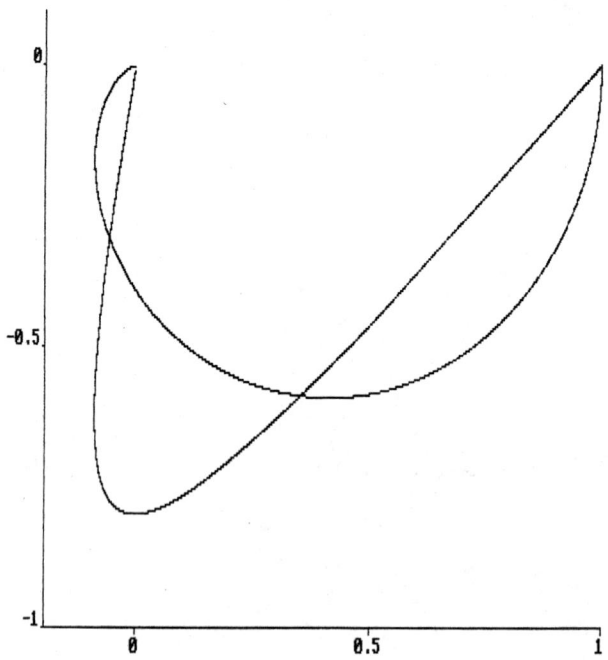

FIGURE 3.34 Polar plots of $-\mathbf{A}_L(j\omega)$ and $\mathbf{P}_L(j\omega)$ for a simple two-real-pole loop gain function. $-\mathbf{A}_L(j\omega)$ starts at 1 ∠0° at $\omega = 0$ and approaches 0 at −180°. $\mathbf{P}_L(j\omega)$ appears to approach 0 at −90°. See text for details.

If $N - M = 1$, then it is evident that:

$$\lim_{\omega \to \infty} \text{Re}\{\mathbf{P_L}(j\omega)\} \to 0 \tag{3.75A}$$

$$\lim_{\omega \to \infty} \text{Im}\{\mathbf{P_L}(j\omega)\} = \omega Y(\omega) \to -b_o \tag{3.75B}$$

Popov's stability theorem states that *the nonlinear system will be absolutely stable if there exists a finite real number, q, such that for all $\omega \geq 0$,*

$$\text{Re}\{(1 + j\omega q)[-\mathbf{A_L}(j\omega)]\} > -1/k \tag{3.76}$$

To see how this relation can be interpreted, let us expand the terms:

$$(1 + j\omega q)[X(\omega) + jY(\omega)] = [X(\omega) - q\omega Y(\omega)] + j[q\omega X(\omega) + Y(\omega)] \tag{3.77}$$

Thus, $\text{Re}\{(1 + j\omega q)[-\mathbf{A_L}(j\omega)]\} = [X(\omega) - q\omega Y(\omega)] > -1/k$. From Equation 3.77, we can write finally:

$$\omega Y(\omega) \leq [X(\omega)/q + 1/(qk)] \tag{3.78}$$

Equation 3.78 describes the Popov line, plotted in the complex plane with the plot of $\mathbf{P_L}(j\omega)$.

Figure 3.35A shows the polar plot of $\mathbf{P_L}(j\omega)$ for a generic minimum-phase $N - M = 3$ system, along with the straight Popov line. It is easy to choose a q value, given k and $\mathbf{P_L}(j\omega)$, so that the $\mathbf{P_L}(j\omega)$ plot lies to the right of the line, signifying a stable system. Figure 3.35B illustrates the polar plot of an unstable system. Note that for the k value obtained from the nonlinearity, there is no q value that will cause the entire $\mathbf{P_L}(j\omega)$ locus to lie to the right of the line.

As an *example* of the application of the Popov criterion, consider the regulator shown in Figure 3.36. The nonlinearity is a linear controller with hard saturation. The loop gain is type 1, with $N - M = 3$. We do not want this regulator to limit cycle; it must operate linearly. To apply Popov, we write the negative loop gain:

$$-\mathbf{A_L}(j\omega) = \frac{1}{j\omega(j\omega + 4)(j\omega + 10)} = \frac{-14\omega^2 - j\omega(40 - \omega^2)}{\omega^2[1600 + 116\omega^2 + \omega^4]} \tag{3.79}$$

The Popov frequency response is

$$-\mathbf{P_L}(j\omega) = \frac{-14}{1600 + 116\omega^2 + \omega^4} - \frac{j(40 - \omega^2)}{1600 + 116\omega^2 + \omega^4} \tag{3.80}$$

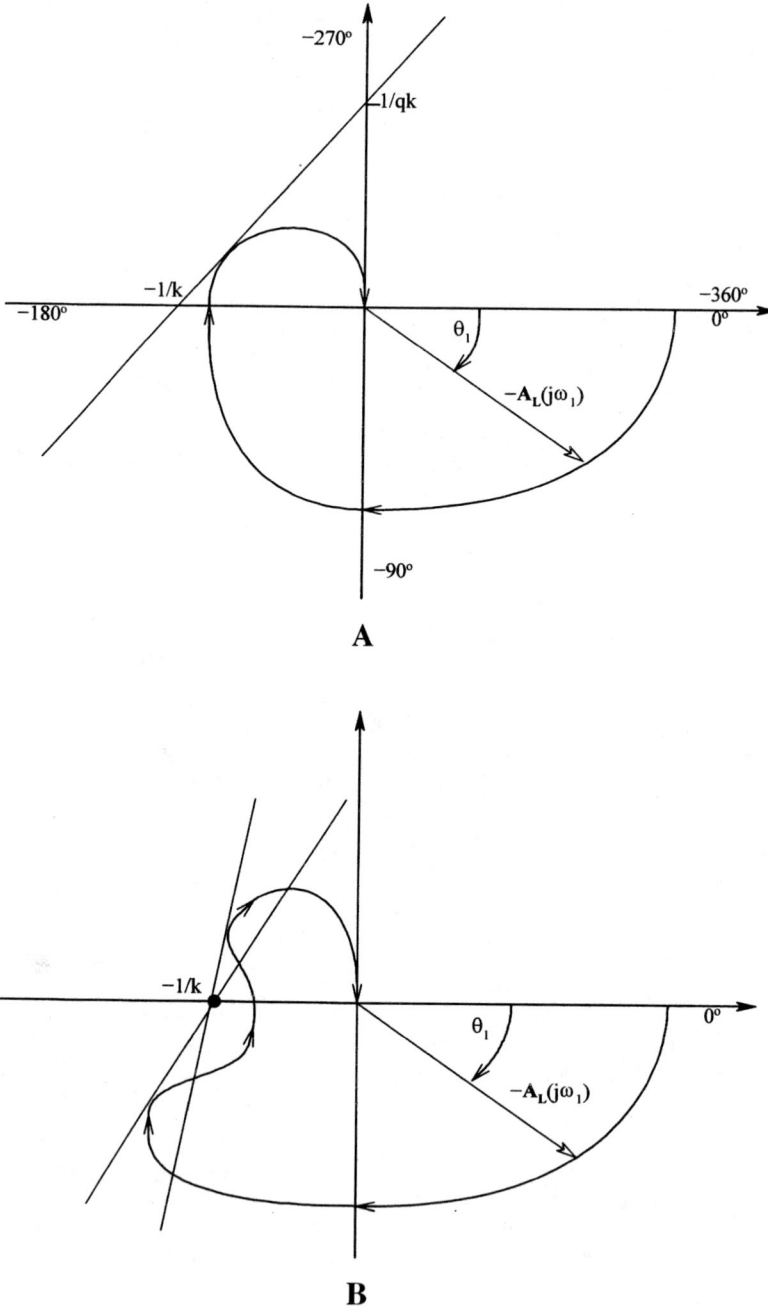

FIGURE 3.35 (A) Polar plot of $\mathbf{P_L}(j\omega)$ for a nonlinear system that can be stable for the Popov k given. (B) Polar plot of $\mathbf{P_L}(j\omega)$ for a nonlinear system that can never be stable for the Popov k given.

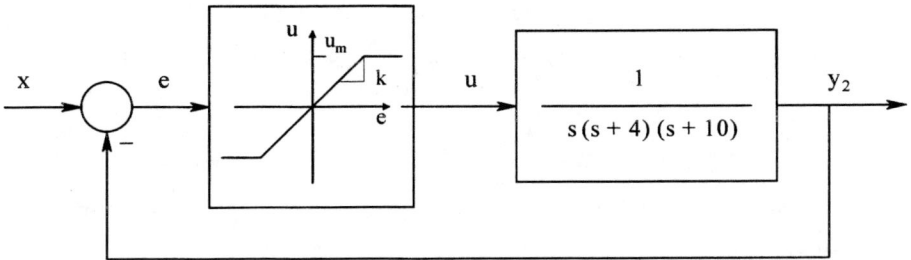

FIGURE 3.36 Block diagram of a type 1 system with a saturating controller.

q is easily chosen so the Popov line lies to the left of $\mathbf{P_L}(j\omega)$. At $\omega = \omega_o = \sqrt{40}$, and $\mathbf{P_L}(j\omega_o)$ is real. For stability, the real $\mathbf{P_L}(j\omega_o)$ must be greater than the point $-1/k$:

$$\frac{-14}{1600 + 116(40) + (40)(40)} = -1.7857 \times 10^{-3} > -1/k \qquad (3.81)$$

So $k < 560$ to ensure stability. If $k > 560$, the system will oscillate with a bounded limit cycle.

By way of comparison, we can examine the stability of the same system with the DF method. The DF for the saturation nonlinearity is well known:

$$N(E) = (2k/\pi)\left[\sin^{-1}(E_m/E) - (E_m/E)\sqrt{1 - (E_m/E)^2}\,\right], \quad E > E_m \qquad (3.82)$$

In the DF method, we plot $\mathbf{A_L}(j\omega)$ and $\mathbf{N}^{-1}(E)$ in the polar plane and require that the two vector loci do not intersect. These plots are shown in Figure 3.37. Note that this system's stability depends on $\mathbf{A_L}(j\omega_o) < 1/k$, exactly the same criterion as in the Popov method. The DF method is probably more useful in this case, however, in that it tells us the amplitude of the limit cycle oscillations.

3.4 LINEAR CLOSED-LOOP SYSTEMS WITH DELAY: STABILITY PROBLEMS

The ideal delay, also called a transport lag or dead-time operation, is often found in the description of physiological systems. For example, because nerve impulses propagate at a finite velocity on axons, delays ranging from 0.01 m/10 m/sec = 1 msec to 2 m/0.1 m/sec = 20 msec may be involved in the transmission of sensory and motor signals in the human body. Other sources of apparent delays in physiological systems are associated with the diffusion of ions, hormones, and neurotransmitters. Diffusion does not generate a pure delay, however, because of the

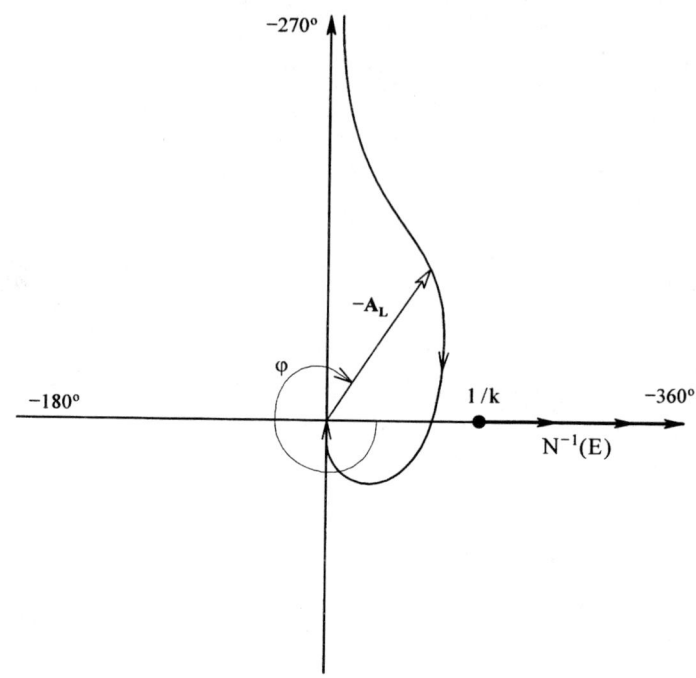

FIGURE 3.37 Polar plot of $A_L(j\omega)$ and $N^{-1}(E)$ illustrating the DF criterion for stability. A complete plot of $A_L(s)$ around the clockwise closed contour C_1 would give a mirror-image frequency response plot around the 0° axis with a closure to $-\infty$. Thus, if $1/k$ lies to the left of the $A_L(j\omega)$ locus, there will be two net encirclements of $1/k$, and thus two CC closed-loop system poles in the right-half s-plane, giving an oscillatory instability.

statistical distribution of velocities and the nonuniform distances involved. When a drug is injected as an intravenous bolus, several cardiac cycles are required before it is uniformly mixed in the blood. Again, this is not a pure delay, but can be approximated by one to simplify modeling.[126] When a hormone activates an intracellular biosynthetic pathway, an apparent dead time is also introduced. This is because of the long sequence of biochemical steps leading to the target effect. For example, the hormone molecules must bind to specific receptors on the target cell surface. Binding leads to the release of a secondary intracellular messenger substance that leads to activation of protein synthesis through the well-known biochemical pathways involving DNA, messenger RNA, etc.[59]

In the time domain, the output of an ideal delay element can be written:

$$y(t) = x(t - \delta) \tag{3.83}$$

where $x(t)$ is the input, $y(t)$ is the output, and δ is the delay. It is easy to show that the Laplace transform of the delay operation is

$$\frac{Y(s)}{X(s)} = \exp(-\delta s) = D(s) \tag{3.84}$$

In terms of frequency response, $\mathbf{D}(j\omega) = 1\,\angle{-}\omega\delta R$. That is, the magnitude of $\mathbf{D}(j\omega)$ is unity, and its phase goes negative linearly with frequency. ($R = 57.3°$/rad.) The pure delay operation can be approximated by rational polynomials. The simplest Padé approximation to $D(s)$ is

$$\hat{D}(s) = \frac{1 - s\delta/2}{1 + s\delta/2} \cong D(s) \tag{3.85}$$

Another approximation to $D(s)$ given by Milsum[100] is

$$\hat{D}(s) = \frac{1}{(1 + s\delta/n)^n} \xrightarrow[\lim n \to \infty]{} D(s) \tag{3.86}$$

A well-known example of a physiological system with delays is the pharmacokinetic/physiological model derived by Slate[126] for the action of injected sodium nitroprusside (SNP) in reducing human mean arterial pressure (MAP). (Details of the closed-loop reduction of MAP are provided in Chapter 8.) Through a statistical process of system identification, Slate modeled the reduction of MAP by SNP as:

$$\Delta P = -\frac{K e^{-s\delta_1}\left(1 + \alpha e^{-s\delta_2}\right)}{\tau s + 1}\left[\dot{\mathrm{SNP}}\right] \tag{3.87}$$

$[\dot{\mathrm{SNP}}]$ is the rate of infusion of SNP (μg/min). Two delays were used to simulate stepwise lags in mixing of SNP in the circulatory system. Typical parameters are as follows: primary delay $\delta_1 = 30$ sec, recirculation delay $\delta_2 = 45$ sec, recirculation fraction $\alpha = 0.4$, time constant $\tau = 50$ sec, and $K = 0.72$ mmHg/(ml/hr) for a 200-μg/ml SNP solution. In terms of SNP mass/time, $K = 0.216$ mmHg/(μg/min).

Slate used a pseudorandom binary noise (PRBN) source to drive the SNP infuser. The MAP/SNP impulse response was found by cross-correlating the measured MAP with the pseudorandom SNP input rate. The impulse response often showed two peaks (minima); hence Slate's use of the second delay term in modeling the MAP/SNP transfer function. Note that Slate's model for the MAP/SNP system has only one pole. The use of one pole is an engineering approximation. Vasodilation by SNP has more than one dynamic step. In fact, in an earlier study of SNP pharmacokinetics, Stoelting[131] gave data for MAP reduction due to an intravenous bolus injection of SNP that Woodruff and Northrop (unpublished) modeled closely by a transfer function with two real poles and one zero. Stoelting's direct impulse response did not show the twin peaks shown by Slate's PRBN method. Slate's model has proven accurate enough, however, for the successful design of a number of closed-loop SNP infusers to reduce postoperative MAP (see Chapter 8).

When an SISO negative-feedback system has a loop gain that includes a delay in the numerator, the system has the propensity to exhibit low damping and to become frankly unstable as the scalar part of the loop gain increases unless some kind of compensatory controller is used. Figure. 3.7 illustrates the root locus diagram of a simple negative feedback system with the loop gain:

$$A_L(s) = -\frac{K e^{-\delta s}}{\tau s + 1} \quad (3.88)$$

In this case, we let $\tau = 1$ and $\delta = 1$. Note that an infinite number of root locus branches originate at poles at $|s| = -\infty$ and progress to join an infinite number of zeros at $|s| = +\infty$ as K is increased from 0 to ∞. The first pair of locus branches to cross the $j\omega$ axis produces frank instability. These branches originate with the pole at $s = -1$ and the pole at $s = -\infty$. The other locus branches also approach the $j\omega$ axis as K is increased. These branches cross the $j\omega$ axis as closed-loop, CC, pole pairs at progressively higher values of K. These branches have asymptotes in the left-half s-plane parallel to the real axis at $s = j0, \pm j2\pi, \pm j4\pi, \pm j6\pi$, etc. In the right-half s-plane, the asymptotes are also parallel to the real axis at $s = \pm j\pi, \pm j3\pi, \pm j5\pi$, etc.

Controller designs for plants with delays can be subdivided into three categories: (1) controllers that introduce zeros to compensate for the nonminimum (excess) phase lag introduced by the delay term, such as PD, PI, and PID; (2) model reference controllers that cancel the effect of the delay with a dynamic model of the plant (e.g., the Smith delay compensator); and (3) self-tuning, adaptive controllers that optimize their performance according to dynamic, on-line estimates of the plant's parameters, including the delay. This third class of controller design is often necessary to preserve robust, accurate controller performance when the plant is a CPK/P system and its parameters are nonstationary.

Isaka and Sebald[63] have described in a review paper the wide variety of controller strategies workers have applied to the reduction of MAP by SNP. Because the MAP/SNP plant has a delay of the same order of magnitude as its time constant (typically 30 sec vs. 40 sec, respectively), it is a challenging system to control without *undershoot* (MAP is *reduced* by SNP). A PI controller is a simple means of controlling a plant with a delay so that it has zero steady-state error. The PI block adds a pole at the origin and a real zero to the loop gain function. The pole at the origin makes the system type 1, having zero steady-state error for constant inputs. Thus the delay system with a PI controller has a loop gain:

$$A_L(s) = -\frac{K_p K_c (s + a) e^{-s T_p}}{s(s + 1/\tau)} \quad (3.89)$$

The zero is chosen so $a > 1/\tau$. The behavior of the PI controller with the first-order delay plant is illustrated with a simulation using Simnon. The program is as follows:

```
CONTINUOUS SYSTEM mappicon      "  07/12/97
STATE Dp v NOISE
DER dDp dv dNOISE
TIME t
"
" 1st ORDER PLANT for BP reduction from IV sodium nitroprusside.
ud = DELAY(ru, Tp)
dDp = -Dp/tau + ud*(Kp/tau)
MAP = MAPo - Dp
"
e = MAP - SP
"
" PI CONTROLLER.
dv = e
u = Kc*(e + a*v)
ru = IF u > 0 THEN u ELSE 0           " u is nonnegative.
sru = ru/10                            " scaled ru.
"
" INPUTS:
MAPo = 185 + Kmap*t + NOISE            " MAP drifts up and is noisy.
dNOISE = -wo*NOISE + SD*NORM(t)        " LPF to filter GWN.
"SP = IF t < 5 THEN (185 - 17*t) ELSE r
"
" CONSTANTS:
a:10                                   " PI zero @.
SP:110                                 " Set point.
Kp:.310                                " Plant gain: mm Hg/(mg/min).
Kmap:1                                 " Rate const. for MAPo increase.
Tp:0.5                                 " Plant delay in min.
Kc:0.2                                 " PD controller gain.
r:100                                  " mm Hg.
tau:0.7                                " Plant TC in min.
SD:50                                  " Noise sd.
wo:0.5                                 " Break freq. of noise
                                         spectrum in r/min.
zero:0
ninety:90
"
END
```

The program adds a linear upward drift and Gaussian noise to the untreated MAP_o. The controller is initially "tuned" by trial and error for the plant so that the zero is at 10 r/min and the controller gain is $K_c = 0.2$. Figure 3.38 shows the system response at turn-on to a set point of 110 mmHg. Note that the untreated MAP, MAP_o, is noisy and drifts upward. The MAP is also noisy and follows the set point with medium error. To make the system more robust, we raise K_c to 2.0 and reduce the zero to $a = 1.5$ to control the undershoot. In Figure 3.39, we see that the steady-state error in the MAP from noise and drift of MAP_o is much less than the low-gain

FIGURE 3.38 Results of a Simnon simulation of the turn-on transient of a closed-loop SNP infusion system using a PI controller. The base MAP (trace 3) is made noisy and drifts upward. Trace 1 is the regulated MAP, trace 4 is the infusion rate of SNP, trace 2 is the set point = 110 mmHg, and trace 5 is 90 mmHg, the lower allowable bound on controlled MAP. Time scale in minutes.

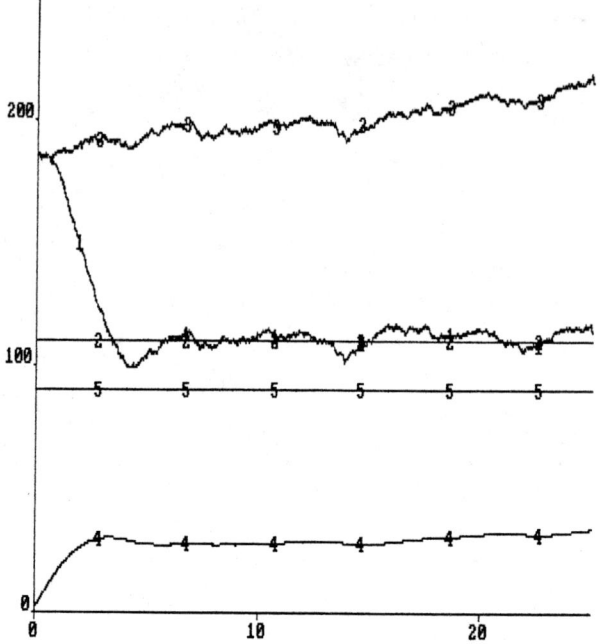

FIGURE 3.39 Simulation of MAP reduction by SNP infusion using the PI controller used in Figure 3.38, except the controller is made quicker by making $a = 1.5$ r/min and $K_c = 2.0$. Same traces. See text for discussion and program.

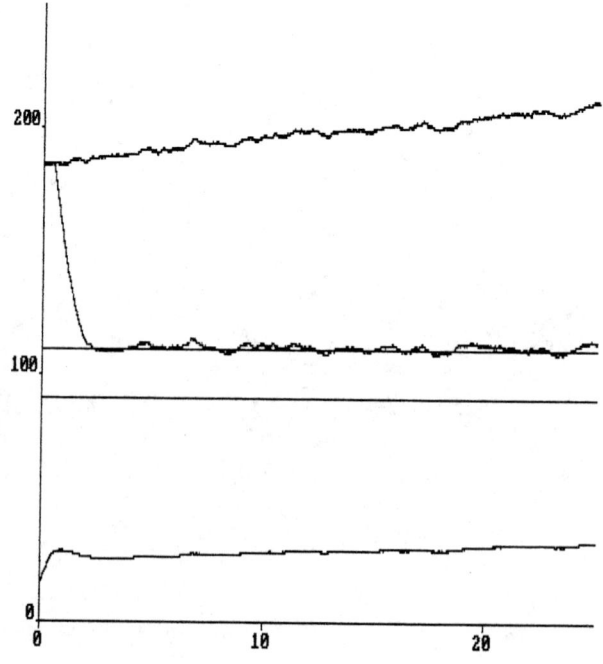

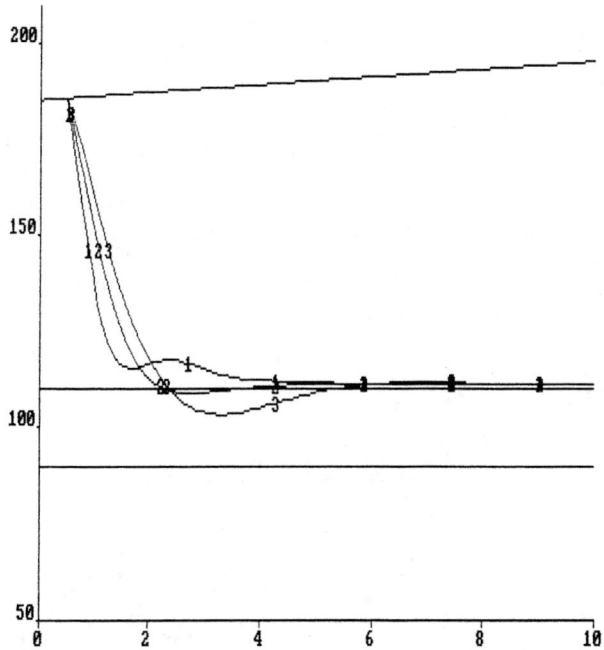

FIGURE 3.40 Simulation of MAP reduction by SNP infusion using the same PI controller used in Figure 3.39, illustrating system robustness to changes in plant time constant. Trace 1, $\tau = 0.4$ min; trace 2, $\tau = 0.7$ min (nominal); trace 3, $\tau = 1.0$ min. Other traces as in Figure 3.38. No noise was added.

system. The system also reaches its set point quicker. It is interesting to note that the simple PI controller is tolerant of plant parameter changes. In Figure 3.40, we see system responses when the plant time constant assumes values of 0.4, 0.7, and 1.0 min. There is a small undershoot for tau = 1 min, and tau = 0.4 min causes "creep" to the set point. In Figure 3.41, the dead time is varied as $T_p = 0.3, 0.4, 0.5, 0.6, 0.7$ min. Values of 0.6 and 0.7 produce a tolerable undershoot in MAP. Other values produce slightly slower settling times as 0.3 is approached.

Another approach to controlling plants with delays uses a plant model reference controller architecture to cancel the destabilizing effect of the dead time. This strategy was first described by Smith[128] in 1959. Figure 3.42A illustrates a continuous Smith delay compensator (SDC) control system for a simple first-order-delay plant with parameters K_p, δ_p, and τ_p. The reference model is split into two parts: a conventional transfer function, followed by a transport lag block. Figure 3.42B illustrates an alternate configuration, in which system error is explicitly added to the model outputs, y_m and $-v_m$. Ideally, the model parameters should be equal to the plant parameters for perfect cancellation of the untoward effects of the plant's dead time. That is, $K_m = K_p$, $\delta_m = \delta_p$, $a_m = a_p$. With reference to Figure 3.42B, we can write an equivalent single-loop loop gain for the SDC control system:

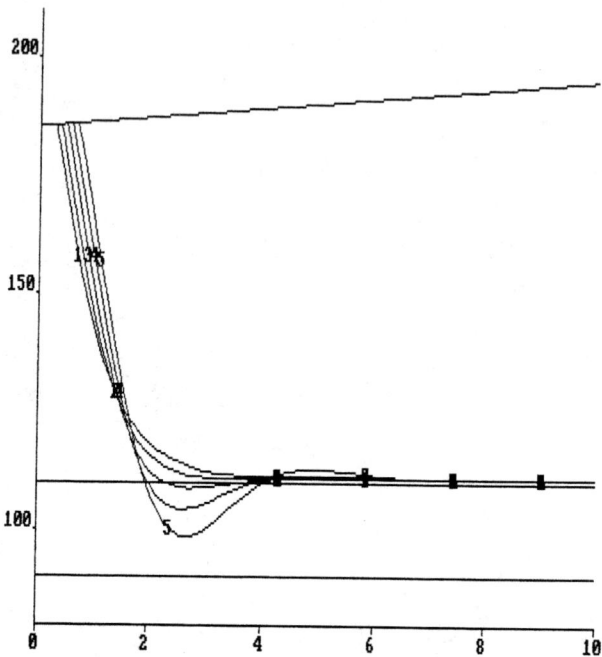

FIGURE 3.41 Simulation of MAP reduction by SNP infusion using the same PI controller as in Figure 3.38, illustrating system robustness to changes in plant dead time, d_p. Trace 1, $d_p = 0.3$ min; trace 2, $d_p = 0.4$ min; trace 3, $d_p = 0.5$ min (nominal); trace 4 $d_p = 0.6$ min; trace 5, $d_p = 0.7$ min. See text for discussion.

$$A_{Leq}(s) = \frac{W}{U}(s) = \left\{ \frac{-K_p e^{-s\delta_p}}{s + a_p} - \frac{K_m}{s + a_m} + \frac{K_m e^{-s\delta_m}}{s + a_m} \right\} K_c \quad (3.90)$$

Putting the terms over a common denominator, we have:

$$A_{Leq}(s) = \frac{-K_p(s + a_m)e^{-s\delta_p} + K_m(s + a_m)\left(e^{-s\delta_p} - 1\right)}{(s + a_p)(s + a_m)} K_c \quad (3.91)$$

Now it is easily seen that if the SDC controller is perfectly "tuned," the loop gain becomes

$$A_{Leq}(s) = \frac{-K_c K_p}{s + a_p} \quad (3.92)$$

Introduction to Linear SISO Control Systems and Systems with Delays

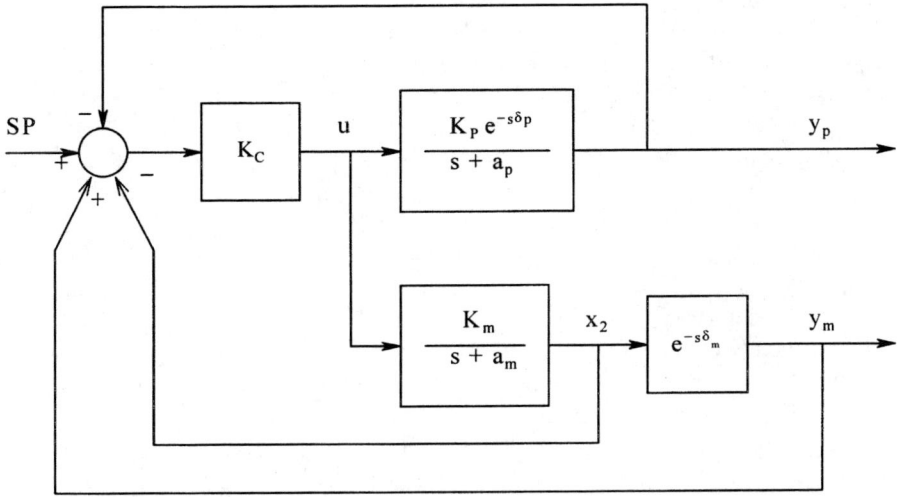

FIGURE 3.42A Block diagram of simple real-pole plant with a delay using a model reference controller architecture called the Smith delay compensator. Note the model outputs, x_2 and y_m, are fed back, as is the plant output, y_p.

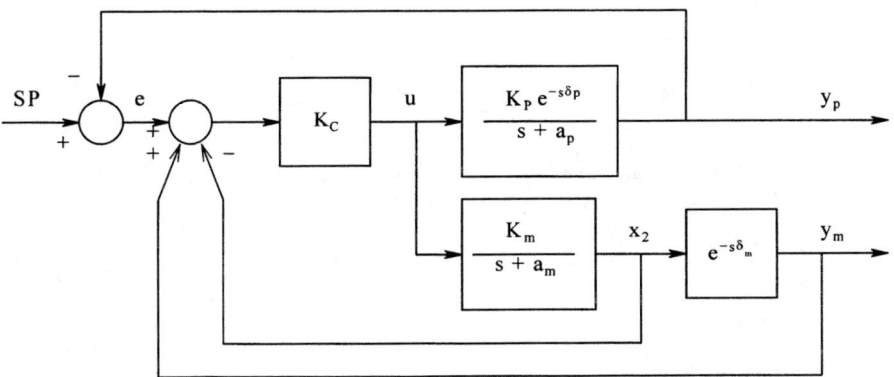

FIGURE 3.42B Same system as in Figure 3.42A, except plant error is shown explicitly.

The cancellation of the delay terms implies that the tuned SDC system's dynamics are that of one with a single-pole loop gain. When we write the tuned closed-loop system's transfer function,

$$\frac{Y}{X}(s) = \frac{K_c K_p / (1 + K_c K_p)}{s / \left[(1 + K_c K_p) a_p \right] + 1} e^{-s\delta_p} \tag{3.93}$$

we see that the dead time is still present, and the closed-loop system's break frequency is raised by (1 + the dc loop gain) of the tuned system. The system is stable even for high K_c.

Because the SDC uses a plant reference model, the question naturally arises as to how well it tolerates model parameter mismatches and noise. The best way to address this question is by simulation. The effects of model parameter mismatch for the simple one-pole plant illustrated in Figure 3.42B are illustrated below. Figures 3.43A, B, and C illustrate the step response of a tuned SDC controller having $K_p = K_m = 0.31$, $K_c = 30$, $a_p = a_m = 1.43$ r/min, $\delta_m = \delta_p = 0.5$ min when each of the SDC reference model parameters is successively detuned by ±20%. In doing such a study, certain trends become evident: As K_c is raised, the system becomes more sensitive to detuning (i.e., the effects of detuning become progressively worse as K_c increases). Second, detuning the parameter a_m has the least effect on spoiling closed-loop-system performance, followed by K_m. Detuning the delay, T_m, is most critical and can easily lead to bounded and unbounded oscillations. A ±20% detuning of T_m will make the system useless when $K_c = 50$. These results tell us that the SDC approach is of limited value if the plant delay is time variable over a wide range or it initially cannot be accurately measured.

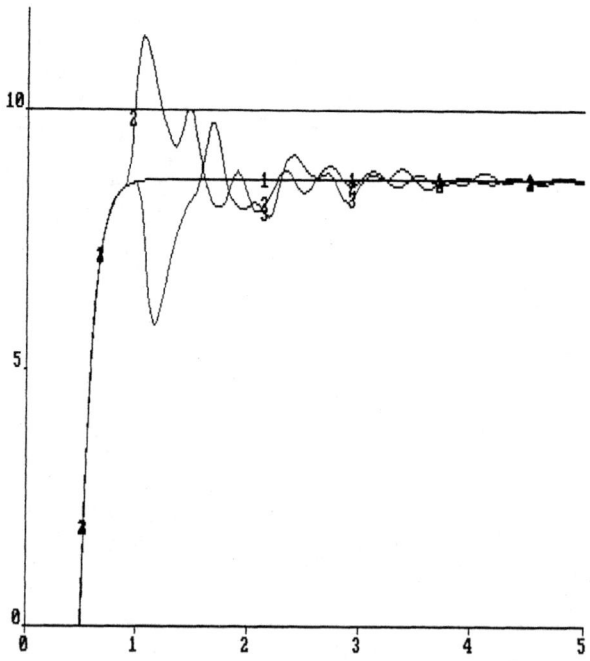

FIGURE 3.43A Simnon simulation showing the effect of detuning the SDC model's delay, d_m, by ±20%. The step response of a simple linear, real-pole plant with delay $d_p = 0.5$ min is shown for model $d_m = 0.5$ (trace 1), 0.4 (trace 2), and 0.6 (trace 3). Note steady-state error due to type 0 loop gain.

FIGURE 3.43B Simulation showing the effect of detuning the SDC model's gain, K_m, by ±20%. Same system as in Figure 3.43A. Trace 1, $K_m = 0.31$; trace 2, $K_m = 0.248$; trace 3, $K_m = 0.372$ min.

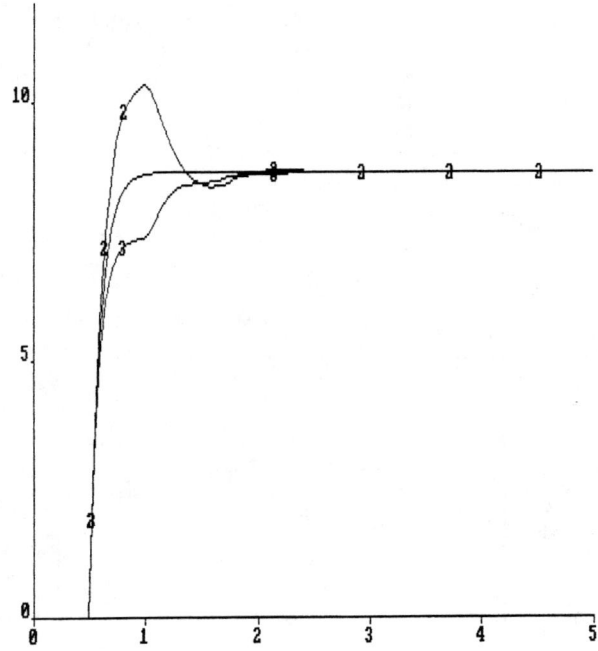

FIGURE 3.43C Simulation showing the effect of detuning the SDC model's corner frequency, a_m, by ±20%. Same system as in Figure 3.43A. Trace 1, $a_m = 1.43$; trace 2, $a_m = 1.144$; trace 3, $a_m = 1.716$ r/min.

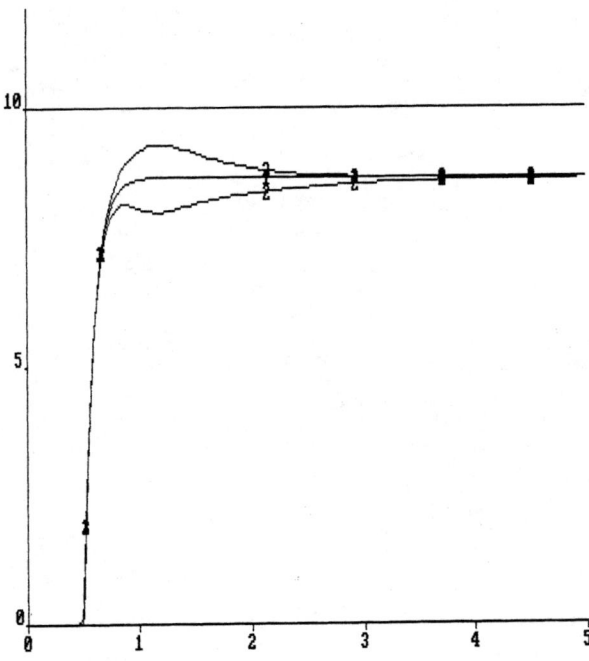

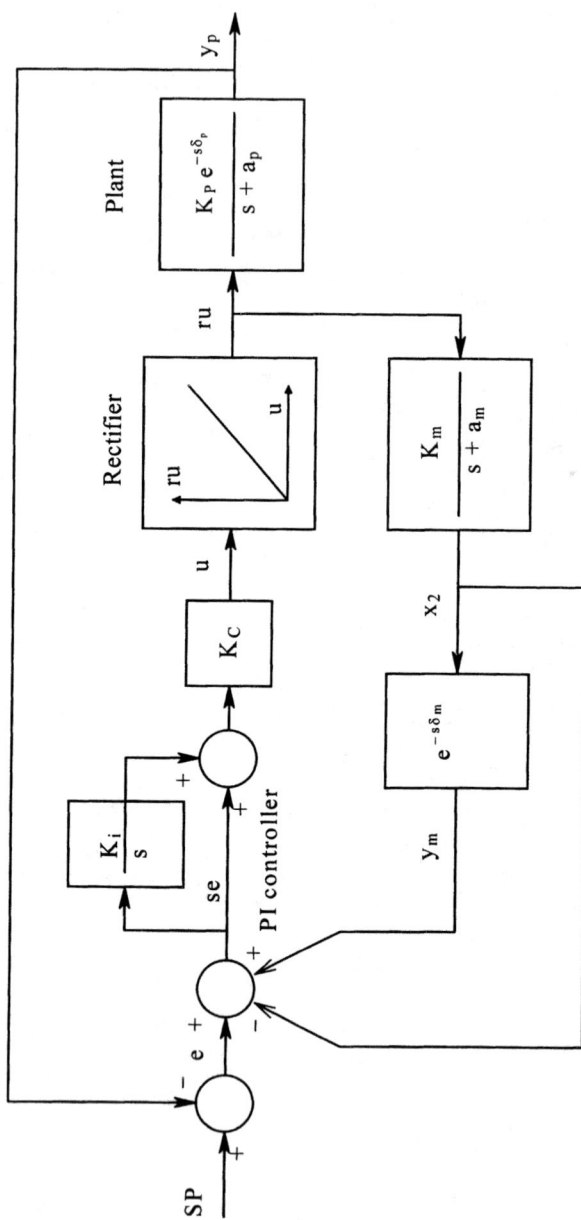

FIGURE 3.44 Block diagram of an MAP regulator using the SDC architecture along with a PI controller to improve system performance. See text for parameter values.

More sophisticated controllers can be used in conjunction with the SDC to control plants with dead times. Figure 3.44 illustrates the use of a PI controller in conjunction with a first-order plant with dead time. The SDC model reference compensates for the plant dead time, while the PI controller makes the closed-loop system type 1, giving zero steady-state error. The PI block also adds a zero beyond the plant pole to improve the delay-compensated system's second-order transient response. In this case, the tuned system has $K_p = K_m = 0.31$, $a_p = a_m = 1.43$ r/min, $\delta_p = \delta_m = 0.5$ min, $K_i = 1.5$, and $K_c = 15$ (controller gain). Figures 3.45A, B, and C illustrate the effects of detuning the SDC ±20% for a_m, K_m, and δ_m. As in the first example, we see that a high K_c makes the closed-loop system more sensitive to detuning. Also, as in the first SDC case, we see that the system tolerates detuning of K_m and a_m better than detuning δ_m. However, the use of the PI compensator makes the closed-loop system more tolerant of detuning than the SDC used alone.

The use of the SDC may be extended to discrete control systems and to integral-pulse frequency-modulation bolus injection systems when the plant has appreciable delay. Because of the individual variability of CPK/P plant parameters, SDC parameters must be initially accurately estimated and actively updated to compensate for nonstationarity which can occur as a result of a diurnal rhythm or due to the side effects of the injected drug. On-line estimation of plant parameters, including the

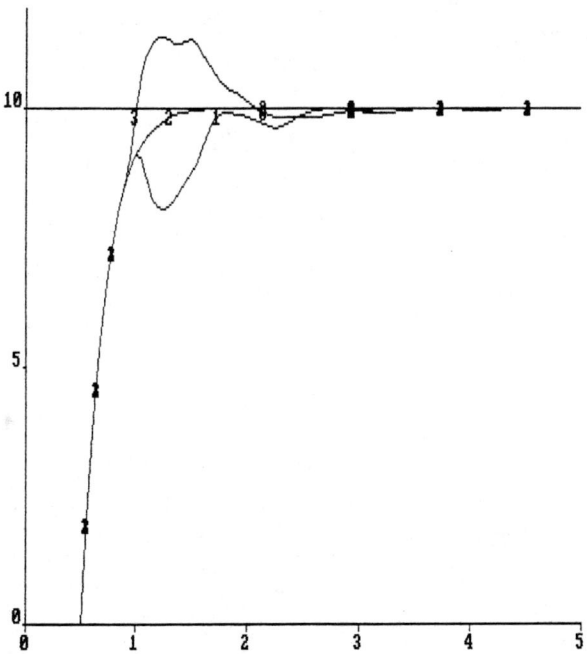

FIGURE 3.45A Effect of detuning the SDC model shown in Figure 3.44. The model plant delay is detuned ±20%. Trace 1, $d_m = 0.6$ min; trace 2, $d_m = 0.5$ min (nominal); trace 3, $d_m = 0.4$ min.

FIGURE 3.45B Effect of detuning the SDC model shown in Figure 3.44. The model dc gain is detuned ±20%. Trace 1, $K_m = 0.31$ (nominal); trace 2, $K_m = 0.248$; trace 3, $K_m = 0.372$.

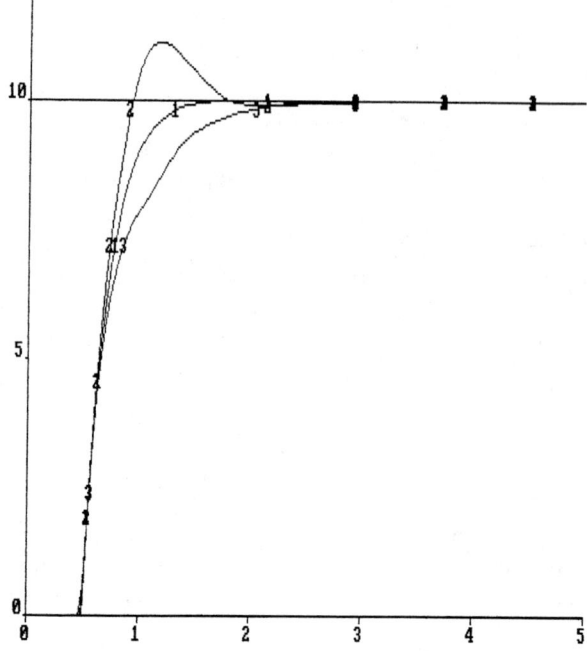

FIGURE 3.45C Effect of detuning the SDC model shown in Figure 3.44. The model real pole is detuned ±20%. Trace 1, $a_m = 1.43$ r/min; trace 2, $a_m = 1.144$ r/min; trace 3, $a_m = 1.716$ r/min.

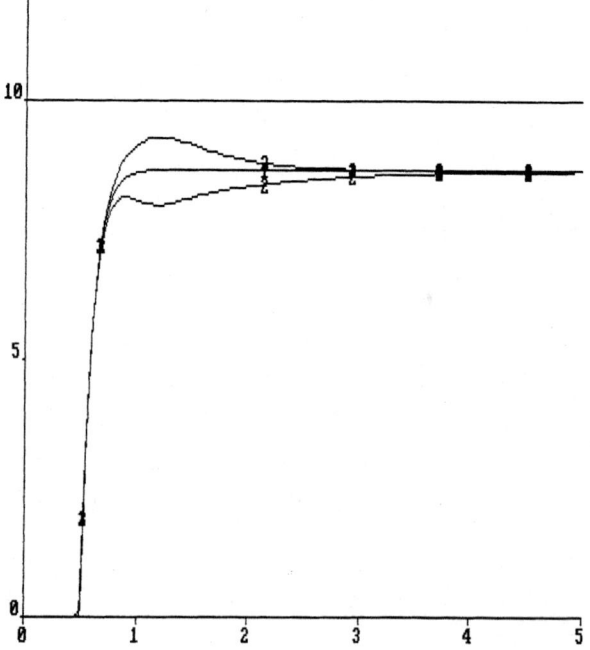

delay, is a topic beyond the scope of this text. The interested reader is referred to texts by Graupe,[57] Gawthrop,[53] Ljung,[84] and the work of Shah.[119,120]

3.5 CHAPTER SUMMARY

We have seen how linear real-pole plants can be effectively controlled using various feedback compensation schemes such as PD, lag, lead, and lag/lead compensation. Closed-loop linear system performance is shown to be markedly improved when the closed-loop system is converted to type 1 from type 0, using PI or PID controllers. Type 1 systems have zero steady-state error.

We have also examined in this chapter various means of determining the stability of linear closed-loop systems from their loop gain functions. In general, stability techniques detect the presence of closed-loop-system poles in the right-half s-plane. This can be done algebraically, as by the RH test, or graphically quantitatively, using the root locus method or the Nyquist stability criterion. The Nyquist method is shown to be extendable to certain nonlinear systems when using the DF method, which replaces a symmetrical, odd nonlinearity in the loop gain with an equivalent amplitude- and frequency-dependent gain. The Popov stability criterion is applicable to the same type of system treated by the DF method; however, it does not give as much information about the possible unstable behavior as does the DF method.

Special stability problems that accrue to linear systems with transport lags in their loop gains are considered by root locus and the Nyquist-based methods. The SDC was introduced as one means of achieving stable control of systems with transport lags. The SDC uses a model reference architecture to nullify the destabilizing effects of the transport lag. The performance of an SDC system depends on how closely the model is tuned to the plant.

PROBLEMS

3.1 A negative-feedback SISO system has the loop gain

$$A_L(s) = \frac{-K(s-1)}{(s+1)}$$

A. Write $A_L(s)$ in vector difference form.
B. Sketch the root locus diagram for the system.
C. Let s assume all values on the clockwise contour, C, in the s-plane. Make a table of $|A_L(s)|$ and $\angle A_L(s)$ for $s = j0, j1, j2, j\infty, \infty, -j\infty, -j2, -j1, -j0$. Plot $A_L(s)$ in the polar plane for these s values.
D. Use the Nyquist criterion to find the range of K for which the system is stable.

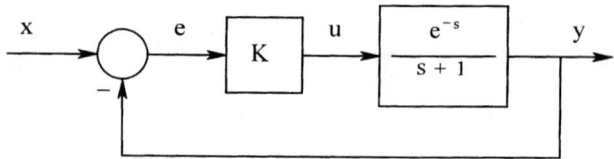

PROBLEM 3.2

3.2 A simple negative feedback system is shown in the figure. The first-order plant has a transport lag. Repeat parts C and D of Problem 3.1 for this system.

3.3 The loop gain of a negative-feedback SISO system with a transport lag δ can be written:

$$A_L(s) = \frac{-K e^{-\delta s}}{s(s+1)}$$

Use the Nyquist criterion to determine the maximum $K = f(\delta)$ for stability. Let $0.5 \leq \delta \leq 5$. Plot K vs. δ. The clockwise contour **s** must follow in the s-plane is the same as Problem 3.1 part C, except it makes an infinitesimal semicircular detour around the origin (and the pole at the origin) in the s-plane. The semicircle goes from $-j\varepsilon$ to $+\varepsilon$ to $+j\varepsilon$, where $\varepsilon \ll 1$. The behavior of $A_L(s)$ in the polar plane for **s** values on this semicircle in the s-plane is very important.

3.4 Consider the functional nonlinearity

$$y(e) = Ae + Be^2 \, \text{sgn}(e)$$

where $A, B > 0$, $\text{sgn}(e) = e/|e|$, $e = E \sin(\omega t)$, $N(E) = Y_1/E$. Sketch and dimension $\mathbf{N}^{-1}(E)$ on polar coordinates.

3.5 Thoroughly discuss the stability of the control system using DF analysis and Nyquist diagrams. What are the conditions under which it is unstable? Do the oscillations grow, or do they reach a stable limit cycle?

3.6 A. Use the Nyquist criterion to find the range of positive K over which the linear open-loop unstable system can be made stable with negative feedback.

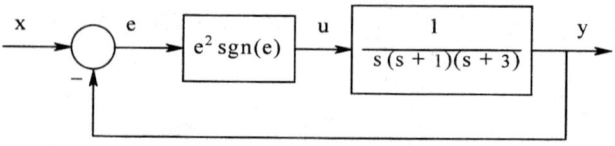

PROBLEM 3.5

Introduction to Linear SISO Control Systems and Systems with Delays

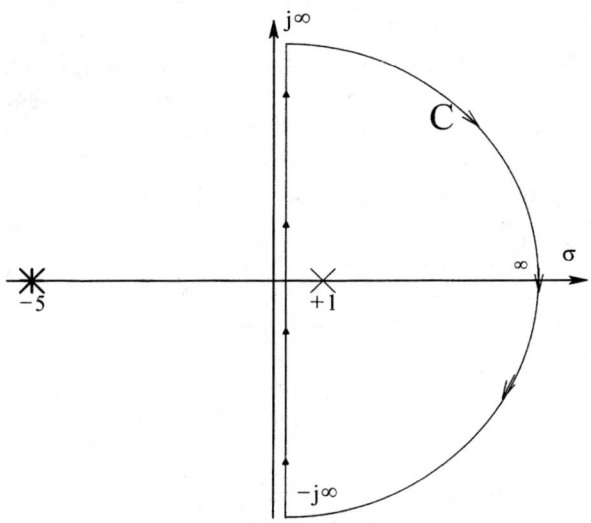

PROBLEM 3.6

$$A_L(s) = \frac{-K}{(s-1)(s+5)^2}$$

- **B.** Clearly sketch the locus of $A_L(s)$ as the vector s traverses the clockwise contour, C, in the s-plane. Show $A_L(s)$ at $s = j0+$, $+j1$, $+j5$, and $j\infty$.
- **C.** Give the frequency, ω_o, that the closed-loop system will oscillate at if K is too large.

3.7 A simple on/off controller for an infusion pump is used to control the level of a cancer chemotherapy agent in the brain's cerebrospinal fluid. The set point is SP = 50 µg/l. The plant dynamics in response to a drug input rate u are

$$Y(s) = U(s)\left[20e^{-s\Delta}/(s+0.5)(s+4)\right]$$

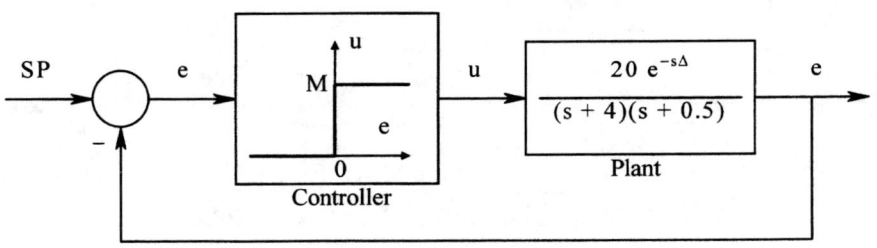

PROBLEM 3.7

Natural frequencies are in radians per hour. $\Delta = 0.1$ hr.
- **A.** Assume a zero-mean, sinusoidal, steady-state error voltage input to the controller. Find the on value of $u(t)$, M, that will yield an average drug level at the output of 50 μg/l.
- **B.** Use the DF method to find the peak value of the oscillations in $y(t)$ in the steady state and the frequency of oscillation in cycles per hour.
- **C.** Model the system's time-domain behavior using Simnon or Simulink. Let SP = 50 μg/l. Show $y(t)$, $u(t)$, SP, and $e(t)$. Show what happens when SP steps from 50 to 75 and then to 25. Allow the system time to reach steady state in each case. Find the steady-state oscillation frequency for each set point.

3.8
- **A.** Make a detailed Nyquist plot for the loop gain:

$$A_L(s) = \frac{-10K(s + 0.5)}{s^2(s + 2)(s + 10)}$$

- **B.** Give the range of positive K over which the closed-loop system is stable.
- **C.** Use the Barkhausen criterion to find the system's frequency of oscillation, ω_o, when it reaches instability.

3.9 A first-order pharmacokinetic plant with a delay is to be controlled with a simple limit-cycling on/off controller. The set point $R = 100$ μg/l.
- **A.** Find u_m (in mg/hr) so that $\overline{D_{SS}} = 100$ μg/l and $\overline{E_{SS}} = 0$.
- **B.** Use the DF method to algebraically find the steady-state limit-cycle frequency and the steady-state peak error amplitude.
- **C.** Simulate the system and compare the DF results with the simulation results.

3.10 A strange nonlinearity converts a sine wave of any amplitude E and frequency ω into a triangle wave with zero mean and fixed peak amplitude M, having the frequency and phase of the input sine wave. Find an expression for the nonlinearities DF, $N(E)$.

3.11
- **A.** Examine the turn-on transient of the SDC-compensated system shown in the figure. Assume zero initial conditions and a set point of SP = x = 10. Assume initially that $K_p = K_m = 40$, $a_p = a_m = 1$, and $\delta_p = \delta_m = 0.2$.

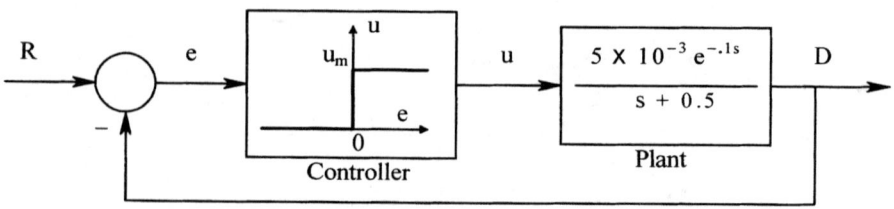

PROBLEM 3.9

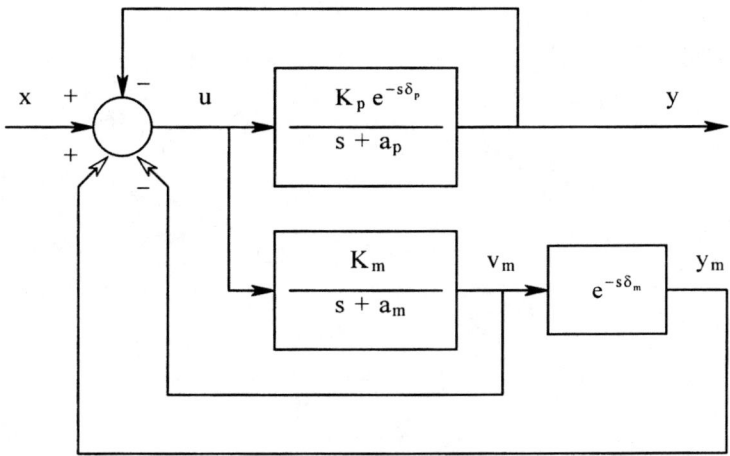

PROBLEM 3.11

B. Now detune the Smith model reference parameters by ±20%, one at a time, and record the results. Which Smith model parameter is the least sensitive to detuning? Which is the most? Why does the tuned system never reach its set point?

3.12 An unstable plant is compensated with a lag/lead feedback filter.
  A. Draw the system's root locus diagram to scale.
  B. Sketch and dimension the locus of $A_L(j\omega)$ for $-\infty < \omega < \infty$ in the complex plane. Give the values of A_L for $\omega = 0$ and $\omega = +\infty$.
  C. Use the Nyquist criterion to find the total range of K over which the system is stable and unstable.

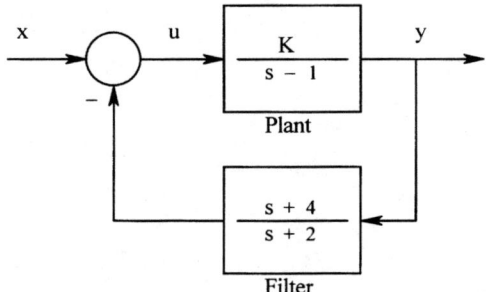

PROBLEM 3.12

4 Introduction to Compartmental Modeling and Pharmacokinetic Systems

4.0 INTRODUCTION

The closed-loop control of a drug concentration or of a physiological parameter has been practically implemented in humans for such diverse systems as postoperative hypertension, postoperative pain, blood glucose concentration, cancer chemotherapy drug concentration,[14,102,105,148] and depth of anesthesia.[81] In every case, to achieve control, a drug must be administered to the body under controlled conditions. The usual input routes used for closed-loop control are intravenous drip (infusion) or bolus injections, intramuscular or subdermal injections, and intraperitoneal infusions or bolus injections. Drugs used in anesthesia are often inhaled as aerosols or gases.

In the case of intravenous administration, the controlling drug or hormone enters the circulatory system, where it is mixed and distributed evenly by the action of the heart. The drug must diffuse out of the circulatory system into extracellular volumes before it reaches the target organ or cells. The controlling drug, as it acts on the target, may also be subject to excretion by the kidneys and intestines, as well as biotransformation and inactivation by organs such as the liver, the renal epithelium, and the intestinal mucosa.[54] It also may be reabsorbed from the intestines or kidneys in the process of excretion. In some cases, the drug may combine with proteins in the blood to form a storage compartment from which it is slowly released. The drug may also be hydrolyzed or broken down by enzymes in the blood, or at cell receptor sites, and these breakdown products are further metabolized or excreted.

In the following sections of this chapter, some of the details of pharmacokinetics and compartmental analysis relevant to closed-loop administration of drugs are described.

4.1 BASIC COMPARTMENTAL ANALYSIS

Central to the formulation of a dynamic description of what happens to a drug in the body, once internalized is the concept of the *compartment* and *compartmental systems*. Godfrey[55] defines a compartmental system as

> ...a finite number of homogeneous, well-mixed, lumped subsystems, called compartments, which exchange [a substance] with each other and with the environment so that the quantity or concentration of material within each compartment may be described by a first order differential equation. A compartmental system may be used to model either the kinetics of one substance, in which case the compartments occupy different spaces and the inter-compartment transfers represent flow of material from one location to another, or the kinetics of two or more substances (such as a drug and its metabolites) in which case different compartments may occupy the same space and some of the inter-compartment transfers represent transformation from one substance to another.

Another definition of compartment, cited by Godfrey, is "a quantity of a substance which has a uniform and distinguishable kinetics of transformation or transport."

Godfrey[55] shows that an N-compartment system is described by N first-order ordinary differential equations (ODEs) and thus has N states. Furthermore, if a chemical reaction occurs within a compartment, first-order ODEs from mass-action kinetics may be added to the compartmental ODEs, raising the total number of states necessary to model the system. Probably the best way to learn about compartmental analysis is to examine several examples, presented below.

In the *first example*, and the simplest, CO_2 gas made by working yeast in bread dough at a rate $\dot{Q}_{DI}$ diffuses out of the dough at a rate $K(Q_D/V)$ mg/min. The ODE describing this process is simply:

$$\dot{Q}_D = \dot{Q}_{DI} - Q_D(K/V) \quad \text{mg/min} \tag{4.1}$$

K is in liters per second, Q_D is in milligrams, and V is the compartment volume in liters. If we Laplace transform the ODE, we can write the transfer function for this simple linear one-compartment system with time constant, V/K (seconds):

$$\frac{Q_D}{\dot{Q}_{DI}}(s) = \frac{1}{s + K/V} \tag{4.2}$$

In the steady state, $Q_{DSS} = \dot{Q}_{DI}(V/K)$ mg.

In the *second compartmental example*, a chemical, C, is synthesized in the mitochondria of a cell at a rate $\dot{Q}_o$. Its concentration is C_m μg/l in the mitochondria. It diffuses out of the mitochondria into the cytoplasm, where its concentration is C_c. It next diffuses through the cell membrane to what is basically zero concentration

Introduction to Compartmental Modeling and Pharmacokinetic Systems

outside of the cell. The two compartments are (1) the mitochondria and (2) the cytoplasm around the mitochondria. The compartmental state equations are based on simple diffusion (Fick's first law):

$$V_m \dot{C}_m = \dot{Q}_o - K_{12}(C_m - C_c) \quad \mu g/min \quad (4.3A)$$

$$V_c \dot{C}_c = K_{12}(C_m - C_c) - K_2 C_c \quad \mu g/min \quad (4.3B)$$

Written in state form, we have:

$$\dot{C}_m = -C_m(K_{12}/V_m) + C_c(K_{12}/V_m) + \dot{Q}_o/V_m \quad \mu g(1 \cdot min) \quad (4.4A)$$

$$\dot{C}_c = C_m(K_{12}/V_c) - C_c(K_2 + K_{12})/V_c \quad \mu g(1 \cdot min) \quad (4.4B)$$

Note that mass-diffusion rates depend on *concentrations* or mass per volume, so the diffusion rate constants, K_{12} and K_2, must have the dimensions liters per minute.

From the ODEs, we see that the system is linear and can be described by a signal flow graph as shown in Figure 4.1. The signal flow graph can easily be reduced by Mason's rule to find the transfer function, $C_c/\dot{Q}_o(s)$:

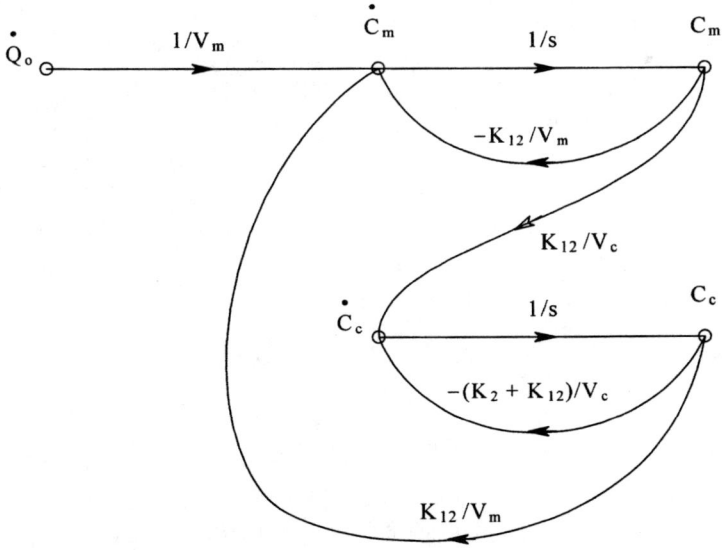

FIGURE 4.1 Signal flow diagram describing the second-order linear compartmental model of example 2.

$$\frac{C_c}{Q_o}(s) = \frac{(1/V_m)(1/s)(1/V_c) K_{12}(1/s)}{1 - \left[-K_{12}/sV_m - (K_2 + K_{12})/sV_c + K_{12}^2/s^2 V_m V_c \right]}$$
$$+ \left[\left(-K_{12}/sV_m \right) \left(-(k_2 + k_{12})/sV_c \right) \right]$$

$$= \frac{K_{12}/V_m V_c}{s^2 + s\left[K_{12}/V_m + (K_{12} + K_2)/V_c \right] + K_2 K_{12}/V_m V_c} \qquad (4.5)$$

$$= \frac{K}{(s+a)(s+b)}$$

Thus the two-compartment system governed by diffusion is seen to have linear second-order dynamics with two real poles.

In considering high-order compartmental systems ($N \geq 3$), there are two major compartmental architectures described by Godfrey.[55] These are shown in Figure 4.2. In the so-called *mammillary model,* a central compartment exchanges material with two or more peripheral compartments. The peripheral compartments do not exchange material with each other; they exchange material only with the central compartment. The *catenary model* architecture assumes a linear array of compartments where each compartment only communicates with its two nearest neighbors. Of course, there are variations on these themes.

Figure 4.3 illustrates a general two-compartment-drug/two-compartment-metabolite model. In terms of states, this is a fourth-order system; it is second order for the drug, however. The describing ODEs are

$$\dot{x}_1 = b_1 u_1(t) + k_{12} x_2 / V_2 - (k_{01} + k_{21} + k_{31}) x_1 / V_1 \qquad \text{mg/min} \qquad (4.6A)$$

$$\dot{x}_2 = k_{21} x_1 / V_1 - k_{12} x_2 / V_2 \qquad (4.6B)$$

$$\dot{x}_3 = k_{31} x_1 / V_1 - (k_{03} + k_{43}) x_3 / V_3 + k_{34} x_4 / V_4 \qquad (4.6C)$$

$$\dot{x}_4 = k_{43} x_3 / V_3 - k_{34} x_4 / V_4 \qquad (4.6D)$$

Compartments 1 and 2 contain a drug, D; compartments 3 and 4 contain its metabolite, M; and k_{31} is the mass-action rate constant governing the transformation of D into M. $\dot{x}_1$ is in milligrams per minute, and x_1 is in milligrams. The intercompartmental mass-transfer rate is proportional to concentration in that compartment (i.e., proportional to x_k/V_k, where V_k is the volume of the kth compartment). The k's are in liters per minute in this example.

Introduction to Compartmental Modeling and Pharmacokinetic Systems

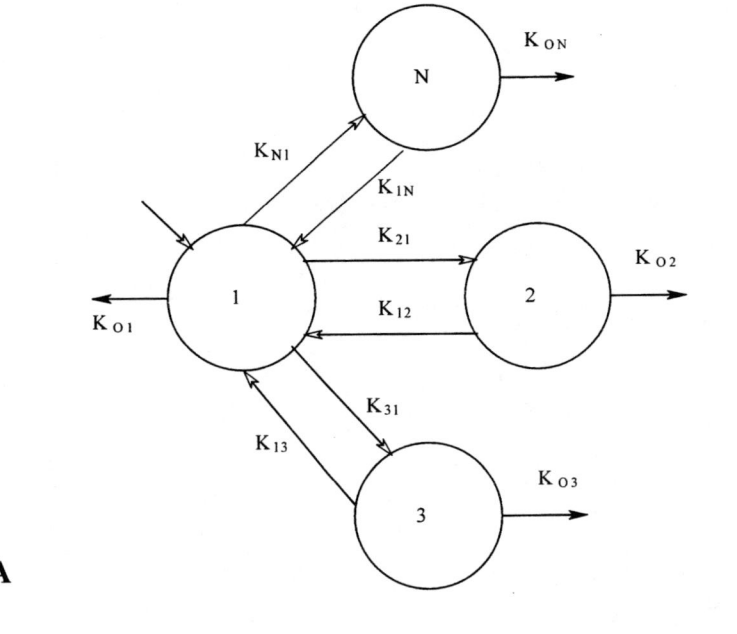

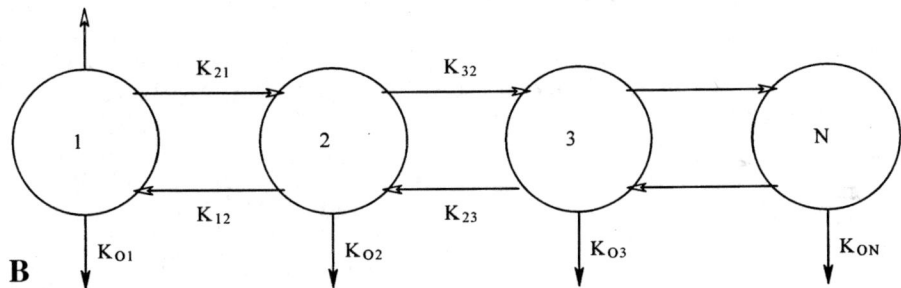

FIGURE 4.2 (A) A mammillary compartmental pharmacokinetic system. (After Godfrey.[55]) (B) A catenary (concatenated) compartmental pharmacokinetic system.

As a *third example,* consider the third-order catenary (concatenated) compartmental system shown in Figure 4.4. Drug input is only to compartment 1; all three compartments have drug loss. The state equations are

$$\dot{x}_1 = -(K_{L1} + K_{21})(x_1/V_1) + K_{12}(x_2/V_2) + 0 + u_1 \qquad (4.7A)$$

$$\dot{x}_2 = K_{21}(x_1/V_1) - (K_{L2} + K_{12} + K_{32})(x_2/V_2) + K_{23}(x_3/V_3) \qquad (4.7B)$$

$$\dot{x}_3 = 0 + K_{32}(x_2/V_2) - (K_{L3} + K_{23})(x_3/V_3) \qquad (4.7C)$$

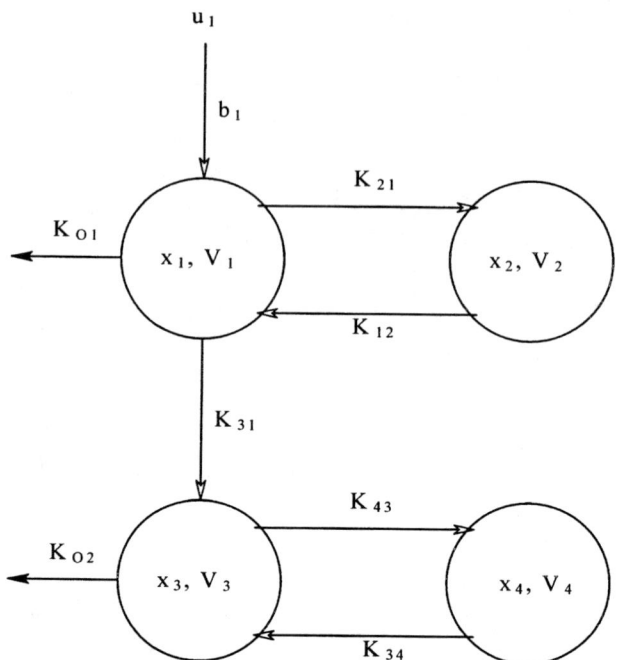

FIGURE 4.3 A fourth-order system consisting of a drug distributed in two compartments (1 and 2) and its metabolite also distributed in two compartments (3 and 4). k_{31} is the metabolic conversion rate constant.

These equations form a linear system: $\dot{\mathbf{x}} = \mathbf{A}\mathbf{x} + \mathbf{B}\mathbf{u}$. The a_{11} element of $\mathbf{A}$ is $-(K_{L1} + K_{21})/V_1$, etc.

Our *fourth example* examines a third-order mammillary system, shown in Figure 4.5. In this case, a central compartment communicates with two independent side compartments. The drug input is to compartment 1; all three compartments lose drug. The state equations are

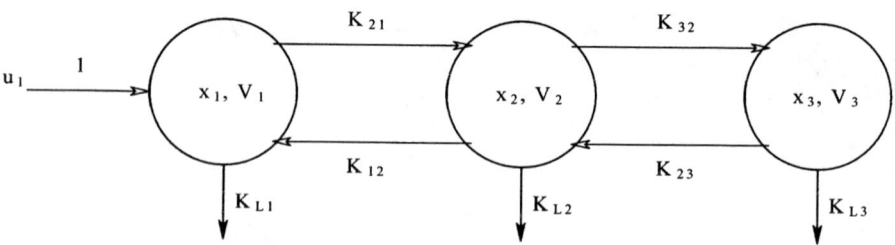

FIGURE 4.4 A third-order linear concatenated compartmental pharmacokinetic system. Drug is lost from all three compartments.

Introduction to Compartmental Modeling and Pharmacokinetic Systems

$$\dot{x}_1 = -(K_{21} + K_{31} + K_{L1})(x_1/V_1) + K_{12}(x_2/V_2) + K_{13}(x_3/V_3) + u_1 \quad (4.8A)$$

$$\dot{x}_2 = K_{21}(x_1/V_1) - (K_{L2} + K_{12})(x_2/V_2) + 0 \quad (4.8B)$$

$$\dot{x}_3 = K_{31}(x_1/V_1) + 0 - (K_{L3} + K_{13})(x_3/V_3) \quad (4.8C)$$

Note that $x_k/V_k \equiv c_k$, the concentration in the kth compartment.

A rapid way of finding the transfer function, $C_2/U_1(s)$, is to draw the signal flow graph for the three linear state equations and use Mason's rule to find C_2/U_1. The signal flow graph for Equation 4.8 is shown in Figure 4.6. In applying Mason's rule, we note that there are a total of five loops, three of them nontouching. There is one forward path that touches all loops but one; thus there is one cofactor term. The first pass at Mason's rule yields:

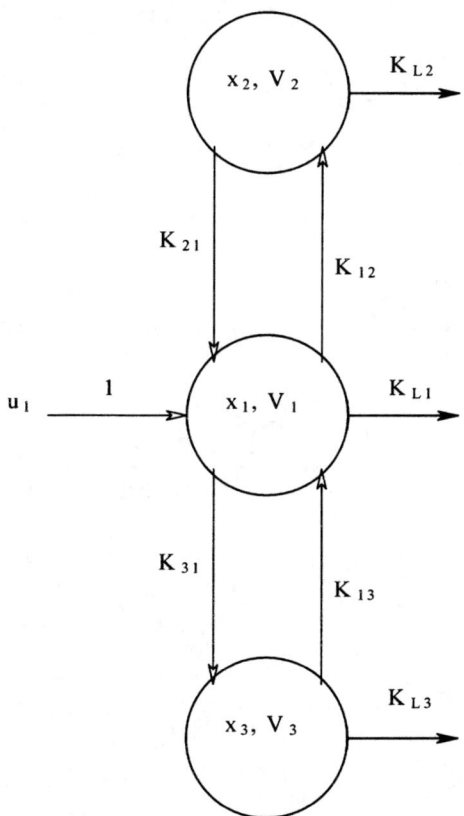

FIGURE 4.5 A third-order linear mammillary compartmental pharmacokinetic system.

$$\frac{C_2}{U_1} = \frac{(1/s)(K_{21}/V_1)(1/s)(1/V_2)\left[1-\left\{-(K_{13}+K_{L3})/sV_3\right\}\right]}{1 - \begin{Bmatrix} -K_{12}(1/s)(1/V_2) - K_{13}(1/s)(1/V_3) \\ -(K_{L1}+K_{31}+K_{21})(1/s)(1/V_1) + K_{12}K_{21}(1/s^2)(1/V_1)(1/V_2) \\ +K_{13}K_{31}(1/s^2)(1/V_1)(1/V_3) \end{Bmatrix}}$$

$$+ \begin{Bmatrix} [-K_{12}(1/s)(1/V_2)][-K_{13}(1/s)(1/V_3)] \\ +[-K_{13}(1/s)(1/V_3)][-(K_{L1}+K_{31}+K_{21})(1/s)(1/V_1)] \\ +[-K_{12}(1/s)(1/V_2)][-(K_{L1}+K_{31}+K_{21})(1/s)(1/V_1)] \\ +[-K_{12}(1/s)(1/V_2)][K_{13}K_{31}(1/s^2)(1/V_1)(1/V_3)] \\ +[-K_{13}(1/s)(1/V_3)][K_{12}K_{21}(1/s^2)(1/V_1)(1/V_2)] \end{Bmatrix}$$

$$- \begin{Bmatrix} [-K_{12}(1/s)(1/V_2)][-K_{13}(1/s)(1/V_3)] \\ [-(K_{L1}+K_{31}+K_{21})(1/s)(1/V_1)] \end{Bmatrix}$$

(4.9)

This result appears awesomely complex, but all that needs to be done is to multiply numerator and denominator by s^3 and collect terms. Now we can write:

$$\frac{C_2}{U_1} = \frac{K_{21}(1/V_1)(1/V_2)\left[s+(K_{13}+K_{L3})/V_3\right]}{s^3 + s^2\left\{K_{12}/V_2 + K_{13}/V_3 + (K_{L1}+K_{21}+K_{31})/V_1\right\}}$$
$$+ s\begin{Bmatrix} -K_{12}K_{21}/(V_1V_2) - K_{13}K_{31}/(V_1V_3) + K_{12}K_{13}/(V_2V_3) \\ +K_{13}(K_{L1}+K_{31}+K_{21})/(V_1V_3) + K_{12}(K_{L1}+K_{31}+K_{21})/(V_1V_3) \end{Bmatrix}$$
$$+ s^0\left\{K_{12}K_{13}K_{L1}/(V_1V_2V_3)\right\}$$

(4.10)

which is of the linear cubic form:

$$\frac{C_2}{U_1} = \frac{(s+a)K_{21}/(V_1V_2)}{(s+b)(s+c)(s+d)} \quad (4.11)$$

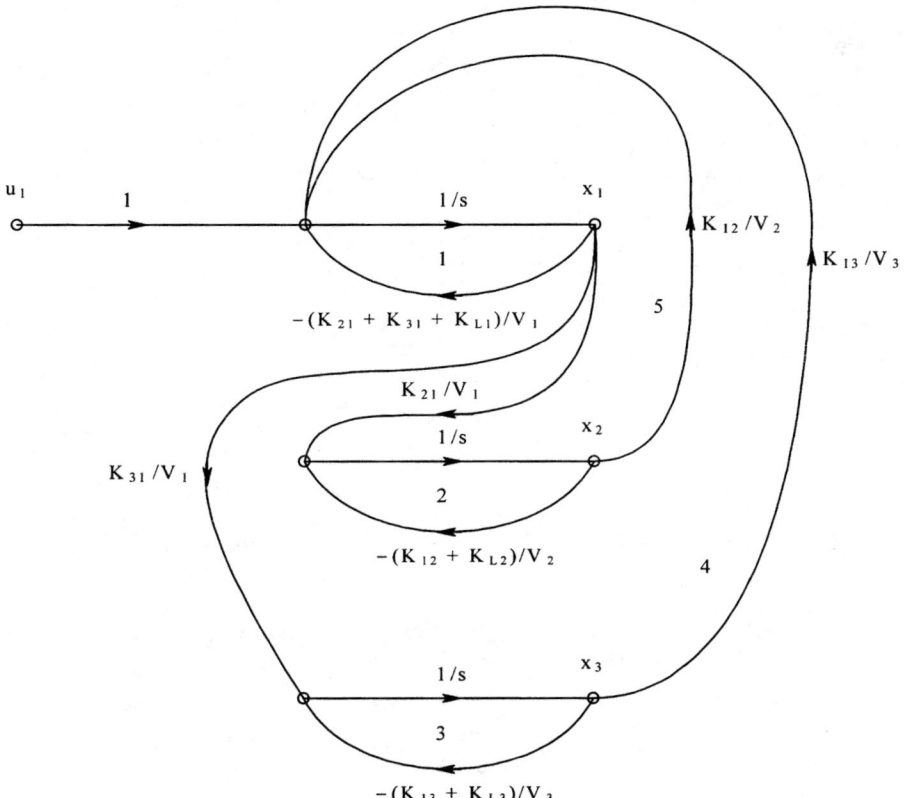

FIGURE 4.6 The third-order signal flow graph describing the state equations governing the mammillary system. There are five loops.

Factoring the real roots of the cubic denominator is best done numerically, using a computer program such as Matlab™.

The diffusion-based compartmental analyses above resulted in linear transfer functions, enabling the use of conventional linear-systems analysis tools such as Laplace transforms and frequency response analysis. Not all compartmental systems are linear, however. The transfer function/Laplace transform approach is not valid for nonlinear and time-variable systems. Instead, we rely on computer solutions of the nonlinear ODEs, using software packages such as Matlab and Simnon™. Several examples of *nonlinear* compartmental systems are presented next.

Godfrey[55] describes a *nonlinear* one-compartment system, shown in Figure 4.7A, in which the rate of loss of a substance from compartment 1 follows a saturating (hyperbolic) relationship similar to the Michaelis–Menten relation (cf. Equation 2.36). The system ODE is

$$\dot{x} = \frac{-V_m}{K_m + x} x + u(t) \qquad (4.12)$$

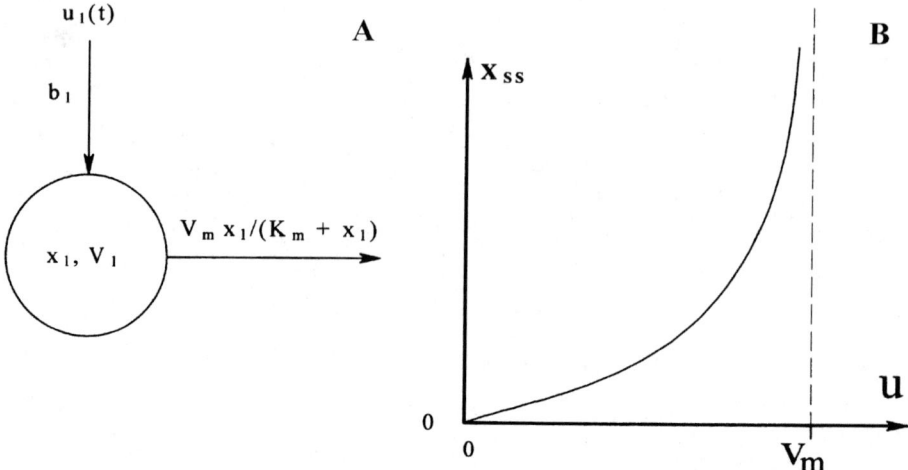

FIGURE 4.7 (A) A nonlinear one-compartment system. Loss rate saturates at V_m. (B) Graph of the steady-state concentration of X, given a constant input rate, U.

The steady-state x, given a constant drug input rate, u (mg/min), is found by setting $\dot{x} = 0$ in the equation above. This yields:

$$\dot{x}_{ss} = \frac{u\, K_m}{V_m - u} \tag{4.13}$$

A plot of this relation is shown in Figure 4.7B. Note that x_{ss} increases rapidly as u approaches V_m. For $u > V_m$, the system cannot eliminate drug fast enough, and it accumulates in the compartment without bound (as long as u is constant).

Godfrey also describes a second-order nonlinear compartmental system, as shown in Figure 4.8. The system equations are

$$\dot{x}_1 = -k_1 x_1^2 - \frac{V_m x_1}{K_m + x_1} + k_{12} x_2 + u_1 \tag{4.14A}$$

$$\dot{x}_2 = \frac{V_m x_1}{K_m + x_1} - k_{12} x_2 - k_{02} \tag{4.14B}$$

Note that the loss rate from compartment 1 appears to be bimolecular (i.e., two molecules of x_1 must combine to exit). Also, transfer of the drug from compartment 1 to compartment 2 has Michaelis-like rate saturation, and the loss of the drug from compartment 2 proceeds at a constant rate k_0, regardless of the drug concentration in compartment 2. Such a constant rate is suggestive of an active pumping mechanism that runs in a rate-limited mode.

Introduction to Compartmental Modeling and Pharmacokinetic Systems

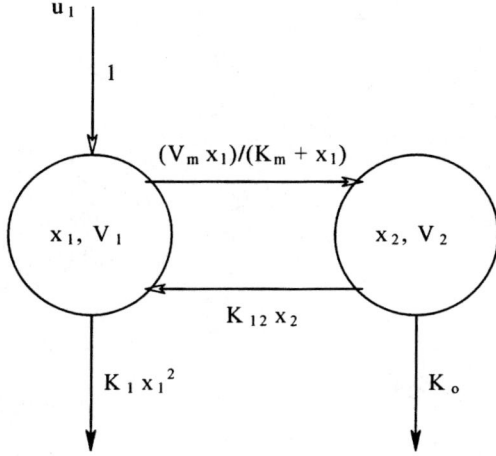

FIGURE 4.8 A second-order nonlinear compartmental pharmacokinetic system taken from Godfrey.[55]

Simnon can be used to obtain an exact solution for this system, given an input rate u_1. The Simnon program is as follows:

```
CONTINUOUS SYSTEM Godfry93    "5/08/97
STATE x1 x2
DER dx1 dx2
TIME t
"
dx1 = -k01*rx1*rx1 + k12*rx2 - Vm*rx1/(Km + rx1) + u1
dx2 = -k12*rx2 - k02 + Vm*rx1/(Km + rx1)
"
u1 = b1*(1 + V12)
V12 = if t > t1 then -1 else 0
"
k02 = IF x2 > 0 THEN k2 ELSE 0    " Can't pump out x2 unless some there.
rx1 = if x1 < 0 then 0 else x1
rx2 = if x2 < 0 then 0 else x2
"
k01:1
k2:.1
k12:1
Vm:1
Km:0.1
b1:1
t1:4
zero:0
"
END
```

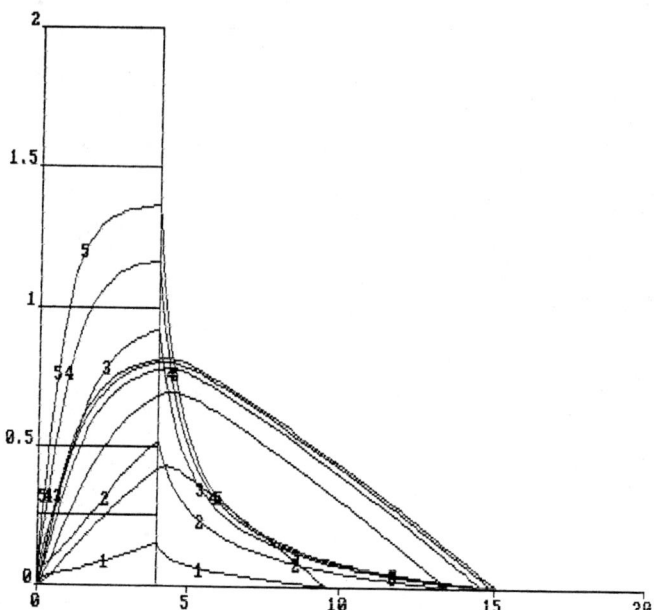

FIGURE 4.9 Results of a Simnon simulation of the nonlinear Godfrey compartmental pharmacokinetic system, "godfry93". The input is applied at a constant rate for 4 hr. Trace 1 = $x1$, trace 2 = $x2$, trace 3 = $u1$ = 0.5. trace 4 = $x1$, trace 5 = $x2$, trace 6 = $u1$ = 1.0. Trace 7 = $x1$, trace 8 = $x2$, trace 9 = $u1$ = 2.0.

Because concentrations (and masses) in compartmental systems are nonnegative, we use "rectified" state variables in the ODEs. Also, the constant loss rate, k_{02}, from compartment 2 is made zero for zero x_2. Results of the simulations in which a pulse of input rate of different values is given are shown in Figure 4.9. We note that as u_1 increases, the rise and fall time of x_1 becomes quicker. Also, x_2's dynamics saturate, proceeding in an almost linear manner to zero concentration at about $t = 15$, regardless of u_1, once $u_1 \to 0$. Without simulation, these details would be difficult to appreciate.

As a final example of a computer solution of compartmental systems, we consider a two-compartment system in which the loss rate coefficient, $k_{02}(t)$, varies periodically with time; such variation might be due to a diurnal rhythm in the body, in which case the period would be 24 hr. The system's ODEs are

$$\dot{x}_1 = u_1 + k_{12} x_2 - \frac{V_m x_1}{K_m + x_1} - k_{01} x_1 \quad \mu g/hr \qquad (4.15A)$$

$$\dot{x}_2 = -k_{12} x_2 + \frac{V_m x_1}{K_m + x_1} - x_2 k_{02}\left[1 + \eta \cos(2\pi t/T)\right], \quad 0 \leq \eta \leq 1 \qquad (4.15B)$$

To make things more interesting, we pose Michaelis-type saturation in the rate of transfer of the drug in compartment 1 to compartment 2. A Simnon program was written, as follows:

```
continuous system NLTV2CPT    "5/08/97
STATE  x1  x2
DER  dx1  dx2
TIME  t
"
dx1 = u1 + k12*rx2 - Vm*rx1/(Km + rx1) - k01*rx1
"
dx2 = - k12*rx2 + Vm*rx1/(Km + rx1) - rx2*k02*(1 + n*cos(2*pi*
t/per))
"
u1 = b1*( 1 + V12)             " Generates pulse input.
V12 = IF t > t1 THEN -1 ELSE 0
"
rx1 = IF x1 < 0 THEN 0 ELSE x1
rx2 = IF x2 < 0 THEN 0 ELSE x2
"
" CONSTANTS
k01:.2
t1:20
Vm:1
Km:0.5
b1:1
n:0.75
k12:.2
pi:3.14159
per:24
k02:.5
"
END
```

Two simulation conditions are shown in Figure 4.10, one with a fixed loss constant from compartment 2 and the other with a time-variable loss constant. In the fixed loss constant case ($n = 0$), the concentration of drug in compartment 2 reaches a saturated value, representing a saturated input rate equal to the output rate. When the loss constant k_{02} is varied sinusoidally around a mean value, the concentration x_2 shows two peaks, and persists in high concentration for about 10 hr longer than when k_{02} is fixed.

In summary, compartmental analysis has been shown to be useful in describing the dynamics of drug distribution in the human body. In many cases, compartmental pharmacokinetic (CPK) systems are linear and are amenable to analysis with linear algebraic tools and Laplace transforms. We have also seen that CPK ODE systems can also be nonlinear and have time-varying coefficients, making pencil-and-paper solutions out of the question. We can learn much about the behavior of these nonlinear and time-varying systems through computer simulations.

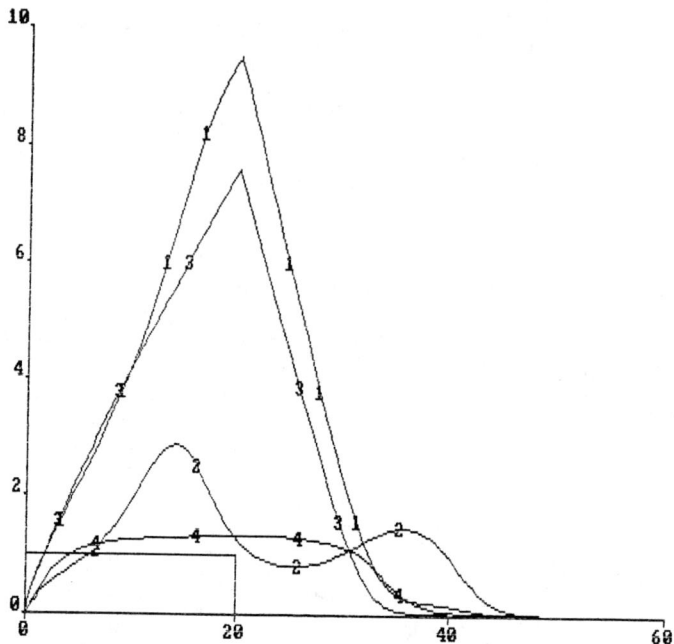

FIGURE 4.10 Results of the Simnon simulation of a two-compartment, nonlinear, time-variable compartmental pharmacokinetic system. Simulation with $k01 = n = 0$: Trace $1 = x1$, trace $2 = x2$, trace $3 = u1$, trace $4 = k02(t)$. Simulation with $k01 = 0$, $n = 0.9$: trace $5 = x1$, trace $6 = x2$, trace $7 = u1$, trace $8 = k02(t)$. Note the final peak in $x2$ at $t = 37$ in the time-variable $k02$ case.

4.2 THE USE OF WASHOUT CURVES AND BOLUS INPUTS TO IDENTIFY THE PARAMETERS OF SIMPLE CPK SYSTEMS

In order to design an external controller for a CPK system, we need to know the system parameters quantitatively, including the number of compartments, their effective volumes, and the rate constants determining linear loss and intercompartmental transfer rates for each compartment. There are two main approaches to CPK parameter identification. One is to assume linearity, a compartmental architecture, and to give a single-bolus injection to one or more compartments, one at a time, and to observe the "impulse response" of concentrations vs. time. From initial values, peak values, and concentration rates, some constants can be assigned for simple one- and two-compartment linear systems. Characterization of higher order and nonlinear systems is more of a challenge and requires special techniques, such as the white noise method,[89] in which details of system architecture are sacrificed to obtain a system description in terms of an overall linear weighting function plus a set of nonlinear weighting functions (the black-box approach).

Introduction to Compartmental Modeling and Pharmacokinetic Systems

The use of a pseudorandom binary noise (PRBN) input which is cross-correlated with the system's output to extract the system's weighting function is useful when the system can be assumed to be linear. Sheppard and Sayers[123] and Slate[126] used the PRBN method to characterize the response of mean arterial pressure (MAP) to the injection rate of sodium nitroprusside (SNP), a blood-pressure-reducing drug. An input "dc bias" of SNP must be used at the input because the infusion rate cannot go negative, and the MAP must stay > 0.

As an example of simple CPK parameter identification, consider the one-compartment linear system characterized by the ODE

$$\dot{x} = -x(K_L/V) + u(t) \quad \mu g/l \tag{4.16}$$

where K_L is the loss rate constant, V is the compartment volume, and $u(t)$ is the input rate. In the first case, let $u(t) = D_o \delta(t)$ (bolus injection). The Laplace-transformed system is

$$\frac{X}{U}(s) = \frac{1/V}{s + K_L/V} \tag{4.17}$$

The drug concentration following the bolus injection is

$$X(s) = \frac{D_o/V}{s + K_L/V} \xrightarrow{\mathcal{L}^{-1}} x(t) = (D_o/V) e^{-(K_L/V)t} \tag{4.18}$$

Since we know D_o, the peak value of $x(t)$ at $t = 0$ gives us the compartment volume, V. The loss rate is found from estimating the system's time constant, $\tau = V/K_L$. In this simple case, τ is the time it takes $x(t)$ to decay to $0.368 \times (0)$.

A *washout curve* is found by infusing drug into the compartment at a constant rate and letting x reach a steady-state equilibrium value, x_{SS}. x_{SS} is found from:

$$X(s) = \frac{U_o}{s} \frac{1/V}{s + K_L/V} \rightarrow x_{SS} = U_o/K_L \tag{4.19}$$

At $t = 0+$, $u(t) \rightarrow 0$, and $x(t)$ decays exponentially from x_{SS}. Thus:

$$x(t) = (U_o/K_L) e^{-(K_L/V)t} \tag{4.20}$$

In this case, the initial value (at $t = 0$) yields K_L, and the time constant estimate yields V.

The complexity of a linear two-compartment pharmacokinetic system, shown in Figure 4.11, is more than doubled. Now we would like to find K_{L1}, K_{12}, K_{21}, V_1, and V_2. We assume we know u_1, and we can measure $c_1 = x_1/V_1$ and $c_2 = x_2/V_2$. The system's state equations are

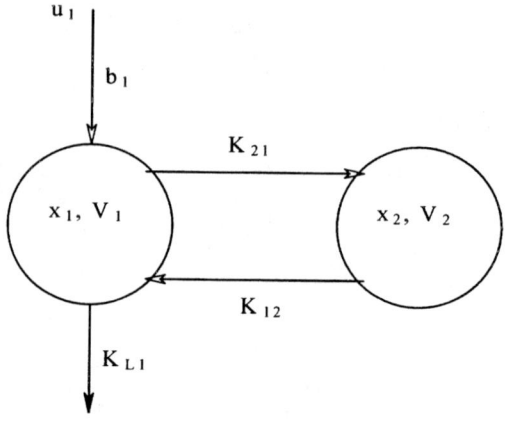

FIGURE 4.11 A two-compartment linear compartmental pharmacokinetic system.

$$\dot{x}_1 = -x_1(K_{L1} + K_{21})/V_1 + x_2(K_{12}/V_2) + u_1 \tag{4.21A}$$

$$\dot{x}_2 = x_1(K_{21}/V_1) - x_2(K_{12}/V_2) \tag{4.21B}$$

An easy way to find the transfer functions for this single-input/two-output (SI2O) system is to use the system ODEs to make a signal flow graph and then reduce it using Mason's rule. Figure 4.12 shows the signal flow graph, and the transfer functions are easily shown to be

$$\frac{C_1}{U_1}(s) = \frac{(1/V_1)(s + K_{12}/V_2)}{s^2 + s\left[K_{12}/V_2 + (K_{L1} + K_{21})/V_1\right] + K_{12}K_{L1}/V_1V_2} \tag{4.22A}$$

$$\frac{C_2}{U_1}(s) = \frac{K_{21}/V_1V_2}{s^2 + s\left[K_{12}/V_2 + (K_{L1} + K_{21})/V_1\right] + K_{12}K_{L1}/V_1V_2} \tag{4.22B}$$

The (factored) quadratic denominator has real roots at $s = -b$ and $s = -c$. If a constant infusion u_{1o} is given, in the steady state it is easy to see that: $c_{1SS} = u_{1o}/K_{L1}$ and $c_{2SS} = u_{1o}K_{21}/(K_{12}K_{L1})$. Thus from steady-state concentrations we can find K_{L1} and the ratio, K_{21}/K_{12}.

The bolus response of the system is

$$C_1(s) = \frac{(D_o/V_1)(s + a)}{(s + b)(s + c)} \tag{4.23A}$$

Introduction to Compartmental Modeling and Pharmacokinetic Systems

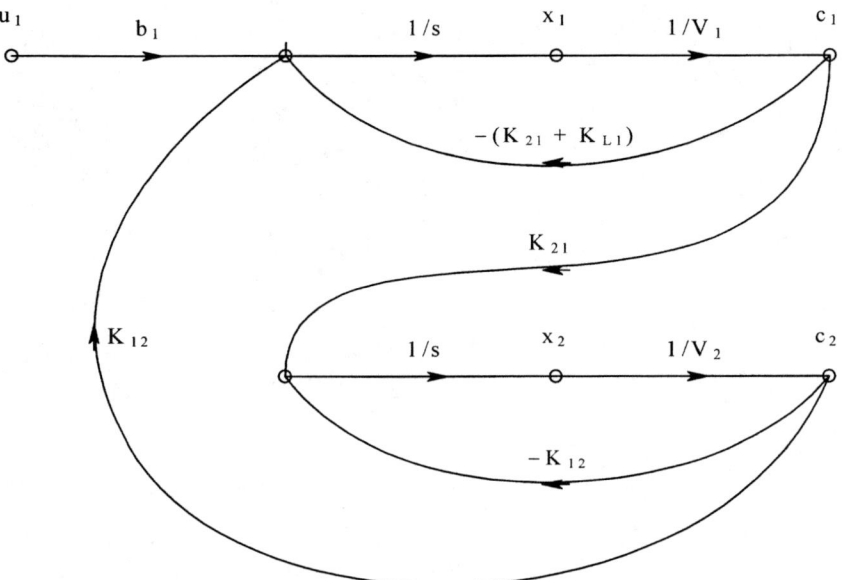

FIGURE 4.12 Signal flow graph for the two-compartment compartmental pharmacokinetic system. Note that it has three loops.

and

$$C_2(s) = \frac{D_o K_{21}/V_1 V_2}{(s+b)(s+c)} \tag{4.23B}$$

Inverse Laplace transforming, we get:

$$c_1(t) = (D_o/V_1)\frac{1}{c-b}\left[(a-b)e^{-bt} - (a-c)e^{-ct}\right] \tag{4.24A}$$

$$c_2(t) = (D_o K_{21}/V_1 V_2)\frac{1}{c-b}\left[e^{-bt} - e^{-ct}\right], \quad c > b \tag{4.24B}$$

From $c_1(0) = D_o/V_1$, we find V_1. $c_2(0) = 0$, however. Factoring the quadratic denominator of the system transfer functions yields the natural frequencies b and c:

$$b, c = -\left[K_{12}/V_2 + (K_{L1} + K_{21})/V_1\right]/2$$
$$\pm \left\{\left[K_{12}/V_2 + (K_{L1} + K_{21})/V_1\right]^2 - 4K_{12}K_{L1}/V_1 V_2\right\}^{(1/2)}/2 \tag{4.25}$$

b and c can be estimated by plotting the natural log of $c_1(t)$ vs. t and examining the slopes. If $c \gg b$, then there will be two distinct linear slope regions. Near $t = 0+$, the c term will dominate, and as $t \to \infty$, the slope is due to the b term. From Equation 4.24A, we can write:

$$c_1(t) = D_o \left[\frac{a-b}{V_1(c-b)} e^{-bt} + \frac{c-a}{V_1(c-b)} e^{-ct} \right] \qquad (4.26)$$

Referring to Figure 4.13, near $t = 0+$, we can write:

$$\ln(c_1) = \ln\left[D_o K_1 e^{-bt} + D_o K_2 e^{-ct} \right] \approx \ln\left[D_o K_2 e^{-ct} + D_o K_1 \right] \qquad (4.27)$$

$$\frac{d \ln(c_1)}{dt} = \frac{-D_o K_2 c e^{-ct}}{\left[D_o K_2 e^{-ct} + D_o K_1 \right]} \approx \frac{\Delta \ln(c_1)}{\Delta t} \qquad (4.28)$$

Solving for the natural frequency c:

$$c \cong \frac{\Delta \ln(c_1)}{\Delta t} \frac{[c_1(0+)]}{D_o K_2} > 0 \qquad (4.29)$$

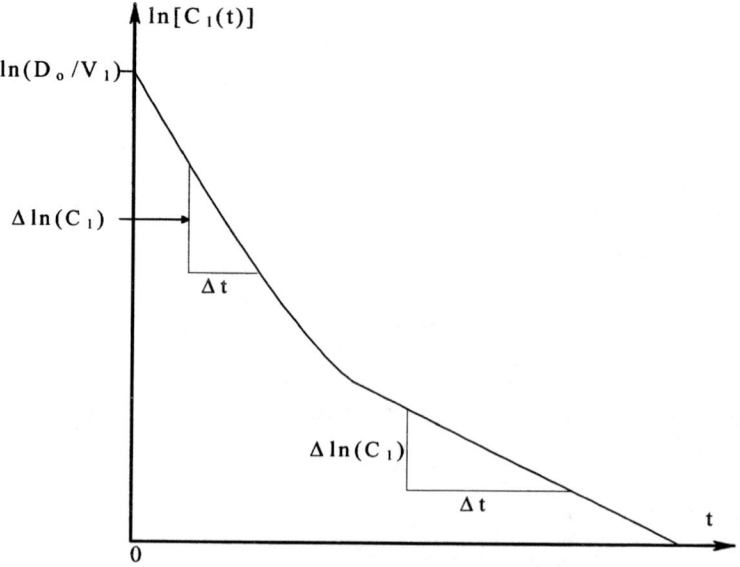

FIGURE 4.13 Bolus response of $c_1(t)$ on a log linear scale to show how natural frequencies can be determined.

Introduction to Compartmental Modeling and Pharmacokinetic Systems

Because the fast exponential term $\to 0$, for $t > 3/c$, we can write:

$$b \cong \frac{\Delta \ln[c_1(t)]}{\Delta t} \tag{4.30}$$

Thus it is possible to estimate the two-compartment pharmacokinetic system's natural frequencies and certain parameters, such as K_{L1}, K_{21}/K_{12}, and V_1.

In summary, it is seen that individual parameter estimation for CPK systems of order $N \geq 3$ can be very complex, whatever method is used. The Laplace initial and final value theorems applied to the states given impulse and step inputs can provide useful information with which to characterize the **A** and **B** matrices. CPK system parameter identification is easiest when there are three or fewer states and their time constants are widely separated.

4.3 OVERVIEW OF DRUG INPUT CONTROLLERS

Perhaps the most common pathway for drug infusion or injection is the intravenous (IV) route. IV administration of drugs provides a rapid means of distributing the drug to various compartments in intimate contact with the circulatory system. Slower pathways for drug infusion include but are not limited to the intraperitoneal, intramuscular, subdermal, and intradermal routes. Drugs can also be directly infused into the cerebrospinal fluid or inhaled as aerosols, in which case they are absorbed by lung tissues. The IV pathway is the most widely used route, however, for closed-loop drug control.

There are several modes of controlling drug infusion rate. The simplest is the on/off mode, effected by a solenoid valve, for example, in an IV drip line. A syringe pump running at a constant rate can also be turned on and off to effect control. If the infusion rate of a syringe pump can be continuously varied between zero and a maximum, then some sort of proportional controller can be designed. Drug input rate is, obviously, nonnegative.

Bolus injectors offer a special challenge.[96] While they can be made mechanically simple, the control dynamics associated with their use can be complex. Control using a bolus injector is accomplished by varying the instantaneous frequency of the injections; the bolus size remains constant. On-line system identification when bolus injectors are used can be easier as the plant output is the superposition of its impulse responses, each one the result of a bolus injection at some time, t_k.

Both on/off limit-cycling controllers and integral pulse-frequency modulation (IPFM) bolus-injecting controllers produce a steady-state ripple on the controlled variable. The peak-to-peak amplitude of the ripple can be designed for and is usually held to 5 to 10% of the steady-state output. Such ripple is not considered to be deleterious in most closed-loop drug infusion systems. Ripple in a fixed-schedule, periodic drug input paradigm generally far exceeds the steady-state ripple from a closed-loop IPFM system or a limit-cycling on/off controller.

4.4 CHAPTER SUMMARY

Pharmacokinetics is the study of how drugs, once given, are distributed throughout the body in time. CPK analysis can be used to predict tissue levels of a drug and to determine how often the drug must be given to obtain an effective concentration at the target organ. The CPK model parameters for a given drug will vary from individual to individual, underscoring the need for parameter identification before *accurate* modeling can take place. Accurate CPK models can be used in the design of closed-loop drug administration systems, designed to maintain the concentration of a drug in a compartment between an upper and lower bound for maximum effectiveness and minimum side effects.

The next chapter treats the design of *nonlinear* controllers for closed-loop drug injection systems. As you will see, there are a variety of design strategies, each with its own costs and benefits.

PROBLEMS

4.1 Design a proportional-plus-integral controller (find values for K_p and K_i) for a continuous drug infusion system for the nonlinear pharmacokinetic plant: x_2 is the output (desired concentration) and u the drug input rate.

$$dx_1 = -K_{01} * x_1^2 + K_{12} * x_2 - V_m * x_1 / (K_m + x_1) + u$$

$$dx_2 = -K_{12} * x_2 - K_{02} * x_2 + V_m * x_1 / (K_m + x_1)$$

Let all constants = 1 except $K_m = 0.1$ and $V_m = 5$. t is in minutes, and $u \geq 0$.
   A. Adjust the controller parameters K_i and K_p so that the desired drug concentration x_2 reaches its set point (SP) in minimum time with no more than 5% overshoot. Give the controller parameters for SP = 1 and SP = 5.
   B. Demonstrate the system's robustness to parameter change by setting SP = 1 and letting the second compartment's loss constant K_{02} jump from 1 to 2 once the system reaches steady state. Demonstrate the validity of your design by simulation. Give plots with appropriate scales. Give programs used.

4.2 A linear, time-invariant, two-compartment pharmacokinetic system is shown in the figure. In it, $K_{12} = K_{21} = K_{L1} = K_{L2} = 2$, $V_1 = V_2 = 4$, $c_1 = x_1/V_1$, $c_2 = x_2/V_2$ µg/ml, $x_1(0-) = x_2(0-) = 0$, $b_1 = 1$.

Introduction to Compartmental Modeling and Pharmacokinetic Systems

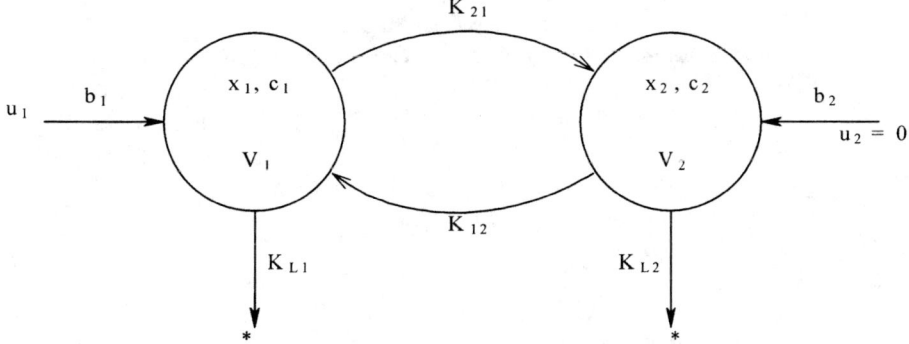

PROBLEM 4.2

A. Write the system's state equations in matrix form.
B. Find $X_1(s)/U_1(s)$ and $X_2(s)/U_1(s)$.
C. Let $u_1(t) = D_o\,\delta(t)$, $u_2 = 0$. Plot and dimension $c_1(t)$ and $c_2(t)$.
D. Consider the system's washout behavior. Assume that $u_1 = 1$ for a long time before $t = 0$. At $t \geq 0$, $u_1 = 0$. Plot and dimension $c_1(t)$ and $c_2(t)$. (Simulate with Matlab or Simnon.)

4.3 The normal physiological rate of secretion of a certain hormone is 2.5 ng/min. The steady-state concentration of the hormone, $[H]_{SS}$, is 2.0 pg/ml in body fluids. When a bolus injection of the hormone is given, the concentration rises abruptly to a peak concentration, $[H]_{PK}$, and then decays with a simple exponential curve to $[H]_{SS}$ again. The half-time for the bolus response to decay to the steady-state value is 20 min.
A. Make a block diagram for the system.
B. Find numerical values for the hormone's loss constant, K_L, and the effective compartment volume.

4.4 A hormone in the extracellular fluid (ECF) at a concentration $[H]$ controls the passive diffusion of a molecule M into a cellular compartment with volume V_C. Let $[M]_o$ be the extracellular concentration of M and $[M]_i$ be the intra-

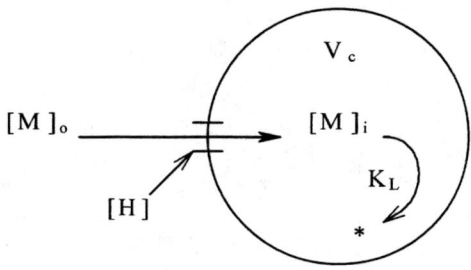

PROBLEM 4.4

cellular concentration of M. The intracellular M disappears at a rate $K_L\,[M]_i$. The transmembrane diffusion constant is a function of $[H]$: $K_D = K_{D0} + \beta[H]^2$.

A. Write the system's state equation for $[M]_i$. (Note that in a diffusion system, the rate of mass flow is proportional to the concentration difference.)

B. Give an expression for the steady-state $[M]_i$, given a constant $[M]_o$.

C. Give an expression for the system's time constant, $\tau = f([H], K_L, K_{D0}, \beta, V_C)$.

4.5 In the two-compartment pharmacokinetic system shown in the figure, the drug is infused at a rate U_1 µg/min into the first compartment. The drug concentration in compartment 2 is the controlled variable. That is, $c_2 = x_2/V_2$ µg/l. Assume $V_1 = 5$ l, $V_2 = 15$ l, $K_{L1} = 1$, $K_{L2} = 1$, $K_{21} = 4$ l/min, and $K_{12} = 14$ l/min.

A. Write the state equations for the system.
B. Make a signal flow graph for the system.
C. Use Mason's rule to find the transfer function, X_2/U_1, in the form:

$$\frac{C_2}{U_1}(s) = \frac{K}{(s+a)(s+b)}$$

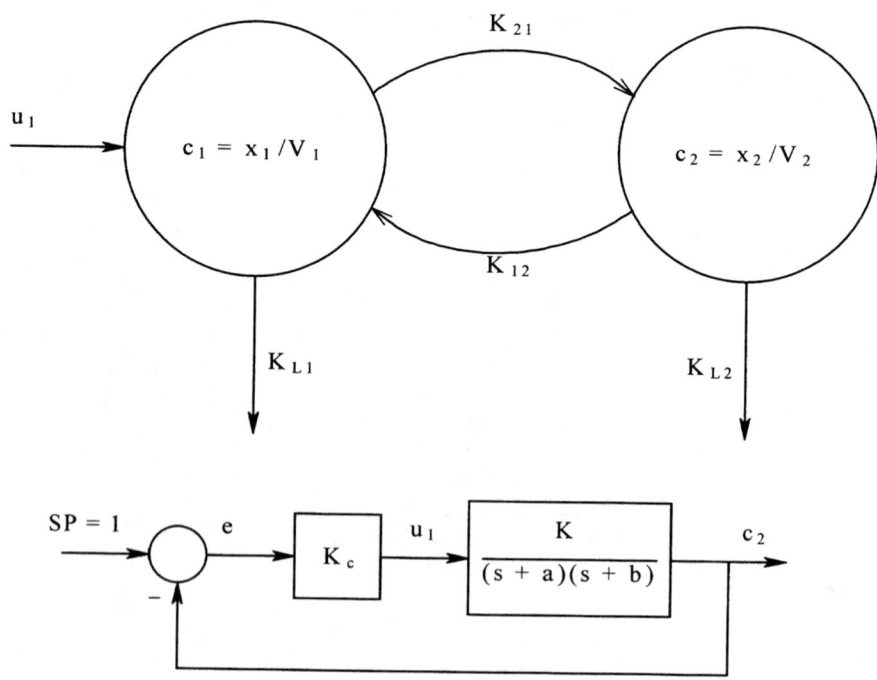

PROBLEM 4.5

That is, find numerical values for K, a, and b. It is now desired to use simple proportional closed-loop control to force c_2 to approach a set point of 1 mg/l. That is, $U_1 = (1 - c_2)K_C$. Note that in a CPK system, $U_1 \geq 0$; there can be no negative drug infusion rate.

D. Under closed-loop conditions, the system's poles can become complex-conjugate for large K_C. Give an expression for C_2/SP. Note that the denominator will be of the form

$$s^2 + s(2\zeta\omega_n) + \omega_n^2$$

Find the numerical value of K_C required to give the closed-loop complex-conjugate poles a damping factor of $\zeta = 0.707$.

E. Find numerical values for the closed-loop system's ω_n and dc gain, given the K_C value found above.

4.6 A linear, time-invariant, two-compartment pharmacokinetic system has $K_{12} = K_{21} = K_{L1} = K_{L2} = 1$, $V_1 = V_2 = 1$, $C_1 = x_1/V_1$, $C_2 = x_2/V_2$, $b_1 = 1$, $x_1(0) = x_2(0) = 0$.

A. Write the state equations for the system in matrix form: $\dot{\mathbf{x}} = \mathbf{A}\mathbf{x} + \mathbf{B}\mathbf{u}$.
B. Determine and plot to scale $x_1(t)$ and $x_2(t)$ for $u_1(t) = \delta(t)$.
C. Repeat part B for $u_1(t) = U(t)$ (unit step).

4.7 An interconnected two-compartment pharmacokinetic system is investigated experimentally by giving a 5-μg bolus IV injection at $t = 0$. That is, $u_1(t) = D_o \delta(t)$, where $D_o = 5$ μg. The concentration of the drug in the blood compartment is modeled by $c_1(t) = C_1 e^{-0.5t}$, where $C_1 = 1$ μg/l. The concentration of the drug in the second compartment, cerebrospinal fluid, is given by $c_2(t) = C_2 (e^{-0.1t} - e^{-0.5t})$, where $C_2 = 0.001$ μg/l. Natural frequencies are in radians per hour.

A. Find the transfer functions:

$$\frac{C_1}{U_1}(s) \quad \text{and} \quad \frac{C_2}{U_1}(s)$$

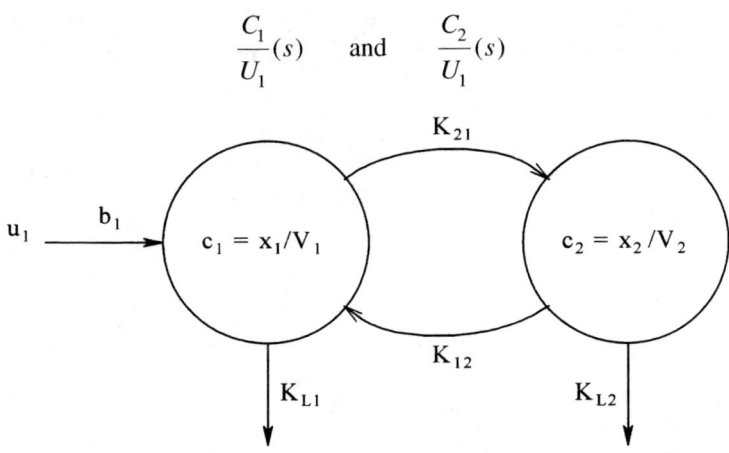

PROBLEM 4.6

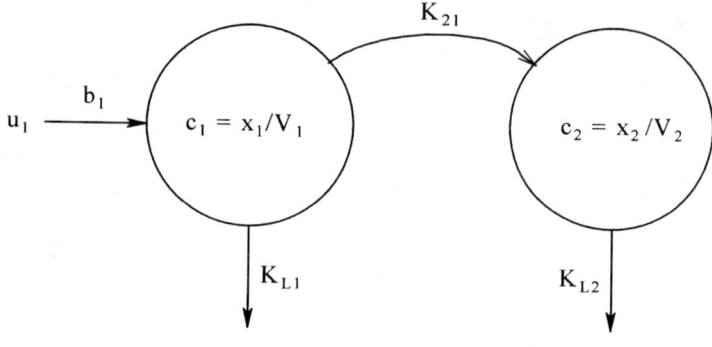

PROBLEM 4.7

B. From the experimental evidence above, evaluate numerically K_{21}, K_{L1}, K_{L2}, V_1, and V_2. It is known that $b_1 = 1$ and $K_{12} = 0$.

4.8 A one-compartment nonlinear CPK system has Michaelis–Menten elimination kinetics.[55] Its ODE is given by:

$$\dot{c} = \frac{V_m c}{K_m + c} + u_1/V_1 \quad \text{mg}/(1 \text{ diagonal min})$$

where c is the concentration of drug in the compartment at time t. The compartment has a known volume V_1 l and can receive a bolus input of D_o mg at $t = 0$ [$u_1(t) = D_o \delta(t)$]. The system can also receive a constant infusion of I_o mg/min [$u_1(t) = I_o U(t)$]. Detail a procedure based on *a priori* knowledge of V_1, $c(t)$, D_o or I_o, and t that will identify V_m and K_m. Steady-state or dynamic values of $c(t)$ may be used. Note that any $I_o < V_1 V_m$ mg/min for a real solution.

4.9. A CPK system has two compartments: compartment 1 (C1) is interstitial fluid with volume V_1 l, and compartment 2 (C2) is the circulatory system with

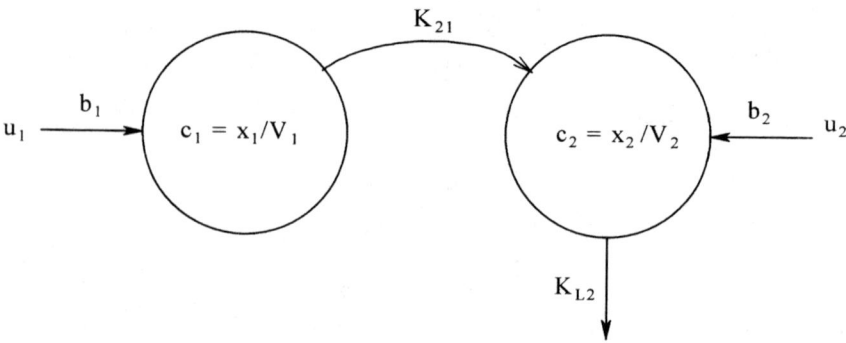

PROBLEM 4.9

volume V_2 l. Drug input can be at rate u_1 to C1 and rate u_2 to C2. Only the drug concentration in C2, c_2, can be measured. There is mass transfer from C1 to C2 at rate $K_{21} c_1$ μg/min. Drug is lost from C2 at rate $K_{L2} c_2$ μg/min. x_1 is the drug mass in C1, and x_2 is the drug mass in C2 (see the figure). b_1 is the bioavailability fraction[55] ($0 < b_1 < 1$). The ODEs describing the system are

$$\dot{x}_1 = b_1 u_1 - K_{21} x_1 / V_1 \qquad \text{μg/min}$$

$$\dot{x}_2 = u_2 + K_{21} x_1 / V_1 - K_{L2} x_2 / V_2 \qquad \text{μg/min}$$

We wish to perform some parameter identification on the system. Inputs u_1 and u_2 will be manipulated, and the concentration $c_2(t) = x_2(t) / V_2$ will be observed.

A. From the ODEs above, derive the Laplace transfer function for $C_2(s)$ for $U_1(s)$ and $U_2(s)$.
B. Let $U_2(s) = u_{20}/s$ (step of height u_{20}). Find a unique expression for K_{L2}.
C. Let $u_2(t) = D_2 \delta(t)$ (a bolus injection of D_2 μg). Find a unique expression for V_2.
D. Let $U_1(s) = u_{10}/s$ (step of height u_{10}). Find a unique expression for b_1, the bioavailability fraction.
E. Assume the natural frequency $K_{L2}/V_2 \gg K_{21}/V_1$. Describe how you can estimate K_{21}/V_1.

4.10 Godfrey[55] gives an example of a nonlinear two-compartment pharmacokinetic system in which compartment 2 has Langmuir saturation. The system is shown in the figure. x_1 and x_2 are the drug masses in compartments 1 and 2, respectively. (Compartment volumes are hidden in K_{L1}, K_{21}, and σ_2.) The system ODEs are

$$\dot{x}_1 = b_1 u_1 - K_{L1} x_1 - K_{21}(1 - \sigma_2 x_2) x_1$$

$$\dot{x}_2 = K_{21}(1 - \sigma_2 x_2) x_1$$

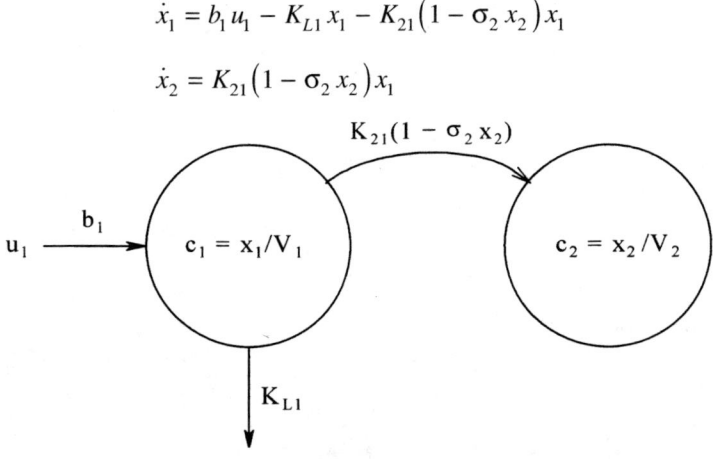

PROBLEM 4.10

- **A.** Investigate the system's dynamics by simulation. Let $\sigma_2 = 20$, $K_{L1} = 1$, $K_{21} = 3$, $b_1 = 0.5$, $x_1(0) = 0$, $x_2(0) = 0$. Generate an impulse input u_1 at $t = 0$ and again at $t = 1$. Impulses can be approximated by narrow, rectangular pulses of width 0.02 and heights u_{10} ranging from 1 to 50. Thus the impulse areas are $D_1 = 0.02\, u_{10}$. At what level does x_2 saturate?
- **B.** Show how the first impulse at $t = 0$ can be used to find b_1.

4.11 After a drug is taken, its concentration decays naturally toward zero. Discuss the various mechanisms by which the active forms of drugs are eliminated from the body. What organs are generally involved?

5 Special Types of Closed-Loop Drug Input Controllers

5.0 INTRODUCTION

The design of closed-loop drug delivery systems presents a special challenge to the biomedical control engineer. The plant parameters often vary markedly in value between individuals, so that individual plant identification is often necessary to achieve accurate control of drug levels. Plant parameters are often nonstationary; the drug itself can alter the physiological states of organs that are involved in a drug's breakdown and elimination. For example, cancer antimetabolites can poison liver cells and/or kidney cells, as well as kill tumor cells. The plants are nonlinear; drugs can be injected, but not uninjected. Excess drug is generally cleared from the body passively, by natural means. Also, the turn-on (step) response of a compartmental pharmacokinetic/physiological (CPK/P) system must show little specified overshoot (or undershoot). This is especially true with blood-pressure-lowering drugs like sodium nitroprusside. The mean arterial pressure must ramp down smoothly to the set point and not shoot past it at system turn-on. Otherwise the patient could faint or go into shock.

Drug input controllers are, in general, nonlinear because no negative inputs are possible. They fall into three broad categories:

1. *Proportional infusion,* where the drug input rate is variable, so $0 \leq u \leq U_{max}$ mg/min. A motor-driven syringe pump is generally used for proportional infusion.
2. *On/off infusion,* where the drug input rate is either 0 or U_{max}.
3. *Pulse frequency-modulated (PFM) bolus injections,* where $u = D_o \sum \delta(t - t_k)$. The times at which the bolus injections are given, t_k, are calculated by the control algorithm. Each bolus is D_o mg of drug. A pulsatile, constant-stroke volume pump is used in this application.

The on/off infusion controller usually operates in a steady-state limit cycle, much as a home heating system is turned on and off by the thermostat. Both the on/off controller and the PFM controller produce a ripple in the controlled drug concentration. The peak ripple can be designed for and in most closed-loop CPK applications does not present a problem.

The types of drugs injected can be antibiotics, cancer antimetabolites, and drugs designed to have specific physiological effects such as pain suppression (opioids), glucoregulation (insulin), blood pressure reduction (sodium nitroprusside), serum calcium regulation (hormones), urine output in burn patients (saline intravenous drip), and depth of anesthesia. General principles of design for proportional infusion control of a CPK/P system with an appreciable delay time or transport lag are discussed first. Controller design for such systems presents a considerable challenge. Following systems with delay time, the design of on/off limit-cycling controllers using describing functions is considered. PFM controllers are then considered, followed by unique parameter-switching controllers.

5.1 DESIGN OF ON/OFF LIMIT-CYCLING CONTROLLERS FOR CPK/P SYSTEMS

On/off controllers are used because they are simple, inexpensive, reliable, and generally do not require a computed control algorithm. Their principal disadvantage is the ripple in the controlled variable, which, under certain conditions, can be objectionably large.

In an on/off controller, the controlled input rate of the drug is either U_{max} or 0. The system runs in a limit-cycling mode, the same as a home heating/cooling system. As a result of the periodic limit cycle, the controlled concentration or physiological parameter cycles between an acceptable maximum and minimum level.

Design of on/off controllers can be made easy through the use of *describing functions*, first described in the U.S. by Kochenburger.[73,107,135,138] (Describing functions were introduced in Section 3.3.4.) Briefly, a describing function is an equivalent amplitude-dependent vector gain which replaces an odd nonlinearity in the system loop gain description. There are two basic assumptions made about the closed-loop system's loop gain architecture when using describing functions. The first is that the linear transfer function plant following the on/off switching controller has an output in the steady state which is a *sinusoid* added to a dc level. (The actual periodic component of the steady-state output will, in practice, have harmonics.) If the linear portion of the plant has natural frequencies below the limit cycle frequency, ω_o, and has a numerator of order s^M and a denominator of order s^N where $N - M \geq 1$, then the assumption that the periodic output of the plant is a sinusoid of ω_o generally can be justified. The second assumption has to do with the odd nonlinearity of the controller. In general, $u = f(e) = -f(-e)$. Because the plant output consists of a dc level plus a sinusoidal ripple, the steady-state system error at the design center is a zero-mean sinusoid of frequency ω_o. That is,

$$e_{ss}(t) = SP - [y] = SP - \left[\bar{y} - E_o \sin(\omega_o t + \phi)\right] = E_o \sin(\omega_o t + \phi) \quad (5.1)$$

The input to the on/off nonlinearity is thus assumed to be a sine wave of frequency ω_o and peak amplitude E_o. This periodic input to the nonlinearity produces a periodic square-wave output, $u(t)$. If we analyze $u(t) = u(t - T)$ and find the fundamental frequency term in its Fourier series, we define the *equivalent gain* for the fundamental frequency, $N(E)$. $N(E)$ is the describing function for the nonlinearity. It is written as a complex number, because there is a phase shift associated with certain *memory* nonlinearities, such an on/off controller with hysteresis. $N(E)$ is defined as:

$$N(E) \equiv \frac{C_1}{E} \tag{5.2}$$

where C_1 is the peak amplitude and phase-first harmonic term in the nonlinearity's output, and E is the peak input sinusoidal error.

Figure 5.1 illustrates the block diagram of a nonlinear system with a simple $\pm M$ controller. For system instability (and oscillations), conditions on E and ω cause the closed-loop system's return difference $\to 0$. That is,

$$F(j\omega_o) = \left[1 - A_L(j\omega_o)\right] \to 0 \tag{5.3}$$

So it follows that for instability,

$$A_L(j\omega_o) = -N(E_o)\, G_p(j\omega_o)\, H(j\omega_o) = 1\angle 0° \tag{5.4A}$$

or

$$-G_p(j\omega_o)\, H(j\omega_o) = N^{-1}(E_o) \tag{5.4B}$$

This so-called instability will produce limit cycle oscillations of finite amplitude and frequency, ω_o.

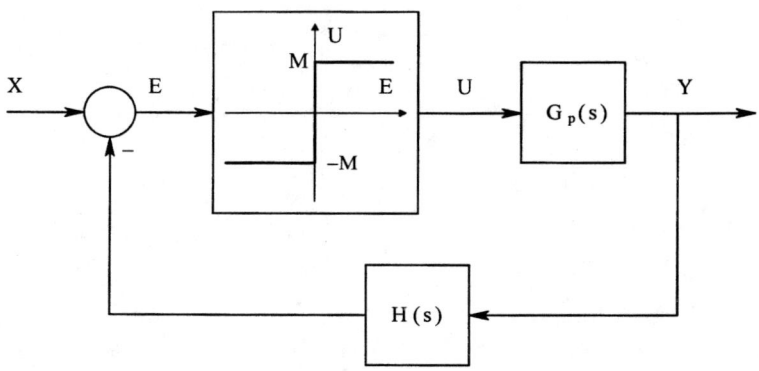

FIGURE 5.1 A simple bang-bang controller.

Equation 5.4B tells us that if we plot $-G_p(j\omega)H(j\omega)$ as a function of ω on polar graph paper, and also the reciprocal describing function vector, $N^{-1}(E)$, their intersection (simultaneous solution) will yield the approximate frequency of steady-state oscillation, ω_o, and E_o, the peak sinusoidal error in the steady state that is the input to the nonlinearity, can be found from $N^{-1}(E)$, Note that $[-G_p(j\omega_o)H(j\omega_o)]$ is the *linear* part of the loop gain.

Let us examine the *first example* of an on/off controller for drug concentration. The system is shown in Figure 5.2A. The plant is a third-order low-pass system with three real poles and no zeros. An on/off controller is to be designed that has specified peak error due to steady-state limit cycling.

First, we must find the U_{max} required. Let us assume that the desired average steady-state drug concentration is SP = $\bar{y}$ = 20 µg/l. We also assume a 50% steady-state duty cycle in the controller when it is limit cycling around the design set point. Thus the controller output in the steady state, $u(t)$, is a 50% duty cycle square wave with minimum value 0 and maximum U_{max}. This means that the average (dc) input

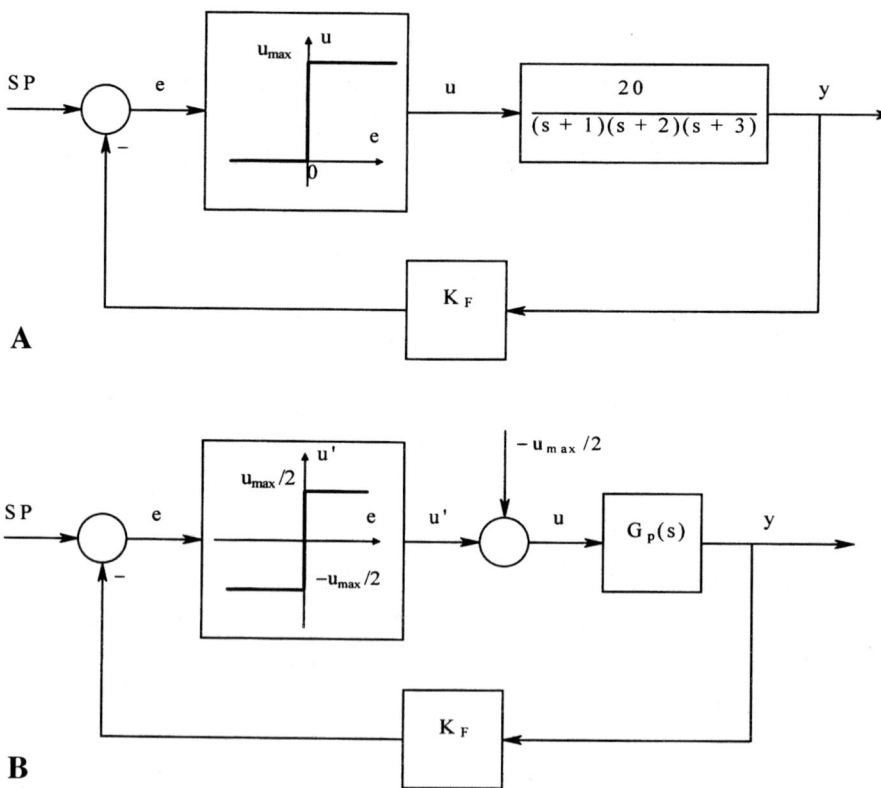

FIGURE 5.2 (A) An on/off controller for drug infusion. The pharmacokinetic plant model has three real poles. (B) The same system as in Figure 5.2A, except the controller is modified to make it an odd function so describing function analysis can be used.

to the plant is $\bar{u} = U_{max}/2$. The dc average plant output is then $(U_{max}/2)(20/6) = 20$, so $U_{max} = 12$ µg/hr.

To use the describing function method, the 0, U_{max} nonlinearity must be made odd, switching between $\pm U_{max}/2$. To do this, we assume that in the steady state, the error is sinusoidal with zero mean $[e(t) = E_o \sin(\omega_o t)]$, and consequently the $u(t)$ square wave will have a 50% duty cycle. Figure 5.2B illustrates the reconfigured system which takes into account $\bar{u}$, $N(E) = (2U_{max})/(\pi E)$ for the symmetrical nonlinearity. However, we plot $\mathbf{N}^{-1}(E)$ in the polar plane along with $-K_F \mathbf{G_p}(j\omega)$, which is

$$\frac{-K_F\,20}{(j\omega+1)(j\omega+2)(j\omega+3)} = \frac{-K_F\,20}{j\omega(11-\omega^2)+(6-6\omega^2)} \qquad (5.5)$$

Now all of $\mathbf{N}^{-1}(E)$ lies on the positive real axis, so where the locus of $-K_F \mathbf{G_p}(j\omega)$ intersects $\mathbf{N}^{-1}(E)$, it is real. This means that the imaginary term in the denominator of Equation 5.5 must equal zero: $j\omega_o (11 - \omega_o^2) = 0$, so $\omega_o = \sqrt{11}$ r/hr. We now consider the magnitudes of the $\mathbf{A_L}(j\omega)$ and $\mathbf{N}^{-1}(E)$ vectors at ω_o (see Figure 5.3).

$$\frac{-K_F\,20}{(6-6\times 11)} = \frac{K_F\,20}{60} = K_F/3 = \frac{\pi E}{2U_{max}} \qquad (5.6)$$

We require that E be 5% of the set point, or 1 µg/l. Substitution of the values of U_{max} and E in Equation 5.6 yields the required value of $K_F = \pi/8$ to meet the 5% error specification. While this simple example was easily solved algebraically, the describing function method is not always so algebraically obliging, and it often is more expedient to do a graphical solution from the polar plot of $\mathbf{N}^{-1}$ and $-\mathbf{G_p}(j\omega)$ or use a simulation.

In the *second example,* we assume a second-order CPK/P plant and an on/off controller with hysteresis. The system is shown in Figure 5.4A. (Hysteresis controllers are widely used in on/off-type temperature regulators, such as found in home heating systems.) In Figure 5.4B, we convert the nonlinearity to a symmetrical (odd) form by adding the dc average controller output to the output of the symmetrical hysteresis controller. As in the previous example, we assume that the steady-state error has zero mean, a 50% duty cycle, and is sinusoidal. Thus the dc steady-state input to the plant is $\bar{u} = U_{max}/2$. We want the average plant output to be equal to the set point, SP. Thus we can find the required U_{max} from:

$$SP = \frac{U_{max}}{2} \mathbf{G_p}(0) = \frac{U_{max}}{2} \frac{K_p}{ab} \qquad (5.7)$$

Thus $U_{max} = SP\,2ab/K_p$.

We next consider the describing function of the hysteresis nonlinearity. If $E > \Delta$, the hysteresis dead zone, u' will be $a \pm U_{max}/2$ square wave with zero mean. There

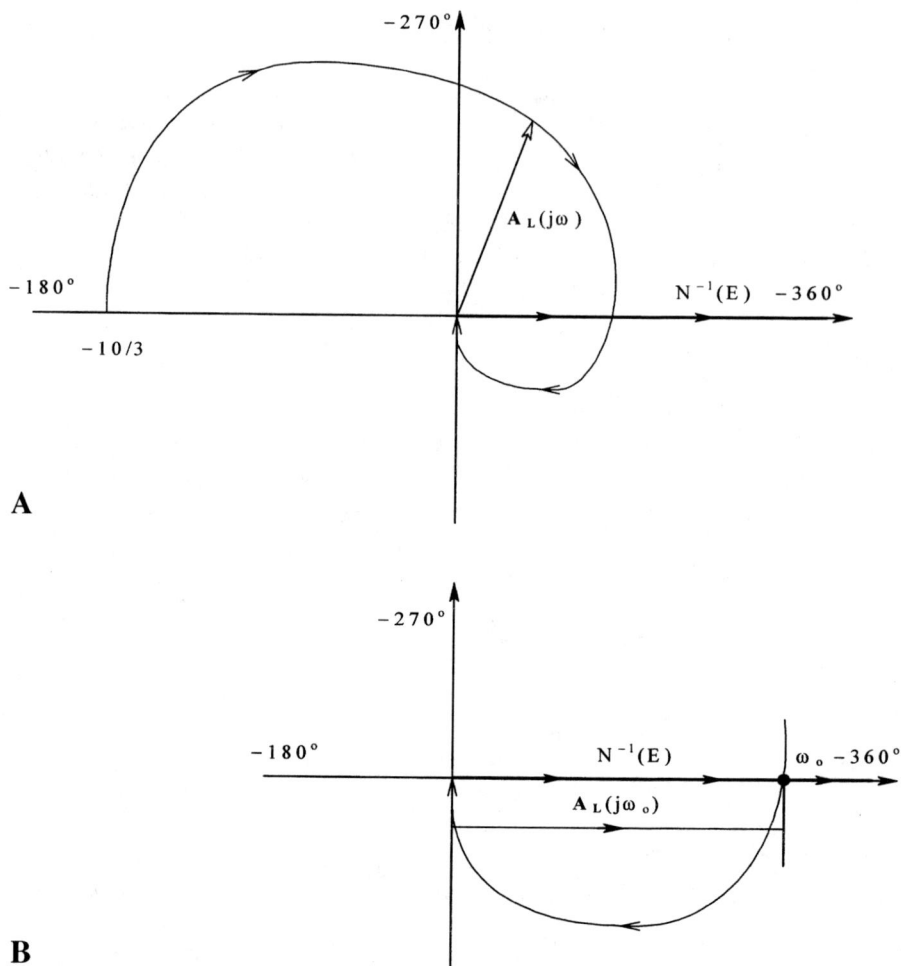

FIGURE 5.3 (A) Polar plots of the $\mathbf{A_L}(j\omega)$ and $\mathbf{N}^{-1}(E)$ loci. (B) Enlarged area of the polar plane of the $\mathbf{A_L}(j\omega)$ and $\mathbf{N}^{-1}(E)$ loci, showing their intersection on the $-360°$ axis at $\omega = \omega_o$.

will also be an amplitude-dependent phase shift between the Fourier series terms of $u'(t)$ and $e(t)$. For the fundamental frequency term in $u'(t)$, this phase lag can be shown to be $\angle N = -\sin^{-1}(\Delta/E)$. Thus the complete describing function for the hysteresis nonlinearity is

$$N(E) = \frac{4(U_{max}/2)}{\pi E} \angle -\sin^{-1}(\Delta/E) \tag{5.8}$$

and its reciprocal is

Special Types of Closed-Loop Drug Input Controllers

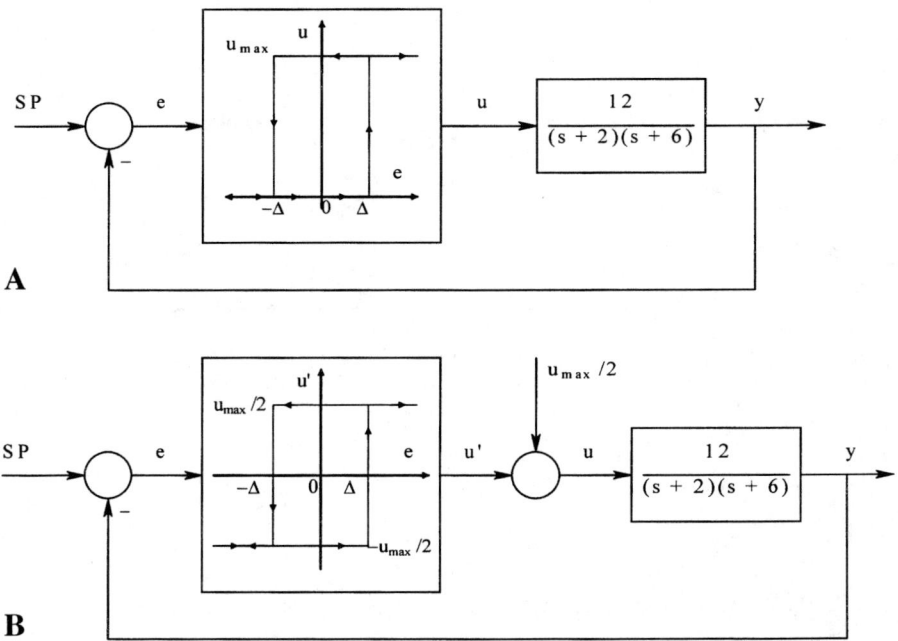

FIGURE 5.4 (A) A second-order CPK plant with a hysteresis on/off controller. (B) The nonlinearity has been converted to an odd function of E so describing function analysis can be used. See text for details.

$$\mathbf{N}^{-1}(E) = \frac{\pi E}{2 U_{max}} \angle + \sin^{-1}(\Delta/E) \tag{5.9}$$

Figure 5.5 shows that $\mathbf{N}^{-1}(E)$ for the hysteresis nonlinearity plots as a straight line parallel to the 0° axis in the first quadrant of the polar graph. $\mathbf{N}^{-1}(E)$ intersects the −270° axis at a radial distance of $\pi\Delta/2U_{max}$ for $E = \Delta$ and extends to the right with increasing E. For the system to produce stable limit cycle oscillations, the vector − $\mathbf{G_p}(j\omega)$ must intersect $\mathbf{N}^{-1}(E)$ at $\omega = \omega_o$, and $E = E_o$. We can design for the intersection within bounds by adjusting Δ and K_F.

To demonstrate the efficacy of a graphic solution, let us define numerical values for the system parameters: SP = 10 µg/l, $E_o \le 5$ µg/l, $K_p = 12$, $a = 2$, $b = 6$ r/hr, $\Delta = 0.2$ µg/l. E_o, U_{max}, and ω_o are to be determined: $U_{max} = (2 \text{ SP } ab)/K_p = 2 \times 10 \times 2 \times 6/12 = 20$ µg/hr. The intersection of $\mathbf{N}^{-1}(E)$ with the −270° axis is at $R = \pi\Delta/2U_{max} = 6.283/40 = 1.571\text{E-}2$. The angle of $-\mathbf{G_p}(j\omega)$ is given by $\varphi = -180° - \tan^{-1}(\omega/2) - \tan^{-1}(\omega/6)$. The magnitude of $-\mathbf{G_p}(j\omega)$ is $K_p/[(\sqrt{\omega^2+2^2})(\sqrt{\omega^2+6^2})]$. In order to limit cycle, the magnitude of $-\mathbf{G_p}(j\omega)$ must exceed 1.571E-2 at $\varphi = -270°$ so there will be an intersection of $-\mathbf{G_p}(j\omega)$ and $\mathbf{N}^{-1}(E)$. By adjusting Δ, we can affect the limit cycle frequency, ω_o, and E_o. Figure 5.6 illustrates the graphical solution for

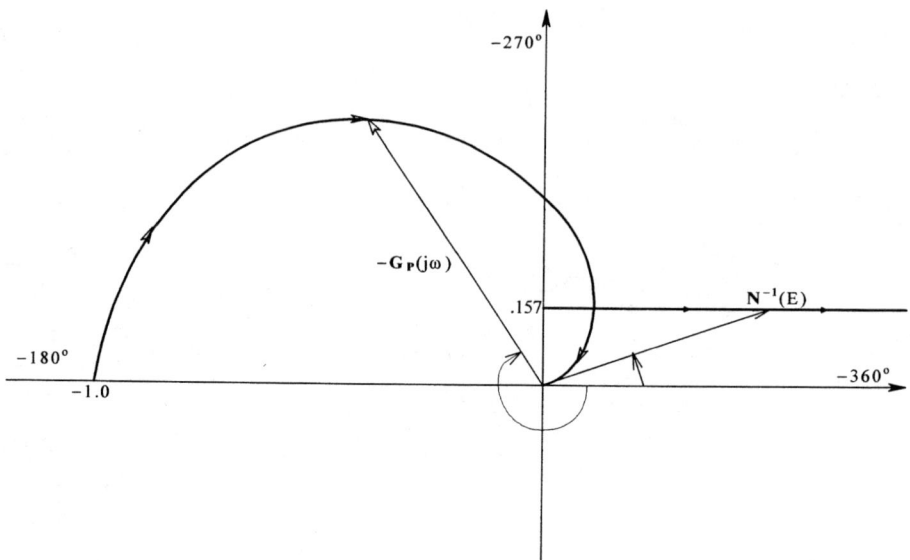

FIGURE 5.5 Polar plots of the $A_L(j\omega)$ and $N^{-1}(E)$ loci. Note that the $N^{-1}(E)$ locus is parallel to the $-360°$ axis.

ω_o and E_o, given the parameters above. We se that the intersection of $N^{-1}(E)$ and $-G_p(j\omega)$ occurs at $\omega_o \cong 16.7$ r/hr. Also, $|N^{-1}(E_o)| \cong 0.0405$, so $E_o = 40 \times 0.0405/\pi = 0.516$ µg/l peak. This represents a 5.2% peak error for the steady-state limit-cycling system.

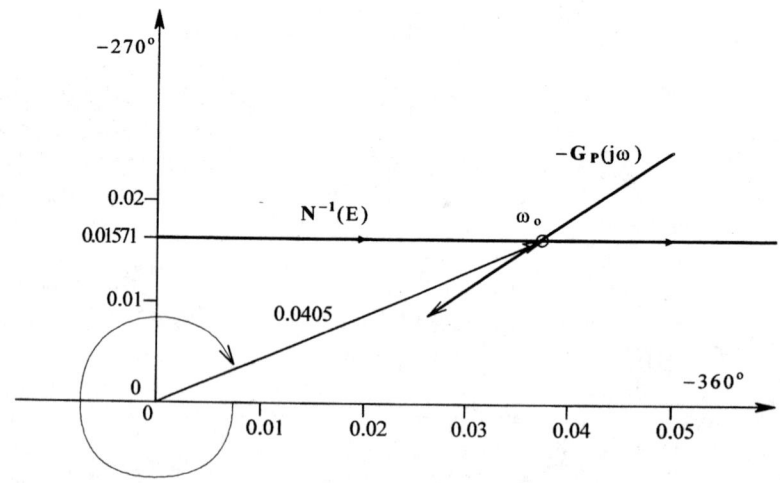

FIGURE 5.6 Enlarged region of the polar plane showing the intersection of the $A_L(j\omega)$ and $N^{-1}(E)$ loci.

Special Types of Closed-Loop Drug Input Controllers

In the *third example* of on/off controller design, we compare the results of a Simnon™ simulation with the describing function graphical solution. We will vary the set point away from the design value to show how the controller changes the duty cycle of the $u(t)$ square wave in order to change the average output level of the plant. The drug infusion regulator in this example is based on a CPK/P plant for blood pressure reduction by a vasodilator drug.[126] The regulated variable is mean arterial pressure (MAP), which is decreased from the unmedicated (high) value of MAP_o according to the dynamics

$$MAP = MAP_o - \Delta P = MAP_o - U(s) \frac{K_p e^{-s\delta}}{(s+a)(s+b)} \qquad (5.10)$$

$u(t)$ is the rate of input of vasodilator drug from the controller (mg/min). The dead time is the sum of the time it takes the drug to reach the venous circulation through the intravenous drip tube plus the time it takes the drug to mix in the circulatory system and reach the capillaries. The closed-loop drug infuser is shown in Figure 5.7. Assume that the desired set point is 110 mmHg MAP. Let $MAP_o = 200$ mmHg. The plant parameters are: $K_p = 2.0$, $a = 1.0$, $b = 0.2$, $\delta = 0.5$ min. The controller rule is $u = U_{max}$ for $E > 0$ and $u = 0$ for $E \leq 0$. To do the describing function analysis, we again plot $-G_p(j\omega)$ and $N^{-1}(E)$ in the polar plane. As before, intersection of the two vectors yields the limit cycle frequency and the peak steady-state error. The limit cycle frequency can also be found analytically by setting the phase of $-G_p(j\omega)$ equal to $-360°$, the angle it has where it intersects the $N^{-1}(E)$ locus. That is,

$$-360° = -180° - \tan^{-1}(\omega_o/a) - \tan^{-1}(\omega_o/b) - \omega_o \delta R \qquad (5.11)$$

Solution of this transcendental equation is best done by trial and error.

The U_{max} required for a 50% duty cycle steady-state oscillation is found from the dc equation:

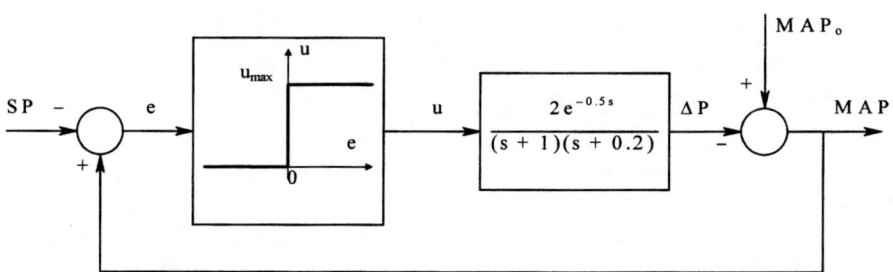

FIGURE 5.7 A limit-cycling on/off drug infusion controller used to reduce MAP. See text for details.

$$\overline{E} = 0 = -\text{SP} + \text{MAP}_o - \Delta P(0) = -110 + 200 - (U_{max}/2)\mathbf{G}_p(0) \rightarrow$$

$$90 = (U_{max}/2)\frac{2.0}{(1)(0.2)} \rightarrow U_{max} = 18.0 \text{ mg/min} \tag{5.12}$$

Therefore, $\mathbf{N}^{-1}(E) = (\pi E)/(2U_{max}) = 0.08727\, E$. $\mathbf{N}^{-1}(E)$ is real and increases linearly with E. Figure 5.8 illustrates the polar plot of $\mathbf{N}^{-1}(E)$ and $-\mathbf{G}_p(j\omega)$. Algebraic trial-and-error solution of the phase equation for the steady-state oscillation frequency yields $\omega_o \cong 1.467$ r/min. Knowing ω_o, we can find the magnitude of $|\mathbf{G}_p(j\omega_o)| = |\mathbf{N}^{-1}(E_o)|$ and then solve for E_o.

$$|\mathbf{G}_p(j\omega_o)| = \frac{2.0}{\sqrt{(\omega_o^2 + 1)}\sqrt{(\omega_o^2 + 0.2^2)}} = 0.761$$

$$0.761 = (\pi E_o)/(2U_{max}) \rightarrow E_o = 8.72 \text{ mmHg} \tag{5.13}$$

Thus the % peak error is $8.72/110 \times 100 = 7.93\%$. This is an acceptable peak error for this type of regulator and CPK/P application.

Figure 5.9 illustrates a Simnon simulation of the MAP regulator. Note that the duty cycle of the steady-state $u(t)$ changes to force the system $\overline{\text{MAP}}$ output to follow the set point. Because the closed-loop system is type 0,[108] there is an average steady-state error for set point values different from the design value (110 mmHg in this case).

The design of on/off controllers using the describing function method is seen to be relatively easy. Analytical, graphic, or simulation means can be used. The use of on/off controllers is acceptable when the ripple level is tolerable by system requirements and slight dc errors have no adverse effect on system performance. In certain applications, where the amplitude of the steady-state ripple in the system output is not acceptable, other controller strategies must be considered.

5.2 NONLINEAR DECOUPLING CONTROLLERS FOR TWO-INPUT/TWO-OUTPUT CPK SYSTEMS

Decoupling is a *model reference control* strategy that allows us to control or regulate one state in a coupled system independently of the other states.[50,64] It has theoretical application in pharmacokinetics as a means of keeping an injected drug's concentration constant in a desired compartment. As you will see, the realization of nonlinear decoupling (NLD) control in CPK systems is generally thwarted by the fact that no negative drug infusion rates are possible in the real world. Obviously, we cannot uninject a drug.

Special Types of Closed-Loop Drug Input Controllers

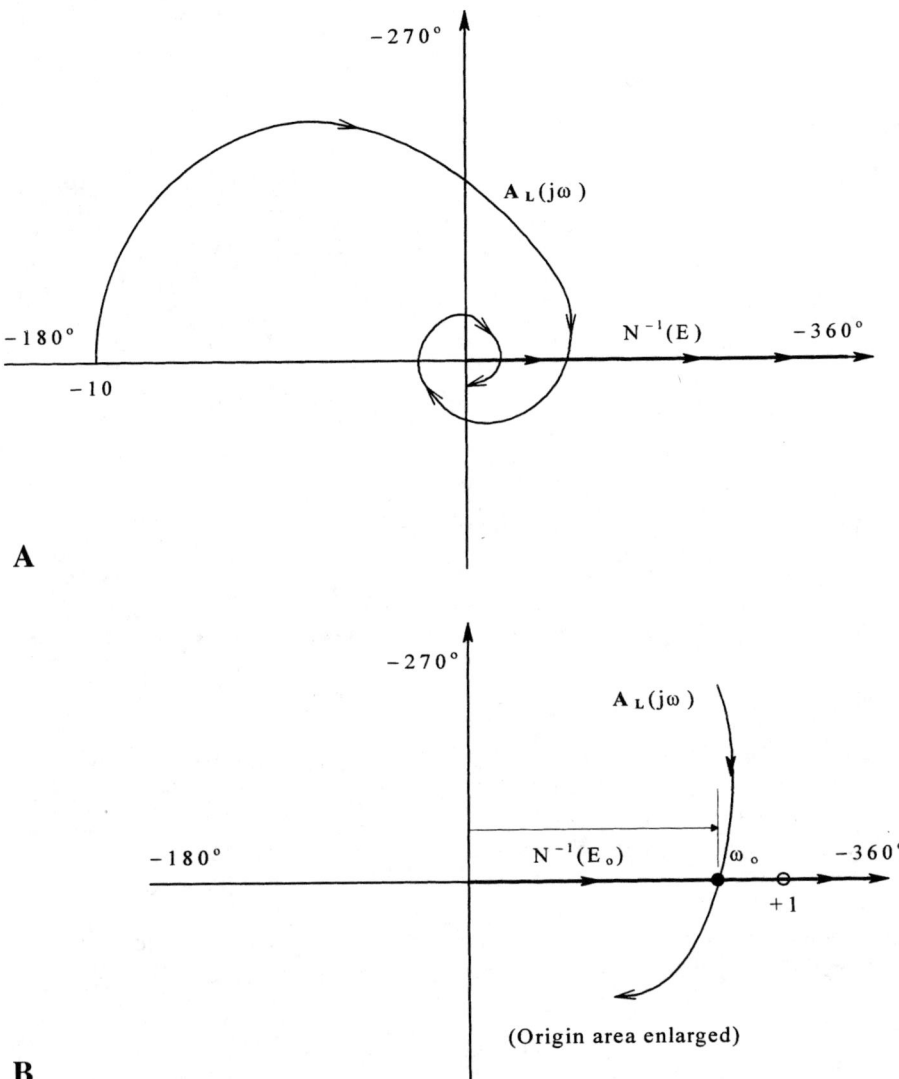

FIGURE 5.8 (A) Polar plots of the $A_L(j\omega)$ and $N^{-1}(E)$ loci. Note that because of the dead time, the $A_L(j\omega)$ locus spirals in around the origin. (B) Detail of the first crossing of $A_L(j\omega)$ and $N^{-1}(E)$ at $\omega = \omega_o$.

Fadali[40] reviewed the use of NLD controllers in continuous drug delivery systems. He gave examples of nonlinear CPK systems that are characterized by two inputs, two outputs, and cross-coupling between the outputs, so that input 1 affects output 2, etc. NLD is a model reference control strategy that requires good estimates of the plant parameters in order to work correctly, not unlike the Smith delay

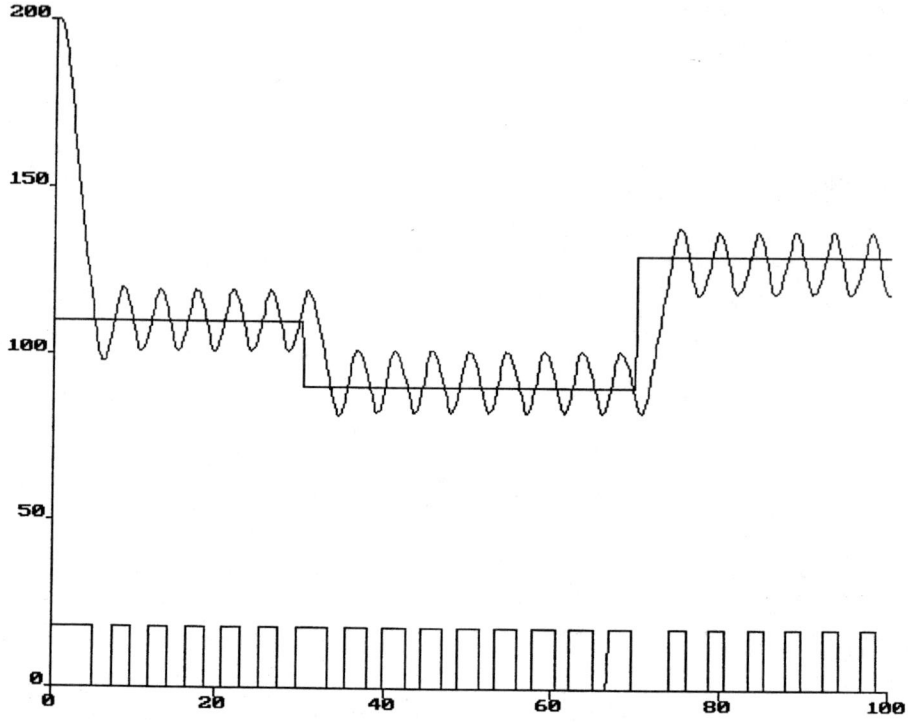

FIGURE 5.9 Simnon simulation of the on/off MAP controller. Note that at the design set point = 110 mmHg, the $u(t)$ pulses have a 50% duty cycle. When the set point is decreased to 90 mmHg, more drug/time is required and the duty cycle of $u(t) > 50\%$. When the set point is made 130 mmHg, the duty cycle is <50%. Note the nearly sinusoidal ripple in MAP(t).

compensator controllers discussed in Section 8.2. We begin by describing a general model of a linear cross-coupled, two-state CPK system:

$$\dot{x}_1 = a_{11} x_1 + a_{12} x_2 + b_1 u_1 \qquad (5.14A)$$

$$\dot{x}_2 = a_{21} x_1 + a_{22} x_2 + b_2 u_2 \qquad (5.14B)$$

Note that the a_{jk} terms can have either sign, the u_k's are inputs ≥ 0, and the outputs are the states **x**. This linear two-compartment pharmacokinetic (2CPK) system is shown as a signal flow graph in Figure 5.10. It is clear that input u_1 affects state x_2 and vice versa.

We wish to design two model reference controllers such that x_2 does not change, given an input to compartment 1, and x_1 remains constant when the input to compartment 2 is changed. Ideally, the decoupling controllers should force the 2CPK plant to behave in an independent first-order manner described by the ordinary differential equations

Special Types of Closed-Loop Drug Input Controllers

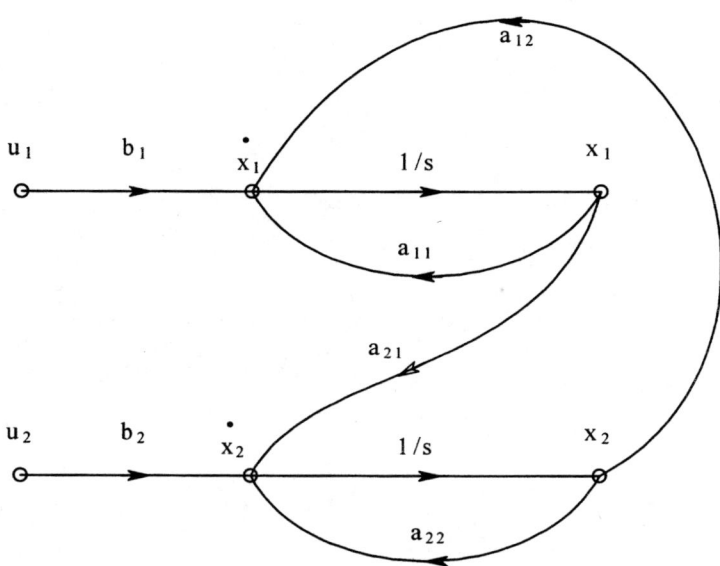

FIGURE 5.10 Signal flow graph describing a cross-coupled 2CPK system.

$$\dot{y}_1 = -P_1 y_1 + c_1 v_1 \qquad (5.15A)$$

$$\dot{y}_2 = -P_2 y_2 + c_2 v_2 \qquad (5.15B)$$

In this notation, $y_1 = x_1$, $y_2 = x_2$, and the independent inputs are now v_1 and v_2. The NLD controllers have the form:

$$\begin{aligned} d_1 &= \left[(\dot{y}_1) - \dot{x}_1\right]/b_1 = \left[(-P_1 x_1 + c_1 v_1) - (a_{11} x_1 + a_{12} x_2)\right]/b_1 \\ &= \left[-(P_1 + \mathbf{a_{11}})x_1 - \mathbf{a_{12}} x_2 + c_1 v_1\right]/\mathbf{b_1} = u_1 \end{aligned} \qquad (5.16A)$$

$$\begin{aligned} d_2 &= \left[(\dot{y}_2) - \dot{x}_2\right]/b_2 \\ &= \left[-(P_2 + \mathbf{a_{22}})x_2 - \mathbf{a_{21}} x_1 + c_2 v_2\right]/\mathbf{b_2} = u_2 \end{aligned} \qquad (5.16B)$$

The boldface parameters are the estimates of the 2CPK plant's parameters used in the NLD controllers. The decoupling controller outputs, **d**, drive the infusion pumps to give the required **u** for decoupling control. As an example, consider the first-state ordinary differential equation with decoupling control:

$$\dot{x}_1 = a_{11} x_1 + a_{12} x_2 + b_1\left[-(P_1 + \mathbf{a_{11}})x_1 - \mathbf{a_{12}} x_2 + c_1 v_1\right]/\mathbf{b_1} = -P_1 x_1 + c_1 v_1 \quad (5.17)$$

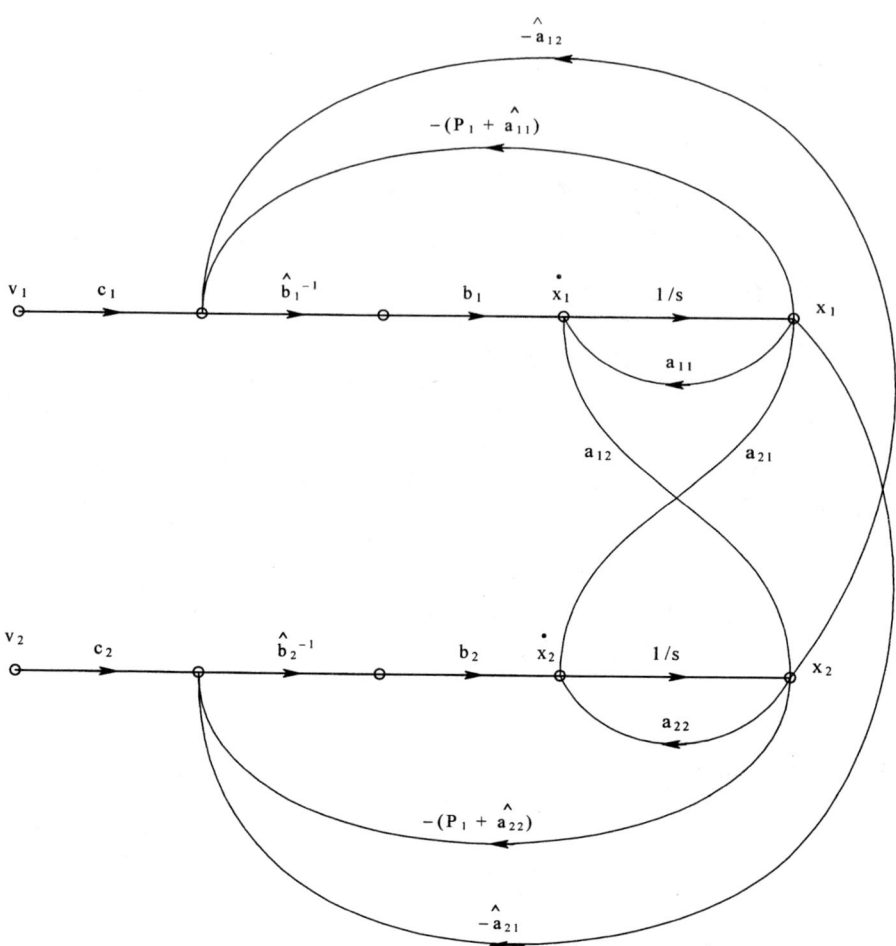

FIGURE 5.11 Signal flow graph showing the structure of the (ideal) decoupling controller applied to the 2CPK system.

Note that $y_1 = x_1$ in this development, and the decoupled x_1 is independent of x_2 and v_2. The complete decoupled-system signal flow graph is shown in Figure 5.11.

We now examine a *nonlinear* 2CPK system suggested by Godfrey[55] in Section 9.3 and give it NLD control:

$$\dot{x}_1 = -K_{o1} x_1^2 + K_{12} x_2 - \frac{V_m x_1}{K_m + x_1} + b_1 u_1 \quad (5.18A)$$

$$\dot{x}_2 = -K_{12} x_2 - K_{o2} + \frac{V_m x_1}{K_m + x_1} + b_2 u_2 \quad (5.18B)$$

We define $\mathbf{y} = [y_1 \ y_2]^T = [x_1 \ x_2]^T$. The NLD controllers are easily seen to be

$$u_1 = \left[-P_1 x_1 + a_1 v_1 + \mathbf{K_{o1}} x_1^2 - \mathbf{K_{12}} x_2 + \frac{\mathbf{V_m} x_1}{\mathbf{K_m} + x_1} \right] \Big/ \mathbf{b_1} \qquad (5.19\text{A})$$

$$u_2 = \left[-P_2 x_2 + a_2 v_2 + \mathbf{K_{12}} x_2 + \mathbf{K_{o2}} - \frac{\mathbf{V_m} x_1}{\mathbf{K_m} + x_1} \right] \Big/ \mathbf{b_2} \qquad (5.19\text{B})$$

Again, the boldface parameters are plant parameter *estimates*. When driven by these (tuned) controllers, the nonlinear compartmental equations become linear. P_1 and P_2 are the closed-loop system's natural frequencies:

$$\dot{y}_1 = -P_1 y_1 + c_1 v_1 \qquad (5.20\text{A})$$

$$\dot{y}_2 = -P_2 y_2 + c_2 v_2 \qquad (5.20\text{B})$$

We now simulate the NLD control of Equations 5.19A and B applied to the nonlinear 2CPK system of Equations 5.18A and B. A Simnon program modified for case 3 is shown below:

```
CONTINUOUS SYSTEM Godfrey9    "Rectified controls 5/27/97
STATE x1 x2
DER dx1 dx2
TIME t
"
" PLANT
dx1 = -k01*x1*x1 + k12*x2 - Vm*x1/(Km + x1) + b1*ru1
dx2 = -k12*x2 - k02 + Vm*x1/(Km + x1) + b2*ru2
"
" DECOUPLING CONTROLLERS
u1 = (a1*v1 + k01*x1*x1 - k12*x2 - P1*x1 + Vm*x1/(Km + x1))/b1
u2 = (a2*v2 + k12*x2 + k02 - P2*x2 - Vm*x1/(Km + x1))/b2
"
K02 = if x2 > 0 then k2 else 0  " prevents loss of x2 if x2 <= 0.
ru1 = if u1 < 0 then 0 else u1  " rectifies drug input rates.
ru2 = if u2 < 0 then 0 else u2
v1 = 1 + V12
V12 = if t > t2 then -1 else 0
v2 = V21 + V22
V21 = if t > t3 then 1 else 0
V22 = if t > t4 then -1 else 0
"
k01:1
k2:.1
k12:1
Vm:1
Km:0.1
```

```
b1:1
b2:1
a1:1
P1:1
a2:.2
P2:.2
t2:25
t3:50
t4:75
zero:0
"
END
```

Three cases are considered: (1) the inputs are applied directly to the system (no NLD control); (2) NLD control given by Equations 5.19A and B is applied to the system, where the controller outputs are allowed to go negative (a nonphysically and realizable condition); and (3) the controller outputs are prevented from going negative (there can be no negative input rates of drugs in reality). The input, v_1, is applied first to compartment 1, then v_2 is applied to compartment 2. Figure 5.12 illustrates the system's responses to case 1. Trace 1 is v_1, trace 2 is x_2, trace 3 is v_2, and trace 4 is x_2. Because the system is well cross-coupled, either input has a profound effect

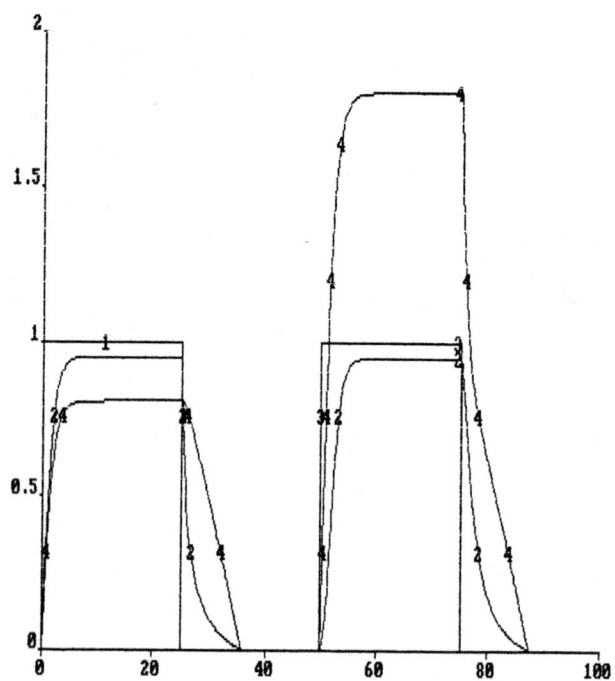

FIGURE 5.12 Simnon simulation of the 2CPK system's response to inputs first to compartment 1, then to compartment 2. No decoupling is used. See text for details.

Special Types of Closed-Loop Drug Input Controllers

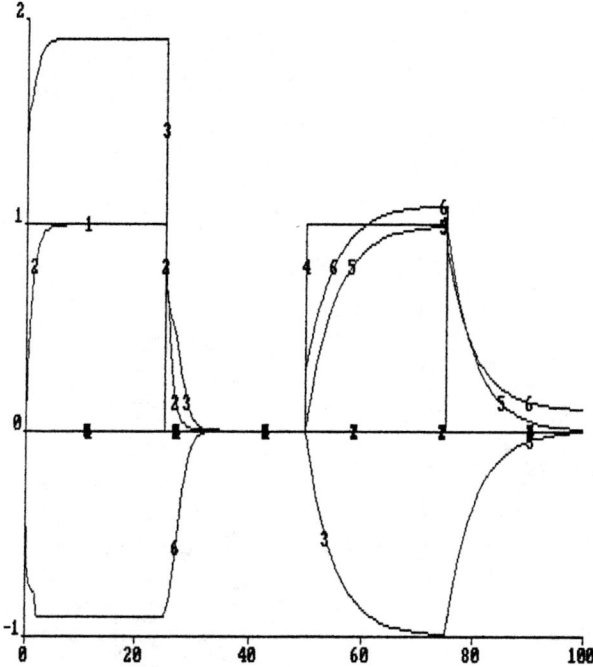

FIGURE 5.13 Results of *ideal* NLD controller action on the 2CPK system. See text for description.

on both states. That is, drug v_1 shows up in both compartments, as does v_2. The effect of NLD control is shown in Figure 5.13, case 2. Trace assignments are $1 = v_1$, $2 = x_1$, $3 = u_1$, $4 = v_2$, $5 = x_2$, $6 = u_2$, and $7 = 0$. Note that the NLD control causes $x_1 \to v_1$ and $x_2 \to v_2$. x_2 remains zero during the v_1 infusion, and x_1 remains zero during the v_2 infusion. However, the decoupling is accomplished by u_2 going negative during the v_1 infusion and u_1 going negative during the v_2 infusion. Note that in a CPK system, drug input rates are nonnegative, so these results represent a physical impossibility. In case 3, we still compute the NLD controls, u_1 and u_2, but prevent them from going negative. In Figure 5.14, all variables are nonnegative. The traces are $1 = v_1$, $2 = x_1$, $3 = ru_1$, $4 = v_2$, $5 = x_2$, and $6 = ru_2$. In this case, x_1 follows v_1 and x_2 follows v_2, as in the nonrectified case 2. However, there is a nonzero x_2 as a result of v_1 and, similarly, a nonzero x_1 as a result of the v_2 infusion. These unwanted responses are smaller than the desired outputs but still represent poor performance of the NLD control strategy under realistic conditions.

In conclusion, it appears that the use of NLD controllers in CPK applications is severely limited because of the restriction on nonnegative control inputs. The NLD control strategy is a model reference control strategy, and thus its success depends on how well the plant's parameters can be measured or estimated. In CPK systems, we often see system parameters change in time, either from diurnal rhythms or the effect of the drug on organs such as the liver and kidneys. However, NLD control

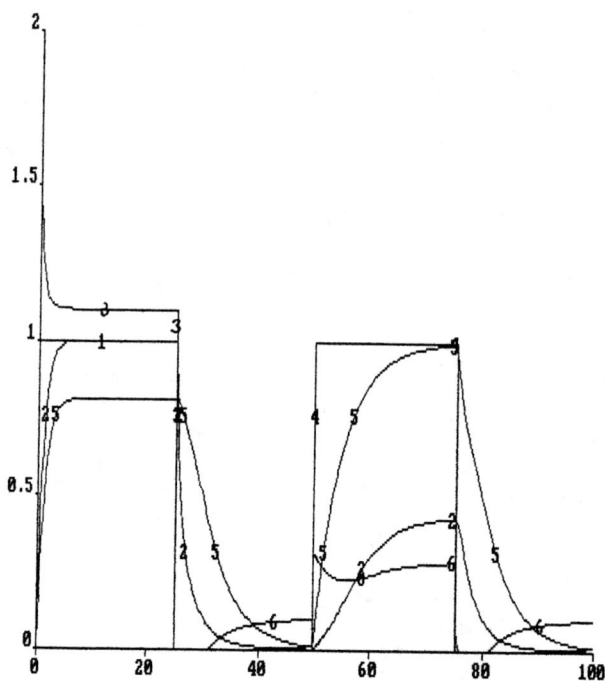

FIGURE 5.14 Results of NLD controller action when drug input rates and states are rectified (prevented from going negative). Note that the NLD control strategy works poorly.

has been applied successfully in other control applications, such as satellite maneuvering, robotics, and flight control, in references cited by Fadali.[40]

5.3 INTEGRAL PULSE-FREQUENCY MODULATION (BOLUS) CONTROLLERS

Integral pulse-frequency modulation (IPFM) bolus controllers achieve CPK/P system control by injecting boluses of the administered drug at appropriate times determined by the controller. Each bolus is a constant amount, D_o. A bolus injection can be considered to be an impulse of drug input rate of area D_o. Thus the CPK/P plant output is the superposition of plant impulse responses.

Properties of IPFM and relaxation pulse-frequency modulation controllers were first described by Meyer,[99] Li,[79] and Pavlidis[112] for conventional nonCPK/P applications such as satellite maneuvering and chemical process control. We will only consider IPFM systems in this section. IPFM controllers can be classified as having two-sided or one-sided outputs. In two-sided IPFM controllers, the output is either positive or negative impulses; in *one-sided IPFM,* which is our concern, the output

Special Types of Closed-Loop Drug Input Controllers

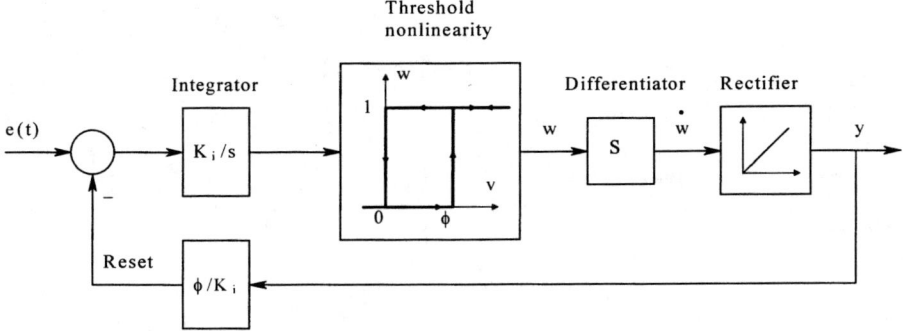

FIGURE 5.15 System for generating a pulsed control using IPFM.

is positive impulses only. The output of a one-sided IPFM drug injection controller can be described mathematically by a sequence of positive impulses:

$$u(t) = \sum_{k=1}^{k=\infty} D_o \delta(t - t_k) \tag{5.21}$$

$\delta(t - t_k)$ is a unit impulse (delta function) occurring at time $t = t_k$. D_o is the area of each output impulse (dimensions, mg). One-sided IPFM is well suited for CPK/P control systems because $u(t)$ can be considered to be a train of bolus injections of a drug. The timing (and rate) of the bolus injections determines the plant output. A bolus injector is physically easy to implement,[96] and cumulative drug dose is obtained by simply counting the number of boluses delivered.

A system for modeling the generation of IPFM is shown in Figure 5.15. (There are several ways IPFM generation can be described mathematically, but this is one of the more intuitive and simple means.) $e(t)$ is the signal driving the integral pulse-frequency modulator. At $t = 0$, $v(0) = 0$, and $e(t) > 0$ is applied to the system and the integrator begins to integrate $e(t)$. The integrator output v grows from zero until it reaches the threshold, φ, at $t = t_1$. The output of the unity-gain hysteresis nonlinearity, w, jumps to 1. This jump is differentiated, forming $\delta(t - t_1)$. The unit impulse at t_1 is passed through a half-wave rectifier, then fed back through a gain block, φ/K_i, where it is given an area φ/K_i. This scaled impulse is subtracted from $e(t_1)$ and by superposition is integrated by the integrator to form a step of height $-\varphi$. Since the integrator output is $v(t_1) = \varphi$, the negative step resets $v(t_1+) \to 0$. The output of the nonlinearity steps back to zero. This generates a negative impulse, $-\delta(t - t_1+)$, which does not pass through the rectifier. Meanwhile, the integrator continues to integrate $e(t)$ starting at t_1+. Its output again rises from $v = 0$ to reach φ at $t = t_2$. The process repeats itself ad infinitum or until $e(t) \to 0$ so that v cannot reach φ. The integration and resetting of the integrator in the IPFM generator can be described mathematically:

$$\varphi = K_i \int_{t_{k-1}}^{t_k} e(t)\, dt, \quad k = 1, 2, 3, \ldots, t_0 = 0 \tag{5.22}$$

The piecewise linear equation above can be rewritten to express the *instantaneous frequency* elements of the output pulse train, r_k:

$$\frac{1}{t_k - t_{k-1}} \int_{t_{k-1}}^{t_k} e(t)\, dt = \frac{(\varphi/K_i)}{t_k - t_{k-1}} = r_k(\varphi/K_i), \quad k = 2, 3, 4, \ldots \tag{5.23}$$

where r_k is defined as the kth element of instantaneous frequency. r_k is simply the reciprocal of the interpulse interval between the kth and $k-1$th output impulses. If $e(t) = E$ (constant), then it is easy to see that $r_k = (K_i/\varphi)\, E$. That is, the frequency of the pulses generated by the integral pulse-frequency modulator is proportional to the dc input. IPFM can be seen as an ideal voltage-to-frequency conversion process or, alternately, frequency modulation generation with zero carrier frequency.

One-sided IPFM works fine if the modulator input is > 0. If $e(t) < 0$, the integrator output can go negative, causing excessive delays when a positive input is integrated to bring $v \to \varphi$ again. To avoid this integrator "wind-up," we often rectify $e(t)$ before applying it to the IPFM generator.

A simple closed-loop, type 0, CPK drug injection control system with an IPFM controller is shown in Figure 5.16. If the system is in the steady state, there will be a constant input pulse frequency from the controller in order to maintain the desired average output level. That is, there will be a net positive average error input to the IPFM controller given by:

$$\overline{e_{SS}} = SP - \overline{y_{SS}} \tag{5.24}$$

Now the average (dc) plant output is the plant's dc gain times the dc average of the constant frequency input pulse train. That is,

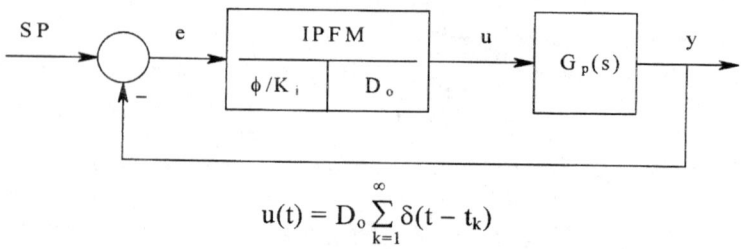

FIGURE 5.16 Block diagram of an IPFM controller used for CPK control by drug infusion.

Special Types of Closed-Loop Drug Input Controllers

$$\overline{y_{ss}} = \frac{D_o}{T_o} G_p(0) \tag{5.25}$$

If we assume the plant has real poles with no repeated roots, then we can write $G_p(s)$ in partial fraction form:

$$G_p(s) = \sum_{i=1}^{i=N} \frac{A_i}{(s + \omega_i)} \tag{5.26}$$

where A_i is the ith residue of the rational polynomial, $G_p(s)$. Thus the dc response of the plant is

$$G_p(0) = \sum_{i=1}^{i=N} \frac{A_i}{\omega_I} \tag{5.27}$$

Also, because the pulses are periodic with period T_o, we can write:

$$\frac{1}{T_o} \int_{t_k}^{t_k + T_o} e(t)\, dt = \frac{\varphi}{K_i T_o}, \text{ or } \overline{e_{ss}} = \frac{\varphi}{K_i T_o} \tag{5.28}$$

Thus from Equation 5.24 we can write:

$$\frac{\varphi}{K_i T_o} = \text{SP} - \frac{D_o}{T_o} \tag{5.29}$$

Equation 5.29 can be solved for T_o, the steady-state period of the controller:

$$T_o = \frac{\varphi/K_i + D_o G_p(0)}{\text{SP}} \tag{5.30}$$

The same relations can be used to obtain an expression for the steady-state average error of the system:

$$\overline{e_{ss}} = \frac{\text{SP}}{1 + K_i D_o G_p(0)/\varphi} \tag{5.31}$$

The periodic output of the IPFM controller in equilibrium will produce a steady-state ripple in the plant output. The peak-to-peak amplitude of the ripple can be found easily if the plant is of order $N - M = 1$. Such plants always respond to an impulse input with a discontinuous or jump output. The size of this jump, ρ, is easily found by the Laplace initial value theorem. We assume the input is $D_o\,\delta(t)$. Thus:

$$\rho = \lim_{s \to \infty} s D_o \sum_{i=1}^{N} \frac{A_i}{(s + \omega_i)} = D_o \sum_{i=1}^{N} A_i \qquad (5.32)$$

If the plant order is $N - M \geq 2$, then the plant output will have a smooth, peaked response for every input impulse. The peaks occur following the impulse input times, t_k. In the steady state, the response starts and ends at the same value. That is, $y(t_k+) = y(t_{k+1}-)$. To find the exact nature of the steady-state $y(t)$ between pulses, we can make use of the *modified z-transform*.[80] The modified z-transform (z,m-transform) allows us to find the exact analog value of the z,m-transformed $f(t)$ at times between the sampling instants. The inverse of a conventional z-transform of $f(t)$ gives us $f(t - nT_o)$. The inverse of the z,m-transform gives us $f(t - [nT_o + mT_o])$, where $0 \leq m \leq 1$. $m = 0$ gives the conventional z-transform; $m = 1$ gives the value

---

TABLE 5.1
Some Common Laplace and z,m-Transforms

$f(t)$	$F(s)$	$F(z, m)$
$U(t)$	$1/s$	$\dfrac{z}{z-1}$
t	$1/s^2$	$\dfrac{mTz}{z-1} + \dfrac{Tz}{(z-1)^2}$
e^{-at}	$1/(s+a)$	$\dfrac{ze^{-mTa}}{z - e^{-aT}}$
$(1 - e^{-at})/a$	$1/s(s+a)$	$(z/a)\left[\dfrac{1}{z-1} - \dfrac{e^{-maT}}{z - e^{-aT}}\right]$
$(e^{-at} - e^{-bt})/(b-a),\ b > a$	$1/(s+a)(s+b)$	$\dfrac{z}{b-a}\left\{\dfrac{e^{amT}}{z - e^{-aT}} - \dfrac{e^{-bmT}}{z - e^{-bT}}\right\}$
$t\,e^{-at}$	$1/(s+a)^2$	$\dfrac{Te^{-amT}\left[e^{-aT} + m\left(z - e^{-aT}\right)\right]}{\left(z - e^{-aT}\right)^2}$
$\sum K_i e^{-\lambda_i t}$ (PF form)	$\sum \dfrac{K_i}{s + \lambda_i}$	$\sum \dfrac{K_i z e^{-m\lambda_i T}}{\left(z - e^{-\lambda_i T}\right)}$

Special Types of Closed-Loop Drug Input Controllers

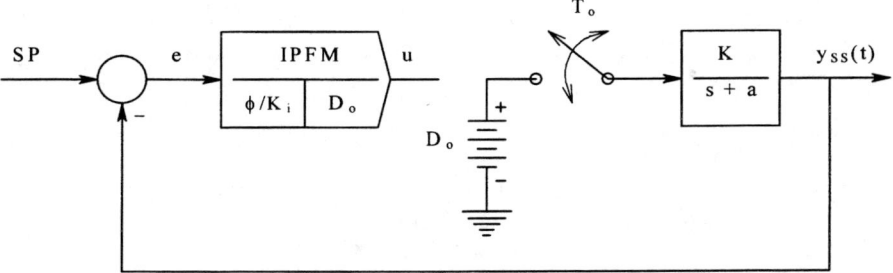

FIGURE 5.17 Block diagram illustrating steady-state operation of an IPFM controller.

of $f(t)$ just before the next sampling instant. Table 5.1 gives a list of time functions, their Laplace transforms, and their z,m-transforms.

The system used to find the between-samples output of the plant in steady-state equilibrium is shown in Figure 5.17. We assume that the closed-loop system is not disturbed in its steady-state equilibrium if the loop is opened at the output of the IPFM controller, and the IPFM controller is replaced by a periodic sampler that provides the plant with pulses of area D_o every T_o seconds, as shown in the figure. Under these conditions, the z,m-transformed plant output can be written:

$$Y(z,m) = \left\{ \frac{D_o z}{z-1} \right\} \left\{ \frac{zKe^{-amT_o}}{z-e^{-aT_o}} \right\} \tag{5.33}$$

The first term in $Y(z, m)$ is the z,m-transform of a sampled step of height D_o; the second term is the z,m-transform of the plant, $K/(s + a)$. Since the system is in the steady state, we can find $y(t)$ between sampling instants (between input impulses) using the *final value theorem* in z:

$$y(mT_o) = \lim_{z \to 1} (z-1) \left\{ \frac{D_o z}{z-1} \right\} \left\{ \frac{zKe^{-amT_o}}{z-e^{-aT_o}} \right\}, \quad 0 \le m \le 1 \tag{5.34}$$

$$y(mT_o) = \frac{D_o K e^{-amT_o}}{1-e^{-aT_o}} \tag{5.35}$$

$$y(0) = \frac{D_o K}{1-e^{-aT_o}}, \quad m-0, \text{ peak } y \tag{5.36}$$

$$y(1T_o) = \frac{D_o K e^{-aT_o}}{1-e^{-aT_o}}, \quad m = 1, \text{ min } y \tag{5.37}$$

To obtain a better feeling for how the equations above can be applied in the design of IPFM controllers, we will consider the *first numerical example* using the first-order CPK plant, $3/(s + 1)$. We set SP = 8 units, $\varphi/K_i = 0.1$, and $D_o = 0.5$.

1. The steady-state equilibrium period is

$$T_o = \frac{\varphi/K_i + D_o G_p(0)}{SP} = \frac{0.1 + 0.5(3/1)}{8} = 1.6/8 = 0.2 \text{ hr} \tag{5.38}$$

2. The average steady-state error is

$$\overline{e_{SS}} = \frac{SP}{1 + (K_i/\varphi) D_o G_p(0)} = \frac{8}{1 + 10(0.5)(3/1)} = 8/16 = 0.5 \tag{5.39}$$

3. The average steady-state output is

$$\overline{y_{SS}} = SP - \overline{e_{SS}} = 8 - 0.5 = 7.5 \tag{5.40}$$

4. The peak output is, by the final value theorem:

$$y_{max} = \lim_{\substack{z \to 1 \\ m=0}} (z-1) \left\{ \frac{D_o z}{z-1} \frac{zKe^{-amT_o}}{z - e^{-aT_o}} \right\} = \frac{0.5 \times 3}{1 - e^{-0.2}} = 8.275 \tag{5.41}$$

5. The minimum output is just before the next input impulse:

$$y_{min} = \lim_{\substack{z \to 1 \\ m=1}} (z-1) \left\{ \frac{D_o z}{z-1} \frac{zKe^{-amT_o}}{z - e^{-aT_o}} \right\} = \frac{0.5(3)e^{-0.2}}{1 - e^{-0.2}} = 6.775 \tag{5.42}$$

6. The peak-to-peak ripple on $y(t)$ in the steady state is

$$\rho = y_{max} - y_{min} = D_o \Sigma A_i = 0.5 \times 3 = 1.5 \tag{5.43}$$

This is a $(1.5/8) \times 100 = 18.75\%$ ripple.

As a *second example* of an IPFM CPK system design, let us consider a second-order plant. We wish to regulate the drug concentration, x_2, in the second compartment. $G_p(s)$ is

$$\frac{X_2}{U}(s) = \frac{400}{(s+10)^2} \tag{5.44}$$

Special Types of Closed-Loop Drug Input Controllers

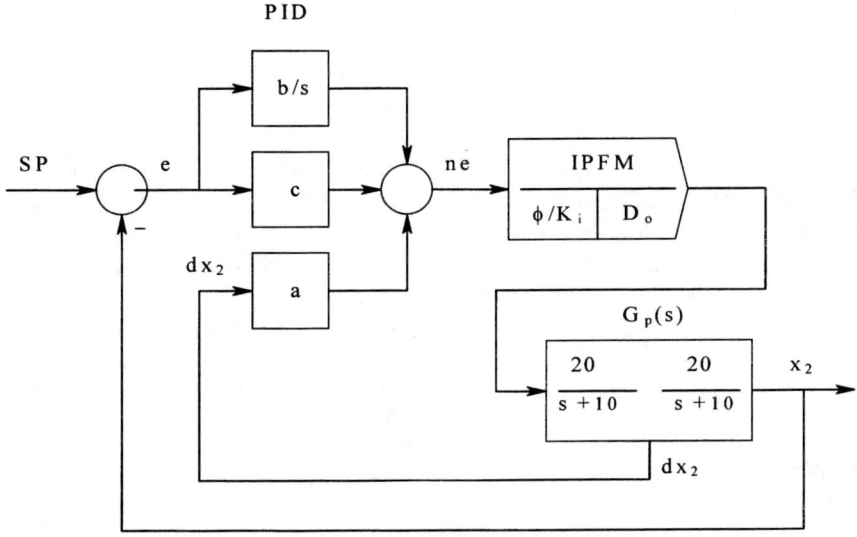

FIGURE 5.18 Block diagram of an IPFM drug injector driven from the output of a PID controller. The output of a two-real-pole CPK plant is controlled.

Furthermore, we want zero average steady-state error. This means that the loop gain of the system must be made type 1 (i.e., it will have one pole at the origin). Simply adding a pole makes the loop gain cubic ($N = 3$) and prone to instability. In this case, we will condition the control system error with a proportional-plus-integral-plus-derivative (PID) unit, the output of which will be the input to the IPFM bolus drug injector. In algebraic terms:

$$E = SP - X_2 \tag{5.45}$$

$$NE(s) = E(s)\frac{b}{s} + cE(s) + asE(s) = a\frac{\left[s^2 + s(c/a) + (b/a)\right]}{s}E(s) \tag{5.46}$$

Because SP is a dc level, it is expedient to use $-dx_2/dt$ for the de/dt term in forming $ne(t)$.

Figure 5.18 illustrates the IPFM/PID control system. The subroutine below illustrates an effective means of simulating a one-sided IPFM generator when modeling the system using Simnon. The Simnon IPFM generator is shown in block diagram form in Figure 5.19.

```
" IPFM Subroutine:
dv = ne - z          " dv is dv/dt
w  = if v > phi then 1 else 0
s  = DELAY(w, tau)
```

```
x = w - s
y = if x > 0 then x else 0
u = y*Do/tau
z = y*phi/tau       " z resets integrator to 0 following an output pulse.
```

v is the integrator output. ne is the input to the IPFM impulse generator. The waveform at y approximates a true delta function. $y(t - t_k)$ can be shown to be a triangular waveform with height 1 and base 2 tau. Thus, instead of unit area, $y(t - t_k)$ has area $A = (1 \times 2\ tau)/2 = tau$. In running the Simnon simulation of the system, we must use fixed-interval Euler integration and make the simulation $\Delta T = $ tau of the IPFM generator. The complete Simnon program is given below:

```
CONTINUOUS SYSTEM ipfmPID    " 11/16/92 mod. 5/28/97
STATE v x1 x2 cr             " Use EULER integration with tau = .001.
DER dv dx1 dx2 dcr
TIME t
"
" QUADRATIC PLANT    " 400/(s + 10)^2
dx1 = 20*u - 10*x1
dx2 = 20*x1 - 10*x2
"
" PID CONTROLLER
trueE = SP - x2              " SP is setpoint, x2 is controlled var.
dcr = trueE                  " Error integrator
ne = c*trueE + b*cr - a*dx2  " PID controller; ne is input to IPFM.
u = Do*y/tau                 " Control input to plant (triangle
                               pulses) with area Do.
```

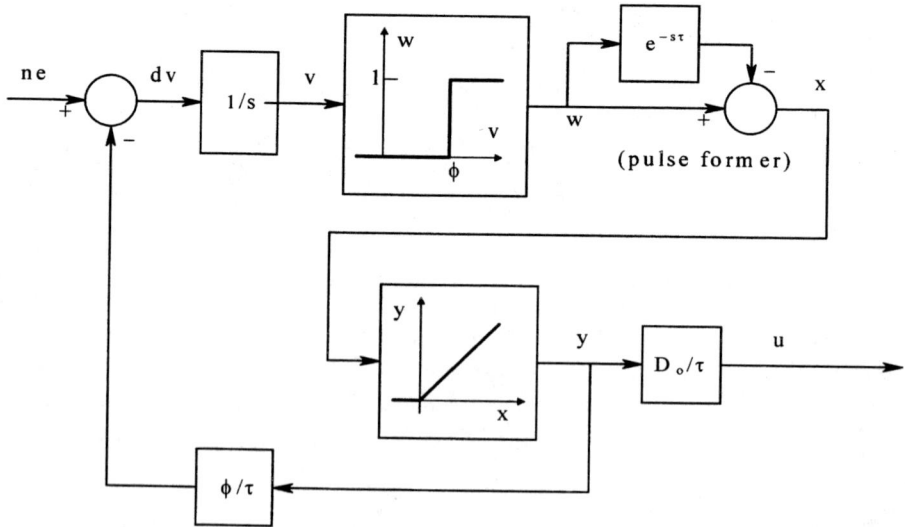

FIGURE 5.19 Block diagram of the mathematical steps used in simulating analog IPFM with Simnon. See text for details.

Special Types of Closed-Loop Drug Input Controllers

```
"
"   IPFM BOLUS DRUG INJECTOR
dv = ne - z                       "  IPFM integrator; z resets v to zero
                                     in tau sec.
w = if v > phi then 1 else 0
s = DELAY(w, tau)                 "  tau is output pulse width.
x = w - s                         "  Pulse former.
y = if x > 0 then x else 0
z = y*phi/tau                     "  z resets IPFM integrator.
"
"   SYSTEM PARAMETERS:
ten:10
zero:0
tau:0.001
phi:0.5
SP:10
Do:0.2              "  Drug bolus size.
a:0.05              "  deriv. gain.
b:4                 "  integral gain.
c:1                 "  proportional gain.
"
END
```

System parameters were chosen to give a quick response without overshoot. Figure 5.20 shows the turn-on transient for the IPFM/PID-controlled system. The spikes at the bottom of the plot are $y(t)$, trace 2 is $x_2(t)$, trace 4 is $ne(t)$, and the set point is trace 1. Note that there is a modest ripple due to the bolus injections from the IPFM controller. Because the closed-loop system is type 1, there is zero average steady-state error. If D_o is made smaller than 0.2, the ripple is smaller and the system is slower. In Figure 5.21, we have set the derivative and integral gains to zero, creating a second-order type 0 system; the proportional gain, c, was left at 1. Note that there is a huge average steady-state error and that the basic equilibrium frequency is lower, giving a large peak-to-peak ripple on x_2. In Figure 5.22, we set $a = b = 0$ and $c = 5$. There is still a large average steady-state error, and the higher gain causes an initial overshoot and erratic behavior of $ne(t)$.

Clearly, combining the PID controller architecture with the IPFM bolus injector allows us to obtain a fast step response with slight overshoot and to design for a tolerable ripple on x_2. It should be pointed out that if the plant has a destabilizing dead-time term, the Smith delay compensation architecture also can be incorporated into the controller with an additional proportional-plus-integral (PI) or proportional-plus-derivative (PD) processor to improve transient response.

Thus, in the *third example of IPFM* control, we use PI compensation along with a Smith delay compensator to regulate a model of postoperative blood pressure reduction with sodium nitroprusside (SNP).[126] Figure 5.23 illustrates the system. The blood-pressure-reduction CPK/P plant is modeled by a dc gain of $K_p = 10$, a real pole at $a_p = 1.33$ r/min, and a dead time of $T_p = 0.5$ min. The Smith delay compensator model reference is assumed to be perfectly tuned (i.e., its parameters are the

FIGURE 5.20 Simnon simulation of the turn-on transient of a PID/IPFM controller operating on a 2CPK plant. See text for details. Trace 1 = set point (SP), trace 2 = plant output (x_2), trace 4 = $ne(t)$ (input to IPFM from PID). The spikes are $y(t)$, the IPFM generator's output. There is zero steady-state error because the system is type 1 because of the PID.

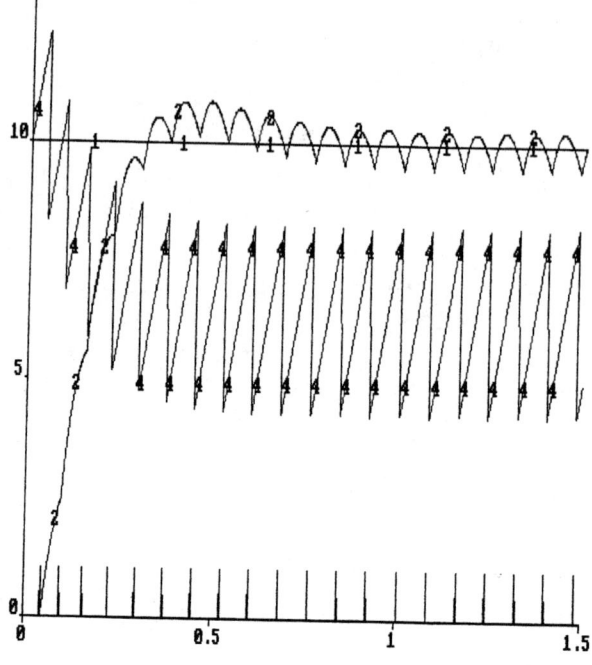

FIGURE 5.21 In this simulation, the "I" and "D" of the PID are set to zero, reducing the system to a simple type 0 form with large steady-state error. Trace 1 = set point, trace 2 = x_2, trace 4 = $ne(t)$ (IPFM input), spikes are IPFM output.

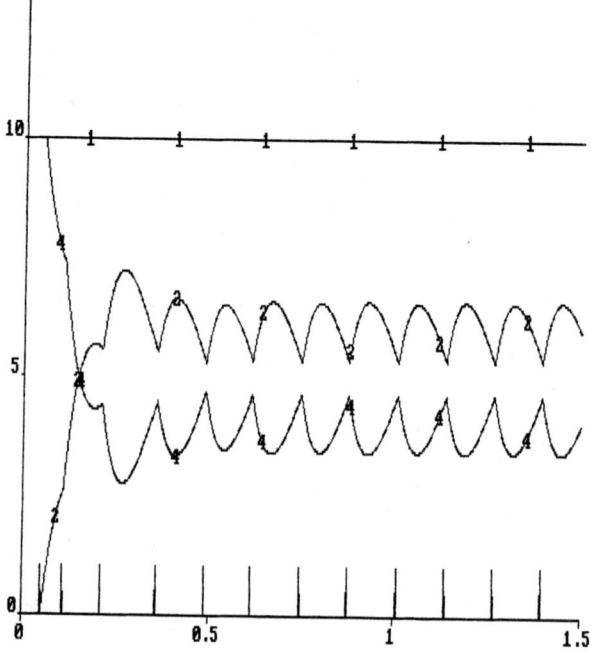

FIGURE 5.22 Now the proportional gain is raised to 5, destabilizing the system. Same trace identification as in Figure 5.21. See text for description.

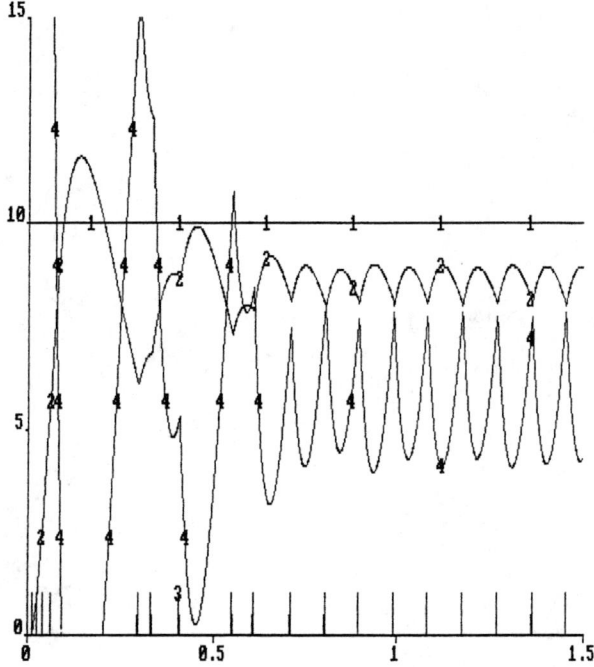

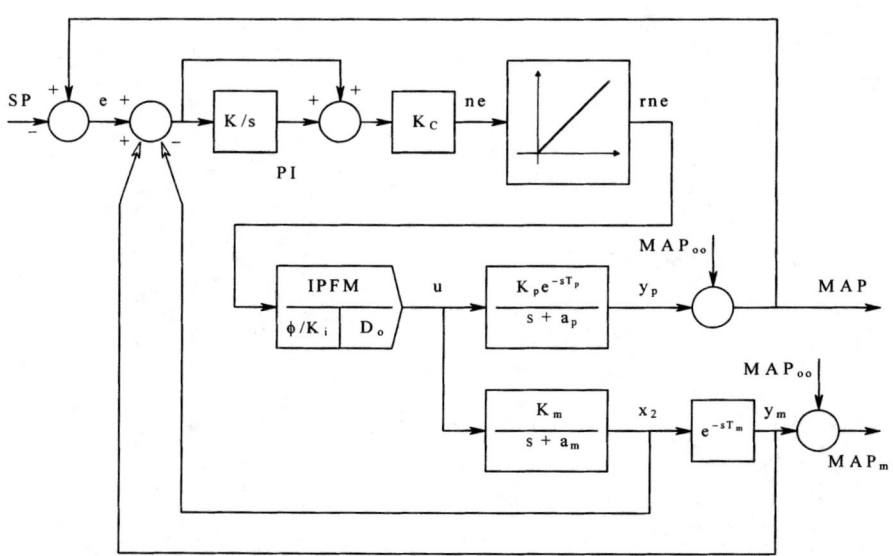

FIGURE 5.23 Block diagram of a model of MAP reduction. IPFM is used to inject SNP. PI control is used in conjunction with a Smith delay compensator to drive the IPFM input, $rne(t)$.

same as the plant's). The output of the "Smith summer," *se*, is the input to the PI block. The integrator gain is $k = 2.0$ r/min, and the overall gain $K_c = 4.0$. The output of the PI block, *ne*, is rectified before being input to the IPFM bolus injector. The Simnon program used for simulation is

```
CONTINUOUS SYSTEM ipfmMAP1      " 10/19/94 v. 5/29/97
STATE v cr x1 x2 "noise         " Use Algor EULER with delT = 0.001
DER dv dcr dx1 dx2 "dnoise      " e.g., SIMU 0 10 0.001
TIME t
"
" 1st ORDER PLANT
"
dx1 = u -ap*x1
yp = Kp*DELAY(x1, Tp)
"
" SMITH REFERENCE MODEL
"
dx2 =Km*u - am*x2
ym = DELAY(x2, Tm)
"
se = e - x2 + ym        " SMITH SUMMER. Fixed SP.
MAP = MAPoo - yp        " MAPo is untreated MAP. Fixed MAPo.
MAPm = MAPoo - ym       " MAP predicted by reference model
e = MAP - SP            " Controller error.
"
"MAPo = MAPoo + Kmap*t + noise  " Add BW-limited noise to MAP drift.
"dnoise = -wo*noise + SD*NORM(t) " BW limiting ODE for noise.
"
dcr = k*se              " PI CONTROLLER; k is integrator gain.
ne = (cr + se)*Kc       " ne is PI controller output to IPFM.
"
rne = if ne > 0 then ne else 0  " Rectify input to IPFM
"
dv = rne - z            " IPFM GENERATOR integrator.
w = if v > phi then 1 else 0    " integrator threshold
s = DELAY(w, tau)               " pulse former
x = w - s
y = if x > 0 then x else 0
z = y*phi/tau                   " IPFM reset
u = Do*y/tau                    " PLANT INPUT
"
SD:200
MAPoo:195
wo:0.5
Kmap:2
ap:1.33      "r/min.
Kp:10
Tp:0.5       "minutes
am:1.33
Km:10
Tm:0.5
k:1.5
```

```
Kc:2
SP:110      "mm Hg set point
phi:15
tau:0.001
Do:1.5      " Bolus size.
upper:130
lower:90
"
```
END

The system's turn-on transient is shown in Figure 5.24. Note that because the system is type 1, there is zero average steady-state error. A slight undershoot is seen. In Figure 5.25, we illustrate system robustness in response to noise and an upward drift in MAP_o. (The noise in MAP_o can be considered to be measurement noise in MAP. The upward drift in MAP_o is typical of long-term treatment with SNP; it occurs over hours, not minutes as shown in the simulation.) In order to make the system "stiff," we have raised K_c to 10.0 and set $k = 1.5$.

In summary, we see that IPFM bolus-injecting controllers create a ripple on the controlled variable. In many pharmacokinetic applications, a small ripple is quite

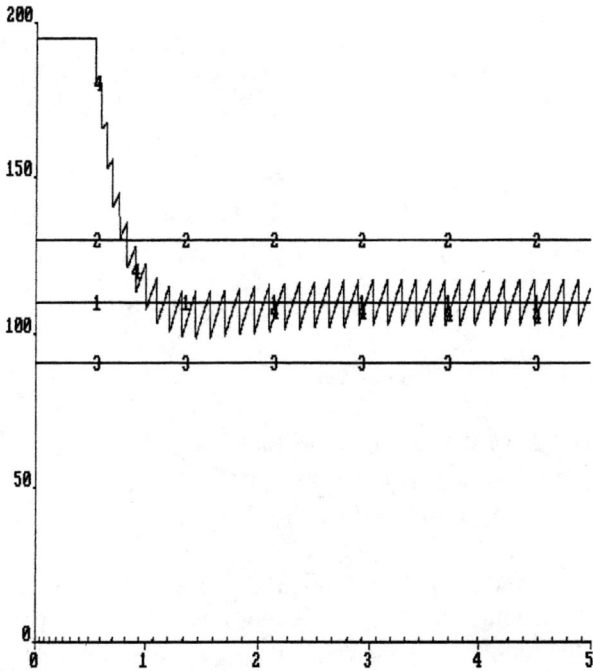

FIGURE 5.24 Simulation of the turn-on transient of the system shown in Figure 5.23. MAP has a sawtooth shape because the SNP/MAP plant has one pole. There is zero average steady-state error. See text for details. Trace #4 is MAP, #1 is SP.

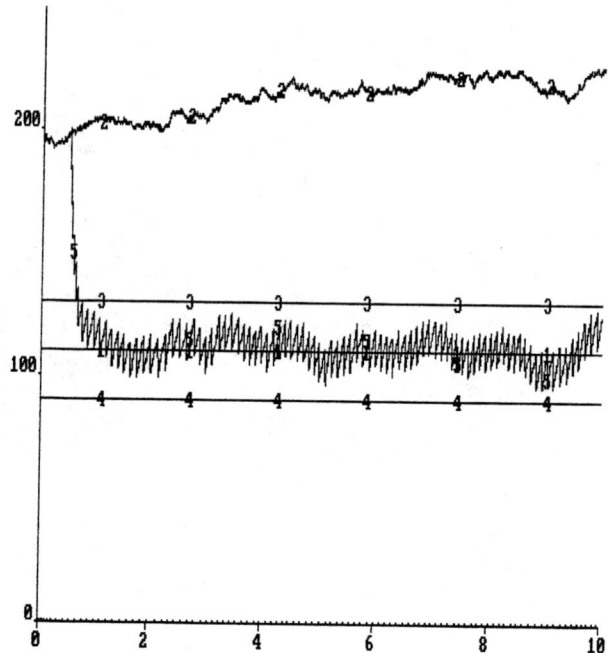

FIGURE 5.25 In this simulation of the system in Figure 5.23, noise is added to the measured MAP and the untreated MAP, MAP_o (trace #2), is allowed to drift upward. In spite of noise and drift, MAP (trace #5) remains within the "safe" bounds, 130 to 90 mmHg, given a set point = 110 mmHg.

tolerable and is a justifiable cost for the electromechanical simplicity of a bolus pump. The IPFM controller effectively inserts a gain, D_o/φ, into the loop gain; however, the IPFM integrator pole does not appear in the loop gain. IPFM bolus injectors can be used alone or with compensation such as PI, PID, PD, and the Smith delay compensator.

5.4 INTRODUCTION TO NONLINEAR PARAMETER-SWITCHING CONTROLLERS: ANALYSIS IN THE PHASE PLANE

Phase plane descriptions of the performance of regulators and control systems provide a useful, qualitative, and quantitative contrast to the conventional frequency-domain and time-domain means of analysis. Phase plane analysis generally assumes that the plant is second order. On Cartesian coordinates, we plot a parameter such as the system error on the x-axis and its derivative on the y-axis, each point being taken at some time, t_k. Thus a phase plane plot, or phase "portrait," parametrically eliminates time. Phase plane plots can be made for (1) initial

Special Types of Closed-Loop Drug Input Controllers

conditions on $(e, \dot{e})$, (2) an impulse input to the system, or (3) a step input to the system. Phase plane plots are useful for describing the behavior of nonlinear parameter-switching or discontinuous control systems and time-optimal controllers where a controller switching rule is devised that will take the system from one set of states to another in minimum time, under certain constraints. The path a system takes in going from one set of states to another in the phase plane is called a *trajectory*.

To become familiar with different types of trajectories, we first examine the phase plane behavior of the well-known linear second-order system described by the ordinary differential equations

$$\ddot{x} + (2\zeta\omega_n)\dot{x} + \omega_n^2 x = 0, \quad \text{(Plus ICs on } x, \dot{x}) \tag{5.47}$$

This second-order linear system can be put in state variable form:

$$\dot{x}_1 = -(2\zeta\omega_n)x_1 - \omega_n^2 x_2 \tag{5.48A}$$

$$\dot{x}_2 = x_1 \tag{5.48B}$$

Note that $x_2 = x$ and $x_1 = \dot{x}$, so we plot x_1 on the vertical axis and x_2 on the horizontal axis. Figure 5.26 illustrates the typical phase portraits for six different root conditions of the characteristic equation for the ordinary differential equation. That is, the roots of $[s^2 + (2\zeta\omega_n)s + \omega_n^2] = 0$ in the s-plane. The straight lines in the nodal phase portraits are called *directrices*; they form the boundaries between trajectories that approach or leave the origin directly and those in which $\dot{x}$ changes sign.

As a *first example* of the application of the phase plane, let us examine the phase plane behavior of a well-known nonlinear ordinary differential equation, van der Pol's equation:

$$\ddot{x} + a(x^2 - b)\dot{x} + x = 0 \tag{5.49}$$

This equation must be put in state variable form in order to model its behavior using Simnon:

$$\dot{x}_1 = -a(x_2^2 - b)x_1 - x_2 \tag{5.50A}$$

$$\dot{x}_2 = x_1 \tag{5.50B}$$

As in the first example, $x_2 = x$, $x_1 = \dot{x}$. The Simnon program follows:

```
continuous system VDPOL
state x1 x2
der dx1 dx2
```

```
time t
"
dx1 = - a*(x2*x2 - b)*x1 - x2
dx2 = x1
"
a:-1
b:1
x2:1
x1:0
"
end
```

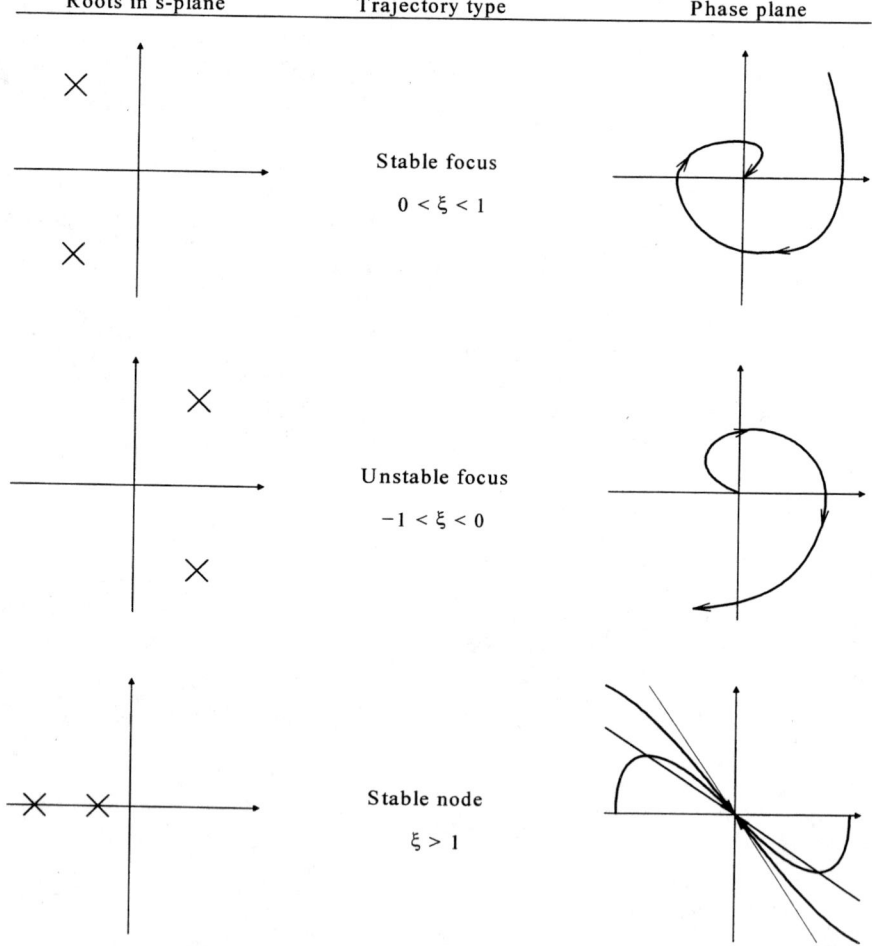

FIGURE 5.26A Typical phase plane trajectories for various quadratic system pole configurations.

Special Types of Closed-Loop Drug Input Controllers

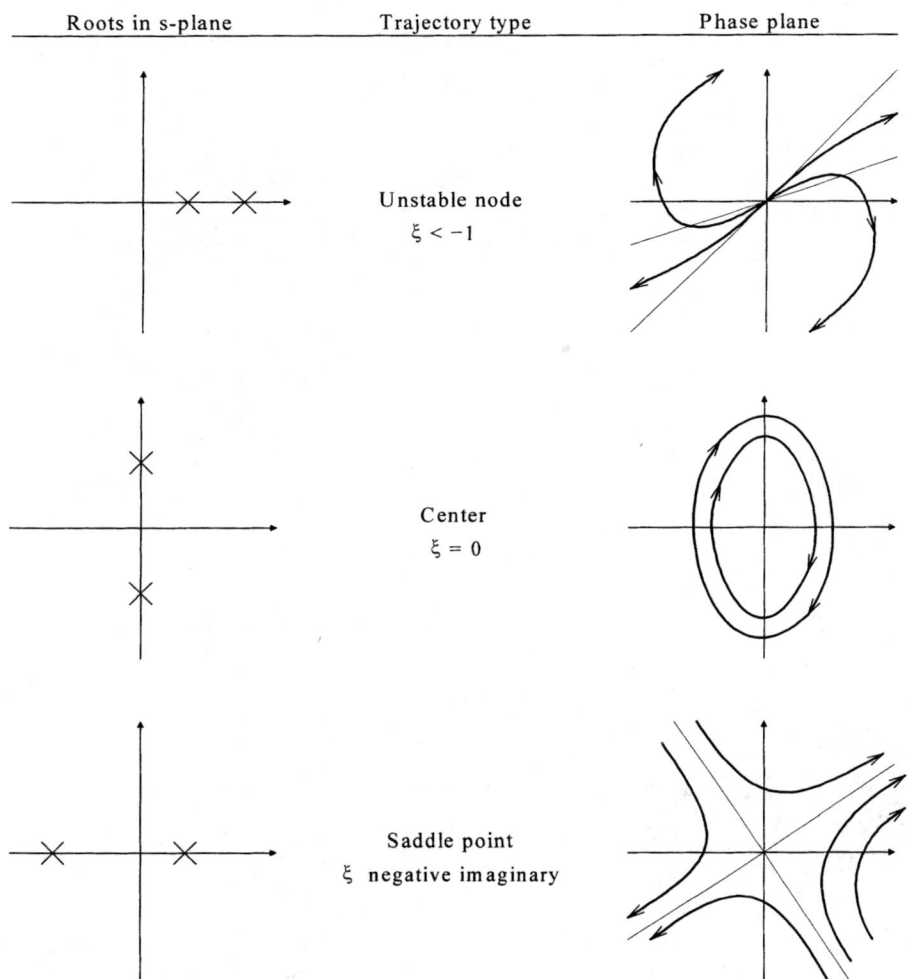

FIGURE 5.26B More phase plane trajectories.

Figure 5.27 illustrates the presence of a stable *limit cycle* (the system is a nonlinear oscillator). In a stable limit cycle, regardless of initial conditions, the trajectories converge on a stable, closed path in the x, $\dot{x}$ plane. Here $a = b = 1$. For several different initial conditions, the phase plane trajectories rapidly converge to the limit cycle trajectory. In Figure 5.28, we set $a = -1$, $b = 1$, and the system becomes conditionally stable. That is, if $x_1(0) = 0$ and $|x_2(0)| < 2$, the trajectories converge to 0,0. For $|x_2(0)| > 2$, the trajectories $\to \infty$.

In the *second example,* we examine the closed-loop type 0 control of a linear second-order CPK plant using PD feedback. Control is by a variable-rate infusion pump. First we examine the pharmacokinetics of the plant. The drug is infused into

FIGURE 5.27 Phase plane trajectories for a stable van der Pol oscillator. Regardless of initial conditions (ICs), all trajectories converge on the stable limit cycle.

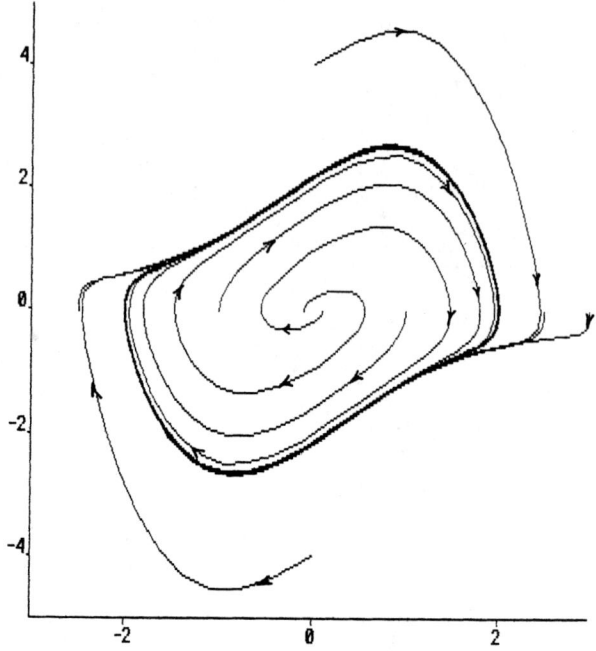

FIGURE 5.28 Phase plane trajectories for a conditionally stable van der Pol system. Trajectories starting inside the critical area converge on the origin; those originating outside the critical area go to infinity, indicating an effective pole in the right-half s-plane. This system does not oscillate.

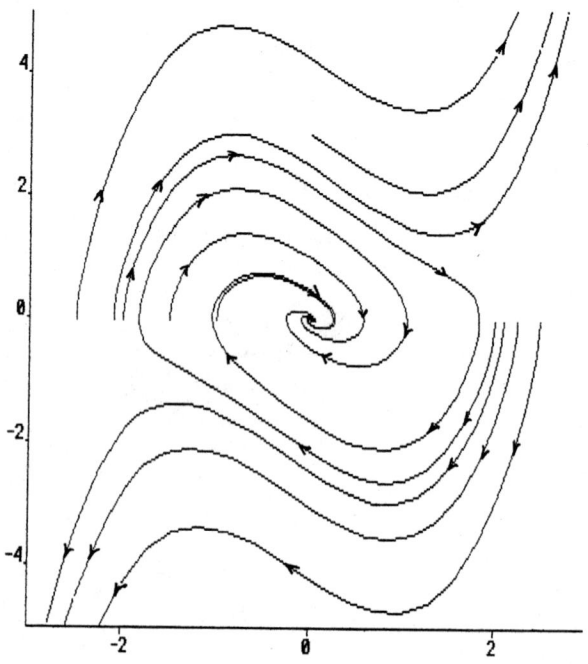

compartment 1; its concentration in compartment 2 is the controlled variable, y_2. Because the system is CPK, the states and infusion rate are nonnegative, thus introducing implied nonlinearity into the system. There is first-order loss of the drug from both compartments and diffusion between them. The system is described by the ordinary differential equations

$$\dot{y}_1 = u - y_1(K_{L1} + K_{21}) + x_2 K_{L2} \tag{5.51A}$$

$$\dot{y}_2 = -y_2(K_{L2} + K_{21}) + x_1 K_{21} \tag{5.51B}$$

These ordinary differential equations are Laplace transformed and written as a transfer function for $K_{L1} = K_{L2} = 1$, $K_{21} = 5$. This gives $a = 11$, $b = 1$ r/hr.

$$\frac{Y_2}{U}(s) = \frac{K_{21}}{(s + K_{L1} + K_{21})(s + K_{L2} + K_{21})} = \frac{5}{(s+11)(s+1)} \tag{5.52}$$

The closed-loop control system is shown in Figure 5.29. A compensatory gain, K_r, is used so that the closed-loop system will have $y_2 = r$ in the dc steady state. From consideration of the closed-loop gain,

$$F_{CL}(s) = \frac{Y_2}{R}(s) = \frac{K_r K_c K_p}{s^2 + s(a + b + K_c K_p) + (ab + c K_c K_p)} \tag{5.53}$$

we have $K_r = (c + ab/K_c K_p)$. We let $c = 20$ r/hr and $K_c = 1$ in the initial design. The zero at 20 r/hr allows the system to have a fast rise time without excessive overshoot as K_c is raised. The Simnon program used to simulate the closed-loop system is

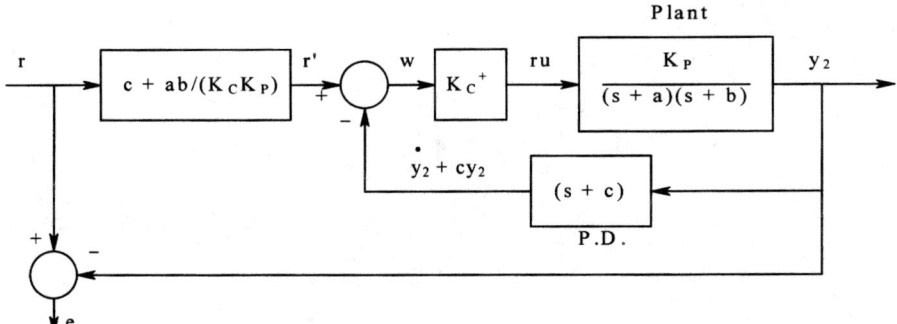

FIGURE 5.29 Block diagram of a quadratic plant controlled by PD feedback. Note that this is a type 0 system with steady-state error to a constant input. The gain $K_r = c + ab/(K_c K_p)$ gives the system zero steady-state error; that is, $y_{2SS} = r$.

```
continuous system PHSPLN02    "6/01/97
state y1 y2
der dy1 dy2
"
dy1 = -6*ry1 + 5*ry2 + ru
dy2 = -6*ry2 + 5*ry1
"
ry1 = if y1 < 0 then 0 else y1
ry2 = if y2 < 0 then 0 else y2
ru  = if u > 0 then u else 0
u = Kc*w
w = r*(c + 11/(5*Kc)) - c*ry2 - dy2    " The input r is normalized so
                                       " that the closed-loop system
e = r - ry2                            " has unity dc gain.
"
zero:0
r:2
c:20
Kc:1
y1:0
y2:0
"
END
```

Note that the states and the input are forced to be nonnegative in the program. The derivative terms can go negative, however.

Figure 5.30A illustrates the system's time-domain output step response for K_c = 1, 2, and 4. Figure 5.30B shows the corresponding phase plane plot for $\dot{y}_2$ vs. y_2. Figure 5.30C shows the rectified control input, $u(t)$, for K_c = 2 and 4. Note that $u(t)$

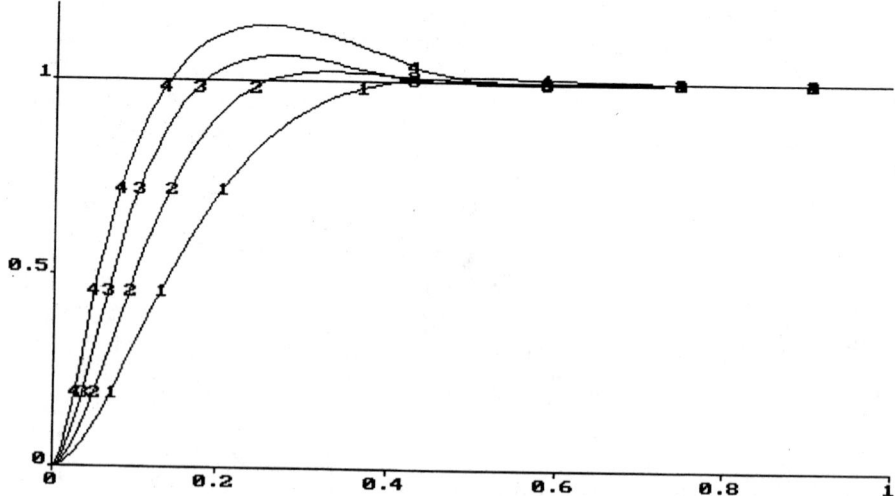

FIGURE 5.30A Step response of the system in Figure 5.29 vs. time. Trace 1, K_c = 1; trace 2, K_c = 2; trace 3, K_c = 4; trace 4, K_c = 8. See text for details.

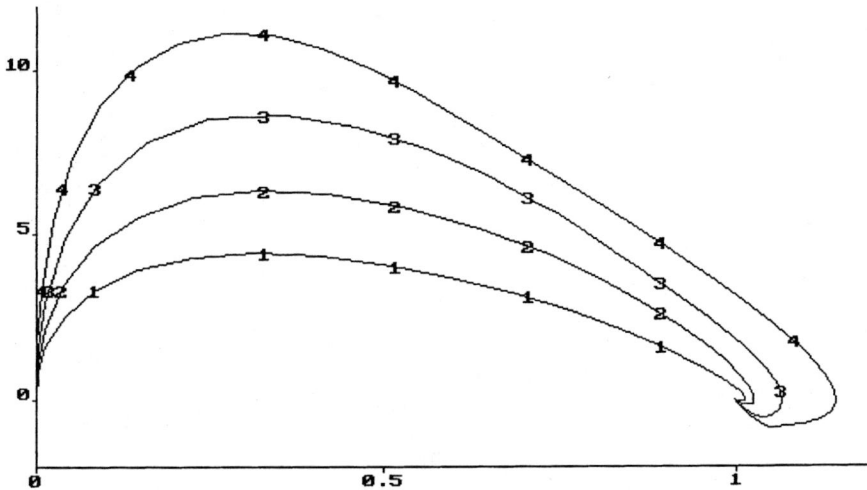

FIGURE 5.30B Corresponding phase plane plots of $\dot{y}_2$ vs. y_2. Same trace numbering as for Figure 5.30A.

tries to go negative for $K_c = 4$, causing the system to become nonlinear. In Figure 5.31A, we see a time-domain "washout curve" for the closed-loop system. The system was given initial conditions $y_1(0) = 0$, $y_2(0) = 2$. Figure 5.31B shows the phase plane trajectory for washout. Curiously, the washout transient is independent of the positive K_c value used. This is because $r = 0$, and the values of y_2 and $\dot{y}_2$

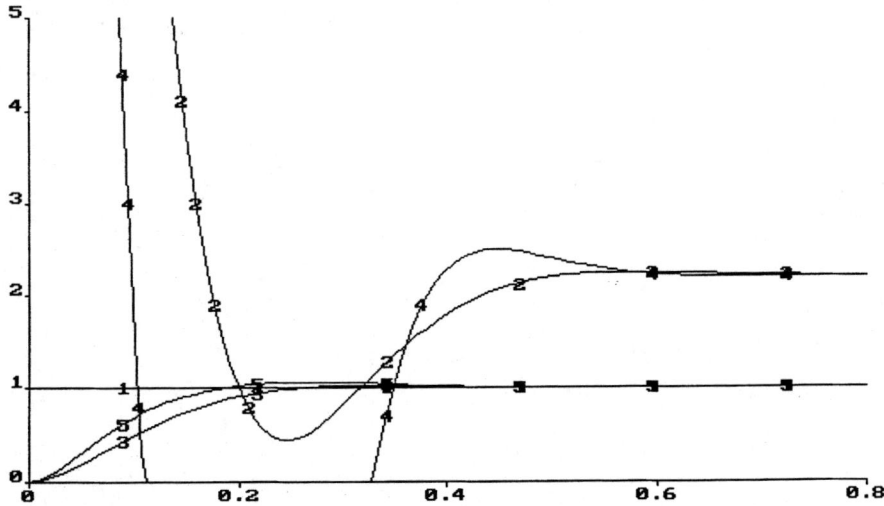

FIGURE 5.30C Plots of rectified control input to the system vs. time. Trace 1 = r (set point), trace 2 = ru for $K_c = 2$, trace 3 = y_2 for $K_c = 2$. Trace 4 = ru for $K_c = 4$, trace 5 = y_2 for $K_c = 4$.

FIGURE 5.31A A simulated time-domain washout curve given the ICs $y_1(0) = 0$ and $y_2(0) = 2$.

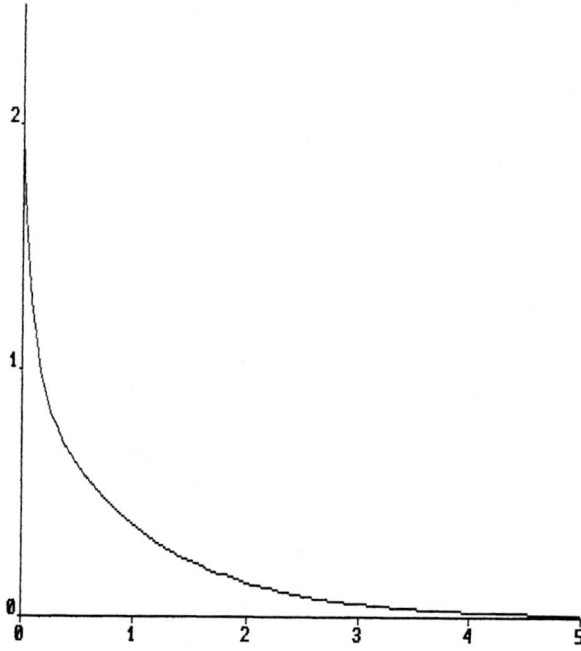

FIGURE 5.31B Phase plane plot of the washout curve in Figure 31A.

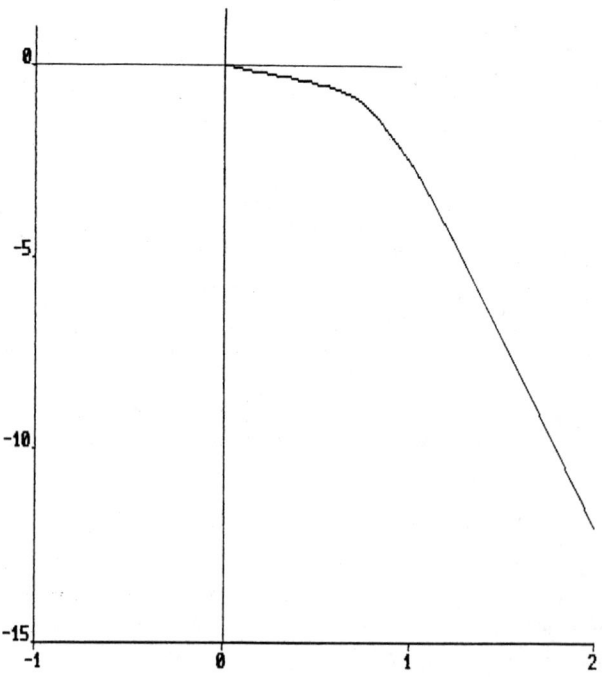

are such that $w < 0$ for $2 \geq y_2(t) \geq 0$; hence $ru = 0$ for $2 \geq y_2(t) \geq 0$. Thus the output decays passively with no control input. The closed-loop nature of the control system is only expressed for conditions on y and $\dot{y}$ such that $w > 0$.

In the *third example*, we again control the 2CPK plant described in example 3, this time using a *parameter-switching controller*. Two proportional feedback gains, c_k, set the position of the compensatory zero in the system's loop gain; hence the closed-loop system's ω_n and ζ are selected by the position of system e, $\dot{e}$ in the error phase plane. The closed-loop system's block diagram is shown in Figure 5.32 It is easy to show that the dc gain of the closed-loop system is 1, $\omega_n = \sqrt{(ab + c_k K_c K_p)}$, and $\zeta = (a + b + K_c K_p)/[2\sqrt{(ab + c_k K_c K_p)}]$. In this system, $a = 11$, $b = 1$ r/hr, $K_p = 5$, and $K_c = 1$. A simple linear-boundary parameter-switching rule was used, illustrated in Figure 5.33. If the e, $\dot{e}$ trajectory is in the sectors bounded by the e-axis ($\dot{e} = 0$), and the line $0 = e + k\dot{e}$, then the gain $c_2 = 110$ is chosen. If the trajectories are outside the sector, then the gain $c_1 = 11$ is selected. In practice, this selection would be done by a digital controller, which we are treating here as a continuous system. From the relations above, we find that when c_2 is selected, $\omega_n = 23.7$ r/hr, $\zeta = 0.359$ (this is the fast system). When c_1 is selected, $\omega_n = 8.12$ r/hr and $\zeta = 1.046$ (slow system). The Simnon program used to simulate the parameter-switching controller uses binary logic to select either c_1 or c_2.

```
continuous system PHSPLN04      "6/01/97 Parameter switching PD controller.
state y1 y2
der dy1 dy2
time t
"
dy1 = -6*ry1 + 5*ry2 + ru
dy2 = -6*ry2 + 5*ry1
"
ry1 = if y1 < 0 then 0 else y1
ry2 = if y2 < 0 then 0 else y2
ru  = if u > 0 then u else 0
```

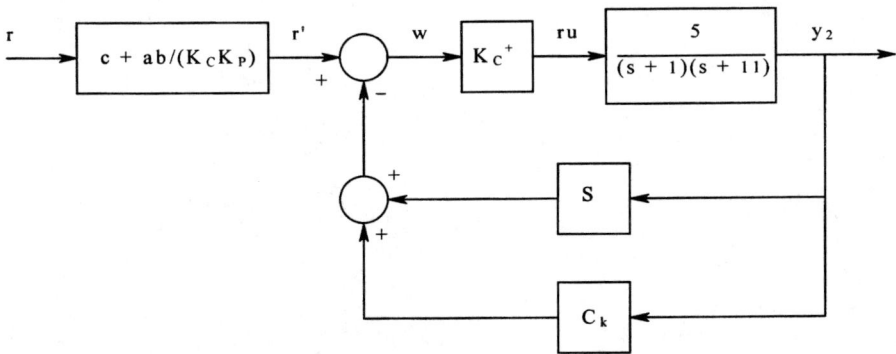

FIGURE 5.32 Block diagram of a type 0 parameter-switching controller using variable PD feedback. A quadratic real-pole CPK plant is used.

```
u = Kc*w
w = r*Fk - Ck*ry2 - dy2
Fk = Ck + 11/(Kc*5)
ZETA = (6 + 2.5*Kc)/SQRT(11 + Ck*Kc*5)
e = r - ry2
"
de = -dy2  "approx for step inputs
SGNdy2 = if -dy2 > 0 then 1 else -1          " Parameter switching logic.
SGNline = if (e - k*dy2) > 0 then 1 else -1
SF = sgndy2*sgnline
Ck = if SF > 0.1 then C1 else C2
"
zero:0
r:2
k:.1
C1:11   " For CL damping = 1
C2:110  " For CL damping = .01
Kc:1
y1:0
y2:0
"
END
```

System response to a step input is shown in Figure 5.34A. Here, the slope of the switching line, k, is varied, using fixed values of K_c, c_1, and c_2. The initial descending arc trajectory on the left is the fast response mode. When the trajectory

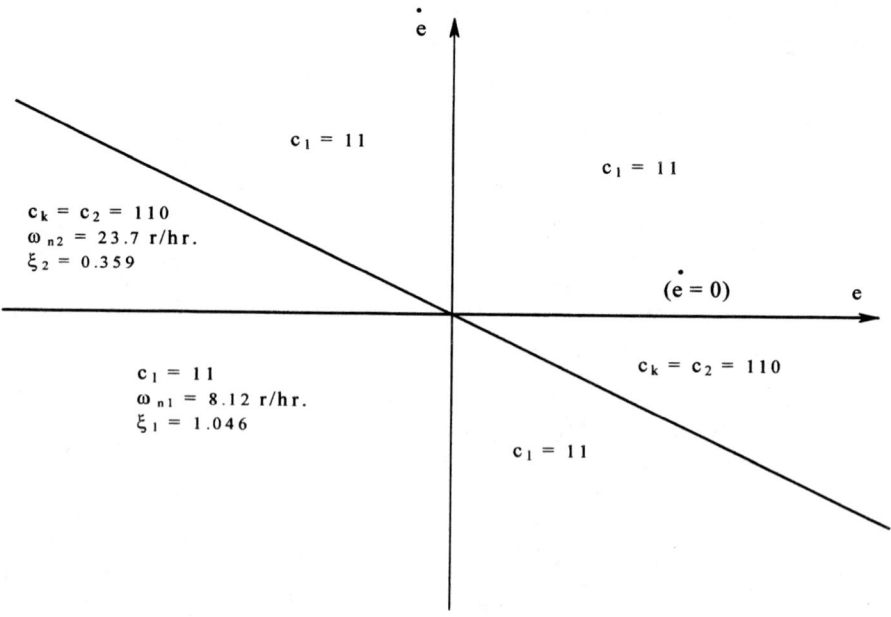

FIGURE 5.33 Switching boundaries in the phase plane for the parameter-switching controller.

Special Types of Closed-Loop Drug Input Controllers

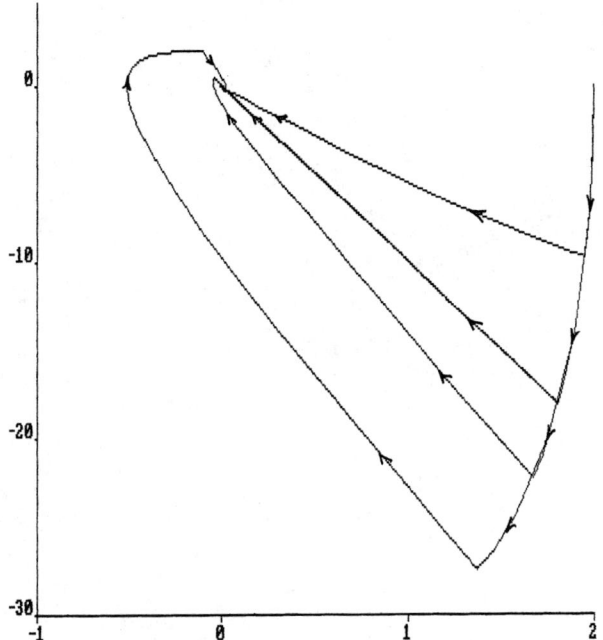

FIGURE 5.34A Phase plane plot of switching system $\dot{e}$ vs. e for a step of two input. As the slope, $-1/k$, of the switching line is varied ($k = 0.2, 0.1, 0.075, 0.05$), the system becomes quicker but develops an overshoot.

reaches the linear boundary, $\dot{e} = -e/k$, c_1 is switched in and the closed-loop system becomes slow with $\zeta \cong 1$. Making $1/k$ too large leads to faster initial response (the negative $\dot{e}$ builds up), but an unpleasant overshoot occurs because the trajectory enters the low damping region in the second quadrant when it crosses the negative e-axis. In Figure 5.34B, $k = 0.075$ for all runs, and c_2 is varied. c_2 values of 110, 440, and 1000 are used. There is little overshoot for all three c_2 values, but the lower ζs and higher ω_ns lead to a steeper initial segment to the trajectory; hence higher peak velocity and faster settling time. When $c_2 = 440$, $\omega_n = 47.02$ and $\zeta = 0.181$; when $c_2 = 1000$, $\omega_n = 70.79$ r/hr and $\zeta = 0.120$. Figure 5.35 shows the time-domain plot of the input step, r; the controlled variable, y_2; and SF. SF selects the feedback gain c_1 or c_2. When SF = 1, $c_k = c_1$ and system damping $\to 1$. When SF = -1 (zero on the plot), $c_k = c_2$ and we have the fast system. The third region of low damping is in response to e, $\dot{e}$ trajectory behavior very near the origin and thus is not resolvable on the y_2 plot.

In the *fourth design example,* we introduce the RADD system architecture. RADD is the acronym for radial damping. Figure 5.36 shows the system block diagram. The same plant is used as in the two previous examples. A normalizing gain acts on the input r so that $y_{2SS} = r$ in the dc steady state. Instead of switching the feedback gain, c, between two values as in the previous example, we let c be

FIGURE 5.34B Phase plane plot of $\dot{e}$ vs. e for a step of two input. Here, $k = 0.075$, and c_2 is varied. The highest velocity is reached for $c_2 = 1000$. Then $c_2 = 440$, and 110 gives the slowest response. Note that final overshoot is the same for all three c_2 values.

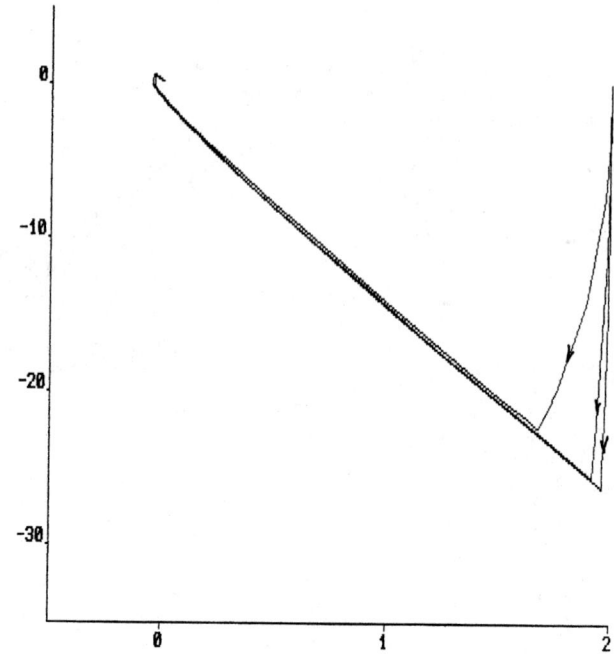

FIGURE 5.35 Time-domain response of the switching system showing $r = 2$, $y_2(t)$, and the switching function, SF. When SF = 1, the system has high closed-loop damping ($\xi \to 1$); when SF = −1, c_k is selected to give low damping. See the program PHSPLN04 for details.

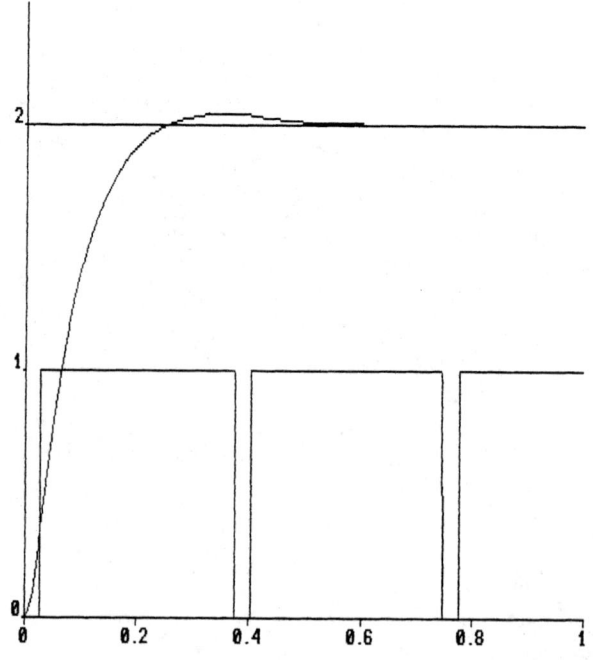

Special Types of Closed-Loop Drug Input Controllers

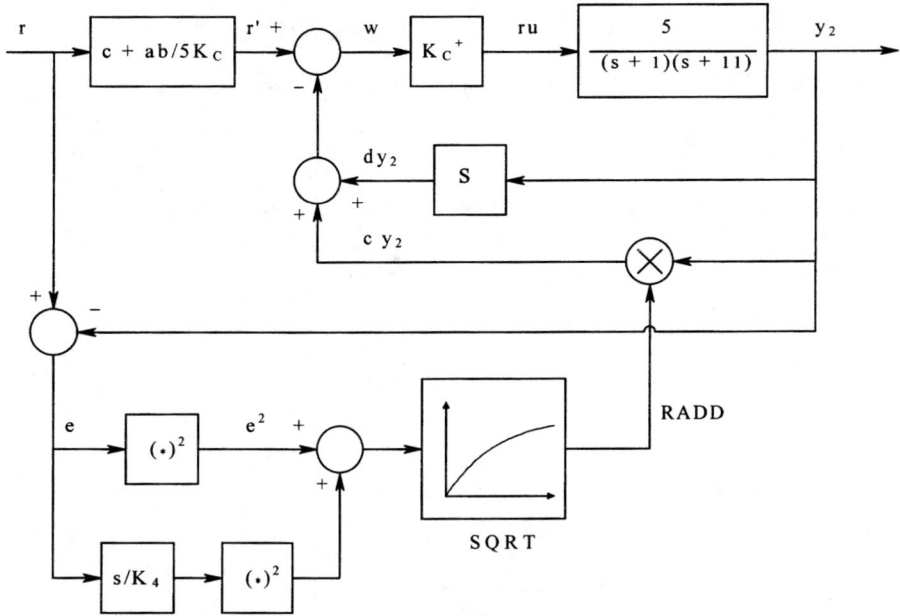

FIGURE 5.36 Block diagram of the RADD system where the proportional gain c is made a function of the trajectory's instantaneous distance from the origin in the error phase plane.

continuously variable as a function of the radial distance a point $(e, \dot{e})$ is from the origin of the $\dot{e}, e$ plane. This distance is

$$\text{RADD}(t) = \sqrt{\left[e^2(t) + \dot{e}^2(t)\right]} \quad (5.54)$$

When a point on a trajectory is relatively far from the origin, RADD is relatively large and we want the closed-loop damping to be low and ω_n to be high. As the trajectory approaches the origin, we want ζ to smoothly increase to unity while ω_n smoothly decreases to a low design value. Contours of iso-zeta are circles or ellipses in the $e, \dot{e}$ plane. One can devise a number of functions to make ζ a *decreasing* function of RADD. For example, in the Simnon program below, we see that c can be expressed by the following *increasing* functions of RADD:

$c = c_1 + k_1 \text{RADD} + k_2 \text{RADD}^2 + k_3 \text{RADD}^3$ (power series)

$c = c_1 + c_2 \left(1 - e^{-\text{RADD} k_2}\right)$ (exponential function)

$c = c_1 + (c_2 - c_1) \dfrac{\text{RADD}}{(\text{RADD} + k_1)}$ (Hill function)

$$c = c_1 \, \text{EXP}\left(2.3 k_1 \, \text{rho}\right)$$ (exponential law 2)

rho = RADD/r (nondimensional radius for step inputs)

The Simnon program is

```
continuous system PHSPLN05    "6/03/97 RADD PD controller.
state y1 y2
der  dy1 dy2
time t
"
" PLANT
dy1 = -6*ry1 + 5*ry2 + ru
dy2 = -6*ry2 + 5*ry1
"
ry1 = if y1 < 0 then 0 else y1
ry2 = if y2 < 0 then 0 else y2
ru = if u > 0 then u else 0
u = Kc*w
w = r*F - C*ry2 - dy2
F = C + 11/(Kc*5)
"
e = r - ry2
de = -dy2     "approx for step inputs
"
RADD = SQRT(SQR(e) + SQR(de/k4))
"
" C is a decreasing positive function of RADD.
"
"C = C1 + k1*RADD + k2*SQR(RADD) + k3*RADD*SQR(RADD)
C = C1 + C2*(1 - EXP(-RADD* k2))
"C = C1 + (C2 - C1)*rho/(rho + k1)
rho = RADD/r                  " rho is RADD normalized to input.
"C = C1*EXP(2.303*rho*k1)
ZETA = (6 + 2.5*Kc)/SQRT(11 + C*Kc*5)   " Zeta is system damping constant.
"
" CONSTANTS:
zero:0
r:2
"F:1
k1:1    " Constants k1, k2, k3, and k4 vary with C relation used.
k2:.01
k3:.1
k4:5
C1:11   " For CL damping = 1
C2:1000 " For low damping
Kc:1
y1:0
y2:0
"
END
```

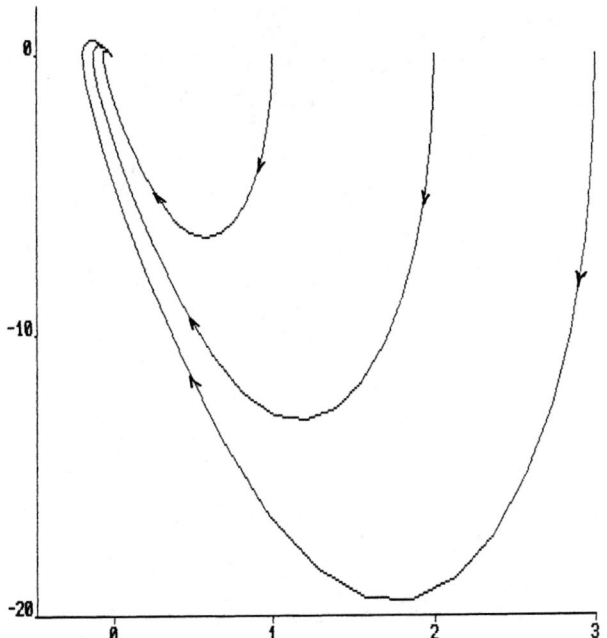

FIGURE 5.37A Phase plane trajectories ($\dot{e}, e$) for the RADD system's step response using a RADD function normalized with respect to the input, r. See text for details.

Figure 5.37A illustrates the $e, \dot{e}$ phase plane trajectories for step inputs of $r = 1$, 2, and 3 units. In this case, the c function is the exponential function:

$$c = c_1 * \exp(2.303 * \text{rho} * k_1) \tag{5.55}$$

Because the argument of the exponential is normalized for step inputs, the damping remains the same for steps of $r = 1$, 2, and 3. This is shown in the time-domain plot in Figure 5.37B, where steps of 2 and 3 are given. Note that the $\zeta(t)$ waveform does not change with increasing step size because the normalized parameter rho is used to calculate c.

Figure 5.38A illustrates the phase plane trajectories for the system's error when the nonnormalized exponential expression for c is used:

$$c = c_1 + k_1 \left[1 - \exp(-\text{RADD}\, k_2) \right] \tag{5.56}$$

Step inputs of 1, 1.5, and 2 units were used. Note that the system becomes progressively more damped as $r = e(0) \to 0$. In Figure 5.38B, the time-domain response of this system is shown. Note that damping increases for the smaller r values, and likewise, overshoot decreases for smaller r values.

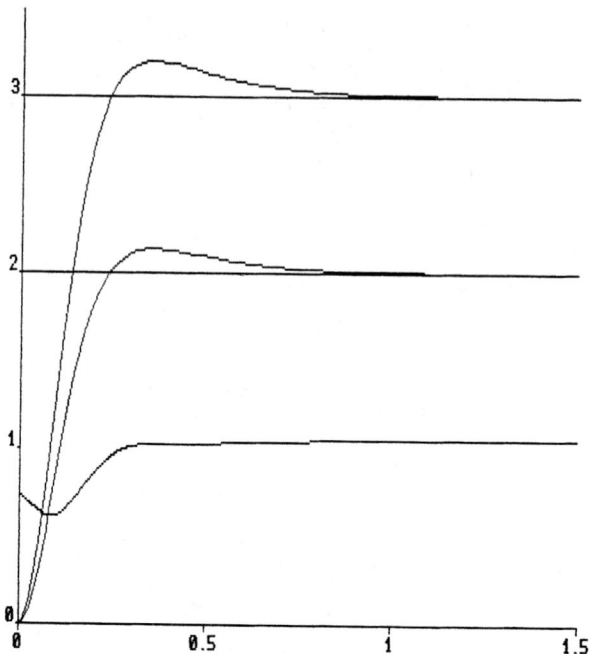

FIGURE 5.37B Time-domain step responses of RADD system for $r = 2$ and $r = 3$. The same normalized RADD function is used. Bottom trace is the "instantaneous damping," $\xi(t)$, for the two responses.

It is of interest to compare a RADD nonlinear control strategy with the two-parameter linear-boundary switching architecture described above on the same 2CPK plant. A Simnon program was written to do this:

```
CONTINUOUS SYSTEM phspln06    "V.6/04/97
STATE  y1 y2
DER  dy1 dy2
TIME  t
"
"  PLANT
dy1 = - 6*ry1 + 5*ry2 + ru
dy2 = - 6*ry2 + 5*ry1
"
"  PLANT INPUT
w = r*F - C*ry2 - dy2
u = Kc*w
F = C + 11/(Kc*5)
e = r - ry2
"
"  RADD CONTROLLER
de = -dy2                       " Approx error deriv. for constant r.
RADD = SQRT(sqr(e) + sqr(de))   " Pythagorean error measure
```

FIGURE 5.38A Phase plane error trajectories for step inputs to the RADD system. Note that as r increases, the system becomes less damped and exhibits an overshoot.

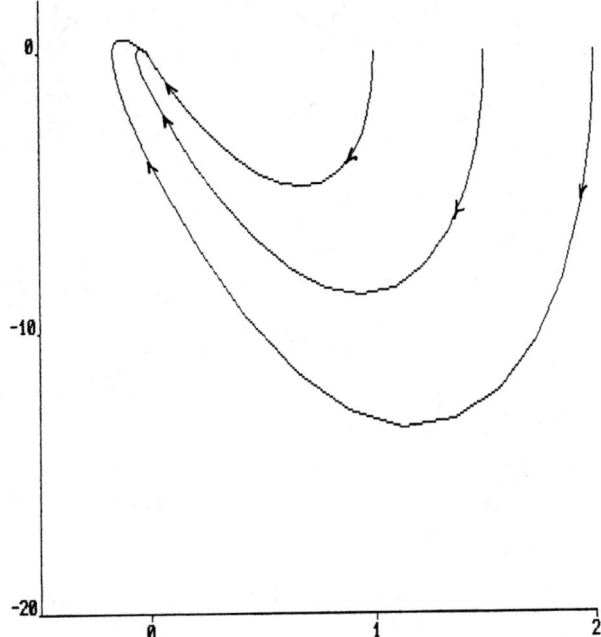

FIGURE 5.38B Time-domain response of the system to $r = 2$. Note the y_2 overshoot. Bottom trace is $\xi(t)$. Compare with $\xi(t)$ for the normalized system in Figure 5.37B.

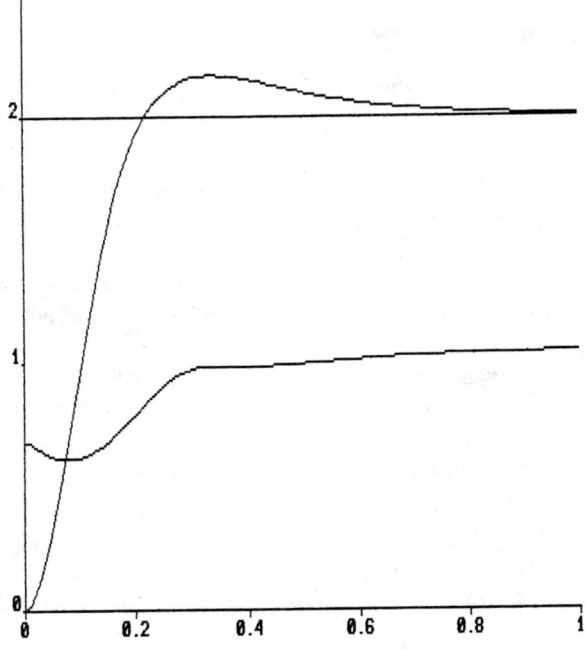

```
P1 = C1 - (C1 - C2)*SQR(RADD)/SQR(RADD + k1)    "RADD Hill function
                                                 controller.
"
" SWITCHING CONTROLLER
SGNde = if de > 0 then 1 else -1
SGNline = if (e + k*de) > 0 then 1 else -1
SF = SGNde*SGNline
P2 = if SF > 0.1 then C1 else C2    " C1 gives hi damping, C2 low.
"
C = if SW > 0 then P1 else P2       " SW selects SWITCHING or RADD
                                      controller.
"
" NONNEGATIVE VARIABLES:
ru = if u < 0 then 0 else u
ry1 = if y1 < 0 then 0 else y1
ry2 = if y2 < 0 then 0 else y2
"
ZETA = (6 + 2.5*Kc)/SQRT(11 + C*Kc*5)
x4 = 8*t - 16                       " To draw axes on phase plane; use simu 0 4.
"
" CONSTANTS
SW:0   " SW = 1 gives RADD control, SW = 0 gives Switching control.
k:.1
k1:1
C1:11
C2:200
r:2
zero:0
Kc:1
"
END
```

Figure 5.39 shows a comparison of the RADD continuously variable damping controller with the two-parameter-switching controller. The switching controller wins! It is faster, better damped, and its behavior does not depend on the magnitude of RADD. The RADD controller, using a nonnormalized RADD, shows progressively more overshoot as $r = e(0)$ increases. The time-domain plot (Figure 5.40) illustrates how the actual system damping varies for the two types of controller. The switching controller has a shorter settling time with little overshoot. It should be stressed that the comparison is made for a step input, which amounts to the turn-on transient for the CPK system. No restraint was assumed for a maximum value of u. In practice, there will be a u_{max}, which will slow the rise times.

In a *fifth and final example* of a parameter-switching system, we examine an application to CPK control by a unique multiparameter-switching control system first proposed by Flügge-Lotz and Wunch[47] in 1955 and implemented by Flügge-Lotz et al.[48] in 1958. See also the text by Flügge-Lotz.[46] In the Flügge-Lotz system architecture, a PD feedback strategy is used in which *both* the P and D gains are switched according to a complex binary combinational switching law. Figure 5.41 illustrates the system. A type 0 2CPK plant is controlled. The derivative feedback

FIGURE 5.39 Error phase plane trajectories for two values of r (2, 2.5) for the RADD PD controller vs. the parameter-switching PD controller. Note that the switching controller has negligible overshoot compared to the nonnormalized RADD system.

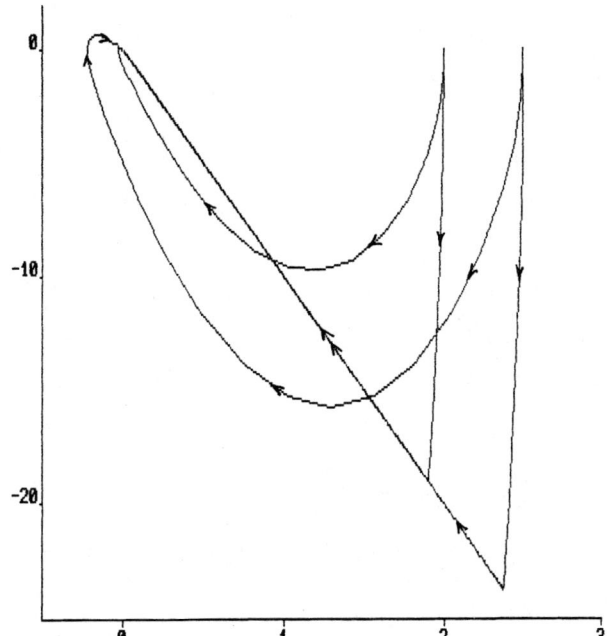

FIGURE 5.40 Time-domain plot of y_2 for an input step of $r = 2.5$ for the RADD system and the switching PD system. Trace 1 = y_2, trace 2 = ξ for the switching controller. Trace 3 = y_2, trace 4 = ξ for the RADD controller.

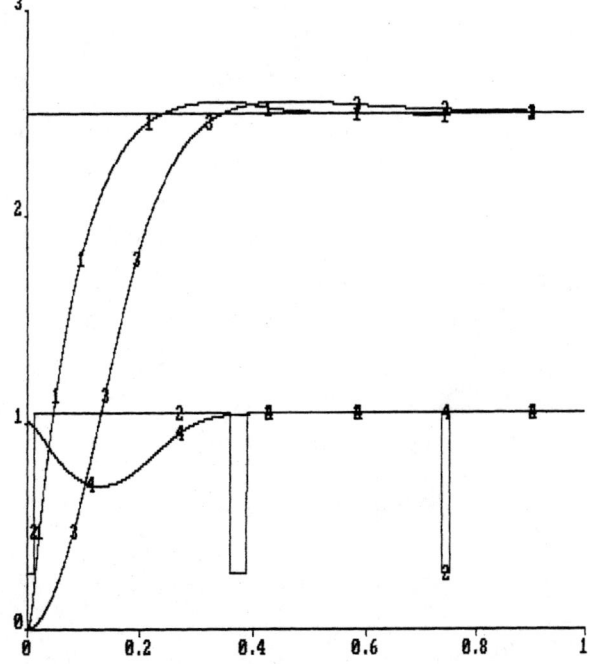

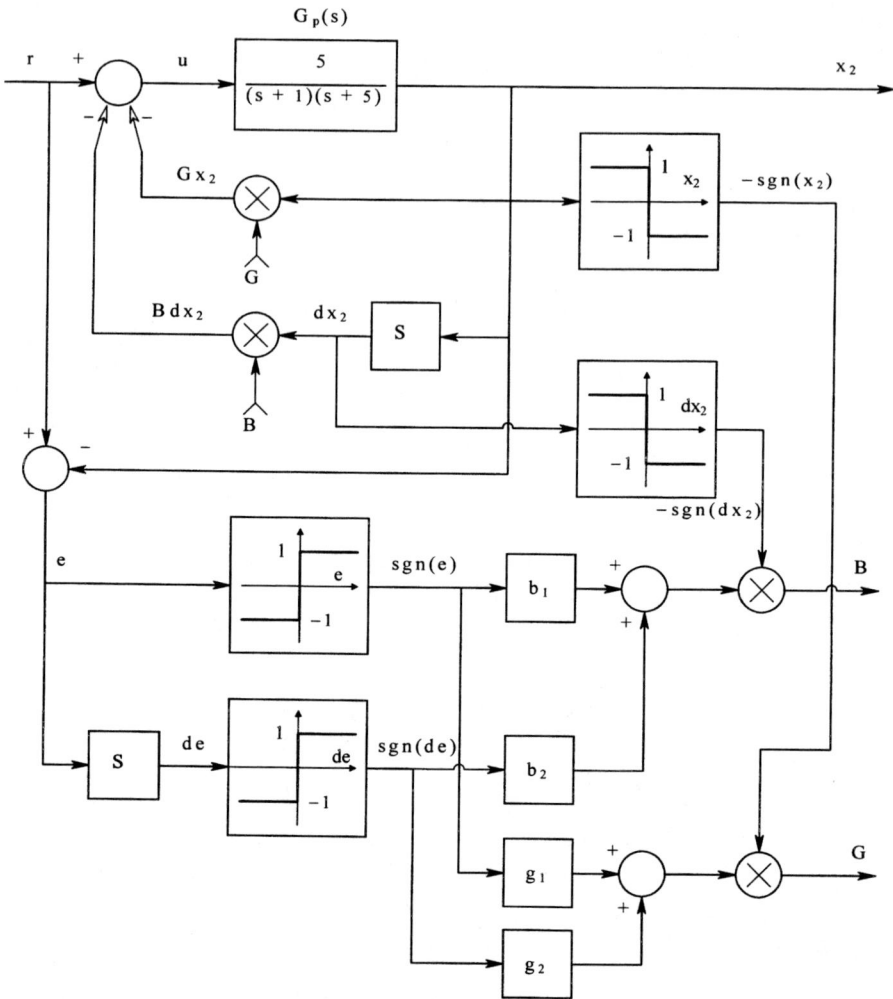

FIGURE 5.41 Block diagram of the multiswitching controller of Flügge-Lotz et al.[48] applied to a two-pole CPK plant. See text for discussion.

gain is B, and the proportional feedback gain is G. The values for B and G are determined by the switching laws:

$$B = -\text{SGN}(\dot{x}_2)\left[b_1\ \text{SGN}(e) + b_2\ \text{SGN}(\dot{e})\right] \quad (5.57\text{A})$$

$$G = -\text{SGN}(x_2)\left[g_1\ \text{SGN}(e) + g_2\ \text{SGN}(\dot{e})\right] \quad (5.57\text{B})$$

Depending on the signs, B and G can each have four numerical values. By judicious choice of b_1, b_2, g_1, and g_2, the closed-loop system can be given different ω_n and ζ values to facilitate its tracking the desired set point. The closed-loop transfer function for the 2CPK system is

$$\frac{X_2}{SP}(s) = \frac{5}{s^2 + s(6 + 5B) + (5 + 5G)} \tag{5.58}$$

Its natural frequency is $\omega_n = \sqrt{5}\sqrt{(1+G)}$, and its damping factor is given by:

$$\zeta = \frac{6 + 5B}{2\omega_n} \tag{5.59}$$

The four constants are chosen so that the six basic root conditions can be realized, as illustrated in Table 5.2. For the example above, we chose $b_1 = 2$, $b_2 = 1$, $g_1 = 0.8$, and $g_2 = -0.8$. Thus G can equal 0, +1.6, or −1.6, and B can equal four values, ±3 or ±1. This system exhibits a "weaving" microbehavior when the output, x_2, is tracking the input, SP(t). Figure 5.42 illustrates this weaving behavior for an input, $SP(t) = 2[1 - \cos(\omega t)]$. Note that a ripple proportional to the height of SP is present as the result of the rapid parameter-switching behavior of B and G by which the closed-loop ω_n and ζ values are selected. The actual control input depends on the ω_n and ζ values. Figures 5.43A, B, and C illustrate the unusual $u(t)$ waveforms as x_2 is forced to track SP(t). Finally, Figure 5.44 shows how G and B values are switched to accomplish control. Perusal of this figure reveals that for the input given, six combinations of B and G values are seen. These are summarized in Table 5.2, along with the resulting closed-loop natural frequencies and damping that result. Note that the choice of b_k and g_k values is flexible. We wish to end up with closed-loop system performance that tracks the input well by microweaving and has low root mean squared (rms) error. Obviously simulation is the only practical means of verifying a design for such a complex time-variable nonlinear system.

The Simnon program used to simulate the controller is given below. It is run with Euler integration with Δt = 0.0001 in order to see crisp switching.

TABLE 5.2
Closed-Loop System ω_n and ζ Values Selected by the Flügge-Lotz and Taylor Controller

B Value	G Value	ω_n	ζ
3	0	2.236	4.696
−1	−1.6	j 1.732	−j 0.289
−3	0	2.236	−2.012
1	1.6	3.606	1.524
1	−1.6	j 1.732	−j 3.175
−1	1.6	3.607	0.139

FIGURE 5.42 Time-domain behavior of the Flügge-Lotz controller. The CPK plant output, x_2, follows the sinusoidal input, SP, with a "microweaving" behavior. The frequency of the input, f, is 0.025 cycles/hr.

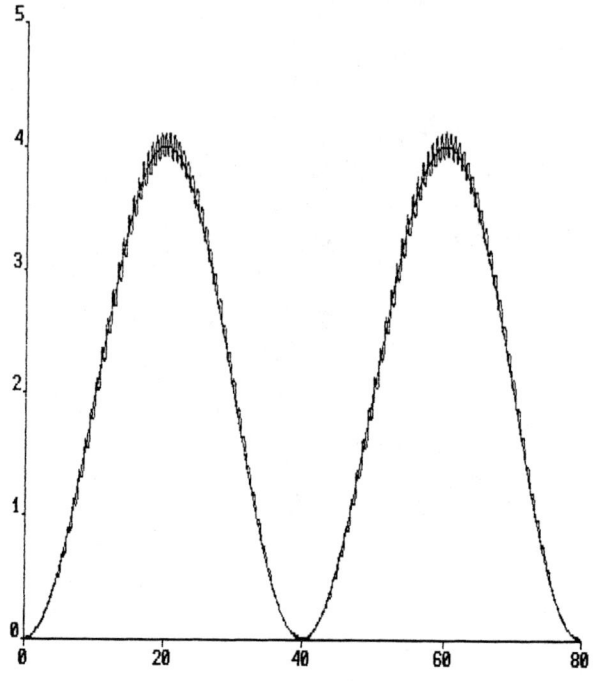

FIGURE 5.43A Plot of x_2, SP, and $u/10$ for $f = 0.1$ cycle/hr. u is nonnegative.

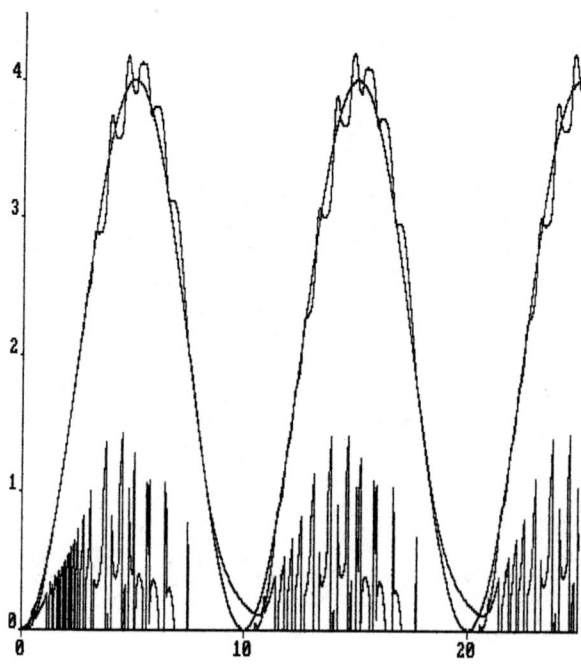

FIGURE 5.43B Enlarged plot of one cycle of x_2, SP, and $u/10$ for $f = 0.1$ cycle/hr. Note detail of $u(t)/10$. Discontinuities are due to switching.

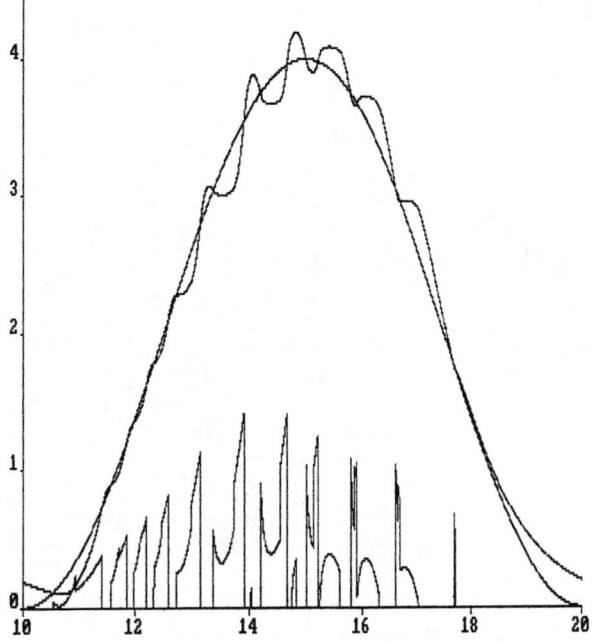

FIGURE 5.43C Another enlarged plot of one cycle of x_2, SP, and $u/10$ for $f = 0.1$ cycle/hr.

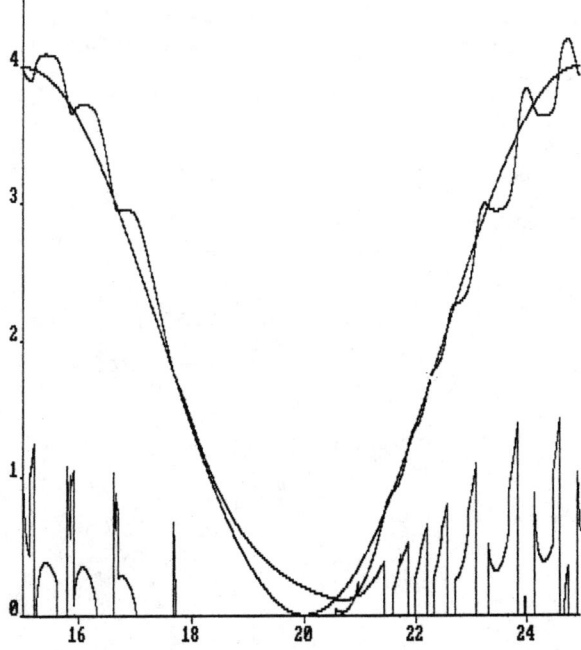

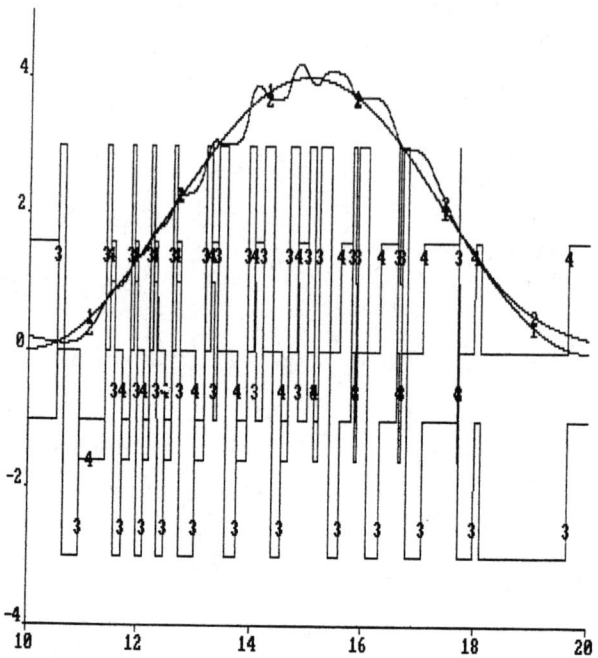

FIGURE 5.44 One cycle of x_2 and SP. G and B are determined by the switching equations. Trace 1 = SP, trace 2 = x_2, trace 3 = B, trace 4 = G. See text for description.

```
CONTINUOUS SYSTEM FLUGGE3     "12/09/92 Rev. 8/8/94, 11/09/94, 6/5/97.
STATE x1 x2 r mse
DER dx1 dx2 dr dmse
TIME t
"
" QUADRATIC PLANT
dx1 = 5*ru - rx1
dx2 = rx1 - 5*rx2
"
w = 2*pi*f
v = A*sin(w*t)         " w is frequency of input in r/hr.
dr = v                 " SP = r*w is actual system input. Integrated v
                         is used to enable calc. of de.
SP = r*w
"
" SWITCHING CONTROLLER:
u = SP - G*x2 - B*dx2
ru = if u < 0 then 0 else u
rx1 = if x1 > 0 then x1 else 0
rx2 = if x2 > 0 then x2 else 0
e = SP - x2
de = dr*w - dx2
us = ru/10
B = -sign(dx2)*(b1*sign(E) + b2*sign(dE))     " Flugge-Lotz & Taylor's
                                                SWITCHING RULES
```

```
G = -sign(x2)*(g1*sign(E) + g2*sign(dE))
"
se = SQR(e)                    " se = error squared
dmse = -.02*mse + .02*se       " LPF to average se, gives mse.
rmse = SQRT(mse)               " rmse = rms error of system.
"
zero:0
pi:3.14159
f:0.025
A:2
b1:2
b2:1
g1:0.8
g2:-0.8
x2:0
x1:0
"
```
END
—

An interesting property of the Flügge-Lotz system is that it runs with a low constant rms error as the input frequency increases. Above a critical input frequency, the rms error increases rapidly. For a comparable CPK plant with a conventional linear controller using fixed PD feedback, the rms error increases monotonically with increasing input frequency, remaining well above that of the Flügge-Lotz system until a high critical input frequency is reached.

Flügge-Lotz analyzes the "microweaving" behavior of the Flügge-Lotz switching controller in the phase plane. The interested reader can pursue this topic further in her text.[46]

5.5 CHAPTER SUMMARY

Four types of nonlinear controllers suitable for regulating the *in vivo* level of a drug or a physiological parameter were introduced in this chapter. The design of such controllers requires special techniques, including modeling for validation. The advantage of certain types of nonlinear control can range from lower implementation cost to improved dynamic performance and robustness.

On/off controllers are the simplest and least expensive to implement. The control is either maximum or zero. They are well suited for high-order real-pole plants and plants with dead-time lags, such as found in many CPK/P systems. On/off controllers have a simple 0,1 architecture, and they may have hysteresis or hysteresis with a dead zone. In the steady state, the closed-loop regulator oscillates around a mean value close to the set point. Design is concerned with specifying the amplitude and frequency of the oscillations; hence system error in the steady state. The well-known describing function method was introduced, and design examples were given.

NLD controllers using a model reference architecture were introduced and examples were given. The NLD method was shown to have limited use in CPK/P systems because of the functional requirement for negative control. Negative drug concentrations and infusion rates are obviously impossible. The possibility of using agonist/antagonist drug pairs in NLD control is a strategy that should be investigated.

IPFM controllers were next examined. IPFM controllers inject drug boluses, rather than a continuous nonnegative infusion. Some simplicity is gained in terms of the mechanical design of the bolus delivery pump; it is perhaps more suitable for implantation, and cumulative dose is easily obtained by counting pump cycles. IPFM controllers produce a ripple in the controlled variable, as do limit-cycling on/off controllers. IPFM controllers can be combined with conventional PD, PI, and PID control elements for improved dynamic performance. Several examples of IPFM CPK/P system design were given.

Finally, in Section 5.4, nonlinear parameter-switching controllers were introduced and analyzed and designed in the phase plane. Phase plane analysis is best suited for second-order systems, although higher orders can be used. The concept of switching boundaries in the $e, \dot{e}$ plane was introduced. That is, when the system's error lies in certain bounded areas in the $e, \dot{e}$ plane, certain controller parameters are switched in, giving the closed-loop system desired pole and zero positions. The RADD controller was seen to continuously adjust closed-loop system poles as a function of the instantaneous states, e and $\dot{e}$. The performance of parameter-switching and RADD controllers was compared.

The behavior of an "extreme" parameter-switching controller, introduced by Flügge-Lotz and colleagues in the mid-fifties,[46–48] was examined. This controller was originally designed for an autopilot controlling an underdamped second-order plant. $SGN(x_2)$, $SGN(e)$, and $SGN(\dot{e})$ are used to determine six combinations of closed-loop system ω_n and ζ. (x_2 is the controlled variable, and ω_n can be real or imaginary; ζ can be real <1, real >1, negative imaginary, or negative real < –1.) For $\dot{x}_2$ below a fuzzy limit, x_2 follows the nonnegative set point with a "microweaving" behavior, oscillating above and below the set point waveform. This contributes to a low steady-state rms error at low input frequencies which remains low with increasing frequency up to a limit, where it abruptly increases. The rms error for a conventional type 0 controller would increase monotonically with input frequency. Thus the type 0 Flügge-Lotz-architecture CPK/P controller exhibits better dynamic performance than a conventional system at low frequencies; negative control is not required.

The nonlinear controllers described in this chapter are easy to implement because of the general availability of microcontrollers with A/D and D/A interfaces. It is easy to examine the performance of nonlinear controller designs through simulation software such as Simnon and Matlab's Simulink®.

PROBLEMS

5.1 Use the describing function method to design an on/off drug infusion controller to regulate the concentration Y of an anticancer drug in the cerebrospinal

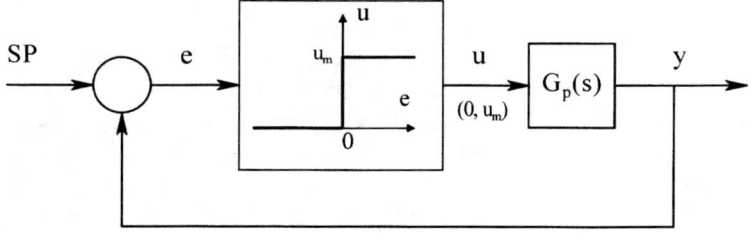

PROBLEM 5.1

fluid. Y is in micrograms per liter, and u is the infusion rate in micrograms per minute; $u \geq 0$. The plant transfer function is

$$G_p(s) = Y/U = 10 \exp\left[-sT_m + (sT_s)^2/2\right]$$

where $T_m = 4.0$ min, $T_s = 3.0$ min, and the set point is SP = 50 μg/l.
A. Find M so that the average error = 0 in the steady-state limit cycle.
B. Draw the system's dimensioned Nyquist diagram to scale in the region of the intersection of $-\mathbf{G_p}(j\omega)$ with the reciprocal describing function, $\mathbf{N^{-1}}(E)$.
C. Find the steady-state limit cycle frequency and the steady-state peak error.

5.2 Design an on/off controller *with hysteresis* to regulate the concentration of a cancer chemotherapy drug in the aqueous humor of the eye. Assume that the drug concentration is measured continuously by a noninvasive optical fluorescence method. The plant transfer function relating chemotherapy drug concentration in the aqueous humor (μg/l) to intravenous infusion rate (μg/min) is

$$G_p(s) = -5.33 \exp\left[-sT_d + (sT_s)^2/2\right]$$

where $T_d = 1.3152$ min, $T_s = 1.00$ min, and the set point is 100 μg/l. In the steady state, $\bar{e} = 0$, and the duty cycle of $u(t)$ is 50%.

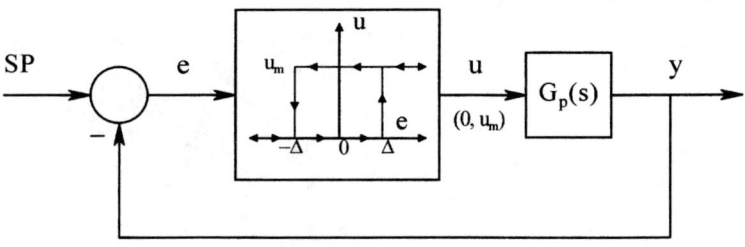

PROBLEM 5.2

A. Find U_M such that the steady-state $\bar{e} = 0$.
B. Draw the system's Nyquist diagram to scale on polar graph paper. Show the intersection of $\mathbf{A_L}(j\omega)$ with the inverse describing function, $\mathbf{N}^{-1}(E)$.
C. Find the controller hysteresis width Δ so that $E_{pkSS} = 10$ μg/l.
D. Find the steady-state limit cycle frequency, given the values of U_M and Δ found above.

5.3 A second-order pharmacokinetic plant is to be controlled by a Flügge-Lotz and Taylor-type switching controller. The plant is described by the ordinary differential equations (use Euler integration with $\Delta t = 0.001$)

$$dx1 = 5 * ru - x1$$

$$dx2 = x1 - 5 * x2$$

$x2$ is the controlled variable. The controller is defined by the equations

$u = SP - x2 * \mathbf{G} - dx2 * \mathbf{B}$ " SP is the input signal $x2$ must follow.

$e = SP - x2$ " True system error.

$de = dr * omega - dx2$ " Not an ODE.

$\mathbf{B} = -sign(dx2) * (b1 * sign(e)$ " Flügge-Lotz and Taylor controller.
$\quad + b2 * sign(de))$

$\mathbf{G} = -sign(x2) * (g1 * sign(e)$
$\quad + g2 * sign(de))$

$ru =$ if $u < 0$ then 0 else u " No negative drug infusions.

In order to have access to the exact error derivative, de, we will integrate the input v,

$$v = A * \sin(omega * t)$$

Thus:

$dr = v$

$SP = r * omega$ " $dSP = dr * omega$

$omega = 2 * pi * f$

$pi: 3.14159$

From the above, we see that $SP = A * [1 - \cos(2 * pi * f * t)]$, a nonnegative, cosinusoidal set point. Use $b1 = 2$, $b2 = 1$, $g1 = 1$, and $g2 = -1$.

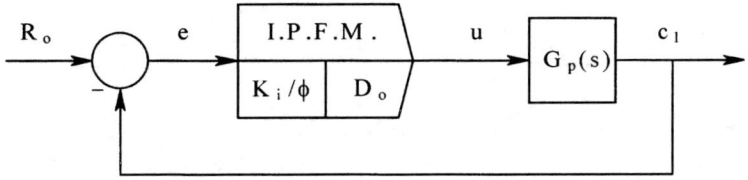

PROBLEM 5.4

A. Draw a block diagram of the system.
B. Find the numerical values for the closed-loop system's ω_n and ξ, given the **G** and **B** values possible. Note that $x2$ is nonnegative.
C. Use Simnon to plot SP, $x2$, and $ru/10$ for several cycles of SP for frequencies of 0.2, 0.1, 0.05, 0.025, and 0.01 cpm. For $f = 0.025$, show the fine structure of SP, $x2$, B, and G to illustrate the switching action of the system.
D. Generate the system's rms error for the frequencies in part C. Note that:

sqe = SQR(e)

dmse = $-0.02 * $ mse $+ 0.02 *$ sqe   " LPF ODE to average sqe and form mse.

rmse = SQRT(mse)

Note that the mse averager requires over 100 min of model time to reach steady state. At the end of a 100-min simulation, type *disp rmse* [enter] to see the steady-state numerical rmse of the system.

5.4 Design an IPFM controller to maintain the drug concentration, C_1, in a one-compartment pharmacokinetic plant with $V_1 = 20$ l, $K_L = 0.2$ l/min.
  A. Find the pharmacokinetic plant's time constant, τ_p, in minutes and dc gain, $G_p(0)$.
  B. Find D_o (μg) so that the peak-to-peak ripple in $C_1(t)$ will be 5% of the set point.
  C. Find K_i/ϕ such that $\overline{E}_{SS}/R_o = 0.01$ (1% steady-state error).
  D. Find T_o, the steady-state equilibrium controller period.
  E. Now assume $K_L' = 0.5 K_L$ as the result of kidney failure (assume SP, V_1, K_i/ϕ, and D_o stay the same). Find τ_p', $G_p'(0)$, T_o', $\overline{E}_{SS}'$, and peak ripple as a percentage of SP.
5.5 Design an IPFM drug injection regulator (i.e., find D_o, ϕ/K_i) so that in the steady-state equilibrium, $|E_{pk}|$ and $|E_{min}| < 0.15$ g/l. The set point SP = 1 g/l. Assume $u_{SS}(t) = D_o \Sigma \delta(t - nT_o)$. Verify your design by calculating T_o, E_{pkSS}, E_{minSS}, and $\overline{E}_{SS}$ for the system.
5.6 A dual-mode rate-feedback switching controller is used to control the drug concentration D for a second-order CPK plant. Rate gain C_1 is selected when

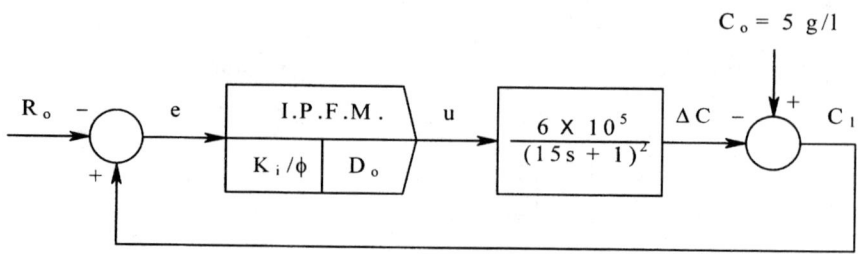

PROBLEM 5.5

$sgn(\dot{e}) * sgn(e + \dot{e}/1.581) = 1$. C_1 gives a closed-loop $\xi = 1$. When $sgn(\dot{e}) * sgn(e + \dot{e}/1.581) = -1$, C_2 is selected, giving $\xi = 0.2$.

A. Calculate the values of C_1 and C_2 to give the closed-loop system $\xi = 1$ and 0.2, respectively.

B. Draw the switching lines on the $e, \dot{e}$ phase plane. Show which sectors have $\xi = 0.2$ and which have $\xi = 1$.

C. Write a program to simulate the closed-loop system. Show the system's step response in both the time domain and in the phase plane. Note that the error derivative is not explicitly available; it must be calculated. Also examine the equilibrium phase plane trajectories for different ICs on e and $\dot{e}$, given $R \equiv 0$.

5.7 An $N - M = 1$ CPK plant is to be controlled with IPFM. The set point SP = 11 µg/l.

$$G_p(s) = \frac{4}{s+1} + \frac{1}{s+4} \quad \text{time in hours}$$

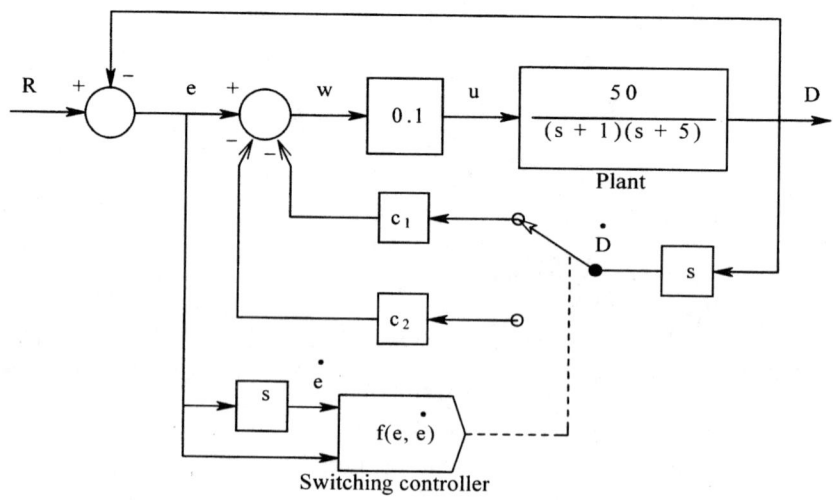

PROBLEM 5.6

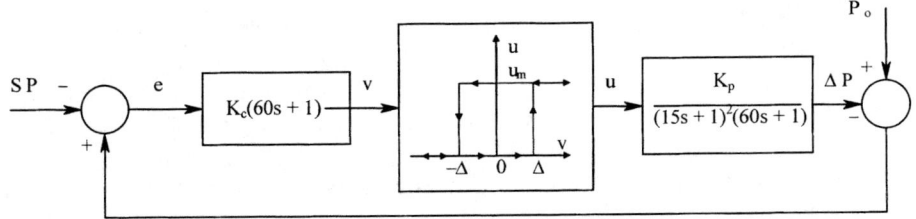

PROBLEM 5.8

A. Find D_o required so that the steady-state peak-to-peak ripple in $D(t)$ is 5% of the set point.
B. Find ϕ/K_i such that $\overline{E}_{SS}/SP = 0.01$ (1% steady-state error).
C. Find the steady-state controller period, T_o.
D. Find e_{pkSS}.

5.8 An on/off controller *with hysteresis* is designed to administer a drug to reduce a physiological parameter from $P_o = 5$ g/l toward the set point SP = 1 g/l. The system is shown in the figure; all time constants are in minutes. Assume in the steady-state $v(t) = V_o \sin(\omega_o t)$. Also, $\Delta = 0.791$, $U_M = 10^{-5}$ g/min.
A. Use the describing function method to find the steady-state $P(t) = P_{oSS} + \Delta P \sin(\omega_o t + \varphi)$.
B. Also find $v(t)$ and $e(t)$.

5.9 After a parathyroidectomy, a dog becomes drastically hypocalcemic and would die in tetany unless drastic procedures are taken to restore the normal blood calcium concentration of [Ca^{++}] = 10 mg/dl. An on/off controller with hysteresis controls a parathyroid hormone (PTH) drip. The PTH acts (along with vitamin D3 and calcium gluconate injections) to raise the [Ca^{++}]. The [Ca^{++}] is sensed on-line and compared to the desired-level SP = 10 mg/dl. The integral of the error signal, $v(t)$, is the controller input.

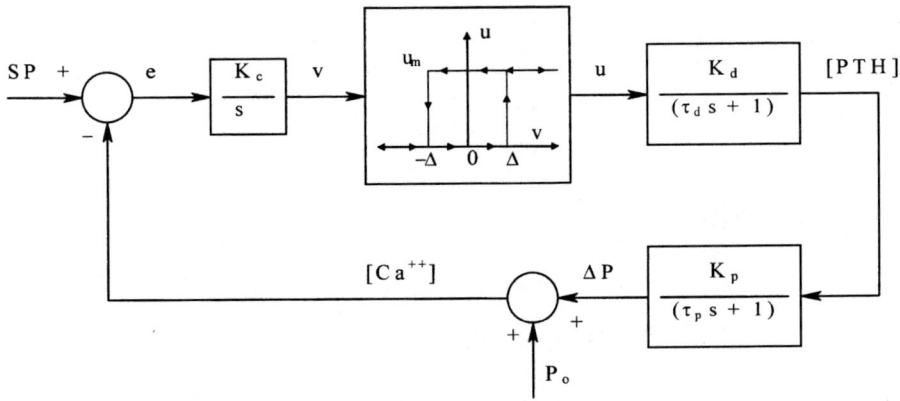

PROBLEM 5.9

A. Use the describing function method and a graphical solution to find the steady-state limit cycle ω_o in radians per minute.
B. Find the peak value of $e(t)$ and [PTH] in the steady state.
C. Design an IPFM controller to operate on $e(t)$ to form a pulsed $u(t)$ which will give a lower max $|e|$ than the hysteresis controller.

5.10 Use simulation to investigate the nonlinear two-input/two-output system of Fadali's example 3.[40]

$$\dot{x}_1 = a_{11} x_2 + b_1 d_1$$

$$\dot{x}_2 = -a_{22} x_1^2 + b_2 d_2$$

$$\mathbf{y} = \begin{bmatrix} x_1 & x_2 \end{bmatrix}^T = \begin{bmatrix} y_1 & y_2 \end{bmatrix}^T$$

Let:

$$b_1 = 2.\text{E-}2 \qquad a_{11} = 1$$

$$b_2 = 2.\text{E-}2 \qquad a_{22} = 1.\text{E-}2$$

A. Let d_1, d_2 be inputs:

$$d_1 = 1 \text{ for } 0 \leq t \leq 5, \text{ and for } t \geq 40, \text{ else } 0$$

$$d_2 = 1 \text{ for } 15 \leq t \leq 20, \text{ and for } t \geq 40, \text{ else } 0$$

Plot $x_1(t)$, $x_2(t)$, d_1, and d_2.

B. Now devise a nonlinear decoupling controller with $a_2 = P_2 = 0.4$, $a_1 = P_1 = 1.0$. Make v_1 and v_2 inputs the same as d_1 and d_2 in part A. Plot x_1, x_2, d_1, d_2, v_1, and v_2.

C. Now let x_1, x_2, d_1, and d_2 be nonnegative (drug concentrations and input rates). Plot x_1, x_2, d_1, d_2, v_1, and v_2.

6 Hormonal Regulation of Sodium, Potassium, Calcium, and Magnesium Ions

6.0 INTRODUCTION

In this chapter, the physiology and biochemistry of four complex, endogenous, hormonally governed, ionic regulatory systems are considered. They are of interest because of their similarities of operation. The kidneys play a major role in the operation of all four regulators. These four systems are not as well defined as the glucoregulatory system because of lack of detailed knowledge about the regulation of the microcirculation of the kidneys. Block diagrams are given to describe these endogenous regulators; however, no computational models are given.

Guyton[59] has observed that the functions of the body are regulated by two major physiological systems: (1) the nervous system and (2) the endocrine system. The endocrine, or hormonal, system shows a wide range of response times in exerting its functions. Certain hormones affect function in seconds; others take minutes, hours, or even days to take effect. Delays in the nervous system's actions are generally on the order of tens of milliseconds. The nervous system and the endocrine system are seen to interact in certain hormonal systems. For example, neurosecretory cells (specialized neurons that secrete hormones when stimulated by other neurons) in the posterior pituitary gland and the cells of the adrenal medulla release hormones into the circulatory system when given specific nervous stimulation.

Hormones are carried to target cells in the body by the circulatory system. They are generally present in extremely minute concentrations, some as low as 1 pg/ml. The molecules of hormones are either proteins (polypeptides) made from chains of amino acids or from single amino acids or are steroids consisting of several joined rings of carbon atoms with appropriate radicals attached. The adrenal medulla and the thyroid gland secrete hormones based on single amino acids. The pancreas and the anterior pituitary secrete polypeptide hormones, while steroids come from the adrenal cortex, the ovaries, and the testes.

Hormones in the blood or in the extracellular fluids combine with *receptor proteins* located either on a cell's surface membrane (receptors for polypeptide hormones and catecholamine hormones), in its cytoplasm (steroid hormone receptors), or in its nucleus (e.g., thyroid hormone receptors). There may be 2×10^3 to 10^5 receptors for a specific hormone per cell.[59] The density of receptors associated with a cell is not fixed. The cell is constantly manufacturing receptors at a rate which is itself under control. In most cases, the binding of a hormone with its receptor has a down-regulatory effect on the rate of receptor production. However, in some cases, there is an opposite effect.[59]

A hormone receptor has a high specific affinity for its corresponding hormone molecule. That is, when the hormone molecule encounters its receptor, a strong chemical bond forms between them, locking them together. The binding of a hormone with its receptor triggers a sequence of chemical events which leads to the hormonal effect on the cell. One such biochemical scenario is that the binding activates the enzyme *adenyl cyclase* in the membrane. The adenyl cyclase molecule exposed to the cytoplasm causes immediate conversion of adenosine triphosphate in the cytoplasm to *cyclic adenosine monophosphate* (cAMP), which then initiates a number of cellular functions including protein synthesis before adenyl cyclase is inactivated. cAMP is known as a *second messenger*. The type of cell determines what the end product of cAMP production will be; for example, in a thyroid cell, cAMP triggers the production of thyroid hormone; in a renal tubule epithelial cell, cAMP increases the membrane permeability to water.

The cAMP system stimulates target effects in the following hormonal systems:[59] (1) adrenocorticotropin, (2) thyroid-stimulating hormone, (3) luteinizing hormone, (4) follicle-stimulating hormone, (5) vasopressin (also known as antidiuretic hormone), (6) parathyroid hormone, (7) glucagon, (8) catecholamines (epinephrine, norepinephrine), (9) secretin, and (10) the hypothalamic-releasing hormones. In some instances, hormone–receptor binding releases another second messenger known as *cyclic guanosine monophosphate* (cGMP), which is a nucleotide like cAMP. Still another second messenger system involves the hormone binding to a receptor that causes the passage of calcium ions into the target cell. These internalized Ca^{++} bind with a protein called *calmodulin*. The Ca^{++}–calmodulin complex activates various intracellular biochemical processes in a manner similar to that of cAMP or cGMP. An example of the Ca^{++}–calmodulin system is found in the induced contraction of smooth muscle.

The various cells of the immune system also produce hormones that largely affect the function of the immune system. These immune system hormones are called lymphokines or, more broadly, cytokines or autacoids. We will examine their actions in Chapter 10.

The secretion rates of all hormones are under feedback control. In some cases, these negative feedback loops are very complex, multistep pathways; in other cases, the causality is relatively simple. Our interest in this chapter is to describe the hormonal control loops that regulate blood potassium ion concentration, blood sodium ion concentration (hence osmotic pressure), blood calcium ion concentration, and blood magnesium concentration. As you will see, the regulation of $[K^+]$, $[Na^+]$, P_{osm}, and $[Mg^{++}]$ is intimately interrelated through the functions of the kidneys. [Ca^{++}] is regulated very tightly (it is a very robust system) compared to [K$^+$], [Na$^+$], P_{osm}, and [Mg^{++}].

6.1 IMPLICIT SUMMING POINTS IN HORMONAL REGULATION

In a conventional control system or regulator, there is a summing point where the desired input or set point is compared with the actual output to generate a system error. In the simplest case, the error is conditioned and used to drive the plant. Hormonal systems, on the other hand, exhibit closed-loop causality but have no specific summing node. This section describes how a hormonal system can be reduced to a conventional closed-loop form by deriving an *implicit summing point*.[70]

As a first example, consider a simple, static, ideal, linear, hormonal loop in which the steady-state (SS) hormone concentration [H] controls the concentration of a cell product, [P]. (See Figure 6.1A.) That is,

$$[P] = K_1[H] \tag{6.1}$$

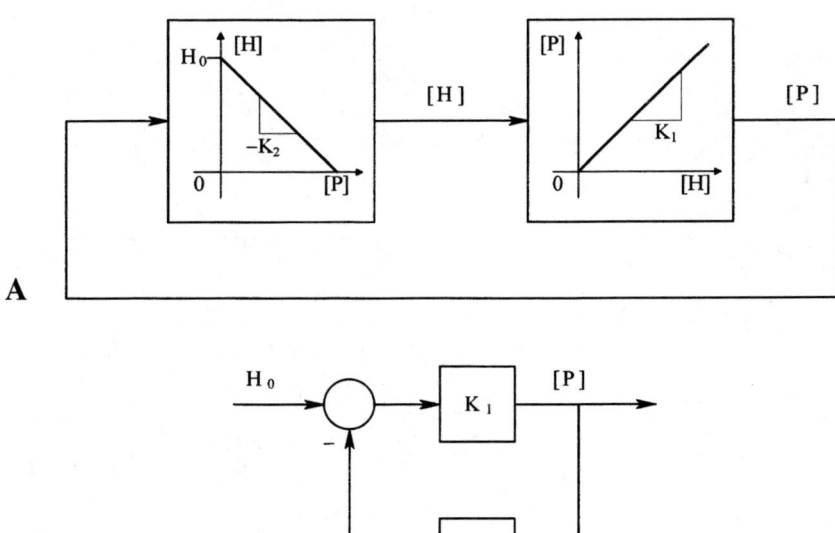

FIGURE 6.1 (A) Simple, static, linear hormonal regulatory model. The defining equations are the equations of the linear input/output (I/O) relations. (B) Equivalent linear block diagram describing the hormonal system as a single-loop negative feedback system with an implicit summing point and virtual set point.

The SS product concentration inhibits or down-regulates the production of the hormone so that:

$$[H] = [H_o] - K_2[P] = K_2[P_o] - K_2[P] \tag{6.2}$$

If we combine these two equations, we can write:

$$[P] = [P_o]\frac{K_1 K_2}{1 + K_1 K_2} = [H_o]\frac{K_1}{1 + K_1 K_2} \tag{6.3}$$

This is the transfer function of a simple negative feedback system, shown in Figure 6.1B. Note that the virtual set point is $[P_o] = [H_o]/K_2$.

As a *second example* of finding an implicit summing point, we consider the quasistatic causal relationship between mean arterial pressure (MAP) and water intake rate. The basic system is shown in Figure 6.2. Here, a specific summing point computes the net water loss or gain rate of the body from water loss rate in urine, insensible water loss rate, and net water input (drinking and/or intravenous

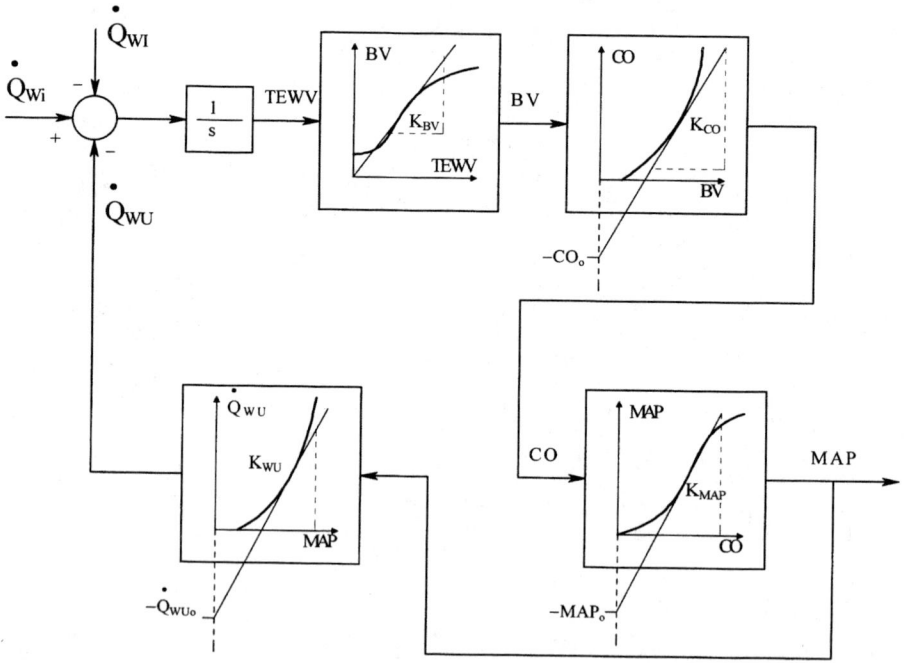

FIGURE 6.2 Nonlinear quasistatic system describing water balance and MAP in the human. One real summing node is used to sum the rate of water loss through the kidneys, the rate of insensible water loss, and the rate of water input from drinking or intravenous injection. The net input rate is integrated (note that there are initial conditions) to form the total extracellular water volume (TEWV). (Adapted from Guyton.[59])

infusion). The summer output is integrated to obtain the total extracellular water volume, TEWV. SS blood volume (BV) is related to TEWV by a saturating curve. SS BV affects SS cardiac output (CO), which in turn determines MAP. The SS loss rate of water in the urine, $\dot{Q}_{wu}$, is seen to be a steeply increasing function of MAP. The curves for BV, MAP, CO, and $\dot{Q}_{wu}$ can be linearized at their operating points. Thus:

$$BV = K_{BV} \text{ TEWV} \tag{6.4A}$$

$$CO = K_{CO} BV - CO_o \tag{6.4B}$$

$$MAP = K_{MAP} CO - MAP_o \tag{6.4C}$$

$$\dot{Q}_{wu} = K_{WU} MAP - \dot{Q}_{wuo} \tag{6.4D}$$

Figure 6.3 illustrates the linearized regulatory system shown as a signal flow graph. Note that the linearized offsets are dc quantities and may be ignored when considering the *changes* in MAP caused by changes in $\dot{Q}_{wi}$:

$$\frac{\Delta MAP}{\Delta \dot{Q}_{Wi}} = \frac{(1/s) K_{BV} K_{CO} K_{MAP}}{1 + (1/s) K_{BV} K_{CO} K_{MAP} K_{WU}} = \frac{K_{BV} K_{CO} K_{MAP}}{s + K_{BV} K_{CO} K_{MAP} K_{WU}} \tag{6.5}$$

Thus the MAP responds to transient changes in $\dot{Q}_{Wi}$ with a time constant of $1/(K_{BV} K_{CO} K_{MAP} K_{WU})$ sec, and given a step input of $\dot{Q}_{Wi}$, the SS ΔMAP can be shown to be

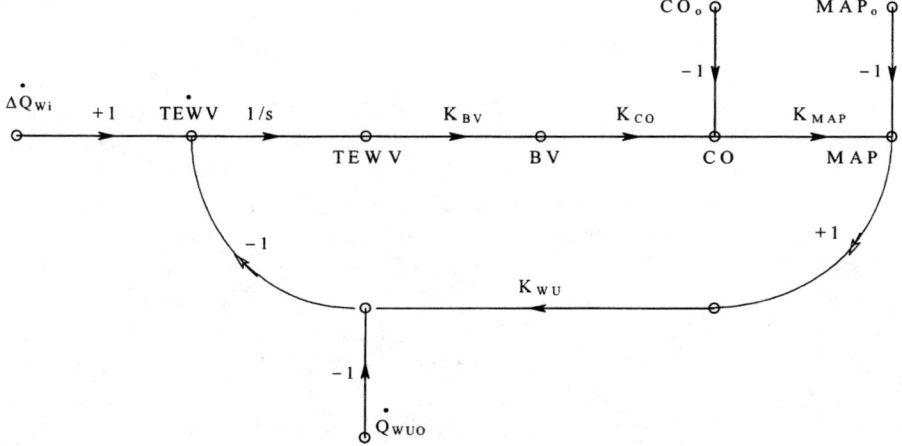

FIGURE 6.3 Linearized water regulatory system. Linearization is done by replacing each nonlinearity with a straight-line approximation tangent to the steady-state operating point on each curve. The straight-line approximations give equivalent gains at each operating point and virtual dc inputs.

$$\text{SS } \Delta\text{MAP} = \dot{Q}_{Wi}/K_{WU} \tag{6.6}$$

The dependence of MAP on K_{WU} underscores the importance of normal kidney function in maintaining normal MAP, given disturbance inputs of ingested water or anomalous water loss, such as by excess sweating.

6.2 INTRODUCTION TO THE PHYSIOLOGY OF THE KIDNEYS

The kidneys are normally associated with the production of urine and the elimination of nitrogenous waste. The kidneys are in fact very complex biochemical engines whose functions are intimately related to not only eliminating urea but also to the *conservation* of many ions and molecules that are initially filtered. From a systems viewpoint, each kidney is a multiple-input/multiple-output plant with massively parallel structure. The inputs are through the renal artery and the nervous system; the outputs are through the ureter and the renal vein. The inputs include filterable substances in the blood (e.g., water, various ions, and low-molecular-weight organic molecules), autonomic nervous signals that control vasoconstriction (and hence blood flow and pressures), and various *hormones* that affect the rate of blood filtration and the rate of absorption of substances such as glucose, water, Na^+, K^+, etc. By simple subtraction, the renal vein carries the difference between the inputs in the renal artery and the substances lost in the urine (e.g., urea, water, Na^+, K^+, glucose, etc.).

Figure 6.4 illustrates schematically the anatomy of a nephron, the basic functional unit of the kidneys. Note that a nephron consists of a *Bowman's capsule* to which the arterial blood is supplied from the *afferent arteriole* which branches off one of the small *arcuate arteries*. The arcuate arteries in turn branch from the *interlobar arteries*, which branch from the *segmental arteries*. All segmental arteries branch from the main *renal artery*. The filtered blood exits the Bowman's capsule in the efferent arteriole, which branches to the *vasa recta* and the peritubular capillaries. Eventually this returning blood reaches a branch of the arcuate vein, etc.

The initial filtrate passes from Bowman's capsule through a thin tubule containing very specialized cells whose diverse functions are under the control of certain hormones. The output of the tubule is filtrate, not urine; it passes through a converging series of collecting ducts leading to the left and right ureters, which supply the urinary bladder. Further processing of the filtrate leaving the tubules occurs in the collecting ducts. Their output is essentially urine. The kidneys have a massively redundant structure; each contains about 1.2 million nephrons. This redundancy underscores the physiological importance of these organs.

Figure 6.4 also shows the mean hydraulic pressures present in a typical nephron (assuming an empty bladder). The arcuate artery sends arterioles to the anastomosing vessels in Bowman's capsule, where the initial filtering of the blood takes place. Water, glucose, urea, creatinine, ions, amino acids, other low-molecular-weight

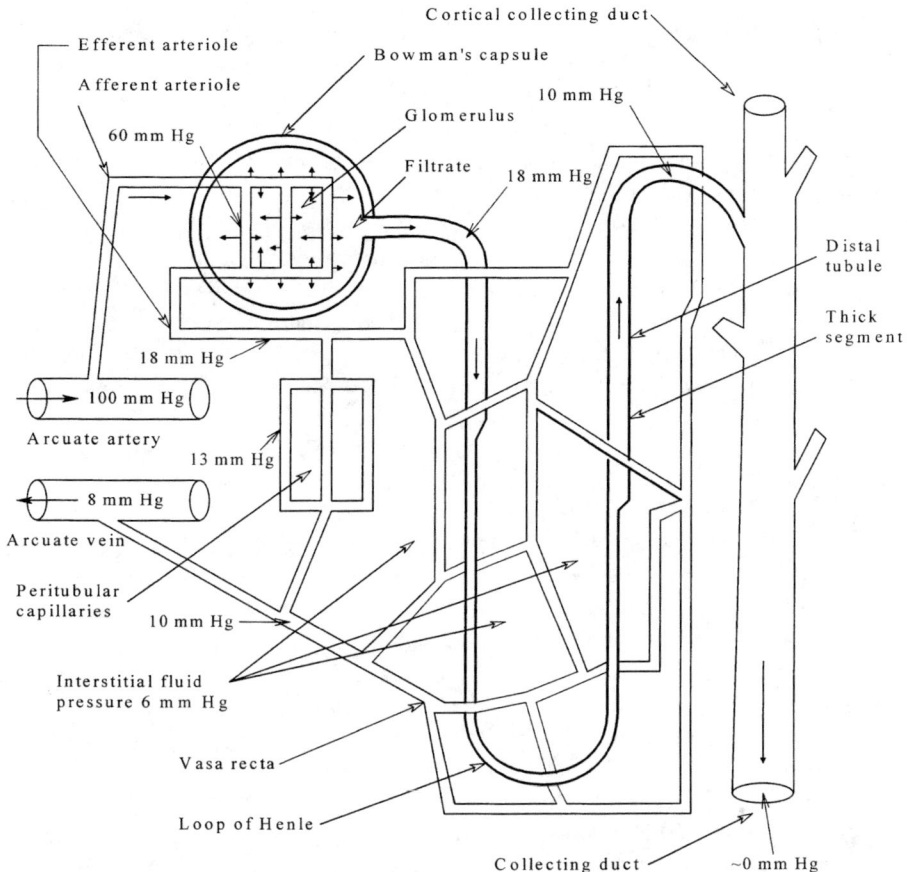

FIGURE 6.4 Schematic structure of a "typical" nephron. Cortical and juxtaglomerular nephrons differ in their size and proportions. Approximate normal hydraulic pressures are shown in the blood vessels of the kidney and the duct carrying filtrate.

substances in the blood form the filtrate. The blood, less the filtrate, next passes through the efferent arteriole, then through the peritubular capillaries, the *vasa recta*, and then to a venule leading to an arcuate vein. Pressures in the nephron system largely determine the rate of filtrate formation. Pressures are regulated by selective vasoconstriction mediated by local hormones and the autonomic nervous system.

A "normal" 70-kg male has a mean CO of 5600 ml/min of blood; of this flow, a *renal fraction* of about 1200 ml/min passes through the kidneys, or 21% of the CO. This renal fraction can be modulated by the body between 12 and 30%.[59] Total glomerular filtrate (GF) is formed at a mean rate of 125 ml/min, but it can be modulated between a few millimeters per minute to over 200 ml/min. The *filtration fraction* is defined by Guyton as the glomerular filtration rate (GFR) divided by the *plasma* flow rate through the kidneys. The normal plasma input to the kidneys is 650 ml/min, so the filtration fraction is 19%. (Note that *blood volume* depends not only on the liquid portion, plasma, but also the red blood cells. [Red blood cells are not filtered.]) Over 99% of the filtrate is absorbed by the tubular cells; otherwise, we would need a water intake of 7.5 l/hr to make up for the water loss alone!

The formation rate of GF depends not only on the hydraulic pressure difference between the inside of the glomerular capillaries and the filtrate pressure in Bowman's capsule but also on the *colloid osmotic pressure* (P_{CO}) of the blood due to high-molecular-weight proteins. The colloid osmotic pressure of the glomerular filtrate is ≈ 0 because no large-molecular-weight proteins normally enter the GF. The effective filtration pressure is $P_{eff} = P_{cap} - P_{CO} - P_{BC}$. Thus, in the normal case, the capillary pressure $P_{cap} = 60$ mmHg, $P_{CO} = 32$ mmHg, and the pressure in Bowman's capsule $P_{BC} = 18$ mmHg, so $P_{eff} = 10$ mmHg. The rate of formation of GF is proportional to P_{eff}.

Guyton[59] shows that the GFR is autoregulated. That is, the GFR normally remains constant at 125 ml/min over a MAP range of 75 to 160 mmHg, with P_{CO} remaining constant. This autoregulation occurs through modulation of the blood pressure in the Bowman's capsule by local hormonal control of the diameters of the efferent and afferent arterioles supplying the capsule. An optimum autoregulated GFR is necessary to optimally filter the blood in the capsule and reabsorb water, ions, glucose, etc. in the tubules at the normal rates. If the GFR is too high, there is not enough time for absorption in the tubules; if it is too low, there is excess absorption, urine output is too low, and the body cannot eliminate urea properly. The mechanism whereby each of the 2.4 million nephrons adjusts its GFR involves their *juxtaglomerular complexes,* which by local hormonal control modulate the hydraulic resistance of the afferent and efferent glomerular arterioles, thus adjusting P_{eff} and hence the individual GFR. A portion of the thick ascending limb of the renal tubule passes in close contact with the glomerular arterioles. Specialized dense epithelial cells in the juxtaglomerular portion of the ascending tubule, forming the *macula densa,* lie against specialized smooth muscle cells in the arteriole walls called juxtaglomerular cells. Guyton states that a low GFR causes excessive absorption of Na$^+$ and Cl$^-$ over the tubule so that the tubular fluid at the region of the *macula densa* is hypotonic. This hypotonic fluid in some way causes the *macula densa* cells to secrete some hormone that causes the juxtaglomerular muscle cells to secrete the protein enzyme renin. Renin causes the local formation of angiotensin II. Angiotensin II causes the smooth muscle cells of the efferent arteriole to contract, decreasing the diameter of the efferent arteriole, which raises its hydraulic resistance and the P_{eff} in the Bowman's capsule. An increase in P_{eff} of course raises the GFR. What we have described is a local type 0 regulator for GFR which operates on each individual glomerulus in the kidneys.

As mentioned above, over 99% of the GF is absorbed into the body through the cells in the walls of the tubules and collecting ducts.. Depending on the molecule or ion in the GF, the absorption mechanism is either active or passive. Tubular epithelial cells do selective *active transport* of glucose, Na$^+$, K$^+$, Ca^{++}, amino acids, Cl$^-$, PO$_4^-$, urate ions, etc.[59] to limit their loss in the urine. Water, on the other hand, is reabsorbed *passively* at a rate under hormonal control.

Different regions of every tubule perform specialized functions in terms of secretion and absorption of various ions and molecules. A tubule can be divided into three regions: (1) the *proximal (thick) segment*, (2) the *loop of Henle* (thin region), and (3) the *distal tubule* (a thick region). The tubule empties into the collecting duct, where some further absorption takes place. The collecting ducts converge on the ducts of Bellini, where the fluid is essentially urine as passed by the body.

First, we will focus on the actions of the brush-border epithelial cells that line the lumen of the proximal tubule. The bases of the brush-border cells face the inside wall of the tubule, the tubular basement membrane. The bases contain molecular pumps powered by the cleaving of the high-energy phosphate bonds of adenosine triphosphate (ATP) by the enzyme ATPase. The biochemical energy so released actively expels three Na^+ ions from the cell in exchange for two K^+ ions for every pump cycle. The expelled Na^+ ions end up back in the circulatory system by diffusing into peritubular capillaries. These sodium pumps create a low intracellular sodium concentration and about a -70-mV internal potential in the brush-border cells. Na^+ ions in the proximal tubule filtrate combine with *Na^+ carrier proteins* located in the brush-border membrane. The -70-mV internal potential and the low intracellular Na^+ concentration cause facilitated inward diffusion of Na^+ from the fluid in the tubular lumen. Thus Na^+ ions enter the brush-border cells passively down a potential *and* concentration gradient and are pumped out of the tubule into interstitial space by ATP-dependent molecular pumps. The Na^+ carrier proteins in the brush-border membrane also participate in *secondary active transport* of amino acids and glucose into the brush-border cells. Every time an Na^+ ion diffuses into the cell, it can carry with it a glucose or amino acid molecule, depending on the type of carrier protein. Other substances important to the body are reabsorbed by active processes in the brush-border cells of the proximal tubule, including certain proteins, vitamins, and acetoacetate ions. Large-molecular-weight proteins are taken up by pinocytosis into the brush-border cells, where they are broken down into amino acids which are released into interstitial space. Another type of Na^+ carrier protein figures in the secondary active transport of hydrogen ions from the cell into the tubular fluid. This third type of carrier protein is mostly found in the cells lining the distal tubule walls.

Substances that are actively absorbed from the tubular lumen exhibit *transport maxima*. This is because there is a finite number of active absorption sites and they work at essentially a fixed rate. Thus, if the concentration of a substance in the filtrate saturates the number of carrier proteins available, further increase in the concentration of the substance will be followed by a corresponding increase in the substance in the urine. A well-known example of a transport maximum occurs with glucose. If a total of 125 ml/min of GF is normally formed in the Bowman's capsules, and the normal average glucose concentration is 100 mg/dl, then the *tubular load* of glucose in the kidneys is 125 mg glucose per minute. Guyton[59] shows that when the tubular load of glucose exceeds 220 mg/min (176 mg/dl glucose in the blood), then significant amounts of glucose appear in the urine. Figure 6.5 illustrates the typical glucose loss rate in urine (in mg/min) vs. the tubular load of

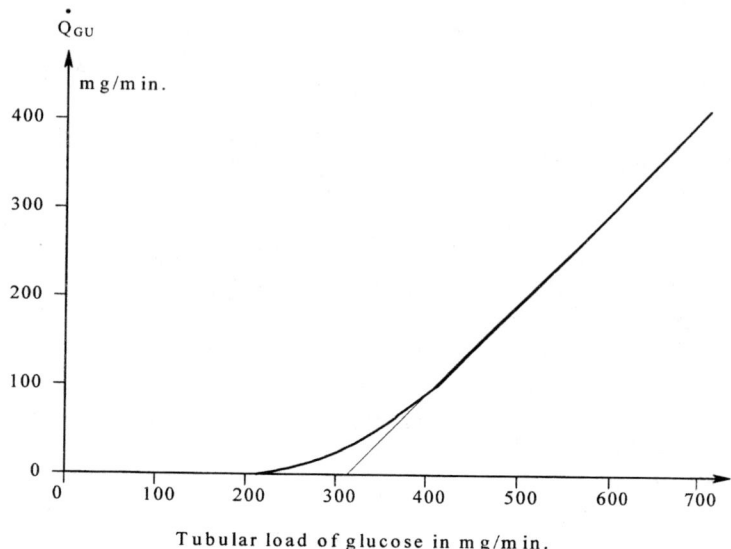

FIGURE 6.5 Typical glucose loss rate in urine as a function of the tubular load of glucose. (See text for explanation.) (From Guyton, A.C. 1991. *Textbook of Medical Physiology*, 8th ed., W.B. Saunders, Philadelphia. With permission.)

glucose. Extrapolating the linear portion of the curve to zero loss, we see that it intersects the tubular load axis at 320 mg/min. Thus the *transport maximum* is defined as 320 mg/min, equivalent to a blood glucose concentration of 256 mg/dl when the GFR is 125 ml/min. The abrupt appearance of glucose in the urine is an important diagnostic sign for hyperglycemia and diabetes mellitus. Table 6.1 summarizes transport maxima for important substances absorbed from tubular fluid.

TABLE 6.1
Tubular Transport Maxima for Important Substances Absorbed from GF

Substance	Transport Maximum	Units
Glucose	320	mg/min
Phosphate	0.1	mM/min
Sulfate	0.06	mM/min
Amino acids	1.5	mM/min
Urate	15	mg/min
Plasma protein	30	mg/min
Lactate	75	mg/min
Acetoacetate	About 30	mg/min
Hemoglobin	1	mg/min

Note that substances absorbed by diffusion do not have transport maxima per se (e.g., Na^+).

While many physiologically important ions and molecules are absorbed from the filtrate in the tubules by specialized tubular epithelial cells, certain low-molecular-weight molecules which are filtered or cleared from the plasma flowing through the glomeruli are not in any way absorbed from the filtrate. Such nonabsorbed substances appear in the urine at the rate at which they are filtered, and thus their concentration in the urine can be used as a measure of the plasma flow rate through the glomeruli when their plasma concentration is known. The polysaccharide *inulin* (not insulin), with a molecular weight of 5200 g, and the monosaccharide *mannitol* are nonabsorbed substances that are used for tests of kidney function (GFR).

For example, assume that inulin in saline is given intravenously so that the SS plasma concentration is 1 mg/ml, or 192 μM. The urine is collected by catheter over a period of 15 min. The cumulative urine volume is analyzed and found to contain 1875 g of inulin, or an average of 1875/15 = 0.125 g inulin per minute. If we divide 0.125 g/min by 0.001 g/ml, we find the GFR is 125 ml/min, assuming that *all* of the inulin passing through the glomeruli is cleared. Note that the low plasma molar concentration of inulin causes a negligible rise in the plasma osmotic pressure.

para-Aminohippuric acid (PAH) is a substance that can be used to estimate the total plasma flow through the kidneys. While inulin is cleared through the glomeruli, PAH is cleared through both the glomeruli and the peritubular capillaries into the tubules, so that about 91% of the plasma PAH is removed from the total volume of plasma passing through the kidneys. For example, assume that the concentration of PAH in the plasma is 1 mg/dl, and 5.85 mg of PAH is passed into the urine per minute. Thus the total plasma flow is 5.85/(0.91 × 1) = 6.43 dl/min.

In general, the ability of the kidneys to "clear" *any substance* from the plasma can be calculated by the formula[59]

$$\text{Plasma clearance (ml/min)} = \frac{\text{Urine flow rate (ml/min)} \times \text{Conc. in urine (mg/dl)}}{\text{Conc. in plasma (mg/dl)}}$$

(*Any substance* includes those with transport maxima.)

In summary, we have demonstrated that the kidneys and their components are indeed remarkably complex biochemical machines. They are indeed multiple-input/multiple-output systems that have both autonomic nervous and hormonal control inputs.

6.3 THE REGULATION OF PLASMA SODIUM ION CONCENTRATION AND OSMOLARITY

First, *osmotic pressure* and *osmolarity* will be defined, and then these parameters will be discussed in the context of physiological regulation. If a nonionic substance

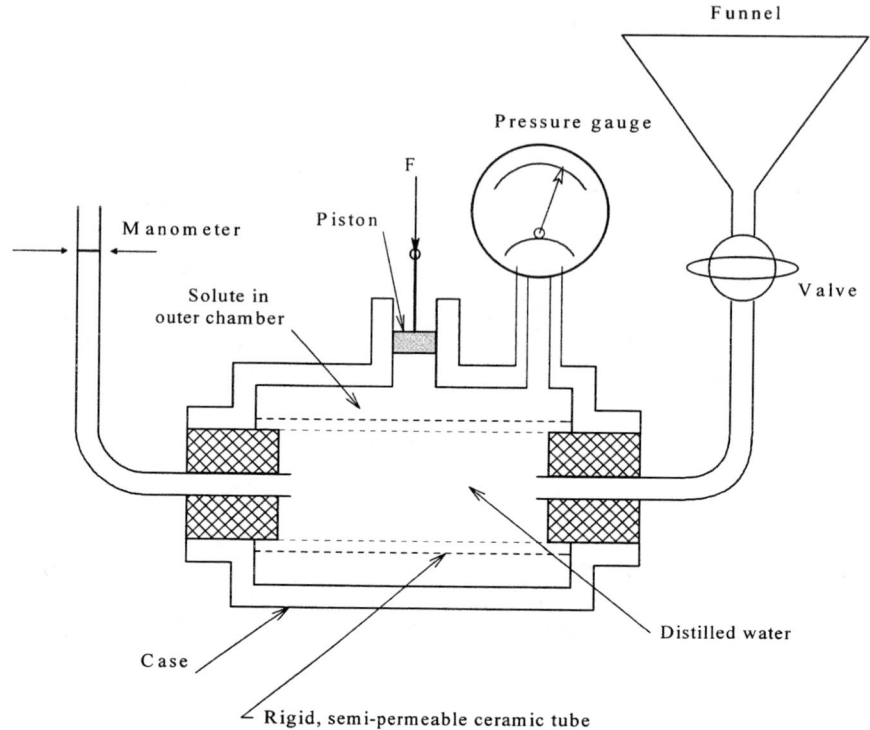

FIGURE 6.6 Apparatus to measure the osmotic pressure of a solution. The porous, rigid, ceramic tube is permeable to water but not solute in outer chamber. The piston of area A applies pressure P_s to solution to counteract the osmotic pressure of water leaving the porous tube.

such as glucose is dissolved in water and placed in a chamber separated from pure water by an ideal *semipermeable membrane* (a membrane which allows water molecules to pass freely in either direction, but not glucose molecules), there will be a tendency for water to pass from the pure water side of the membrane into the solution. Now, if hydraulic pressure is applied to the solution, it will counteract the tendency of water to diffuse through the membrane into the solution. The exact pressure which balances the diffusion tendency is called the *osmotic pressure* of the solution. Figure 6.6 illustrates an apparatus used to measure osmotic pressure. The solution pressure is raised until the water manometer is brought back to its initial value, and the osmotic pressure (hydraulic pressure on the solution) is read.[90] It can be shown that the osmotic pressure of a solution of one solute is given by $P_{osm} = RCT$. P_{osm} is the osmotic pressure in atmospheres, R is the gas constant = 0.0821 l atmosphere per mole degree, C is the molar concentration of the solvent (g mol/l), and T is the Kelvin temperature. If an ionizable solute is used, both + and −ions exert osmotic pressure, so $C = 2M$, where M is the molar concentration of NaCl, for example. If more than one molecular species is present, then $C = M_1 + M_2 + M_3$, etc., where M_k are the molar concentrations of each nonionizable molecule. Thus the total osmotic pressure of a mixed solution can be broken down to the sum of the partial osmotic pressures of the various solutes. Note that 1 mol/l of glucose solution has the same osmotic pressure as $1/2$ mol/l of NaCl solution, or 1 mol/l of albumen with a molecular weight of 70,000 g/mol.

Osmolarity is defined as the number of osmoles per liter of solution; *osmolality* is defined as the number of osmoles per liter of water (solvent). The *osmole* is 1 g molecular weight of a nonionizable solute. For example, 1 osm of glucose is 180.16 g; 1 osm of sodium chloride is 58.44 ÷ 2 g. Finally, at physiological temperatures (37°C = 310 K), the osmotic pressure in millimeters of mercury can be written: $P_{osm} = 19.3 \times$ mosm/kg H_2O.

Data given by Guyton[59] show that the total osmotic pressure at 310 K is 5450 mmHg for plasma, 5430 mmHg for *interstitial fluid* (IF), and 5430 mmHg for intracellular fluid. P_{osm} is higher for plasma because it contains more high-molecular-weight proteins than does IF. Most of the P_{osm} of plasma and IF is due to Na^+ and Cl^- ions. In fact, Na^+ and Cl^- account for 248 mosm/l out of a total of 281.3 mosm/l in IF and 249 mosm/l out of a total of 282.5 mosm/l in plasma. Thus regulation of the body's sodium ion concentration has a large effect on the plasma and IF P_{osm}.

The mean concentration of sodium ions in interstitial fluid is about 140 mosm/l of H_2O. $[Na^+]$ is regulated so that it normally ranges from 135 to 147 mosm/l of H_2O. The nonlethal range of $[Na^+]$ is 115 to 175 mosm/l of H_2O.[59] Sodium ions are essential for the normal generation of nerve action potentials, the proper flow rate for aqueous humor in the eye, etc.

There are two major components of the $[Na^+]$ and P_{osm} regulatory system: (1) *osmoreceptor neurons* located in the *supraoptic nucleus of the hypothalamus*, the terminals of which are located in the posterior half of the pituitary gland, and (2) when stimulated by high P_{osm}, the hormone vasopressin, also known as antidiuretic hormone (ADH), is secreted from the osmoreceptor nerve endings. ADH is a novapeptide, that is, a small protein made from nine amino acids:

$$\text{CYS-TYR-PHE-GLN-ASN-CYS-PRO-ARG-GLY-NH}_2$$

The released ADH enters the capillary blood and is transported by the circulatory system to target cells with ADH receptors in the kidneys. The cells most affected by ADH are the epithelial cells in the cortical collecting tubules and the collecting ducts. A raised concentration of ADH causes the increased permeability of the collecting tube cells to water. Thus, raised ADH concentration increases the rate of passive diffusion of water from the collecting tube lumen into extracellular space. This increased water retention causes IF P_{osm} to decrease and is seen to be a type 0 hormonal negative feedback loop. Under high [ADH] conditions, the urine passed is very concentrated due to the water absorption. Note that the number of Na^+ ions in the urine is unaffected by ADH; it only acts on water.

The concentration of ADH in the blood can vary widely, depending on circumstances. The "normal" concentration of ADH is 1 to 2 ng/l. Stress (trauma, pain, anxiety), morphine, nicotine, low MAP due to blood loss or dehydration, tranquilizers, and anesthetics all act as antidiuretics and promote water retention in the body through an enhanced secretion rate of ADH. Alcohol, on the other hand, is a diuretic; it promotes water loss. It is known that a BV loss greater than 25% will cause a 20- to 50-fold increase in the rate of ADH secretion.[59] The ADH–P_{osm} system operates around a set point or bias level. That is, if P_{osm} decreases (e.g., due to excessive water intake), the osmoreceptors receive less stimulation and the ADH concentration falls, causing less water to leave through the collecting duct walls and the production of greater quantities of dilute urine.

Because many other factors affect IF P_{osm}, such as dietary water intake and abnormal levels of proteins or glucose, and other physical and psychogenic factors affect the secretion rate of ADH, the Na$^+$–P_{osm} system is truly multiple input/multiple output and is effectively cross-coupled with systems that regulate [K$^+$] and MAP. We will develop a comprehensive block diagram describing the functional interrelationships between the IF concentrations of K$^+$, Na$^+$, P_{osm}, and dietary inputs of K$^+$, Na$^+$, and water. Figure 6.7 shows a simple systems block diagram of the Na$^+$–P_{osm} portion of this regulator. The action of the posterior pituitary's osmoreceptor cells is illustrated on the left. The rate of secretion of ADH as a function of P_{osm} follows a typical sigmoid curve, showing saturation in Q_{ADH} at high P_{osm}. We assume that ADH is broken down (inactivated) at a rate proportional to its concentration (this assumption follows simple first-order mass-action kinetics). The IF

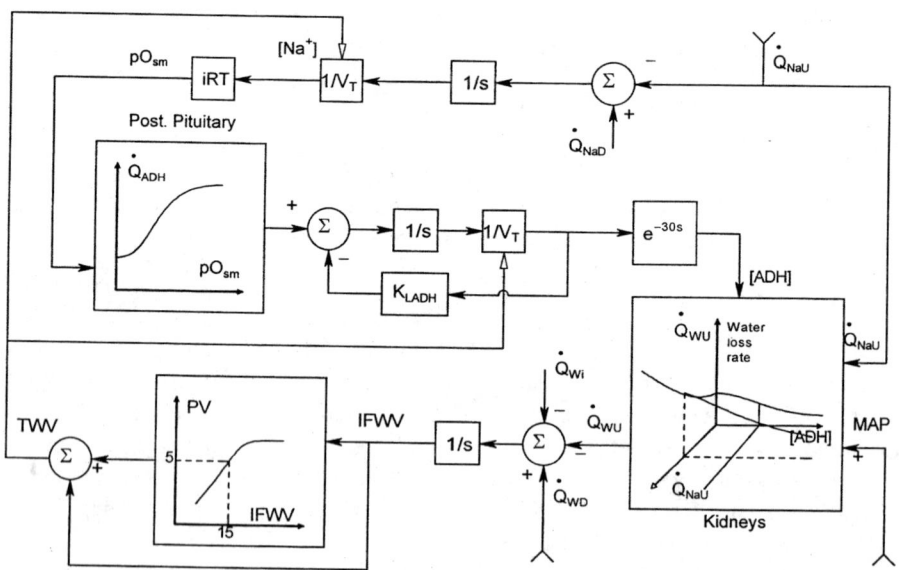

FIGURE 6.7 Systems block diagram of the plasma sodium ion concentration/osmotic pressure regulatory system. Note that the total (extracellular) water volume (TWV) is the sum of plasma water volume and IF water volume. TWV = V_T in the diagram. See text for description of operation.

concentration of ADH affects the water loss rate in the urine, as described above. A high [ADH] produces low water loss rate, a high water retention rate, Q_{WR}, and concentrated urine. Water retention rate is summed with insensible water loss rate and water input rate. The net water balance rate is integrated to give the *total interstitial fluid water volume* (IFWV). BV follows IFWV with a saturating curve. Normal IFWV is about 15 l in a 70-kg man, yielding a BV of 5 l. Total (extracellular) water volume (TWV) in the body is assumed to be the sum of IFWV + water in the blood. Water in the blood is estimated as 0.6 BV. Thus TWV ≅ 18 l. The P_{osm} is seen to be due to the total concentrations of ions and nonionizable solutes in the IF. A fraction of 140/281.3 mosm/l H_2O is the normal component of P_{osm} due to Na^+ in the IF. Thus IF [Na^+] contributes about 50% of the total IF P_{osm}. K^+ normally contributes 4/281.3 → 1.6% of the total IF P_{osm}, and other positive ions contribute less. If we assume the remaining solutes have fixed concentration, and neglect the effect of K^+, Mg^{++}, and Ca^{++} on P_{osm}, then we can write:

$$P_{osm} = 19.3\left(140.65 + [Na^+]\right) \tag{6.7}$$

Note that the concentrations are in milliosmoles per liter H_2O, and P_{osm} is in pressure units of millimeters of mercury. P_{osm} can also be written in terms of TWV:

$$P_{osm} = 19.3\left(\{2.53E3\} + \{Na^+\}\right)/TWV \tag{6.8}$$

{Na^+} is the total extracellular milliosmoles of sodium. The total water volume (TWV) is in liters. {Na^+} is found by integrating the net sodium rate balance. The net sodium rate balance is ($\dot{Q}_{NaD} - \dot{Q}_{NaU}$) mosm/min. $\dot{Q}_{NaD}$ is the dietary input, and $\dot{Q}_{NaU}$ is the loss rate of Na^+ in the urine (assumed constant in this case). Note that the integrators in the diagram are state integrators, required to simulate first-order ordinary differential equations.

To examine heuristically how the Na^+–P_{osm} system responds to an *increase* in insensible water loss rate, it is immediately apparent that the IFWV responds to a step increase in $\dot{Q}_{WI}$ by ramping down. This decrease in IFWV leads to a decrease of TWV. The *decrease* in TWV causes an *increase* in P_{osm}, other factors remaining constant. Increased P_{osm} causes an increase of Q_{ADH} and, after a first-order lag, an increase in blood [ADH]. Increased [ADH] causes a decreased water loss rate in the urine and hence an *increase* in IFWV and BV. Thus a negative feedback causality has been demonstrated for the Na^+–P_{osm} regulator.

Note that a sudden *increase in blood volume* (as from a transfusion of whole blood) causes an *increase* in MAP. Increased MAP in turn causes *increased water loss rate* in the urine (part of the slow blood pressure regulator system), *decreasing* IFWV and hence BV and TWV. Decreased TWV causes an *increase* in P_{osm} and hence an increase in [ADH]. Increased [ADH] decreases the water loss in the urine and hence *increases* IFWV and BV. Guyton[59] observes:

> One mechanism above all others dominates the control of BV and extracellular fluid volume. This is the effect of blood volume on arterial pressure, on the one hand, and

the effect, on the other hand, of arterial pressure on urinary excretion of sodium and water. This interplay between the kidneys and the circulatory system is so powerful in controlling blood volume — and secondarily extracellular fluid volume as well — that it dominates over the other mechanisms that help to control salt and water intake or salt and water excretion.

The dominance of the BV regulator lies in the central nervous system. If BV is caused to increase abruptly, there is a corresponding increase in venous return pressure to the heart. This increase in pressure stimulates stretch receptors in the walls of the left and right atriums. The stretch receptors send signals to the autonomic central nervous system, which in turn sends sympathetic inhibitory signals to the kidneys' arterioles, which increase the blood pressure in Bowman's capsules, thus increasing the GFR. Increased GFR leads to increased urine output (increased water *and* salt loss). The central nervous system also sends signals to the neurosecretory cells that secrete ADH, reducing the rate of production of ADH. Reduced ADH, of course, leads to increased water loss in the urine. Causality in the BV regulator is summarized in Figure 6.8. This is a very high-gain type 0 system. Using the implicit summing point approach described in Section 6.1, we see that the steady-state sensitivity of the BV system is very low to changes in rate of fluid intake. It is easy to show:

$$\Delta BV = \Delta \dot{Q}_{WI} \frac{1}{K_{CO} K_{MAP} K_{WU}} \tag{6.9}$$

Guyton observes that because the gains in Figure 6.8 are so high, the static BV regulation vs. $\dot{Q}_{WI}$ curve shown in Figure 6.9 is flat over a wide range. It should be pointed out that at the high end of the curve, BV regulation can be at the expense of water storage in extracellular fluid space in the form of a gel, creating the

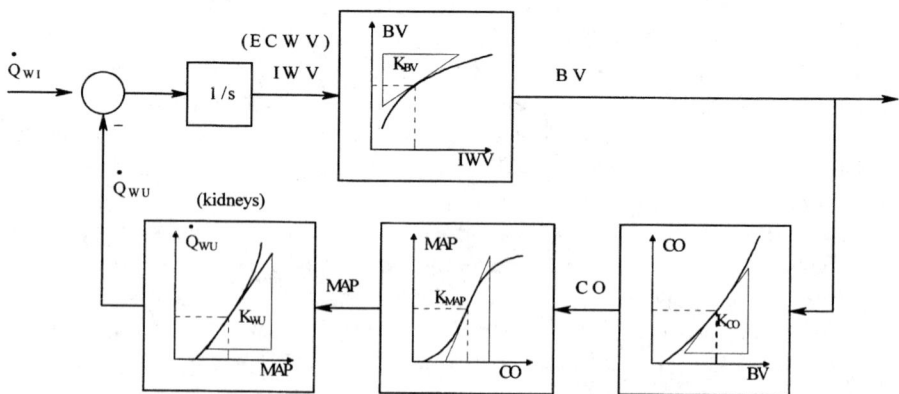

FIGURE 6.8 The BV/average water input rate regulator. See text for description of this high-gain system.

Hormonal Regulation of Sodium, Potassium, Calcium, and Magnesium Ions

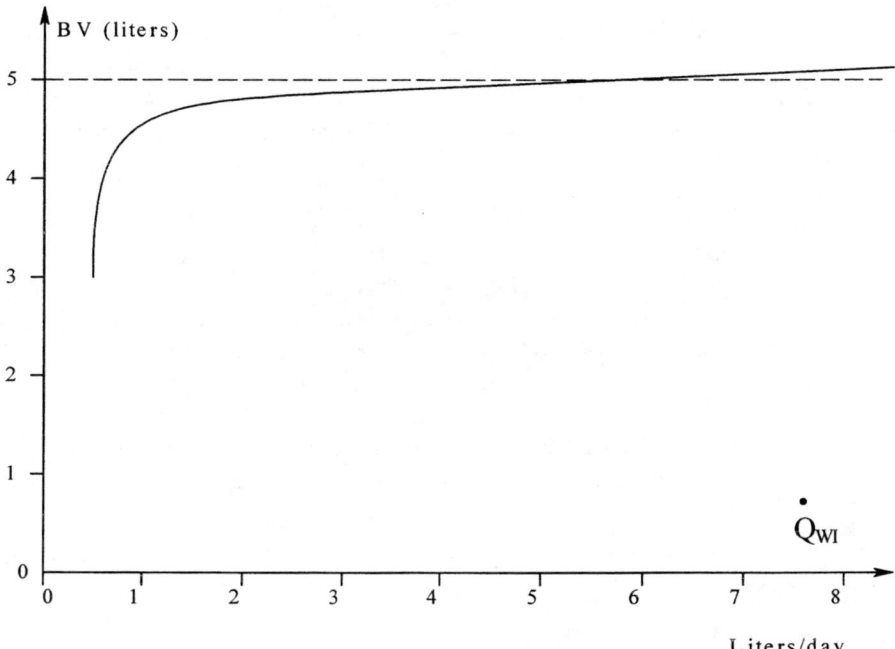

FIGURE 6.9 Curve showing static BV regulation vs. average daily water input. The relatively low slope of the curve from 3 to 8 l/day input is indicative of a high-gain type 0 control system. (From Guyton, A.C. 1991. *Textbook of Medical Physiology*, 8th ed., W.B. Saunders, Philadelphia. With permission.)

condition known as *edema*. Edema normally does not occur in healthy persons because the kidneys shed excess water before the conditions for edema formation occur.

6.4 REGULATION OF PLASMA POTASSIUM IONS BY THE ALDOSTERONE SYSTEM

The action of the K^+–aldosterone control system is to maintain plasma and IF [K^+] within narrow physiological bounds in spite of dietary inputs and other renal events, such as changes in the GFR. The normal (regulated) range of [K^+] in IF is 3.5 to 5.0 meq/l; the nonlethal range is 1.5 to 9.0 meq/l. There is about 3500 meq K^+ in the body, of which only 60 meq is in the IF and extracellular fluid. Potassium is concentrated in the cells by ATP-driven, active transport "pumping" in which Na^+ is expelled from inside cells and K^+ is pumped in, both ions against concentration and potential gradients.

The adrenal cortical hormone *aldosterone* (ALDO) is involved in a complex manner in both the regulation of blood volume and IF [K^+]. Even though Na^+ ions figure in these coupled regulatory systems, aldosterone concentration has little effect on IF [Na$^+$], which is regulated by ADH.[59] Conversely, [ADH] has little effect on IF [K$^+$] if the ALDO system is intact. The cells in the *zona glomerulosa* of the adrenal cortex secrete ALDO primarily in response to IF [K$^+$]. The mechanism by which IF [K$^+$] affects the ALDO secretion rate is not known. An increase in [angiotensin II] in the blood causes an increase in the rate of ALDO secretion as well, but the angiotensin II mechanism is about 1/100 as effective as the [K$^+$] stimulus. Adrenocorticotropic hormone (ACTH) is important for ALDO secretion in that if it is absent, no ALDO will be secreted! Some [ACTH] is required for normal IF [K$^+$]-controlled secretion of ALDO. Guyton[59] states that a 10 to 20% decrease in IF [Na$^+$] can on rare occasions double Q_{ALDO}. It is generally agreed that IF [Na$^+$] and [ACTH] have minor effects on the secretion of ALDO by the adrenal cortex. The "normal" Q_{ALDO} = 150 μg/day, or 104.2 ng/min (average). The mean IF concentration of ALDO is 100 ng/l. ALDO is broken down in the liver, and by-products are secreted in the feces and urine. A simple first-order loss model for ALDO gives the ordinary differential equation

$$[\dot{ALDO}] = -\left(TWV \cdot K_{LALDO}\right)[ALDO] + TWV \cdot \dot{Q}_{ALDO} \quad \text{ng/min} \quad (6.10)$$

where TWV = total water volume in which ALDO is dissolved $\cong$ 18 l. Laplace transforming the ordinary differential equation, we can write:

$$\frac{[ALDO]}{Q_{ALDO}}(s) = \frac{1/TWV}{s + K_{LALDO}/TWV} \quad (6.11)$$

Using the Laplace final value theorem and the values for [ALDO] and $\dot{Q}_{ALDO}$, we find that K_{LALDO} = 1.042 l/min. The washout (loss) time constant for ALDO is

$$\tau = TWV/K_{LALDO} = 18/1.042 = 17.3 \text{ min} \quad (6.12)$$

We now examine the actions of ALDO in the body. Its principal effect is in the kidneys, but it also targets cells in the salivary glands, sweat glands, and the intestinal mucosa in the colon. The action of ALDO on these cells causes them to absorb Na$^+$ back into the body. Water generally follows the absorption of Na$^+$ in the intestines, so P_{osm} is preserved, and BV increases. The effect of ALDO in the kidneys is mainly on the epithelial cells of the collecting ducts. Here, raised [ALDO] causes an increase in the rate of pumping where potassium is exchanged for sodium. Specifically, high [ALDO] increases the rate at which K$^+$ is transported into and Na$^+$ is transported out of the outer borders of the collecting duct's epithelial cells. K$^+$ then diffuses out of the inner borders of the duct's epithelial cells into the duct's lumen, and sodium diffuses out of the duct's fluid into the epithelial cells. Figure 6.10 illustrates the effect of IF [K$^+$] on $\dot{Q}_{ALDO}$. (Note that the $\dot{Q}_{ALDO}$ vs. [K$^+$] curve is really a hypersurface, because $\dot{Q}_{ALDO}$ is affected by angiotensin II and adrenocorticotropin as well as [K$^+$].) The IF concentration of potassium *increases* the loss rate

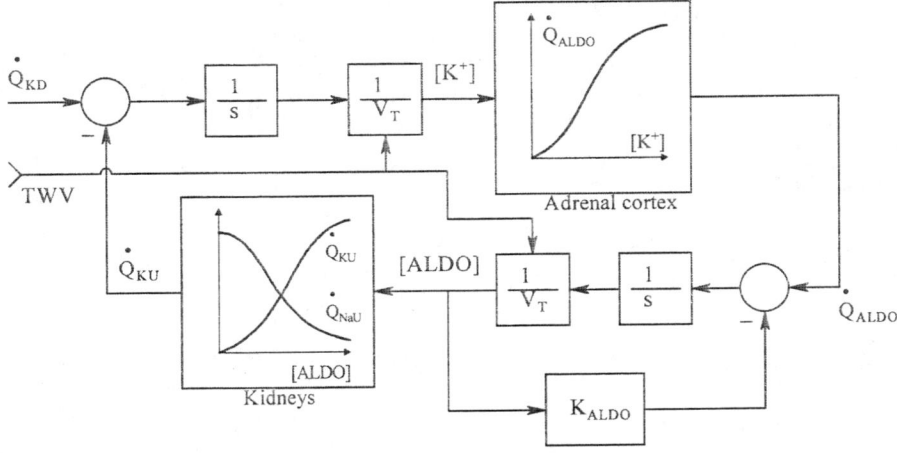

FIGURE 6.10 Block diagram of the [K$^+$]–ALDO. Note that TWV enters parametrically. First-order loss dynamics are assumed for ALDO.

of K$^+$ in the urine and *decreases* the loss rate of Na$^+$ in the urine. This means more Na$^+$ is retained in the body when [ALDO] is high. Paradoxically, the aldosterone-induced retention of Na$^+$ does not significantly raise IF [Na$^+$] or P_{osm}. Apparently water is reabsorbed following the sodium ions, keeping P_{osm} constant. Thus nature has to a large degree decoupled the K$^+$ and Na$^+$ regulatory systems, even though they share a common pump.

A comprehensive schematic for the regulation of IF [Na$^+$], [K$^+$], and extracellular fluid volume is shown in Figure 6.11. At the top of the diagram we see the [K$^+$] regulator. Note that there is an inverse relationship between potassium and sodium loss rates. Total water volume enters the system parametrically, determining K$^+$ and Na$^+$ concentrations and P_{osm}. Note that P_{osm}, derived from [Na$^+$], also activates the subject's *drinking behavior*, described here by a hysteresis function. Nature has provided man and other mammals with a drinking *satiety* behavior mode in which just enough water is drunk to restore a normal P_{osm} and then the urge to drink disappears. There is a 1/2 to 1-hr delay for all of the water drunk to become uniformly distributed in the body and give a steady-state lowered P_{osm}. If the hysteresis were not present, this lag would cause a tendency to overdrink, yielding a hypotonic P_{osm}. Of particular interest in the diagram is the relation between water loss rate and the sodium loss rate and [ADH]. It is clear that high [ADH] increases the rate of water absorption back into the IF and thus reduces water loss in the urine. At high IF [K$^+$] levels, IF [ALDO] is elevated and ATPase pumping is stimulated, so there is an increased K$^+$ loss rate in the urine, and as a result of the K$^+$–Na$^+$ coupling in the pump, there is a reduced rate of Na$^+$ loss in the urine. That is, Na$^+$ ions from the collecting duct fluid are pumped back into the IF at an elevated rate. Water follows these sodium ions, keeping P_{osm} relatively constant in the IF around the collecting ducts. Thus high IF [K$^+$] leads to reduced water loss rate in the urine. This

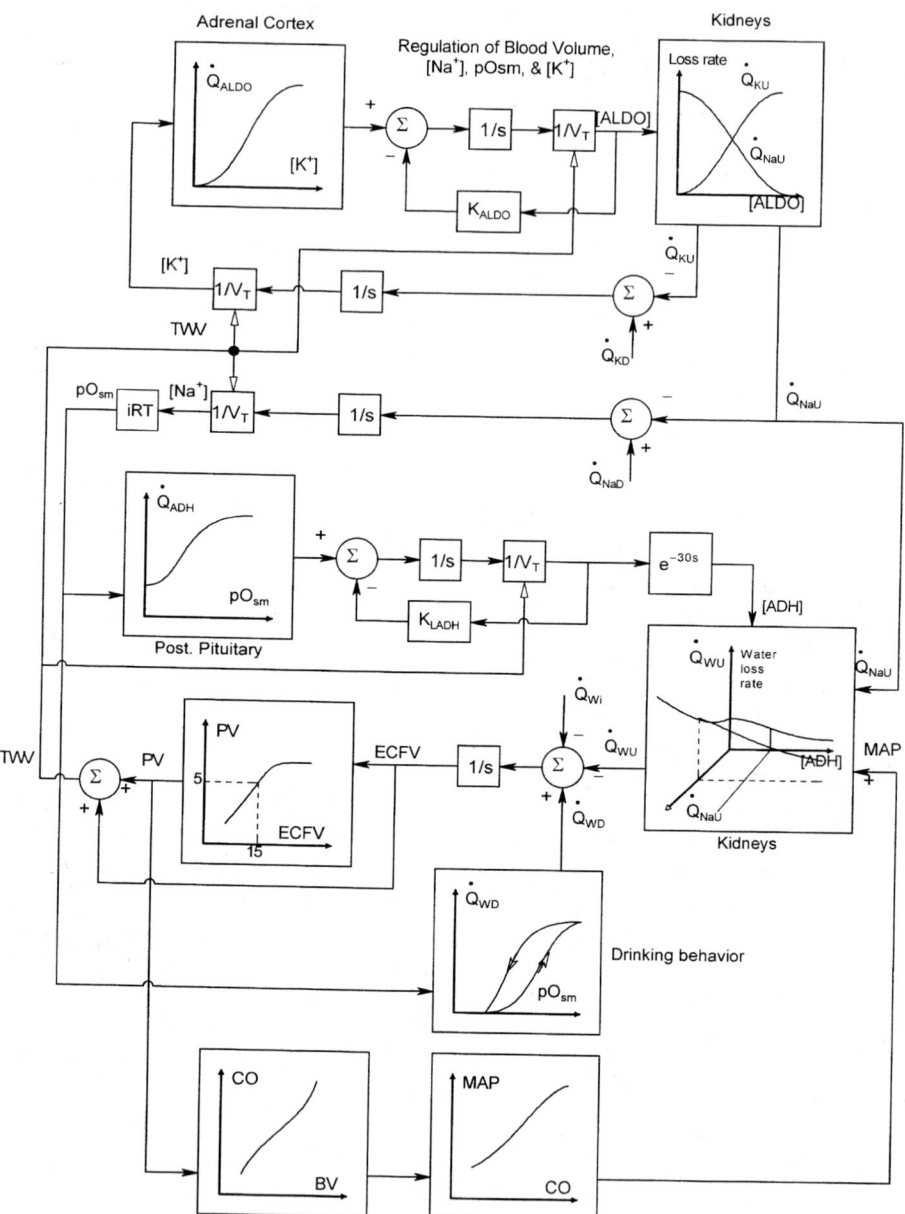

FIGURE 6.11 Block diagram fusing the [K⁺]–ALDO regulator, the plasma sodium ion–osmotic pressure regulator, and the extracellular fluid volume (ECFV) regulator. The parameter V_T is critical in determining concentrations.

mechanism makes the $[Na^+]/P_{osm}$–ADH system relatively insensitive to the operation of the $[K^+]$–ALDO regulator. The net water balance includes insensible water loss rate (Q_{WI}), water input from drinking (Q_{WD}), and water loss rate through the urine affected by ADH- and ALDO-induced sodium pumping (Q_{WU}). In addition (not shown), an 18-l impulse input of water is needed to set the TWV initial condition in the model. Similarly, appropriate input impulses to the Na^+ and K^+ summers are required to set the normal SS $[Na^+]$ and $[K^+]$.

In summary, the regulators for IF $[K^+]$, IF $[Na^+]$ (P_{osm}), and BV are intimately coupled through the mechanisms of the kidneys' excretory and absorption mechanisms. Nature has effectively decoupled the $[Na^+]$–ADH system from the $[K^+]$–aldosterone system so that if the ALDO system is blocked, $[Na^+]$ is still regulated.

6.5 CALCIUM ION REGULATION

Calcium is another very important ion in the body because of its role in muscle contraction and relaxation, nerve synaptic function, cardiac function, control of hormone synthesis, etc. Normal total plasma calcium ranges from 9 to 11 mg/100 ml of plasma, with a mean of 9.5 mg/dl (equivalent to 2.4 mM/l or 4.8 meq/l). Plasma calcium is distributed in three compartments: (1) 50% is ionized and is diffusible through capillary walls into extracellular fluid volume, (2) 40% is bound to large protein molecules which normally cannot diffuse into extracellular fluid volume, and (3) 10% of plasma calcium is *chelated* with a high-affinity bond to citrate ions. In addition to the three plasma calcium compartments, calcium is also present in two IF (extracellular) compartments, as ionized Ca^{++} and as chelated calcium. The active portion of bones in the body also provides an important buffer compartment for calcium storage and release. Hormones regulate the storage and release of calcium in bone and the uptake and excretion of dietary calcium.

Hypocalcemia (serum $[Ca^{++}]$ < 7 mg/dl) causes hyperexcitability in the central nervous system and peripheral nervous system (PNS). There is increased neuronal membrane permeability. Muscle cramps occur. At $[Ca^{++}] \approx 6$ mg/dl, muscles go into tetanus, and there is laryngospasm and airway obstruction. The diaphragm cannot function. Death usually occurs at $[Ca^{++}]$ < 4 mg/dl. When *hypercalcemia* (serum $[Ca^{++}]$ >12 mg/dl) occurs, the muscles are weak and sluggish, and the nervous system is slow and depressed. There is an increased Q-T interval in the ECG, with slow systole. Hypercalcemia also causes constipation due the loss of gut motility and general loss of appetite. Above 17 mg/dl, there is precipitation of $CaPO_4$ in the tissues (the "Lot's wife" syndrome). It is apparent that hypocalcemia is a far more immediately life-threatening condition than is hypercalcemia.

Other calcium-related conditions are *rickets* and *osteoporosis*. Rickets is generally seen in children who are fed a diet deficient in vitamin D. In the absence of vitamin D, calcium input from diet falls, excess PO_4 ions are excreted, and serum $[Ca^{++}]$ falls. Low serum $[Ca^{++}]$ causes elevated parathyroid hormone, which promotes increased osteoclast activity and weakened bones. Treatment of rickets includes vitamin D and a high-calcium diet.

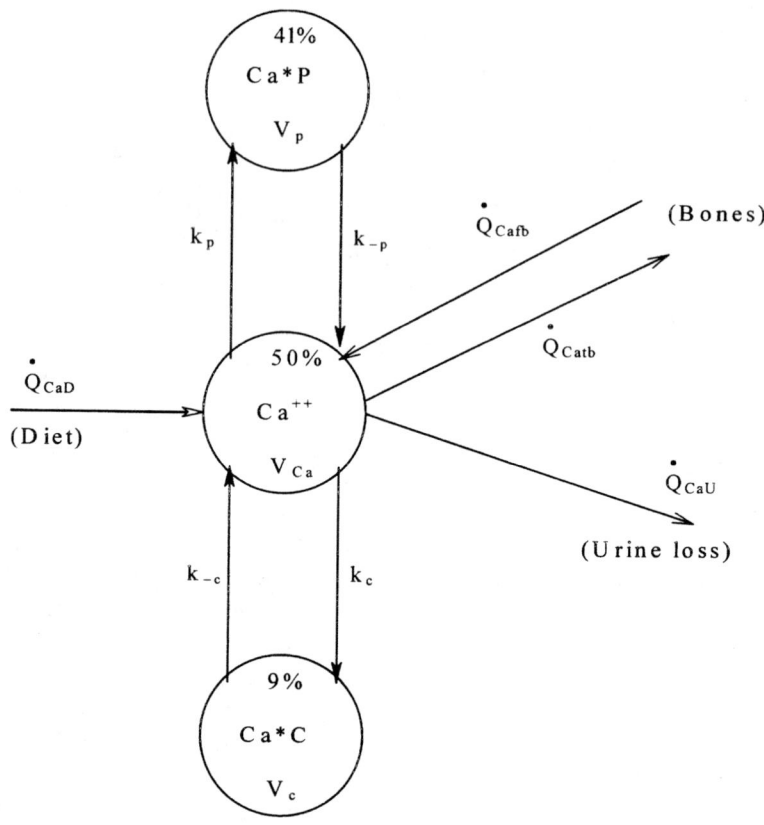

FIGURE 6.12 The three calcium compartments: central, extracellular, and extraosteo. Free ionic calcium and calcium bound to proteins and as chelate are shown.

Osteoporosis is not a calcium-regulator-related disease, even though weakened, brittle bones are its principal symptom. Osteoporosis is not a "single-cause" disease. Many factors can contribute to its onset and progress. It is associated with (1) a lack of use of the bones, (2) malnutrition including lack of vitamin C and protein in the diet, (3) estrogen decline in menopause, (4) Cushing's disease, (5) lack of insulin, and (6) excess adrenocortical hormones.

The three central calcium compartments are illustrated in Figure 6.12. Note that the protein-bound calcium acts as a buffer. It exchanges calcium with the central ionized calcium compartment, as does the citrate chelate compartment. The plasma Ca^{++} compartment also exchanges calcium mass with the bones, the gut (digestive system), and the kidneys. The calcium regulatory system has two inputs (dietary calcium and vitamin D), several storage forms (bone, protein complex, chelate, intracellular calcium), and two outputs (in urine and feces). Three state equations can be written to describe the mass transfers among the three plasma compartments. Diffusion and mass-action kinetics are the controlling processes. We assume that citrate- and calcium-binding protein are present in excess. The equations are

$$[\dot{Ca}*P] = -K_{-p}[Ca*P]/V_P + K_P[Ca^{++}]/V_{Ca} \qquad (6.13A)$$

$$[\dot{Ca}*C] = -\left(K_{-C} + K_{LC}\right)[Ca*C]/V_C + K_C[Ca^{++}]/V_{Ca} \qquad (6.13B)$$

$$[\dot{Ca}^{++}] = -\left(K_C + K_P\right)[Ca^{++}]/V_{Ca} + K_{-C}[Ca*C]/V_C + K_{-p}[Ca*P]/V_P$$
$$- \dot{Q}_{CaU} - \dot{Q}_{Catb} + \dot{Q}_{Cafb} + \dot{Q}_{CaD} \qquad (6.13C)$$

where [Ca * P] is the concentration of calcium * protein complex in the circulatory system fluid volume, V_P; [Ca * C] is the concentration of calcium * citrate chelate in the chelate volume, V_C; [Ca^{++}] is the concentration of ionized calcium in the circulatory system and extracellular fluid volume, V_{Ca}; $\dot{Q}_{CaU}$ is the loss rate of Ca^{++} in the urine; $\dot{Q}_{Catb}$ is the loss rate of Ca^{++} into the bones (storage); $\dot{Q}_{Cafb}$ is the input rate of Ca^{++} released from the bones; and $\dot{Q}_{CaD}$ is the input rate of Ca^{++} from diet.

The calcium system is interesting because it is regulated through the action of two hormones: *parathyroid hormone* (PTH), which is an 84-amino-acid protein, and *calcitonin*, a 32-amino-acid polypeptide secreted by the thyroid glands' parafollicular "C" cells. The rate of release of both hormones is controlled by IF [Ca^{++}]. The rate of release of PTH is *decreased* by high [Ca^{++}]. Conversely, high [Ca^{++}] *increases* the release rate of calcitonin from the thyroid. The normal concentration of PTH ranges from 100 to 300 ng/l, and calcitonin concentration in the blood is normally less than 100 ng/l, with a half-life of about 10 min.[141] Of the two hormones, PTH is the most potent; the bones and the kidneys are its principal target organs. Figure 6.13 illustrates the major relationships in the [Ca^{++}] regulatory system. Figure 6.14 shows how PTH acts on certain kidney tissues to cause them to increase the rate of conversion of 25-hydroxycholecalciferol (HCC) to the active hormone 1,25-dihydroxycholecalciferol (DHCC), which acts on the intestinal epithelium to increase the level of calcium-binding protein and the enzymes calcium-stimulated ATPase and alkaline phosphatase. The net result of these actions is an increase in the rate of [Ca^{++}] absorption from the gut, $\dot{Q}_{CaD}$. Note that vitamin D (cholecalciferol [CC]) can come either from diet or the action of ultraviolet light on the skin to convert 7-dehydrocholesterol to CC. CC is converted in the liver to HCC. There is a high-gain, biochemical, negative feedback loop whereby the concentration of HCC inhibits the liver conversion of CC to more HCC. Figure 6.15 illustrates how this local feedback loop stabilizes the concentration of HCC against variation in the rate of vitamin D intake. Only at very low vitamin D intake rates does the regulation fail.

Further evidence for the high gain in the IF [Ca^{++}] regulator is illustrated in Figure 6.16. The linear portion of the plasma concentration of the active form of vitamin D, DHCC, vs. plasma [Ca^{++}] is very steep. The normal "operating point" of plasma [Ca^{++}] vs. DHCC is shown to be in the linear range of the curve. From a nonlinear control point of view, one might argue that there is a "bang-bang" relation between DHCC concentration and plasma [Ca^{++}]. That is, the concentration of DHCC becomes very small for plasma [Ca^{++}] > 10 mg/dl, drastically reducing $\dot{Q}_{CaD}$.

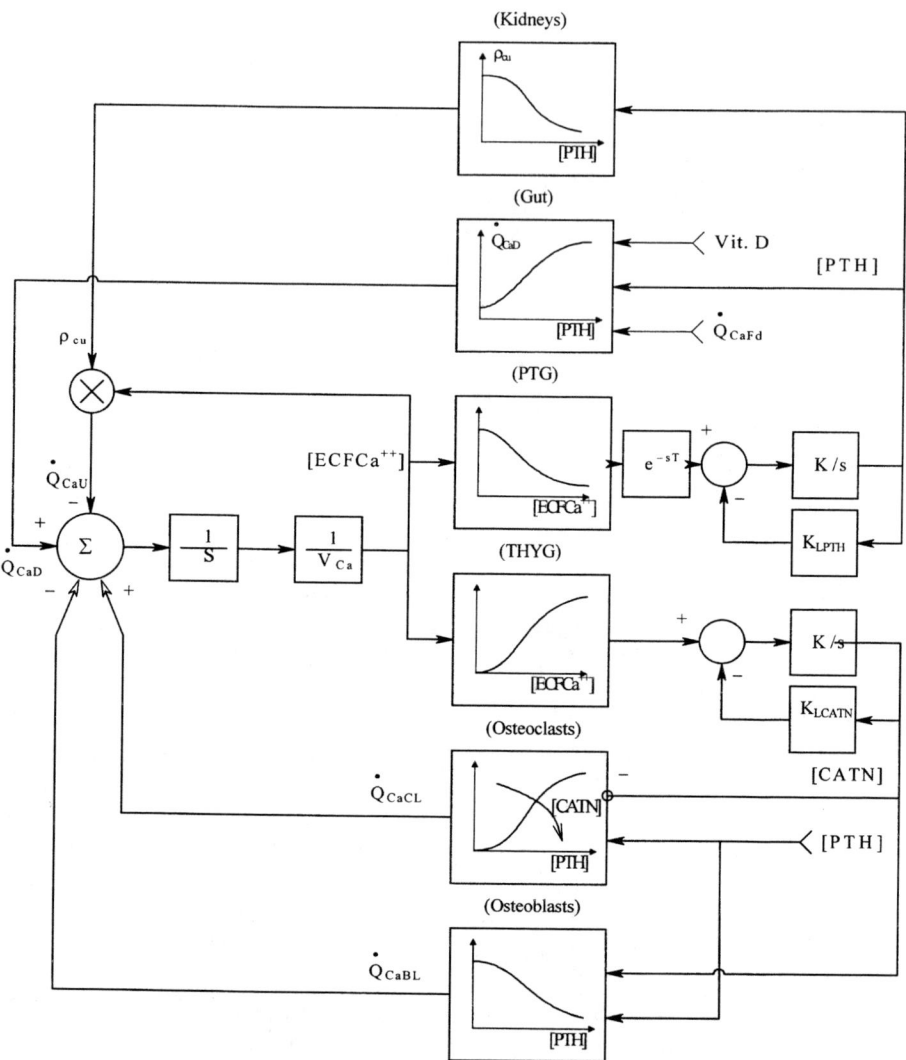

FIGURE 6.13 Block diagram describing calcium metabolism. (Extracellular Ca^{++} exchange with protein and chelate is neglected.)

Elevated [PTH] acts directly in the kidneys to *decrease* the rate at which PO$_4$ ions are reabsorbed in the proximal tubules, thus increasing the rate at which PO$_4^-$ is lost in the urine. [PTH] also acts on the epithelial cells lining the ascending limbs of the loops of Henle, the distal tubules, and the collecting ducts to cause *increased* Ca^{++} reabsorption, decreasing the rate of Ca^{++} loss in the urine. At the same time, elevated [PTH] also causes an *increased* rate of Mg^{++} and H$^+$ reabsorption, as well as *decreased* reabsorption of Na$^+$, K$^+$, and amino acids in the same cells.

Hormonal Regulation of Sodium, Potassium, Calcium, and Magnesium Ions 241

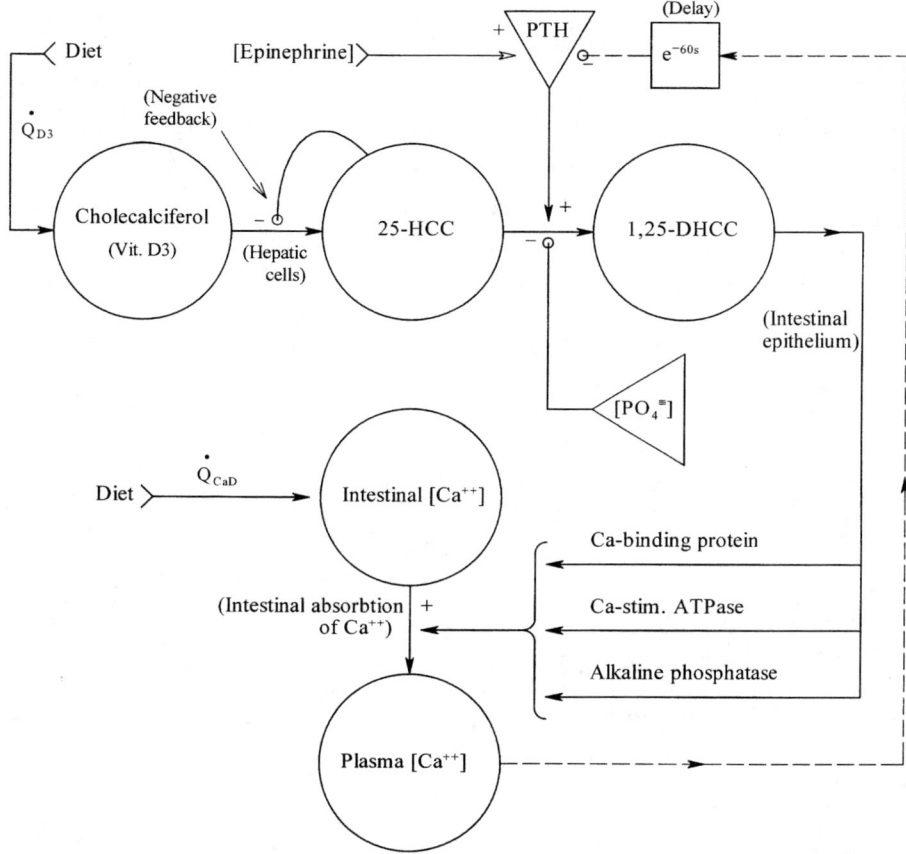

FIGURE 6.14 The role of vitamin D_3 and PTH in regulating plasma [Ca^{++}]. Causal negative feedback paths are shown by circles.

The action of PTH on bone cells is complex. There are three major types of bone cells involved with the control of [Ca^{++}]: osteoclasts, which break down bone salts to release calcium and PO_4 ions, and osteoblasts and osteocytes. Osteoblasts build up bone, forming crystalline bone "salts" in an organic cell matrix. The principal bone salt is *hydroxyapatite*, with the formula $Ca_{10}(PO_4)_6(OH)_2$. Other bone salts can contain Mg, Ma, K, and HCO_3^-.[59] Osteoblasts also lay down a collagen fiber matrix (osteoid) into which the bone salts are deposited. Some osteoblasts become entrapped in the interior of the osteoid; these are the osteocytes of mature bone. The osteocytes and osteoblasts form a thin, single-cell-thick membrane around each bone, called the *osteocytic membrane system*. The osteocytic membranes form a barrier between the bone and the extracellular fluid space. The small amount of fluid between the bone and the osteocytic membranes is called *bone fluid*. The osteocytic membrane system cells normally pump calcium ions from the bone fluid into the IF.

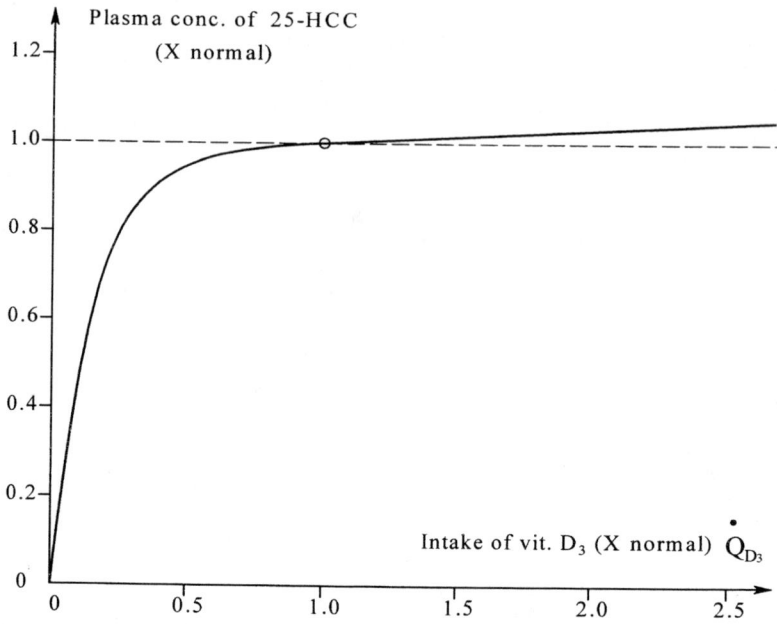

FIGURE 6.15 Plasma concentration of HCC vs. rate of input of vitamin D_3. The flatter part of the curve is the typical response of a high-gain type 0 feedback system. (From Guyton, A.C. 1991. *Textbook of Medical Physiology,* 8th ed., W.B. Saunders, Philadelphia. With permission.)

Raised [PTH] in the IF binds with receptors on the osteocytic membrane system cells and increases the rate of calcium pumping from the bone fluid through the osteocytic membrane into the extracellular fluid. This action depletes the bone fluid [Ca^{++}] and causes dissolution of bone salts (osteolysis) around the osteocytes and osteoblasts; the collagen matrix is left intact, however. For conditions of low [PTH], the rate of calcium pumping slows, the bone fluid [Ca^{++}] rises, and bone salts are redeposited in the collagen matrix.

Interstitial fluid [PTH] is elevated for a long time, the activated osteoblasts and osteocytes evidently send a chemical signal (unknown) to the osteoclast cells, which become activated to break down bone, releasing Ca^{++} into the IF. Osteoclasts do not have receptors for PTH. The secondary activation of the osteoclast cells causes immediate release of Ca^{++} from existing osteoclasts and the proliferation of new osteoclasts that further break down the bone.

Calcitonin (CATN) has a weak effect on plasma [Ca^{++}] in adult humans.[59] When the thyroid gland is removed, there is little effect on the overall regulation of plasma [Ca^{++}]; the PTH system dominates the regulation. When plasma [Ca^{++}] rises above about 9 mg/dl, the rate of CATN secretion increases in a linear fashion. In addition to elevated plasma [Ca^{++}], CATN secretion is stimulated by high levels of gastrin,

Hormonal Regulation of Sodium, Potassium, Calcium, and Magnesium Ions

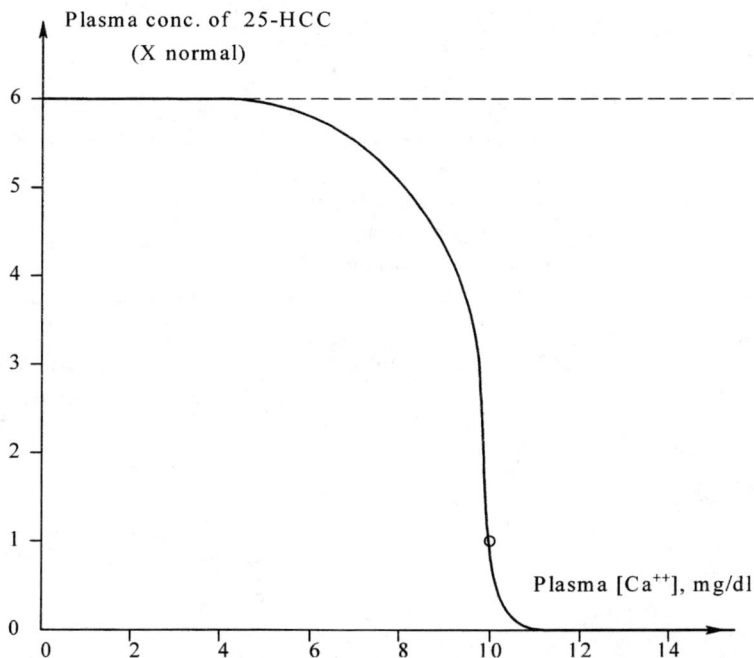

FIGURE 6.16 Relative concentration of DHCC vs. plasma [Ca++]. (From Guyton, A.C. 1991. *Textbook of Medical Physiology*, 8th ed., W.B. Saunders, Philadelphia. With permission.)

cholecystokinin, glucagon, and β-adrenergic agonists.[141] CATN appears to be a short-term down-regulator for high plasma [Ca^{++}] which is the result of a high-calcium diet. CATN is most effective in young animals. CATN reduces bone salt breakdown by inhibiting osteoclast function. CATN also increases the urinary excretion rate of Ca^{++}, Mg^{++}, Na$^+$, K$^+$, and HPO$_4^{2-}$ when present in the blood, contributing to its short-term hypocalcemic effect. CATN's peak action is felt less than an hour after an injection. PTH, on the other hand, reaches its peak effects in 3 to 4 hr.

In summary, high [PTH] *increases* $\dot{Q}_{CaD}$, the rate of dietary input of calcium. Vitamin D is also necessary for $\dot{Q}_{CaD}$ to increase. High [PTH] also *decreases* $\dot{Q}_{CaU}$, *decreases* $\dot{Q}_{Catb}$, and *increases* $\dot{Q}_{Cafb}$. CATN is seen to have a short-term effect in decreasing plasma [Ca^{++}] which is elevated from diet. The CATN system is overridden by the PTH system in long-term regulation.

No exogenous closed-loop control systems have been developed for calcium. Imbalances in plasma calcium are detected through blood analysis and corrected by diet and/or exogenous hormone injections. Extreme cases of hypocalcemic tetany are corrected by intravenous (IV) infusion of calcium gluceptate solution (0.9 meq Ca^{++} per milliliter). Calcium gluconate, carbonate, or phosphate pills are given as oral Ca dietary supplements.

6.6 MAGNESIUM ION REGULATION

Magnesium is another very important ion. It is the second most plentiful intracellular cation. Magnesium is a cofactor of all enzymes mediating phosphate transfer reactions with ATP and other energy-supplying nucleotide triphosphates. Many other enzyme systems also require this ion for operation. The binding of messenger RNA to ribosomes, a key step in protein synthesis, also requires Mg^{++}. Mg^{++} is also required for ribosomal integrity.[61] Exogenous magnesium is given orally or by IV infusion to promote normal ciliary transport of mucous in the bronchial airways of asthma patients.

The central nervous system and neuromuscular system are depressed by high plasma $[Mg^{++}]$. Acetylcholine (ACh) release is suppressed at motor end plates by high plasma $[Mg^{++}]$, as is the subsynaptic membrane sensitivity to released ACh. High plasma $[Mg^{++}]$ slows the action of the heart by increasing conduction time. Plasma $[Mg^{++}] > 15$ meq/l can cause cardiac arrest. Conversely, low plasma $[Mg^{++}]$ causes increased central nervous system irritability, convulsions, and psychotic behavior. The ECG changes with *hypomagnesemia* are similar to those seen with hypercalcemia. Low plasma magnesium causes excess ACh release at motor end plates and can result in muscle tetany.[61]

There is about 2000 meq of Mg^{++} in a 70-kg human adult. About 50% of the Mg^{++} is in bone salts; 45% is found intracellularly, and only 5% is in the extracellular fluids. Normal dietary intake of Mg^{++} is 20 to 40 meq/day. Intestinal absorption of Mg^{++} occurs mostly in the upper small intestine by an active process closely related to the transport system for calcium.[61] Most of the excretion of Mg^{++} from the body occurs by the kidneys. It is reabsorbed from the filtrate by the proximal tubule cells, so that about 3 to 5% of that Mg^{++} filtered is secreted in the urine. Magnesium loss in the urine is increased by many diuretic agents, including alcohol. Consequently, hypomagnesemia can occur as a side effect of diuretic therapy. Magnesium deficiency can also result from diarrhea and can occur in persons with poorly treated diabetes mellitus. Hypermagnesemia is usually the result of renal insufficiency, which can have many causes.[61]

Because the exact hormonal control mechanisms for plasma $[Mg^{++}]$ is unknown, exogenous hormonal intervention to correct for $[Mg^{++}]$ imbalance is not possible. Low plasma $[Mg^{++}]$ is generally corrected by administering magnesium orally as a citrate, sulfate, gluconate, oxide, or hydroxide. In acute cases, an $MgSO_4$ or $MgCl_2$ solution is given slowly by IV.[61] Treatment for hypermagnesemia is basic: The patient is caused to undergo diuresis by drinking water or by receiving an IV infusion of nonmagnesium saline. All ionic concentrations in the plasma are generally monitored during any exogenous adjustment of a plasma cation concentration.

Because only 5% of the body's magnesium is in blood and extracellular fluid, it appears that the settling time for an impulse input of IV Mg^{++} to the blood and extracellular fluid compartment should be slow, with time constants on the order of hours for the plasma $[Mg^{++}]$ to enter the bone and intracellular compartments. Plasma $[Mg^{++}]$ loss through the urine appears to provide the fastest loss rate, if the kidneys are functioning normally.

6.7 CHAPTER SUMMARY

The highlights of the endogenous hormonal regulation of sodium, potassium, calcium, and magnesium ions in the human were reviewed. These systems were seen to be made more complex because of the role of total water volume in determining all ionic concentrations. An increase in total water volume increases BV, which increases blood pressure, which in turn increases the flow rate of GF, including water. The reabsorption rate of water is affected by the osmotic pressure of the filtrate as it passes through the tubules, by the flow rate itself, and by the concentration of ADH acting on the tubular epithelial cells. The rate of ADH release is an increasing function of the osmotic pressure of IF (nearly the same as plasma P_{osm}). P_{osm}, in turn, is largely determined by the plasma sodium ion concentration. Figures 6.8, 6.10, and 6.11 illustrate the complexity of these ionic regulators. In spite of the complexity, and the parametric coupling through total water volume, nature has achieved decoupling for the sodium/P_{osm} and potassium regulators. This is illustrated by the fact that the concentration of the hormone ALDO, which regulates plasma [K^+], has only a very slight effect on plasma [Na^+], which is regulated by manipulation of water reabsorption by ADH. This decoupling is more remarkable when one recalls that ALDO manipulates the coupled pumping of potassium and sodium in kidney tubular epithelial cells.

In spite of the various pharmacological means available to correct hypo- and hypernatremia, -kalemia, -calcemia, and -magnesemia, no exogenous closed-loop therapy for imbalances of these ions has been attempted. In fact, an endogenous physiological control system has yet to be defined for magnesium.

PROBLEMS

6.1 Describe, using block diagrams, the sequence of physiological events the body uses to restore normal blood and extracellular fluid [Na^+] when a person is given an IV bolus of saline solution such that the person's blood [Na^+] rises initially to 225 meq/l. The person is allowed to drink water and urinate as needed.

6.2 Describe, using block diagrams, the sequence of physiological events the body uses to restore normal blood and extracellular fluid [K^+] when a person is given a sublethal IV bolus dose of concentrated KCl solution. The person is allowed to drink water and urinate as needed.

6.3 Describe the physiological events that happen when a dog is euthanized with an IV bolus dose of concentrated KCl solution.

6.4 Describe, using block diagrams, the sequence of physiological events the body uses to restore normal BV, MAP, and ion concentrations when a person drinks an enormous amount of water.

6.5 Other than the obvious central nervous system effects of inebriation, describe how ingestion of ethanol affects the blood concentrations of [K^+], [Na^+], [Ca^{++}], [Mg^{++}], and BV.

6.6 Calculate the osmotic pressure (in mmHg) at 37°C of "normal saline solution" (9 g/l NaCl).

6.7 A recipe for artificial aqueous humor contains the following substances dissolved in water to make 1 l of solution: 0.77 g dextrose, 0.26 g urea, 0.0504 g lactic acid, 0.185 g ascorbic acid, 131 mM/l NaCl, 20 mM/l $KHCO_3$. Calculate the osmotic pressure of this solution (in mmHg) at 37°C.

6.8 Find the "normal" values of the following physiological parameters for a normal 70-kg adult male: total interstitial fluid water volume (IFWV), BV, CO in liters per minute, MAP (brachial) in millimeters of mercury, and water loss rate in the urine ($\dot{Q}_{WU}$) in liters per minute.

6.9 Describe the hormonal and biochemical events present in the condition diabetes insipidus. What procedures are taken to correct this form of diabetes?

6.10 Describe the structure and function of a dialysis machine (artificial kidney). How does it differ from a real kidney in terms of transport substances?

7 Regulation of Blood Glucose

7.0 INTRODUCTION

In this chapter, the regulation of blood glucose in normal humans will be described and mathematically modeled. The roles of the pancreatic hormones insulin and glucagon in maintaining normoglycemia will be described, and the sources and sinks of glucose in the body, including storage in the liver as glycogen, will be examined. Dynamics will be stressed wherever possible. Next, the condition type I diabetes mellitus and insulin therapy will be considered and modeled. Finally, the designs of various "artificial beta cells" that attempt to regulate blood glucose by monitoring its concentration and computing an exogenous insulin infusion rate will be reviewed.

The reason glucose is important in the body is that it is the principal source of energy for cell metabolism. There are a number of biochemical pathways whereby other sugars, starches, fats, and proteins in the diet can be converted to glucose. Once in the circulatory system, there are many sinks for glucose:

1. It is stored in the liver, and to a lesser degree in muscle cells, as a high-molecular-weight polymer, glycogen.
2. It can be lost in the urine if the blood glucose concentration rises above a threshold (about 1.8 g/l).
3. It diffuses into *insulin-sensitive cells* (such as muscle cells) at a rate determined by the concentration gradient of glucose across the cell membranes. This gradient is the glucose concentration difference between extracellular (EF) and intracellular fluid (IF). If a person is exercising, muscle cells metabolize glucose at an increased rate. The diffusion constant for glucose on insulin-sensitive cells increases monotonically with the concentration of insulin. Insulin binds with receptor sites on the insulin-sensitive cells and activates the glucose transport protein in their cell membranes, increasing glucose permeability to a maximum of 10- to 20-fold.
4. Glucose diffuses into *noninsulin-sensitive cells* (such as nervous tissue) at a rate proportional to the glucose concentration difference between extracellular fluid and intracellular fluid.

There are three sources of glucose in the body: (1) *diet*; (2) *gluconeogenesis*, where biochemical entities such as glycerol, lactate, alanine, pyruvate, oxaloacetate, dihydroxyacetone, etc., are converted in complex, multistep processes to glucose; and (3) breakdown of liver glycogen to glucose. (The latter two steps are stimulated by the pancreatic hormone glucagon.)

The glucoregulatory system is regulated by two hormones, *insulin* and *glucagon*. It has one external (diet) and two internal inputs (from the liver and gluconeogenesis) and four sinks. In the absence of insulin, little glucose can enter the insulin-sensitive cells or be stored in the liver as glycogen. Thus the blood glucose concentration rises, raising the osmotic pressure of the blood. Water leaves the cells, and a condition of intracellular dehydration occurs. The high osmotic pressure triggers the need to drink water, a major symptom of diabetes.

If excess insulin is present, too much glucose enters the insulin-sensitive cells and the liver, and the blood glucose concentration falls. If blood glucose concentration falls too low, the central nervous system cells are deprived of their energy source, and a person can experience nervous irritability, convulsions, lose consciousness, or die.

7.1 MOLECULES IMPORTANT IN NORMOGLYCEMIC REGULATION

The pentose sugar glucose has two isomers. That used in nature is D-glucose, or dextrose. It is called D-glucose because when linearly polarized light is passed through a length L of an aqueous solution of D-glucose, the polarization angle is rotated *clockwise* (as seen from the source) by an angle $\theta = LC[\alpha]$, where C is the concentration of the solution, and $[\alpha]$ is the specific rotation of glucose in degrees/ (meter × grams per liter) evaluated at a known temperature and wavelength. If the L-glucose isomer is ingested, it cannot be metabolized by the body! Figure 7.1 illustrates a D-glucose molecule; it has a molecular weight of 180.12 g/mol.

The hormones insulin and glucagon are proteins manufactured in the pancreas, insulin by the beta cells and glucagon by the alpha cells. Insulin consists of two sulfur-linked amino acid chains with a total of 51 amino acids, and it has a molecular weight of 5808. Glucagon is a single chain of 29 amino acids, with a molecular weight of 3485. The molecular structures of both insulin and glucagon are known

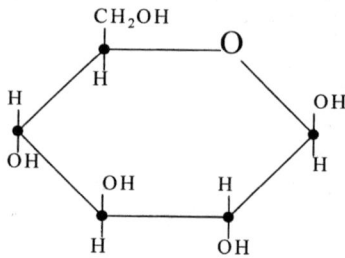

FIGURE 7.1 A D-glucose molecule. Glucose is a pentose monosaccharide.

exactly.[59] Both hormones are secreted into the hepatic portal venous system via the pancreaticoduodenal vein. Thus liver cells are the first to receive the secreted insulin and glucagon and generally see higher concentrations of these hormones than are present in the systemic plasma.

Glycogen is a polymer made in the liver and muscle cells by linking many uridine diphosphate glucose (UDG) molecules together enzymatically. A typical glycogen molecule can have a molecular weight of 5 million or more. UDG is made from glucose 1-phosphate (G1P), which in turn is made from glucose 6-phosphate (G6P). G6P is made directly from glucose that diffuses into the liver cells. Why does nature store glucose in the liver cells as a polymer? The answer lies in osmotic pressure: 1 mosm of glucose weighs 0.180 g, 1 mosm of glycogen may weigh as much as 5E3 g. This means that 28,000 glucose molecules joined as one glycogen molecule will raise the intracellular P_{osm} the same amount as a single, free, intracellular glucose molecule. Thus intracellular storage of glucose as a very-high-molecular-weight polymer does not impose any excessive rise in liver cell intracellular P_{osm}.

Muscle glycogen provides a local, fast source of energy for sudden bursts of muscle activity. If muscles work anaerobically, lactate (lactic acid) is produced. This diffuses out of the muscle cells into the blood, then to the liver cells which take it up and convert it to pyruvate, then to G6P, then to free glucose.[141] The chemical kinetics of glycogen storage and glucose release from liver cells are examined in the next section.

In 1994, a new protein hormone called *leptin* was described[24,30,103,149] that has far-reaching implications in the control of the amount of excess dietary calories stored as fat in fat cells (adipocytes) vs. the amount of glucose stored as glycogen in the liver and muscles. Leptin was first discovered in congenitally fat mice. Leptin has 146 amino acids with a complex helical, folded structure not unlike interleukins 2 and 4 and growth hormone.[149] Leptin is evidently secreted by *all* adipocytes in the body. There are known leptin receptors on certain hypothalamic neurosecretory cells and on the pancreatic beta cells that secrete insulin. There may be leptin receptors on certain hepatocytes as well. Apparently, the grand purpose of leptin is to regulate the amount of stored body fat. Obesity is a condition that affects about one third of the American population. It can lead to many health problems; for example: diabetes, heart disease, high blood pressure, stroke, arteriosclerosis, etc. Exogenous leptin has been used successfully to cause obese mice genetically lacking the ability to produce the hormone to lose weight. Thus there is hope that it can be used in conjunction with other drugs to effect fat reduction in humans. The leptin feedback system is described in the next section.

7.2 THE NORMAL BLOOD GLUCOSE REGULATION SYSTEM

The description of glucoregulation begins by first discussing the protein hormone *insulin*. Insulin is produced by the pancreatic beta cells in response to elevated

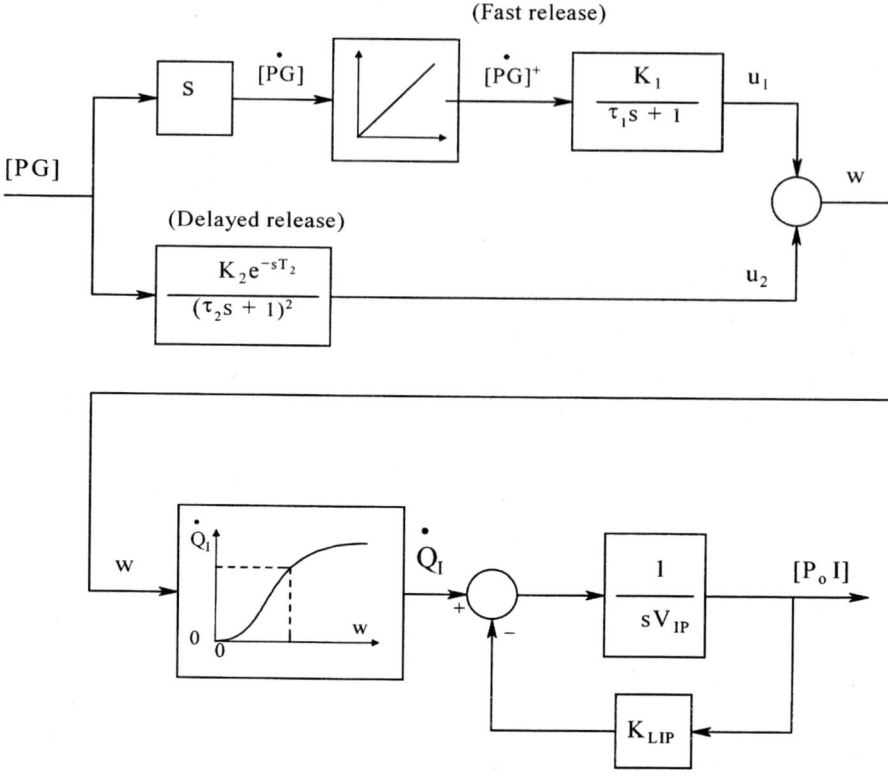

FIGURE 7.2 Block diagram modeling the nonlinear dynamics of insulin release from pancreatic β-cells. [PG] = plasma glucose concentration, [PoI] = portal insulin concentration. The upper, fast-release path obeys *unidirectional rate sensitivity*.

plasma (blood) glucose concentration. The release rate of insulin into the portal circulation can be modeled by the system shown in Figure 7.2. A fast, rectified, single time constant path is in parallel with a slow, two-pole lag path. The former path describes the dynamics of immediate, stored insulin release in response to a sudden *increase* in blood glucose concentration [BG]. The two-pole, slow path with the transport lag term models the metabolic activation of beta cells to synthesize new insulin for release. At a normal, resting [BG] = 90 mg/dl, the beta cells secrete insulin at a rate Q_I = 10 ng/min/kg body weight of insulin. At very high [BG] > 500 mg/dl, Q_I saturates at about 20 times normal, or 200 ng/min/kg body weight.[59] A number of substances associated with eating other than [BG] increase the insulin secretion rate. These substances include the amino acids lysine and arginine as well as the gastrointestinal hormones gastrin, secretin, cholecystokinin, and gastric inhibitory peptide. Also, the hormones glucagon, cortisol, and growth hormone increase Q_I or potentiate the action of [BG] on insulin secretion.[59]

Once secreted, insulin is inactivated by several biochemical processes. It is broken down by proteolytic enzymes when it binds to insulin receptor proteins on

Regulation of Blood Glucose

cell surfaces. Liver and kidney cells also contain enzymes that destroy free insulin. The liver is the principal site of insulin inactivation. Liver *glutathione–insulin transhydrogenase* (GIT) cleaves the three S–S bonds in the insulin molecule, rendering it inactive and subject to further proteolysis.[65]

If an intravenous (IV) bolus injection of insulin is given to normal subjects, the blood insulin concentration, [BI], falls from its peak value at $t = 0+$ to the normal concentration with an exponential decay that can be described by three exponential terms. That is,

$$[BI](t) = I_o + Ae^{-at} + Be^{-bt} + Ce^{-ct} \qquad (7.1)$$

This is considered to be good evidence for a three-compartment pharmacokinetic model for insulin; the three compartments are probably the blood (plasma), the extracellular fluid, and the liver cell volume. The fast phase of insulin decay is seen to take about 20 min for the concentration to go from an initial 400 µU/ml to 20 µU/ml.[124]

Insulin has other hormonal effects in addition to its major one of facilitating the diffusion of extracellular glucose molecules into insulin-sensitive cells. Insulin inhibits gluconeogenesis; it does this by decreasing the quantities and activity of the liver enzymes necessary for gluconeogenesis. Insulin acts to decrease the rate of release of amino acids from muscle and other nonliver tissues, thus reducing the available pool of precursor molecules used in gluconeogenesis. Insulin also promotes the conversion of excess intracellular glucose into fatty acids. These fatty acids are converted to triglycerides in low-density lipoproteins, which are transported to adipose tissues where they become fat.[59] High portal insulin concentration [I] inhibits the enzyme *glucose phosphatase* in liver cells, an enzyme which removes the phosphate group from G6P, converting it to glucose which can diffuse from the liver back to the blood. High portal [I] also activates the liver enzyme *phosphorylase*. Phosphorylase makes G6P from glucose which has diffused into liver cells, thereby trapping the glucose inside the liver cells. G6P is enzymatically converted to *G1P*, then to *UDP*, then to glycogen. Thus elevated insulin increases the rate at which liver glycogen is formed and inhibits its breakdown.

The hormone glucagon (GLN), as previously discussed, is made by the pancreatic alpha cells of the islets of Langerhans. Its release rate is stimulated by low [BG]. Normal plasma concentration of glucagon is between 100 and 200 ng/l. One of the major effects of glucagon is to raise the rate at which glycogen is broken down, releasing glucose stored in the liver cells back into the blood and extracellular fluid. Glucagon also stimulates the process of gluconeogenesis, which also raises [BG] over the long term. The steps by which glucagon stimulates the breakdown of glycogen are well known.[59] After combining with glucagon receptor proteins on the hepatic cell membrane, the enzyme *adenyl cyclase* is activated. Adenyl cyclase catalyzes the formation of cyclic AMP, which activates the following series of reactions: protein kinase regulator protein → protein kinase → phosphorylase *b* kinase → phosphorylase *b* → phosphorylase *a*. *Phosphorylase a* is the active enzyme that cleaves a glycogen unit, forming G1P. G1P is converted to G6P by

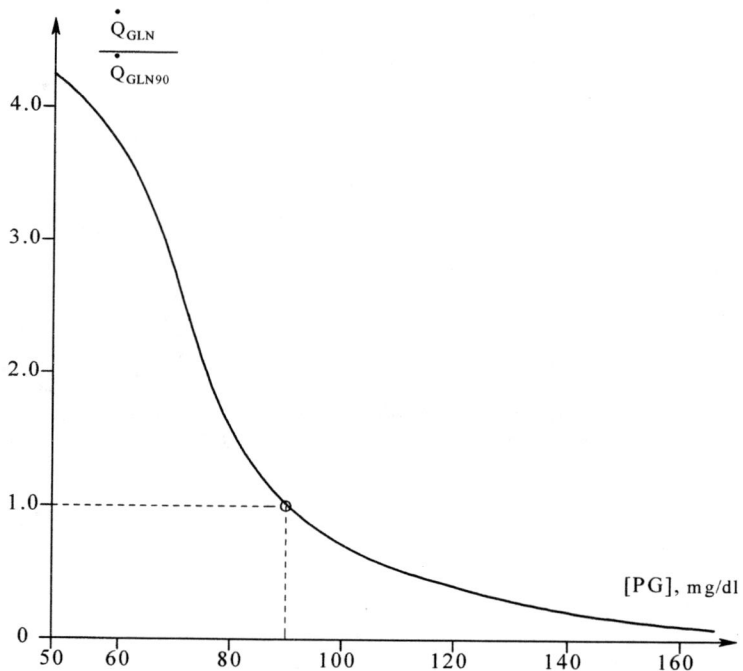

FIGURE 7.3 Normalized rate of glucagon secretion as a function of plasma glucose concentration, [PG]. The half-life of free glucagon is about 10 min.[141] (Adapted from Guyton.[59])

another enzyme, and then phosphatase, activated by glucagon, removes the phosphate group and the resulting D-glucose diffuses out of the liver cell. Note that the outward flow of glucose occurs during periods of fasting as the glucose regulator attempts to maintain normoglycemia.

Gluconeogenesis is activated by high [GLN]. Several metabolic pathways are involved, and many enzymes are necessary for the multistep reactions. The details of these reactions are beyond the scope of this text. However, high [GLN] causes free amino acids in the blood to be taken up by liver cells where they are converted to glucose. High [GLN] also activates the enzyme that converts pyruvate to phosphoenolpyruvate; this conversion is a key step in gluconeogenesis. Note that the gluconeogenesis processes are catabolic. Under extreme conditions of starvation, the body, in an attempt to maintain normoglycemia, first converts fats and then proteins to glucose through gluconeogenesis pathways. In extreme cases of starvation, what is left is "skin and bones"; we have all seen pictures of victims of prison camps, etc., that underscore this grim but life preserving process.

Figure 7.3 illustrates the basic relationship between [PG] and $\dot{Q}_{GLN}$, the rate of release of glucagon from the pancreatic alpha cells. We assume simple, first-order, loss kinetics for the breakdown of [GLN] in the absence of detailed pharmacokinetic data. As in the case of insulin, there are several factors that modulate $\dot{Q}_{GLN}$ other than [PG]. High concentrations of amino acids (especially arginine and alanine) in the blood after a high-protein meal stimulate the release of GLN. GLN promotes the rapid conversion of the amino acids to glucose, raising [PG].

The *delta cells* of the pancreatic islets of Langerhans secrete the 14-amino-acid peptide hormone *somatostatin*. Somatostatin is secreted in response to almost all factors related to eating. It has a very short half-life of about 3 min. Somatostatin acts to *suppress* the secretion rates of both insulin and glucagon. It also has a general slowing effect on all aspects of the digestion process. Guyton[59] speculates that somatostatin's role is to extend the period of time over which food nutrients are digested and assimilated into the blood. By suppressing the release of glucagon and insulin, somatostatin slows utilization of the absorbed nutrients by the tissues, making them available over a longer period.

Enough is known about the newly discovered protein hormone *leptin*, introduced in the previous section, to crudely describe its control loops. The regulation of carbohydrate and lipid metabolism involves complex interaction between many hormones, receptors, appetite, and the diet. Normally, carbohydrates and fats are digested and then processed in several interacting biochemical pathways. An excess of carbohydrates leads to the synthesis of fatty acids in liver cells. These fatty acids are then transformed into triglycerides which are released into the blood in the form of lipoproteins. Under the influence of *insulin*, the lipoproteins are converted into fatty acids in the capillary walls of adipose tissue by the enzyme *lipoprotein lipase*. These fatty acids are taken up by the fat cells and again stored as triglycerides. It is the excess of triglycerides stored in fat cells that produces obesity. The stored triglycerides can be broken down to free fatty acids and released into the blood under the influence of certain hormones and the enzyme *hormone-sensitive triglyceride lipase* (HSTL). Insulin actively *suppresses* the action of HSTL in fat cells. Other hormones *activate* HSTL, however, raising blood fatty acids which can be used for energy. These include epinephrine, norepinephrine, cortisol (a glucocorticoid), growth hormone, and triiodothyronine. (All of these hormones do not necessarily act directly on HSTL.)

The rate of leptin release from fat cells is *raised* by insulin, neuropeptide Y (NPY), triiodothyronine, epinephrine, cortisol, and the amount of triglycerides stored in each cell. Raised leptin concentration in turn *lowers or suppresses* the rate of release of insulin by pancreatic β-cells (at a given plasma glucose concentration), forming a negative feedback loop. A lower insulin secretion rate means a lower insulin concentration than normal. A lower insulin concentration means less glucose enters the liver cells and other insulin-sensitive cells. Thus less glycogen is stored, and less fatty acids are made. Lower insulin concentration leads to *increased* secretion of the hormone *glucagon*. Higher glucagon means that more fatty acids are released from adipocytes into the blood. Elevated glucagon also inhibits the storage of triglycerides in the liver.

Elevated leptin also *decreases* the rate of *neuropeptide Y (NPY)* release from hypothalamic neurosecretory cells in the brain. Thus there are at least two negative feedback loops regulating leptin secretion, leptin/insulin and leptin/NPY. Although the regulated variable appears to be leptin, leptin controls the average dietary calorie input through NPY. Excess calories over time, if not balanced by metabolic loss due to exercise, certainly lead to excess fat, as we all know. Hypothalamic NPY is reported to stimulate hunger and eating behavior and hence caloric intake. It also stimulates the secretion of insulin, although this stimulation may be indirect. NPY is also found in large vesicles in the endings of sympathetic nerve fibers in the body. NPY from this source evidently acts as a potent, long-acting vasoconstrictor, having synergistic action with norepinephrine.[61] These dual actions of NPY are a good example of hormonal pleiotropy.

In mice, chronic obesity appears to be related to either the lack of ability to produce leptin or the lack of leptin receptors in the hypothalamic neurons that secrete NPY. Similar conditions may occur in humans. It remains to be seen if exogenous leptin can aid humans in weight reduction.

7.3 A MODEL FOR NORMAL BLOOD GLUCOSE REGULATION

The human glucoregulatory system consists of *sources and sinks* for glucose, some of which are regulated by the hormones insulin and glucagon which are made in the pancreatic beta and alpha cells, respectively, and secreted into the hepatic portal blood supply.

First, we examine two of the sinks for plasma glucose. The dynamics of glucose entry into insulin-sensitive and noninsulin-sensitive cells will be modeled. These sinks are called *glucose utilization* and are modeled by the systems shown in Figure 7.4. In Figure 7.4A, the diffusion parameter, K_{DISC}, is a function of insulin concentration. Extracellular insulin binds with receptor proteins on insulin-sensitive cell surfaces and within seconds causes the rate at which glucose enters insulin-sensitive cells, $\dot{Q}_{GISC}$, to increase. Insulin effectively increases the diffusion constant, K_{DISC}. Insulin also causes increased cellular permeability to many amino acids, K^+, Mg^{++}, and PO_4^-, as well as slower changes in intracellular enzymes.

To describe *glucose utilization* by insulin-sensitive cells, $\dot{Q}_{GISC}$ is subtracted from the central extracellular glucose rate summer. To derive the dynamics of glucose utilization by insulin-sensitive cells, we subtract the intracellular rate of glucose consumption by metabolism from $\dot{Q}_{GISC}$, then divide the difference by the *intracellular water volume*, V_{IW}, then integrate to obtain the intracellular glucose concentration, [GIC]. We assume simply that the rate of intracellular metabolism is equal to K_{MIC} [GIC] mg/min. The metabolic rate constant, K_{MIC}, is a function of temperature, thyroid hormone concentration, epinephrine concentration, and the mechanical work load if the cells are muscle; for simplicity, we will treat it as a constant. Reduction of the block diagram yields a pseudo-linear transfer function relating $\dot{Q}_{GISC}$ to [PG]:

$$\frac{\dot{Q}_{GISC}}{[PG]}(s) = \frac{H_i K_{DISC} (s\tau_c + 1)}{\left(1 + H_i K_{DISC}/K_{MIC}\right)\left[s\tau_c/\left(1 + H_i K_{DISC}/K_{MIC}\right) + 1\right]} \quad (7.2)$$

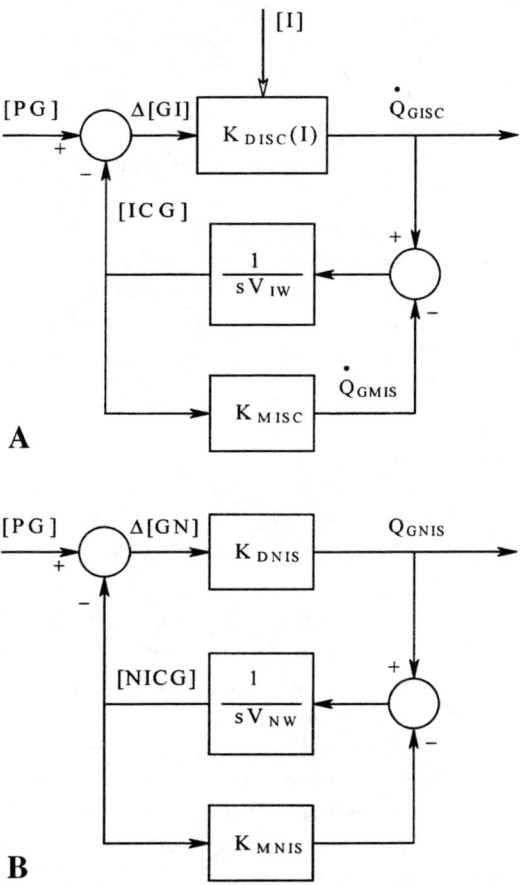

FIGURE 7.4 (A) Block diagram modeling glucose utilization by insulin-sensitive cells with first-order linear dynamics. The insulin concentration increases K_{DISC}. See text for description. (B) Block diagram modeling glucose utilization by noninsulin-sensitive cells with first-order linear dynamics.

where $\tau_c = V_{IW}/K_{MIC}$ min. Note that insulin acts parametrically to increase K_{DISC}, and a number of factors can influence K_{MIC} and hence intracellular metabolism. H_i is the Hill function:

$$H_i = c_o + \frac{c_m K_i [PI]}{1 + K_i [PI]} \tag{7.3}$$

This is the Hill function for insulin activation of K_Ds and enzymes. [PI] is the plasma insulin concentration.

For the case of *noninsulin-sensitive cells* (including central nervous system neurons, red blood cells, intestinal epithelial cells, and kidney tubule epithelial cells), illustrated in Figure 7.4B, the architecture of the block diagram is the same as for insulin-sensitive cells, except that the diffusion constant, K_{DNIS}, is a constant; i.e., it is *not* a function of [I]. $\dot{Q}_{GNIS}$ is the rate at which extracellular glucose enters the noninsulin-sensitive cells; it also must be subtracted from the central extracellular glucose rate summer. After reduction, we find the transfer function:

$$\frac{\dot{Q}_{GNIS}}{[PG]}(s) = \frac{K_{DNIS}\left(s\tau_{cn}+1\right)}{\left(1+K_{DNIS}/K_{MNIS}\right)\left[s\tau_{cn}/\left(1+K_{DNIS}/K_{MNIS}\right)+1\right]} \quad (7.4)$$

where $\tau_c = V_{NW}/K_{MNIS}$ min. The constants K_{DNIS} and K_{MNIS} are considered constants in this case, as the central nervous system's metabolism is relatively constant, and is independent of insulin and glucagon.

Figure 7.5 shows a basic systems block diagram for normal blood glucose regulation. The system producing the hormone glucagon is modeled by simple, first-order loss kinetics. A static, nonlinear function provides the glucagon rate input to the glucagon loss dynamics ordinary differential equation.

The normal pancreatic insulin system is modeled by a single time-constant fast-release component that is only responsive to positive rates of increase of glucose concentration. A second, slow, insulin component has a damped, second-order response to glucose. The sum of the fast and slow insulin release rate systems is further conditioned by a first-order lag system.

A single central glucose compartment, pooling the plasma volume of the circulatory system and the extracellular fluid volume, is used for simplicity. The inputs or sources to the glucose rate summer are from diet, from the liver, and from gluconeogenesis. The sinks of glucose include the liver, the kidneys, and cellular utilization (uptake by the insulin-sensitive cells and uptake by the noninsulin-sensitive cells).

The difference between the outward flow rate of glucose from the liver ($\dot{Q}_{HGR}$) and the inward flow rate from the plasma ($\dot{Q}_{HGS}$) is called the *net hepatic glucose balance* (NHGB). NHGB can be positive, zero, or negative. Negative NHGB occurs when the liver is actively storing glucose as glycogen. (The convention NHGB > 0 for $\dot{Q}_{HGR} > \dot{Q}_{HGS}$ follows Cobelli and Mari.[26]) Note that liver cells can store glycogen up to 5 to 8% of their weight. Liver glycogen has an average molecular weight of 5 million. Muscle cells can also store glycogen, up to a maximum of 1 to 3% of their weight.[59] Glucose from muscle glycogen is generally metabolized inside ¨(e muscle cells, however.

The dietary glucose input rate, $\dot{Q}_{GD}$, for a complex meal containing carbohydrates, sugars, and proteins is generally bimodal. Sugars and carbohydrates are digested first and converted to glucose, followed by the amino acids released by the digestion of proteins. Some amino acids are also converted to glucose.

The loss rate of glucose in the urine can be approximated by a simple linear relation. If we assume a constant glomerular filtration rate of 125 ml/min, the kidneys reabsorb *all* blood glucose molecules in the filtrate up to a [PG] of about

Regulation of Blood Glucose

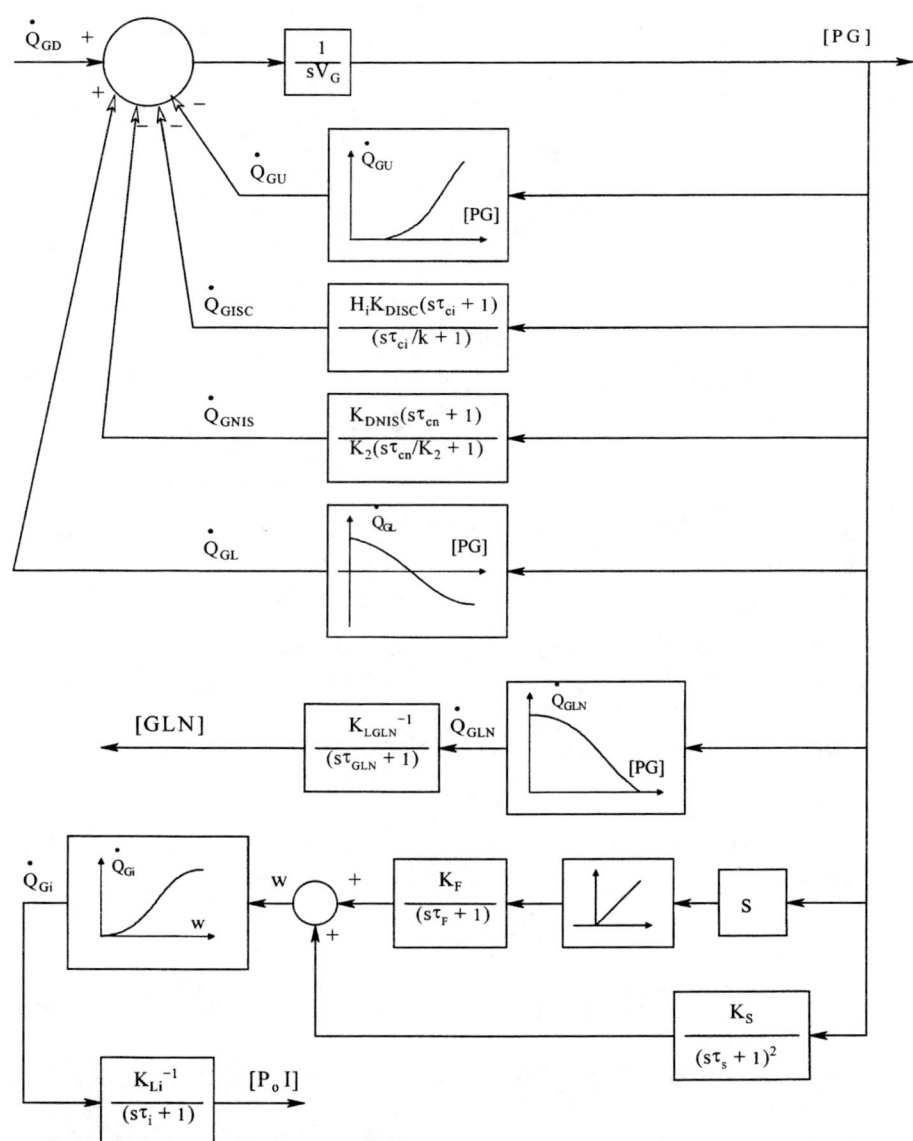

FIGURE 7.5 Block diagram of a single, central glucose compartment with glucose loss rates in the urine, insulin-sensitive cells, noninsulin-sensitive cells and the hepatic glucose flux which depends not only on [PG] but the hormones insulin, glucagon, and leptin. The production of plasma glucagon and portal insulin concentration are also simply modeled. A glucose input rate from diet is shown.

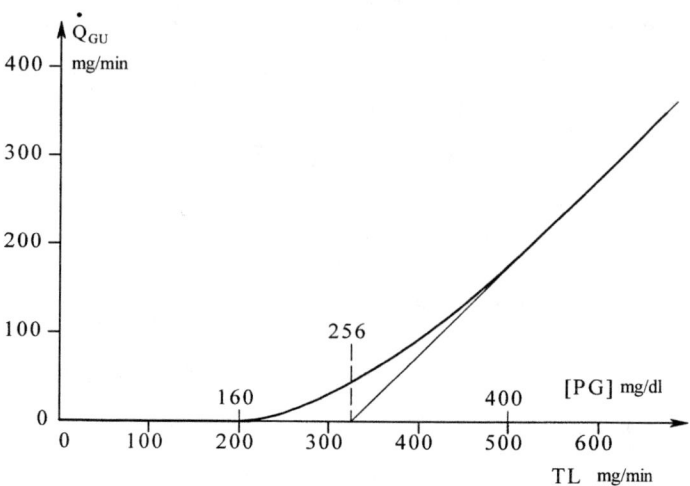

FIGURE 7.6 Rate of glucose loss in the urine as a function of tubular load and also [PG], assuming a constant glomerular filtration rate.

160 mg/dl. Above 160 mg/dl, the kidneys lose more and more glucose in the filtrate as [PG] increases. For [PG] above 320 mg/dl, the loss rate of glucose in the urine, $\dot{Q}_{GU}$, increases linearly with the *tubular load* of glucose (see Figure 7.6). The tubular load (TL) of glucose in milligrams per minute is simply:

$$\text{TL (mg/min)} = [PG] \text{ (mg/dl)} \; 1.25 \text{ (dl/min)} \tag{7.5}$$

The loss rate of glucose through the kidneys, $\dot{Q}_{GU}$, can thus be approximated by the linear equation

$$\dot{Q}_{GU} = -320 + [PG] \text{ mg/min for } [PG] \geq 320 \text{ mg/dl}$$

$$= 0 \text{ for } [PG] < 320 \text{ mg/dl} \tag{7.6}$$

This relation is shown in Figure 7.5.

The dynamics of NHGB can be described using mass action kinetics to describe the liver reactions that form and break down glycogen. Five nonlinear ordinary differential equations are written whose rate constants are functions of the concentrations of insulin and glucagon. The ordinary differential equations are nonlinear because all states are nonnegative. Positive NHGB is defined here as the rate of mass diffusion of glucose from the liver cells compartment to the plasma compartment. That is, positive NHGB makes [PG] rise.

$$\text{NHGB} = K_{DL} (I) \{[LG] - [PG]\} \quad \text{mg/min} \tag{7.7}$$

[LG] is the concentration of free glucose inside liver cells. The five ordinary differential equations describing the dynamics of hepatic glycogen storage and release are

Regulation of Blood Glucose

$$[\dot{LG}] = K_{DL}(PI)\{[PG] - [LG]\} + K_{PH}(GLN)[G6P] - K_{GK}(I)[LG] \quad (7.8A)$$

$$[\dot{G6P}] = K_{GK}(PI)[LG] - K_{GLS}[G6P] - K_{PH}(GLN)[G6P] \\ - K_{61}[G6P] + K_{16}[G1P] \quad (7.8B)$$

$$[\dot{G1P}] = K_{PHL}(GLN)[GN] - K_{16}[G1P] + K_{61}[G6P] - K_{1U}(I)[G1P] \quad (7.8C)$$

$$[\dot{UDG}] = K_{1U}(I)[G1P] - K_{UG}[UDG] \quad (7.8D)$$

$$[\dot{GN}] = K_{UG}[UDG] - K_{PHL}(GLN)[GN] \quad (7.8E)$$

The notation is described in Figure 7.7. Not only are the five states nonnegative, but there is a saturation limit on [GN], the glycogen concentration, at 8% of the liver mass. Note that the rate constants for certain enzymes are generally *increasing*

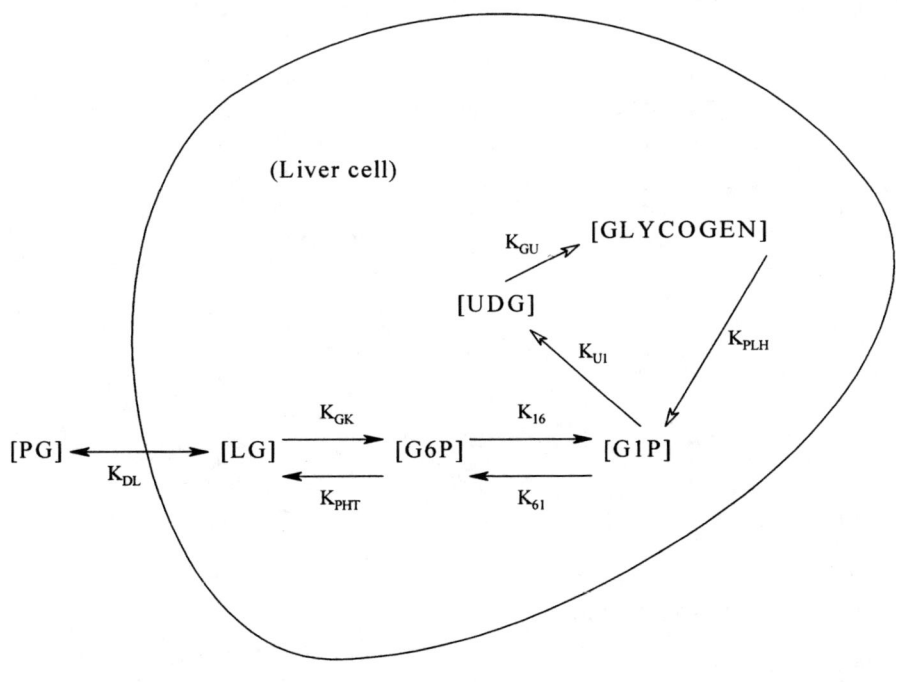

FIGURE 7.7 NHGB in a liver cell.

functions of the concentrations of the respective hormones that modulate them. The precise nature of the parametric modulation of the rate constants is not known, but we expect them to be sigmoid in nature. For example, the rate constant for glucokinase, K_{GK}, is increased by insulin. Therefore we can model K_{GK} with a type of Hill function:

$$K_{GK}(I) = \left\{ K_{GKmin} + \frac{(K_{GKmax} - K_{GKmin}[PI])}{\Phi + [PI]} \right\} \quad (7.9)$$

The exact values for the threshold, Φ, and the minimum and maximum must be determined experimentally. A number of other algebraic forms giving a sigmoid modulation of K_{GK} with [I] are available; these include hyperbolic tangent functions and exponential functions.

While the five ordinary differential equations above are physiologically accurate, they are complex and involve rate constants the values of which are unknown. Thus we seek to represent NHGB by a simpler, heuristic, description. Figure 7.8 illustrates a simplified, second-order nonlinear system that can be used to describe NHGB. The simplified, ordinary differential equations and relations describing NHGB are

$$H_i = c_o + \frac{c_m K_i [PI]}{1 + K_i [PI]} \quad \text{Hill function for insulin activation of } K_D\text{s and enzymes,} \quad (7.10A)$$
[PI] is plasma insulin concentration

$$H_{gln} = b_o + \frac{b_m K_{G14}[GLN]}{1 + K_{G14}[GLN]} \quad \text{Hill function for glucagon activation of enzymes} \quad (7.10B)$$

$$\dot{LG} = H_i K_{DL}\{[PG] - [LG]\} + H_{gln} \text{ rgn} - f_i K_S [LG] \quad \begin{array}{l} \text{g/min} \\ \text{Free liver glucose} \end{array} \quad (7.10C)$$

$$[LG] = LG/V_l \quad \begin{array}{l} \text{g/l} \\ \text{Concentration of free liver glucose} \end{array} \quad (7.10D)$$

$$\dot{GN} = [LG] H_i K_S - \text{rgn } H_{gln} \quad \begin{array}{l} \text{g/min} \\ \text{Rate of glycogen formation} \end{array} \quad (7.10E)$$

rgn = IF gn < 0 THEN 0 ELSE gn   g
Glycogen is nonnegative   (7.10F)

$$NHGB = H_i K_{DL}\{[LG] - [PG]\} \quad (7.10G)$$

As noted previously, the normal pancreas releases insulin in response to elevated plasma glucose concentration in two steps. There is a rapid, transient release of

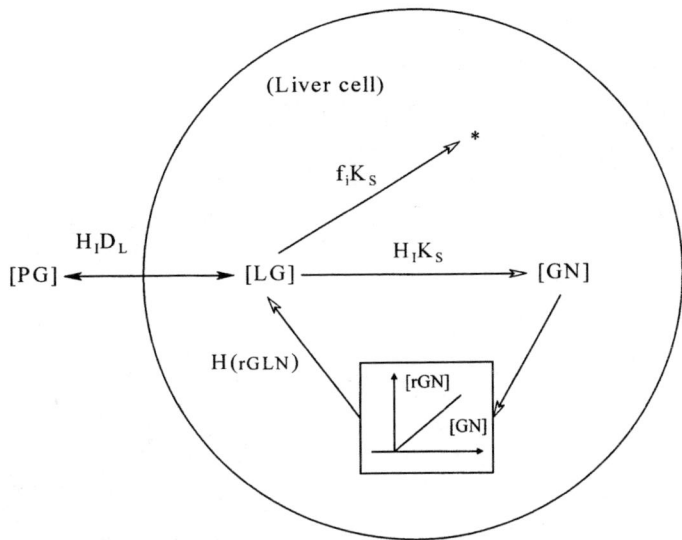

FIGURE 7.8 Simplified, nonlinear, second-order model for NHGB. See text for description.

insulin stored in beta cells in response to a positive rate of increase in [PG], and there is a slower, sustained release involving intracellular synthesis of insulin in response to elevated [PG]. Figure 7.2 illustrates a block diagram describing the fast and sustained release dynamics for insulin. The final ordinary differential equation describes simple, first-order destruction of insulin. Insulin distribution dynamics are ignored.

Glucagon release can be modeled more simply by a first-order loss process. Glucagon release increases with low plasma glucose concentrations. The release rate of glucagon is modeled simply by:

$$\dot{GLN} = K_{G11}/(1 + K_{G12}\, H_i[PG]) - h[GLN] \quad \mu g/min \quad (7.11A)$$

$$[GLN] = GLN/V_{GLN} \quad mg/l \quad (7.11B)$$

It is of interest to simulate the normal glucoregulatory system described above. The following Simnon™ program uses 12 states and allows one to investigate the "normal" responses to an IV glucose tolerance test (IVGTT). Many of the constants were chosen by trial and error to validate the system's responses to known physiological conditions. In this model, three compartments have been added describing insulin distribution, after Cobelli and Mari.[26] Normal pancreatic insulin is secreted into the hepatic portal venous system, from which it directly affects liver cells. Insulin then diffuses into the circulatory system and then to interstitial volume, where it can bind with the receptors on insulin-sensitive cells.

```
CONTINUOUS SYSTEM GLUCOSE5    " Creation date 11/01/94. Mod. 8/25/97.
" Normal glucoregulatory model of 70 kg human. 12 States. 1:43pm
"
STATE g nisG iscG lG gn gln bI ifI pI x1 x3 x2
DER   dg dnisG discG dlG dgn dgln dbI difI dpI dx1 dx3 dx2
TIME  t
"
" GLUCOSE PLANT
"
dg = -dgu - dgnis - dgisc + nhgb + dGin    " dg is rate of glucose
"                                                mass input, g/min.
"            dGin is dietary, gluconeogenesis, and gtt sources.
"
" Utilization Uptake by Non-Insulin-Sensitive Cells.
"
dgnis = Kdnis*(rpG - cnisG)   " dgnis is rate of pG uptake by
"                               non-insulin sens. cells, g/min.
dnisG = -nisG*Kmnis + dgnis   " nisG is mass of G inside NIS cells.
"
cnisG = nisG/Vnis             " Conc. Glucose in NIS cells, g/l
"
" Utilization Uptake by Insulin-Sensitive Cells.
"
discG = -iscG*Kmisc + dGisc       " iscG is total mass of glucose
"                                   inside insulin-sensitive cells.
dGisc = Hi*Kdisc*(rpG - ciscG)    " rate of mass diffusion into
"                                   iscs, g/min. from blood.
ciscG = iscG/Visc                 " Conc. isc Glucose, g/l
"
"            OTHER EQUATIONS:
"
Hi = K6 + (K7*Ki*bI)/(1 + Ki*bI)   " Hill function modelling effect
"                                    of insulin on diffusion constant
R = rpG*Hi                         " R function.
"
pG = g/Vol                         " Plasma glucose conc., g/l.
"
rpG = IF pG < 0 THEN 0 ELSE pG
"
dpG = dg/Vol                       " Rate of change of plasma
"                                    glucose conc.
dgu = IF pG < 1.8 then 0 ELSE Kk*(-1.8 + pG)   " Kidney spillover
"                                                nonlinearity
"           NHGB EQUATIONS:
"
NHGB = Hi*Kld*(clG - pG)           " NHGB in g/min.
"
dlG = -NHGB - Ks*Hi*clG + fgn*Hgln    " lG is hepatic glucose mass
"
dgn = Ks*Hi*clG - fgn*Hgln         " Glycogen stored in liver,
"                                    gms.
clG = lG/Vl                        " Conc. glucose in liver.
"
```

```
Hgln = bo + bm*Kgl4*gln/(1 + gln*Kgl4)    " Hill fctn for glucagon
"                                           action
fgn = IF gn < 0 THEN 0 ELSE K16           " Glycogen switch.
"
"           GLUCAGON EQUATIONS:
"
dgln = - h*gln + Kgl3*fgln        " Glucose release from pancreatic
"                                   alpha cells.
fgln = Kgl1/(1 + Kgl2*R)          " glucagon release rate inverse to R
"                                   parameter.
"
"           PANCREATIC INSULIN DYNAMICS:
"
dpaI = x1 + x3              " Insulin output (fast + slow), microg/min.
rdpG = IF dpG < 0 THEN 0 ELSE dpG
dx1 = - beta*x1 + Kf*rdpG   " Fast insulin
dx2 = -a*x2 + Ksi*a*HpGI    " Slow insulin, eq. 1.
dx3 = -a*x3 + Ksi*a*x2      " Slow insulin, eq. 2.
rdpaI = IF dpaI < 0 THEN 0 ELSE dpaI
HpGI = K14*pG*pG/(K15 + pG*pG)    " Hill function for insulin
"                                   release as a function of pG.
"
"           3 INSULIN COMPARTMENTS: (K8, K9, K10, K11, KLpi, KLifi)
"
dbI = - (K10 + K11)*bI + K9*ifI + K8*pI + dUxin   " Blood insulin
difI = - (KLifi + K9)*ifI + K11*bI        " Interstitial fluid
"                                           compartment.
dpI = - (KLpi + K8)*pI + K10*bI + rdpaI   " Portal compartment
"
"           EXOGENOUS INSULIN INPUT:
"
dUxin = IF t > tins THEN Iin*EXP(-(t - tins)) ELSE 0
"
"           GLUCOSE IVGTT:
"
dGin = Gin1 + Gin2 + Gneo
Gin1 = IF t > tgluc THEN Gin ELSE 0
Gin2 = IF t > (tgluc + 5) THEN -Gin ELSE 0
"
" CONSTANTS
"
tins:800
tgluc:400
beta:0.1
Kf:.333
Ksi:1.7
K4:.001
K5:0.333
K6:.05
K7:15
K8:.7
K9:.7
K10:.4
```

```
K11:.75
KLpi:.2
KLifi:.2
K14:.3
K15:2.
K16:12
gneo:.5
Gin:30    "  g/min
Iin:5
Kdnis:.25
Kdisc:3
Kmnis:.4
Kmisc:.05
Vnis:1
Visc:50
Vl:5
Ki:0.05
Kk:5.5
Ks:1.75
a:.025
h:.1
bo:.05
bm:1
Kld:10
Kgl1:.25
Kgl2:50
Kgl3:3
Kgl4:.2
Kc:1
Vol:75
Kin:.2
point8:.8
zero:0
point9:.9
"
"  INIT
g:75    "  grams gives pG(0) = 1 g/l.
gn:1000
gln:.3
"
END
```

Figure 7.9 illustrates the model's response to an IVGTT. The traces are 1 = [pG], 2 = [bI], 3 = [gln], 4 = [–NHGB], 5 = zero, 6 = 0.8 g/l [pG]. [pG] is to scale; other quantities are not to scale in order to fit them on the pG vertical axis. A negative [–NHGB] signifies glucose release by the liver in response to glucagon (gln). Positive NHGB means the liver is taking up glucose and storing it as glycogen (gn). In the model above, the NHGB equations have been simplified to two nonlinear ordinary differential equations, rather than the five described above. Also, the rate of breakdown of glycogen has been made independent of [gn],

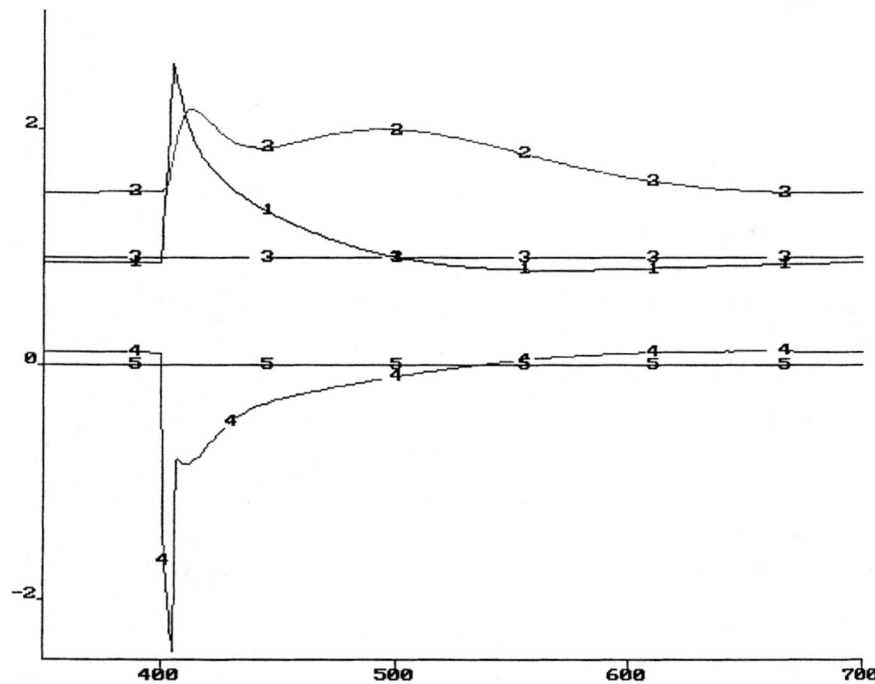

FIGURE 7.9 Response of the 12-state nonlinear model of normal glucose regulation to an IVGTT given at $t = 400$. Vertical axis: grams per liter for [PG]. Horizontal axis: time in minutes. Trace 1 = [PG], trace 2 = blood insulin [bI], trace 4 = NHGB, trace 3 = 0.9, trace 5 = 0.

assuming an excess of glycogen stored. Figure 7.10 illustrates the modeled distribution of endogenous (pancreatic) insulin between the plasma, portal, and interstitial fluid compartments in response to the IVGTT. Also plotted as trace 5 is the rate of glucose loss through the urine. Figure 7.11 shows the effect of an IV bolus dose of (exogenous) insulin on [pG], [bI], [gln], and [–NHGB]. Note that the insulin causes the [–NHGB] to go transiently positive under fasting conditions. The rate of glucose loss into the liver and insulin-sensitive cells causes the [pG] to drop transiently, and paradoxically, [gln] also drops in response to the insulin bolus.

Although the Simnon model above for normal glucoregulation is parsimonious, it gives realistic results. The model includes a switch that turns off hepatic glucose release during fasting when liver glycogen is exhausted. This is shown in Figure 7.12. Note that at $t \cong 360$ min, [gn] $\rightarrow 0$, and consequently NHGB $\rightarrow 0$. [pG] drops rapidly, and [gln] rises in response. [pG] plateaus at about 0.6 g/l because of a fixed gluconeogenesis rate of 0.5 g/min. (It would be more realistic to have gluconeogenesis a lagged, increasing function of [gln], but a fixed value was used in the model for simplicity.)

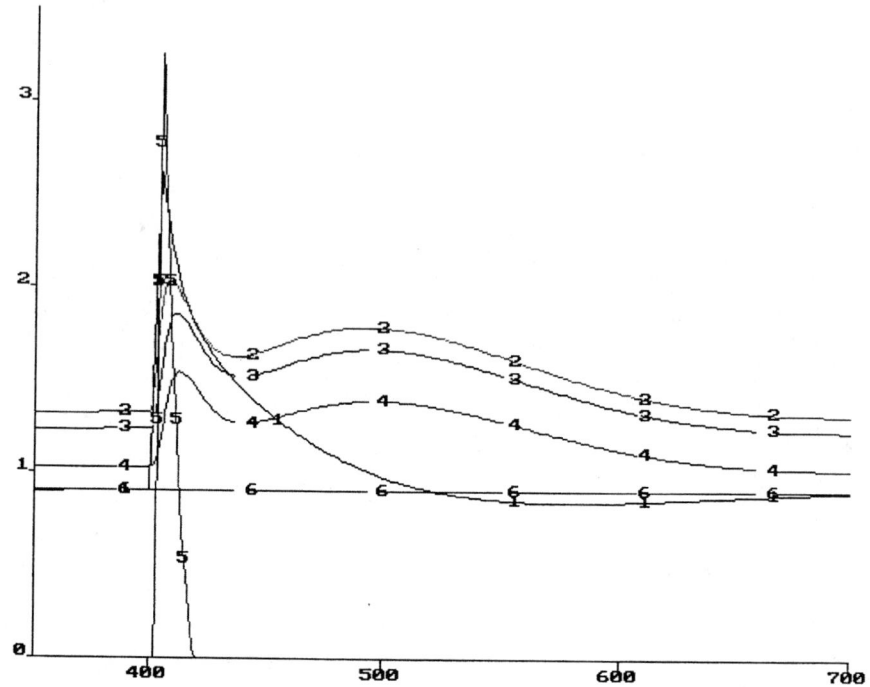

FIGURE 7.10 More results of the simulation of Figure 7.9. Same axes. Trace 1 = [PG], trace 2 = portal insulin concentration [PoI], trace 3 = blood insulin concentration [bI], trace 4 = interstitial insulin concentration [ifI], trace 5 = rate of glucose loss in urine, trace 6 = 0.9 g/l.

7.4 THE COBELLI AND MARI MODEL FOR GLUCOREGULATION

A comprehensive mathematical model for glucoregulation, following that of Cobelli and Mari,[25] is described in Section 5.7.2 of the text by Carson et al.[19] This model uses seven nonlinear state equations. It is presented here because it represents a major, detailed, validated contribution to the mathematical description of glucoregulation. Note that instead of hyperbolic Hill functions to represent saturation phenomena, Cobelli and Mari used hyperbolic tangent functions. The states are

x_1 = quantity of glucose in the plasma and extracellular fluid (mg)
u_{1p} = quantity of pancreatic *stored insulin* (μU)
u_{2p} = quantity of pancreatic promptly *releasable insulin* (μU).
u_{11} = quantity of *insulin in plasma* (μU)
u_{12} = quantity of *insulin in liver* (μU)
u_{13} = quantity of *insulin in interstitial fluid* (μU)
u_2 = quantity of *glucagon* in the plasma and interstitial fluids (pg)

Regulation of Blood Glucose

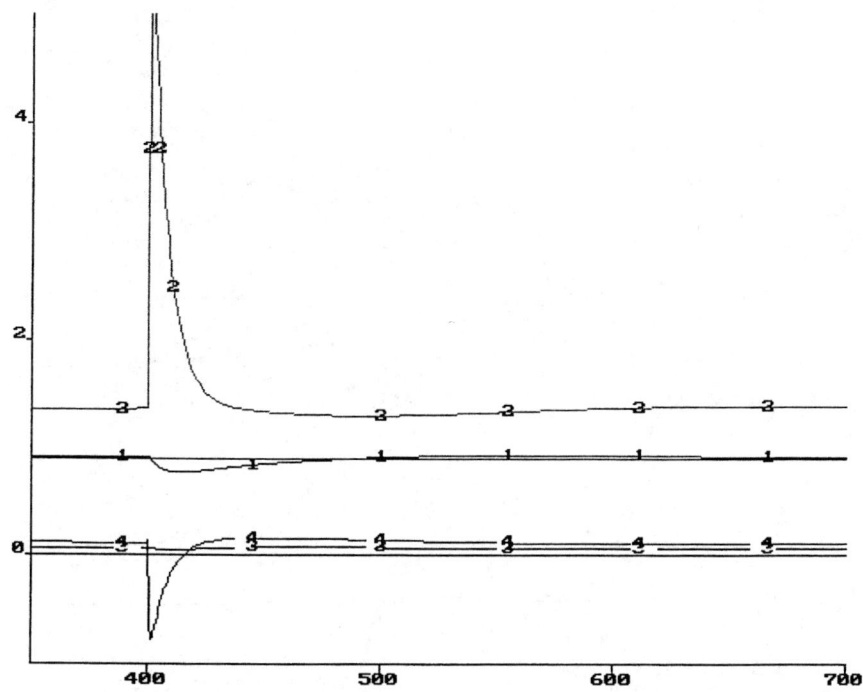

FIGURE 7.11 Simulated results of an IV bolus dose of insulin given at $t = 400$. Trace 1 = [PG], trace 2 = [bI], trace 3 = glucagon concentration [gln], trace 4 = NHGB. Note that the exogenous insulin causes the liver to take up glucose.

Also:

w and $F_1 \ldots F_7$ are nonlinear functions
I_x and I_u are test inputs of glucose and insulin, respectively
m_{ij}, h_{ij}, and k_{ij} are rate constants (min^{-1}); k_{02} is a function of x_1

The seven state equations are

$$\dot{x}_1 = \mathbf{NHGB}(x_1, u_{12}, u_2) - \mathbf{F}_3(x_1) - \mathbf{F}_4(x_1, u_{13}) - \mathbf{F}_5(x_1) + \mathbf{I}_x \quad (7.12\text{A})$$

$$\dot{u}_{1p} = -k_{21} u_{1p} + k_{12} u_{2p} + w(x_1) \quad (7.12\text{B})$$

$$\dot{u}_{2p} = k_{21} u_{1p} - \left[k_{12} + k_{02}(x_1)\right] u_{2p} \quad (7.12\text{C})$$

$$\dot{u}_{11} = -\left(m_{01} + m_{21} + m_{31}\right) u_{11} + m_{12} u_{12} + m_{13} u_{13} + I_u \quad (7.12\text{D})$$

$$\dot{u}_{12} = -\left(m_{02} + m_{12}\right) u_{12} + m_{21} u_{11} + k_{02}(x_1) u_{2p} \quad (7.12\text{E})$$

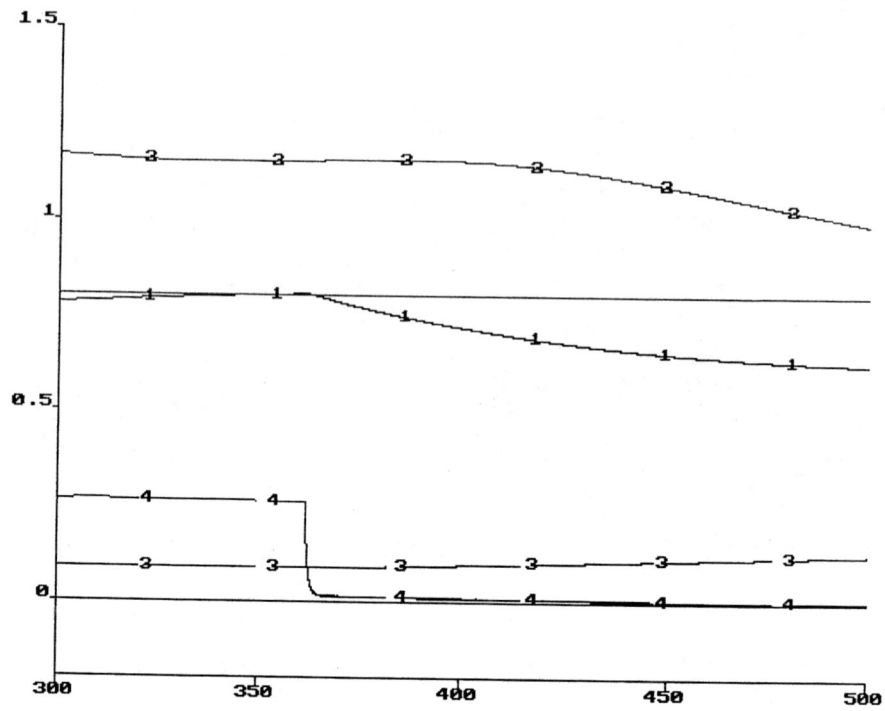

FIGURE 7.12 Simulated results of liver glycogen exhaustion at $t = 360$ min under fasting conditions. Note that [PG] (trace 1) begins to decay. [bI] (trace 2) also decays, while [gln] (trace 3) increases. NHGB drops abruptly to zero as no further glycogen can be converted to glucose and released.

$$\dot{u}_{13} = -m_{13} u_{13} + m_{31} u_{11} \tag{7.12F}$$

$$\dot{u}_2 = -h_{02} u_2 + F_7(x_1, u_{13}) \tag{7.12G}$$

The initial conditions are $x_1(0) = x_{10}$, $u_{1p}(0) = u_{1p0}$, $u_{2p}(0) = u_{2p0}$, $u_{11}(0) = u_{110}$, $u_{12}(0) = u_{120}$, $u_{13}(0) = u_{130}$, $u_2(0) = u_{20}$. NHGB is modeled by:

$$\mathbf{NHGB} = \mathbf{F}_1(x_1, u_{12}, u_2) - \mathbf{F}_2(x_1, u_{12}) \tag{7.13}$$

where

$$\mathbf{F}_1(x_1, u_{12}, u_2) = a_{11} G_1(u_2) H_1(u_{12}) M_1(x_1) = \text{liver glucose production rate, g/min} \tag{7.14A}$$

$$G_1(u_2) = 0.5\left\{1 + \tanh\left[b_{11}(e_{21} + c_{11})\right]\right\} \tag{7.14B}$$

$$H_1(u_{12}) = 0.5\left\{\left(1 + \tanh\left[b_{12}\left(e_{12} + c_{12}\right)\right]\right) + \left(1 + \tanh\left[b_{13}\left(e_{12} + c_{13}\right)\right]\right)\right\} \quad (7.14\text{C})$$

$$M_1(x_1) = 0.5\left\{1 + \tanh\left[b_{14}\left(e_x + c_{14}\right)\right]\right\} \quad (7.14\text{D})$$

$$\mathbf{F}_2(x_1, u_{12}) = H_2(u_{12}) + M_2(x_1) = \text{liver glucose uptake rate, g/min} \quad (7.14\text{E})$$

$$H_2(u_{12}) = 0.5 a_{21}\left\{1 + \tanh\left[b_{21}\left(e_{12} + c_{21}\right)\right]\right\} \quad (7.14\text{F})$$

$$M_2(x_1) = 0.5 a_{22}\left\{1 + \tanh\left[b_{22}\left(e_x + c_{22}\right)\right]\right\} \quad (7.14\text{G})$$

Also, e_{21}, e_{12}, and e_x are the differences between the *actual* plasma (e_{21}, e_x) and hepatic (e_{12}) concentrations of glucagon, insulin, and glucose and their basal or "normal fasting" concentrations, respectively.

$\mathbf{F}_3(x_1)$ is the mass loss rate of glucose through the kidneys (urine) in milligrams per minute per kilogram body weight. It is modeled by:

$$\begin{aligned}\mathbf{F}_3(x_1) &= M_{31}(x_1) M_{32}(x_1) && \text{If } x_1 \geq 2.52 \times 10^4 \text{ mg} \\ &= 0 && \text{If } x_1 < 2.52 \times 10^4 \text{ mg}\end{aligned} \quad (7.15\text{A})$$

$$M_{31}(x_1) = 0.5\left\{1 + \tanh\left[b_{31}\left(x_1 + c_{31}\right)\right]\right\} \quad (7.15\text{B})$$

$$M_{32}(x_1) = \left(a_{31} x_1 + a_{32}\right) \quad (7.15\text{C})$$

"The parameters a_{31}, a_{32}, b_{31} and c_{31} are chosen so that renal excretion is almost zero if plasma glucose concentration is below 180 mg/100 ml and then increases at a constant rate for glucose concentrations in excess of 180 mg/100 ml."[19]

$\mathbf{F}_4$ is insulin-dependent glucose utilization by insulin-dependent cells (mostly muscles and adipose tissues). $\mathbf{F}_4$ is modeled by:

$$\mathbf{F}_4(x_1, u_{13}) = a_{41} H_4(u_{13}) M_4(x_1) \quad (7.16\text{A})$$

$$H_4(u_{13}) = 0.5\left\{1 + \tanh\left[b_{41}\left(e_{13} + c_{41}\right)\right]\right\} \quad (7.16\text{B})$$

$$M_4(x_1) = 0.5\left\{1 + \tanh\left[b_{42}\left(e_x + c_{42}\right)\right]\right\} \quad (7.16\text{C})$$

where e_{13} is the difference between the actual and basal values of interstitial insulin concentration.

The rate of glucose utilization by insulin-independent cells (red blood cells and neurons) is given by $\mathbf{F}_5$:

$$\mathbf{F}_5(x_1) = M_{51}(x_1) + M_{52}(x_1) \tag{7.17A}$$

$$M_{51}(x_1) = a_{51}\tanh\left[b_{51}\left(e_x + c_{51}\right)\right] \tag{7.17B}$$

$$M_{52}(x_1) = \left(a_{52}\,e_x + a_{53}\right) \tag{7.17C}$$

"The parameters are chosen to produce a quasi-constant uptake of approximately 1.5 mg/min/kg."[19]

The insulin subsystem includes insulin *secretion, distribution, and breakdown*. The term $k_{02}(x_1)u_{2p} = \mathbf{F}_6(u_{2p}, x_1)$ is the insulin secretion rate:

$$\mathbf{F}_6(u_{2p}, x_1) = 0.5 a_6 \left\{1 + \tanh\left[b_6\left(e_x + c_6\right)\right]\right\} u_{2p} \tag{7.18}$$

where a_6, b_6, and c_6 are parameters, e_x is the difference between the actual plasma glucose concentration and its basal value (0.915 g/l). Insulin synthesis is described by $w(x_1)$:

$$w(x_1) = 0.5 a_w \left\{1 + \tanh\left[b_w\left(e_x + c_w\right)\right]\right\} \tag{7.19}$$

The model for insulin distribution consists of five compartments, described in the ordinary differential equations of Equation 7.12 and illustrated in Figure 7.13.

Glucagon dynamics are described by one ordinary differential equation with state u_2. The glucagon secretion rate is described by $\mathbf{F}_7$:

$$\mathbf{F}_7(x_1, u_{13}) = a_{71} M_7(x_1) H_7(u_{13}) \tag{7.20A}$$

$$M_7(x_1) = 0.5\left\{1 + \tanh\left[b_{72}\left(e_x + c_{72}\right)\right]\right\} \tag{7.20B}$$

$$H_7(u_{13}) = 0.5\left\{1 + \tanh\left[b_{71}\left(e_{13} + c_{71}\right)\right]\right\} \tag{7.20C}$$

Note that the secretion rate of glucagon is inhibited by high concentrations of glucose *and* insulin.

Parameter values for the model above of Cobelli and Mari are given in their paper.[25] *Liver glucose production:* $a_{11} = 6.71$, $b_{11} = 2.15$, $b_{12} = 1.18\ \text{E-2}$, $b_{13} = 0.15$,

Regulation of Blood Glucose

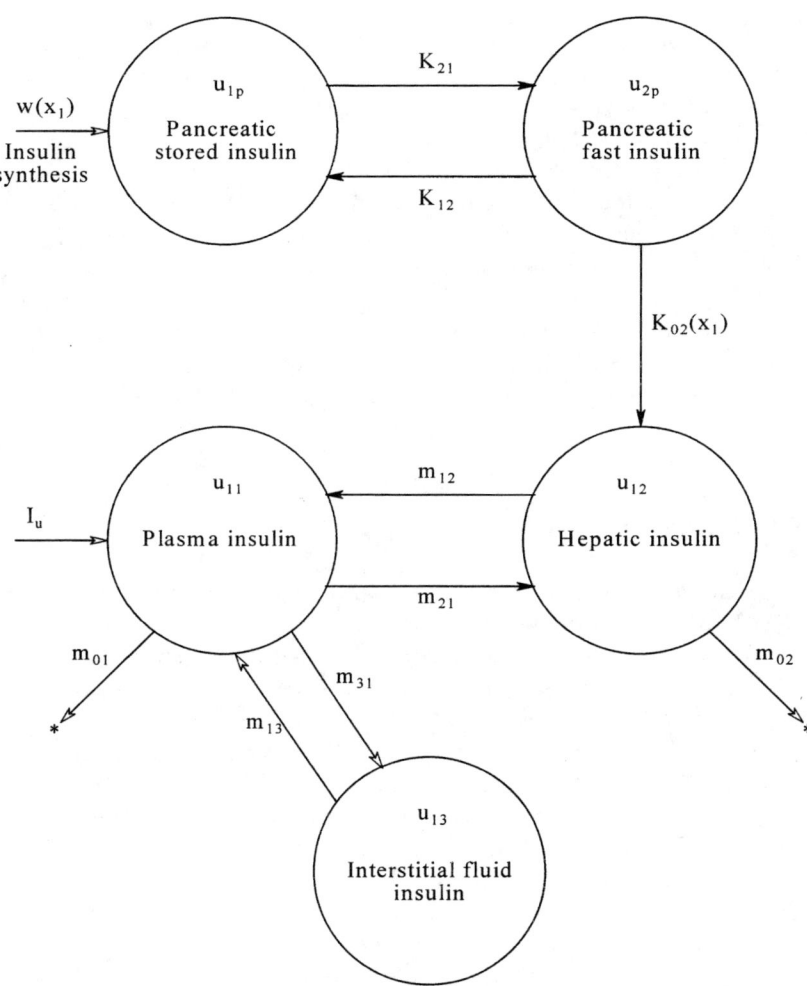

FIGURE 7.13 The five-compartment model for insulin production and distribution of Cobelli and Mari.

$b_{14} = 5.9$ E-2, $c_{11} = -0.854$, $c_{12} = -100$, $c_{13} = 9.8$, $c_{14} = -20$. *Liver glucose uptake*: $a_{21} = 0.56$, $a_{22} = 0.105$, $b_{21} = 1.54$ E-2, $b_{22} = 0.1$, $c_{21} = -172$, $c_{22} = -25$. *Renal excretion*: $a_{31} = 7.14$ E-6, $a_{32} = -9.15$ E-3, $b_{31} = 0.143$, $c_{31} = -2.52$ E4. *Insulin-dependent cellular uptake*: $a_{41} = 3.54$, $b_{41} = 1.4$ E-2, $b_{42} = 1.4$ E-2, $c_{41} = -146$, $c_{42} = 0$. *Insulin-independent cellular uptake*: $a_{51} = 8.4$ E-2, $a_{52} = 3.22$ E-4, $a_{53} = 3.29$ E-2, $b_{51} = 2.78$ E-2, $c_{51} = 91.5$. *Beta cell secretion of insulin*: $k_{12} = 3.13$ E-2, $k_{21} = 4.34$ E-3, $a_w = 0.747$, $a_6 = 1$, $b_w = 1.09$ E-2, $b_6 = 5$ E-2, $c_w = -175$, $c_6 = -44$. *Insulin distribution and breakdown*: $m_{01} = 0.125$, $m_{02} = 0.185$, $m_{12} = 0.209$, $m_{13} = 0.02$, $m_{31} = 0.268$, $m_{31} = 0.042$. *Glucagon secretion and breakdown*: $a_{71} = 2.35$, $b_{71} = 6.86$ E-4, $b_{72} = 0.03$, $c_{71} = 99.2$, $c_{72} = 40$, $h_{02} = 0.068$.

Concentrations are defined as $y_1 = x_1/V_1$ (glucose, in mg/dl), $y_2 = u_{11}/V_{11}$ (plasma insulin), $y_3 = u_{12}/V_{12}$ (hepatic insulin), $y_4 = u_{13}/V_{13}$ (interstitial fluid insulin; all insulin in µU/ml), $y_5 = u_2/V_2$ (blood glucagon in pg/ml). V_1, V_{11}, V_{12}, V_{13}, and V_2 are the distribution (compartmental) volumes, assumed to be 20, 4.5, 3.1, 10.6, and 20% of body weight, respectively. Thus, in a 70-kg person, $V_1 = V_5 \cong 0.2 \times 70 = 14$ l (plasma + extracellular fluids), and $V_2 = 0.045 \times 70 = 3.15$ l (plasma). A "normal" 70-kg individual is assumed to have a plasma glucose concentration = 91.5 mg/dl, plasma insulin concentration = 11 µU/ml, and blood glucagon concentration = 75 pg/ml.

The normoglycemic model of Cobelli and Mari has been validated using physiological data. By excluding pancreatic insulin production, the model was used to simulate type I diabetes and to test control algorithms for the administration of exogenous insulin.[19]

7.5 DIABETES MELLITUS

Type I diabetes mellitus is also called *juvenile-onset diabetes*. It is characterized by severely reduced insulin output from the pancreas. This reduction of endogenous insulin can be due to destruction of pancreatic beta cells by infection or autoantibodies. Outward symptoms of type I diabetes are (1) elevated plasma glucose concentration; (2) *polydipsia,* the desire to consume large amounts of water and other fluids; (3) *polyuria,* the production of abnormally large amounts of urine; and (4) *glucosuria,* or glucose in the urine. Type I diabetics are treated by managing diet, exercise, and injecting exogenous insulin in response to blood glucose measurements and information about diet.

Type II diabetes mellitus (adult-onset diabetes) is characterized by normal insulin levels, but poor response of insulin-sensitive cells to that insulin. Thus plasma glucose is elevated, and the symptoms of polydipsia, polyuria, and glucosuria are present as well. There is evidently loss of insulin receptors, or the receptors lose their affinity for insulin, or the molecular mechanisms whereby bound insulin increases glucose permeability fail in type II diabetes.

Laboratory diagnosis of type I diabetes generally shows a blood pH below 7 with low plasma bicarbonate (metabolic acidosis). Acidosis is the result of high levels of acetoacetic acid and β-hydroxybutyric acid which come from abnormal lipid metabolism. Also, lab work shows very high plasma triglyceride level due to accumulation of very-low-density lipoproteins and chylomicrons (hyperlipemia). Plasma electrolytes, [K$^+$] and [Na$^+$], are low; plasma free fatty acids are elevated.[141]

There are many harmful, long-term, systemic effects from untreated or poorly treated diabetes mellitus. For example, because liver cells do not receive enough glucose, they attempt to make up the difference by the catabolic metabolic process called gluconeogenesis. In gluconeogenesis, fats and proteins are broken down and certain amino acids are converted to glucose. The long-term effect of this process is wasting or emaciation. The liver, however, becomes fatty because of excessive uptake of free fatty acids by liver cells. Because of low intracellular glucose in non-liver cells, there is poor protein metabolism, which leads to conditions such as poor wound healing, cardiovascular disease, kidney disease, retinal degeneration, and osteoporosis.

Type I diabetes mellitus is treatable, because exogenous insulin can be injected by several routes, depending on the speed of response desired. The effectiveness of the treatment increases as the frequency of sampling plasma glucose increases. As will be discussed, many workers have devised control algorithms to automatically inject insulin in response to measurements of blood glucose.

In order to mathematically model type I diabetes, the submodel for the beta cell release of insulin is deleted. The balance of the glucoregulatory model is kept intact, however, including the compartmental distribution and degradation of exogenous insulin. In the model of Cobelli and Mari[25] in Section 7.4 and in the Simnon model in Section 7.3, there are three compartments for insulin: plasma, interstitial fluid, and portal (liver). Cobelli and Ruggeri[27] have shown that the portal route of exogenous insulin infusion is the most effective insulin input route in achieving normoglycemia by an artificial beta cell. Unfortunately, the portal route is not available for outpatient diabetics, and we must work with the IV route.

7.6 EXOGENOUS GLUCOREGULATION: THE ARTIFICIAL BETA CELL

7.6.1 INTRODUCTION

It long has been a bioengineering goal to design a cybernetic "artificial beta cell," or "artificial endocrine pancreas," to replace normal pancreatic function lost in type I diabetics. Ideally, such a system would be implantable, similar to a pacemaker, and would *continuously* monitor the plasma glucose concentration and calculate the required insulin dose to effect normoglycemia. Such an implantable artificial beta cell would consist of the following components: (1) an internal glucose sensor capable of maintaining its calibration over many months; (2) a controller that uses the plasma glucose readings to compute the required insulin dose; (3) a refillable, implanted, insulin reservoir; and (4) a delivery pump. The pump can be of the continuous infusion type or a pulsatile, bolus delivery pump.[96]

A major problem in realizing an implantable, long-term, artificial beta cell has been the long-term stability of the glucose sensor. Many approaches have been taken to design an implanted glucose sensor that maintains its calibration over a long period; none has been successful enough to permit long-term implantation. Glucose sensors can be broken down into several broad categories. The first design is electrochemical. Plasma or interstitial fluid glucose is oxidized to gluconic acid and hydrogen peroxide. The reactions can be written:

$$\text{Glucose} + O_2 \longrightarrow \text{Gluconic acid} + H_2O_2 \qquad (7.21\text{A})$$

$$H_2O_2 \xrightarrow{\text{Catalyst}} 2H^+ + O_2 + 2e^- \qquad (7.21B)$$

The initial oxidation is by the enzyme glucose oxidase or a suitable catalyst. The oxidation releases heat proportional to the limiting reagent (glucose); hence the temperature rise can be used to indicate glucose concentration. Another widely used approach is to use the polarimetric Clark oxygen electrode to sense the local *depletion* of O_2 caused by the oxidation reaction. Still another approach uses the Clark electrode to polarimetrically sense the concentration of H_2O_2 produced by the oxidation of glucose.[142] An example of the latter type of glucose sensor is illustrated in Figure 7.14. The inner membrane is cellulose acetate, and the outer one is cuprophan. This type of sensor is used in the extracorporeal benchtop model 23 glucose analyzer made by Yellow Springs Instruments, Inc.[142]

Optical means can also be used to estimate glucose concentrations. Glucose concentration in a clear solution can be measured by sensing its optical rotation using linearly polarized light. At physiological concentrations, and using 633-nm light, the optical rotation produced by D-glucose in a 1-cm optical path is about 4.5 millidegrees. Several workers[17,31,32,88] have proposed that the aqueous humor of the eye, which contains glucose proportional to plasma glucose, can be used to estimate plasma glucose concentration by optical rotation measurement. Another noninvasive optical means that has been tried uses the percutaneous infrared absorption/reflection or transmission spectrum to measure glucose concentration in tissues and peripheral capillaries.[98]

The problem of long-term stability for implantable electrocatalytic glucose sensors lies in the long-term stability of the enzyme or catalyst used and the long-term integrity of the membrane(s) under *in vivo conditions*. The role of the membranes is to exclude immune system cells, bacteria, viruses, large-molecular-weight

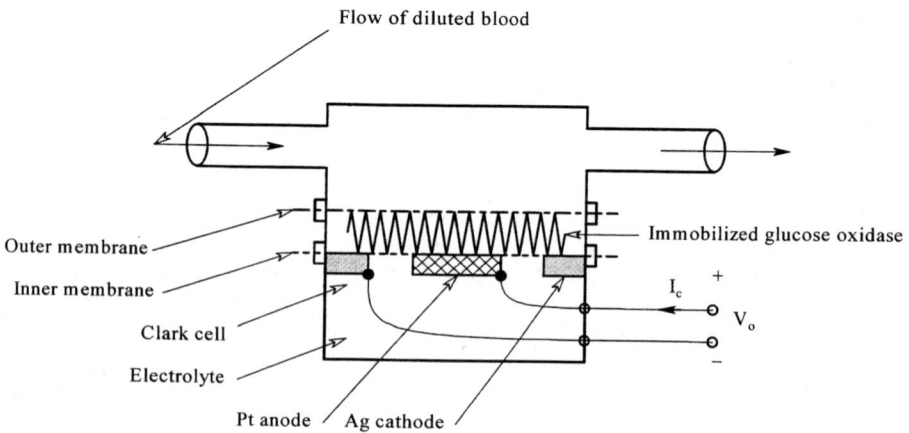

FIGURE 7.14 Cross section through a Clark cell-type glucose sensor using bound glucose oxidase.

Regulation of Blood Glucose

proteins, enzymes, etc., and to permit glucose and O_2 to diffuse inward and gluconic acid to escape outward. Over time, membranes develop defects and pores which allow large proteins, etc., to invade the space where the oxidative catalyst is immobilized. The glucose oxidase or other catalyst is "poisoned" by these proteins, losing its effectiveness, so sensor calibration is gradually lost.

While the design of an implantable, long-term-stable glucose sensor remains an ongoing challenge, many designs for the control algorithm and insulin pump have been set forth in the past 25 years. Effective, robust, controller design for insulin infusion is challenging because of the nonlinearity and high order of the glucoregulatory plant and plant parameter differences among individuals. Many early controller designs were of the cut-and-try variety.

7.6.2 EARLY TYPE 0 ARTIFICIAL BETA CELLS

An early review paper by Broekhuyse et al.[16] discusses and compares four nonlinear artificial beta cell algorithms. These algorithms are not adaptive and presumably would have to be "tuned" to each individual diabetic patient. These four controllers for insulin injection will be described and discussed in this section.

The *first controller* is the Clemens–Biostator–Miles (CBM) controller, first described in U.S. Patent #4,055,175, entitled "Blood Glucose Control Apparatus," dated October 25, 1977. The CBM controller has had several modifications since the original patent. It is a nonlinear, discrete-data controller in which the patient's plasma glucose is sampled periodically, and this information is used to calculate the insulin infusion rate, *IR*. The three infusion modes are

$$IR = RI\left(\frac{G + GD - BI}{QI} + 1\right)^4 \quad \text{for} \quad \left(\frac{G + GD - BI}{QI} + 1\right) > 0 \quad (7.22\text{A})$$

$$IR = 0 \quad \text{for} \quad \left(\frac{G + GD - BI}{QI} + 1\right) \leq 0 \quad (7.22\text{B})$$

$$IR = RI\left(\frac{G - BI}{QI} + 1\right)^4 \quad \text{for} \quad \left(\frac{G - BI}{QI} + 1\right) > 0 \quad (7.22\text{C})$$

$$IR = 0 \quad \text{for} \quad \left(\frac{G - BI}{QI} + 1\right) \leq 0 \quad (7.22\text{D})$$

$$IR = RI\left(\frac{G + GD - BI}{QI} + 1\right)^4 - RI\left(\frac{G - BI}{QI} + 1\right)^4 \quad (7.22\text{E})$$

The original CBM system also had a glucose infusion mode, presumably to compensate for hypoglycemic episodes induced by the nonlinear insulin infusion rate algorithms. The exogenous glucose infusion rate is given by:

$$DR = RD\left(\frac{BD - G}{QD} + 1\right)^4 \quad \text{for} \left(\frac{BD - G}{QD} + 1\right) > 0 \quad (7.23A)$$

$$DR = 0 \quad \text{for} \left(\frac{BD - G}{QD} + 1\right) \leq 0 \quad (7.23B)$$

Note the fourth power exponent of the control polynomials. Not shown in the control algorithms is the saturation of DR at 200 mg/min and IR at 600 mU/min. G = present plasma glucose reading = G_o. GD is a complicated function of the plasma glucose concentration, best illustrated in the block diagram of Figure 7.15. Note that $F(z)$ is an unspecified finite impulse response (FIR) smoothing (low-pass filtering) routine using the present and four or five past samples of plasma glucose [pG]. $D(z)$ is a discrete estimator of [pG], possibly a three-point central difference algorithm.[136] BI

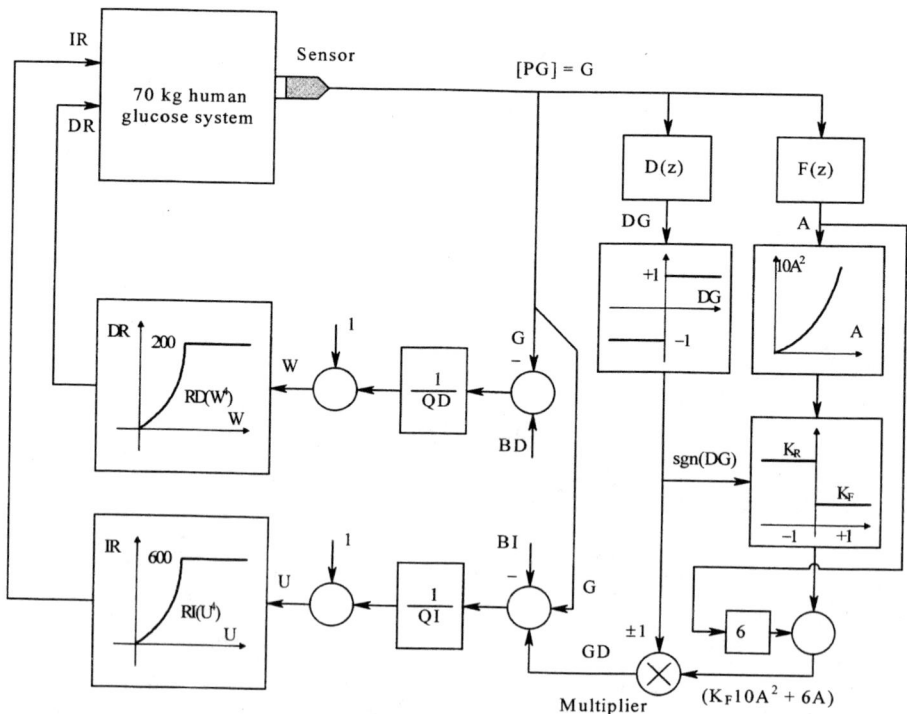

FIGURE 7.15 Block diagram of the highly nonlinear CBM glucose controller. All unmarked inputs at summers are +; − inputs noted.

is the "basal pG concentration," generally taken as 80 mg/dl. *QI* is a gain constant, e.g., 85. *RI* is another constant, e.g., 7.58.

The parameters for glucose infusion are *RD* is about 10 mg/min. *BD* is the basal pG concentration, e.g., 70 mg/dl. The constant *QD* is about 20 mg/dl glucose. (If *BD* = 80 mg/dl, *DR* = 0 at *G* = 100 mg/dl.)

In a later paper by Cobelli and Ruggeri[27] describing the CBM controller, it has evolved into the Biostator II, a trimodal system. The *first mode* is for static hyperglycemia:

$$IR = RI\left(\frac{G-BI}{QI}+1\right)^2 \quad \text{for} \left(\frac{G-BI}{QI}+1\right) > 0 \quad (7.24A)$$

$$IR = 0 \quad \text{for} \left(\frac{G-BI}{QI}+1\right) \leq 0 \quad (7.24B)$$

The *second mode* is the so-called *dynamic mode*:

$$IR = RI\ K\ m(G-BI)10^{-3} \quad \text{for } G > BI \quad (7.25A)$$

$$IR = 0 \quad \text{for } G \leq BI \quad (7.25B)$$

The *third mode* is the static *plus* the dynamic mode. Note that the *IR* argument is now squared, rather than being raised to the fourth power.

Here m is a numerical estimate of [pG], found by computing the slope of the regression line that fits the present and past four measurements of [pG]. The other parameters are defined as above. *K* takes on one of two values (*KR, KF*) depending on the sign of m (+, −). Dextrose (glucose) infusion exactly follows relations 7.23 above. Cobelli and Ruggeri list "optimized" CBM controller parameters from Christiansen, et al.:[21] *BI* = 91.5 mg/dl, *QI* = 40, *RI* = 16.4 (with IR_{po}), *RI* = 21.2 (with IR_{pe}), *KR* = 70, *KF* = 45, *BD* = 60, *QD* = 30, *RD* = 15.

The CBM controller is clearly a kludge, apparently derived with little regard for the physiology of glucoregulation or control theory. In the CBM II system, the insulin infusion polynomial is squared, rather than raised to the fourth power. The potential need to inject compensatory dextrose underscores the poor regulatory dynamics of this system.

The *second glucose controller* is the Albisser/Toronto (A/T) algorithm, developed in 1974.[25] This algorithm uses the hyperbolic tangent function to obtain a soft saturating nonlinearity in the insulin infusion rate, quite the opposite of the power law with hard saturation approach used in the CBM system. The insulin infusion rate over one sample period is given by:

$$\dot{Q}_I = 1/2\ M\left[1 + \tanh\left\{S(G_o + DF - B)\right\}\right] \quad (7.26)$$

where G_o = present [PG] measurement, G_1 = [PG] measured 1 min (sample period) previously, G_2 = [pG] measured 2 min previously, etc., M = maximum insulin infusion rate (400), S = 0.04, B = [pG] for half-maximum infusion rate = 140 mg/dl, $DF = K_1(A^3) + K_2(A)$, $K_1 = 1$, $K_2 = 10$, $A = (4G_o - G_1 - G_2 - G_3)/10$. The A/T algorithm also had a dextrose infusion mode:

$$\dot{Q}_D = 1/2 N\left[1 + \tanh\left\{P(G_o - GD)\right\}\right] \tag{7.27}$$

where N = 200 mg/min, P = 1/31, GD = 50 mg/dl. In 1976, Botz[15] demonstrated that the dextrose infusion was unnecessary with the A/T insulin infusion control algorithm.

The **third algorithm** is the 1977 Kraegen, et al. modification of the A/T controller.[74] This system is referred to as the KAT controller. It is a good example of improvement by nonlinear "tinkering." The insulin infusion rate is given by:

$$\dot{Q}_I = 1/2 M\left[1 + \tanh\left\{S(G_o + DF - B)\right\}\right] \tag{7.28A}$$

$$DF = K_1\left[\exp(RC/K_2) - 1\right] \tag{7.28B}$$

$$RC = (8G_o - 4G_1 - 2G_2 - G_3)/15 \tag{7.28C}$$

where M = 401.84, S = 0.04, B = 140 mg/dl, K_1 = 40, K_2 = 4.

The **fourth algorithm** is a *linear* insulin infusion rate algorithm devised by Fischer et al.:[42]

$$QI = a_0 + a_1\left[G_1 - PG_b\right] + a_2 \Delta pG \tag{7.29A}$$

$$\Delta pG = G_o - G_1 \tag{7.29B}$$

where ΔpG estimates [pG], G_o is the present [pG] sample, G_1 is [pG] measured one sample period ago, a_0 = 37.53, a_1 = 1.09, a_2 = 7.16, PG_b = 90 mg/dl.

Broekhuyse, et al.[16] actually tested the A/T and CBM controllers on human type I diabetics. Performance was poor compared to that of normal nondiabetic subjects with no external insulin infusion. Table 7.1 is taken from Broekhuyse, et al. It summarizes the performance of the A/T and Biostator controllers.

In a 1983 paper, Cobelli and Ruggeri[27] used the validated glucoregulatory system model described in Section 7.4 to evaluate the performance of the CBM-II, the A/T, and the Fischer algorithms in regulating plasma glucose, given one of two insulin input pathways: (1) peripheral (IV) or (2) portal vein injection. The system input was (simulated) oral glucose tolerance tests. Note that the Cobelli and Mari model for a type I diabetic uses *three* interconnecting, concatenated, insulin compartments: (1) liver (into which exogenous *portal insulin* is input directly), (2)

TABLE 7.1
Comparison of Performance of the Toronto and Biostator II Controllers with Normal Human Subjects

	Toronto	Biostator I	Biostator II	Normal
Mean plasma glucose	110 ± 5	128 ± 16	106 ± 4	93 ± 3
Mean amplitude of glycemic excursion	62 ± 10	74 ± 31	71 ± 11	35 ± 4

plasma (into which exogenous *peripheral insulin* is input directly), and (3) interstitial fluid. The important result from Cobelli and Ruggeri's modeling study was that the portal input route for insulin gave greatly improved performance in achieving normoinsulinemia over the peripheral (IV) input route. When the peripheral insulin infusion route is used, hyperinsulinemia results, causing a greatly enhanced peripheral uptake of glucose by insulin-sensitive cells. Peak peripheral glucose uptake rate by model nondiabetics and by type I diabetics having portal insulin input was about the same, 2.5 mg/min/kg. On the other hand, the excess insulin in the peripheral input case resulted in a peak glucose flux of about 7.5 mg/min/kg, a threefold increase. The modeling results of Cobelli and Ruggeri argue strongly for a portal insulin input route. This route is out of the question for self-treating type I diabetics for obvious reasons, but possibly could be surgically implemented for a chronic, implanted, artificial beta cell.

In a further modeling study, Cobelli and Mari[26] used their glucoregulatory model of a type I diabetic to derive the optimum insulin injection rate into the peripheral (plasma) compartment that will create a normal plasma glucose vs. time response to an oral glucose tolerance test (OGTT). Although the glucose response was normal, the plasma insulin levels were elevated, causing excess glucose diffusion into nonhepatic insulin-dependent cells. They also found that in the normal (nondiabetic) case, the liver uptake was about half of the mass of the glucose in the OGTT. When the peripheral insulin route was used with the optimum insulin injection profile, the liver uptake was only about 10% of the OGTT mass. This suggests that although plasma glucose follows a "normal" profile, little glucose is stored in the liver as glycogen which can be released later as plasma glucose under the influence of glucagon. This means that hepatic gluconeogenesis is higher than normal even though plasma normoglycemia is approached using an optimum insulin infusion into the plasma compartment.

In 1986, Northrop and Woodruff[105] introduced the design for a novel, nonlinear, integral-pulse frequency-modulation (IPFM) controller for IV bolus insulin injections in type I diabetics. These authors demonstrated the effectiveness of their controller in a model study. Their glucose plant is shown in Figure 7.16. Although simpler mathematically than the models of Northrop, Cobelli, and Mari described in Sections 7.3 and 7.4, respectively, this model contains the basics of glucose utilization, renal loss, NHGB, glucagon synthesis, and compartmental dynamics for insulin

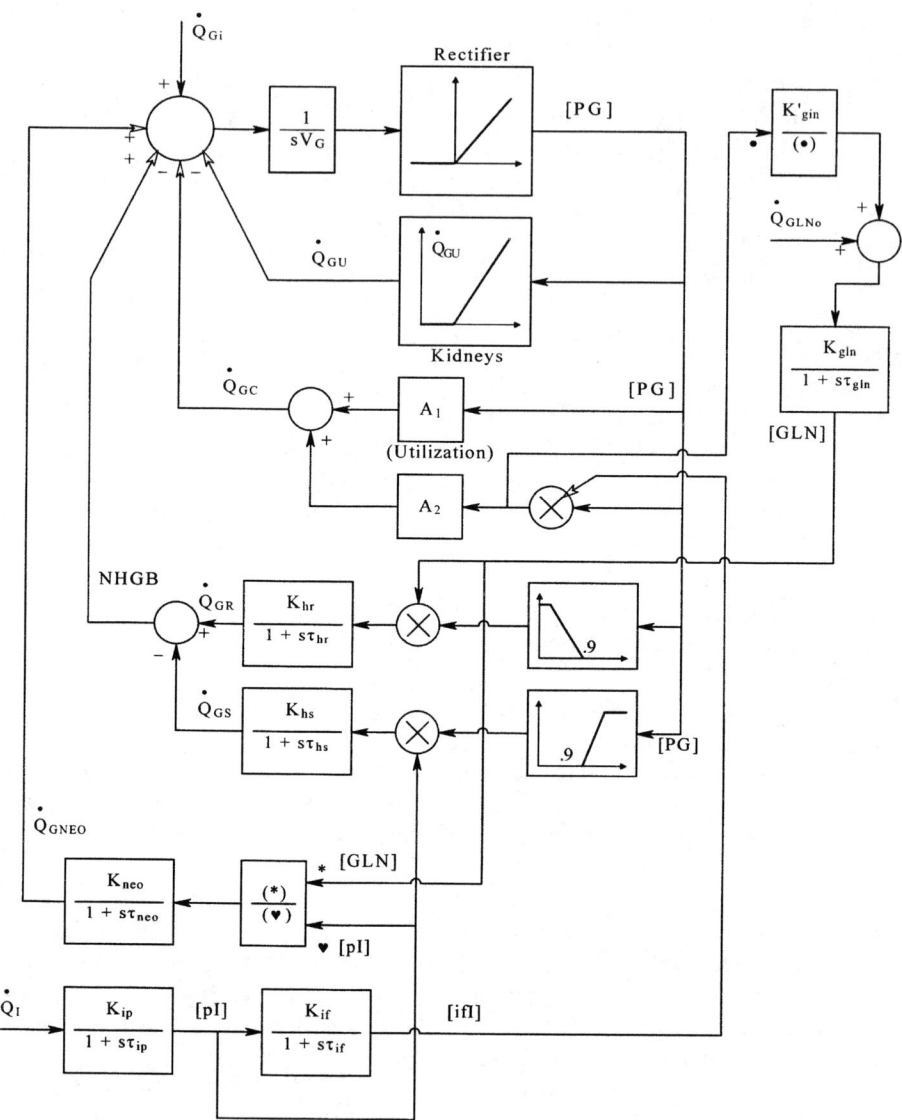

FIGURE 7.16 Block diagram of the simple model of glucose metabolism used by Northrop and Woodruff in 1986 to test their controller performance.

injected peripherally (in plasma), portally, or into the interstitial fluid compartment. The slope, m, of the renal loss function is 5.56 (mg/min)/(g/l). IVGTTs and OGTTs were used in the model tests.

The IPFM controller is a nonlinear digital system that injects small boluses of insulin, if required, at the system's sampling instants. A bolus injection pump has a simple structure that is easily adapted to long-term implantation.[96] The boluses are

Regulation of Blood Glucose

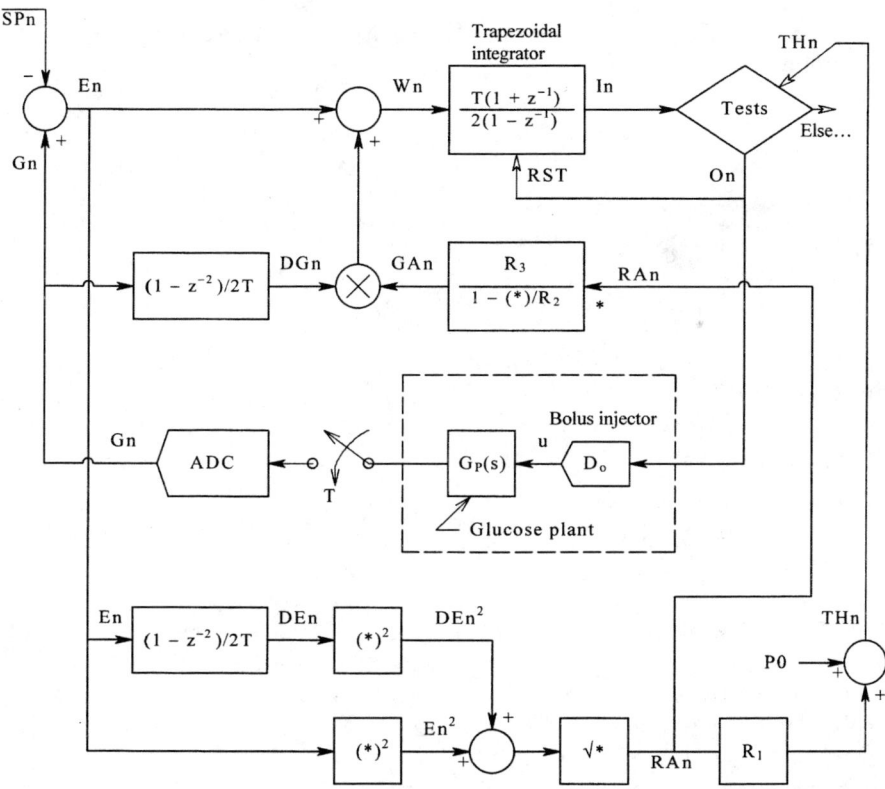

FIGURE 7.17 Block diagram of the discrete, radial-damping-modulation, proportional-plus-derivative, IPFM controller devised by Northrop and Woodruff. The analog plant, $G_P(s)$, is the system in Figure 7.16. Even though the IPFM generator has an integrator, the closed-loop system is type 0.

all the same size or are zero. Figure 7.17 illustrates the discrete IPFM controller used by Northrop and Woodruff. A discrete signal, W_n, is integrated by trapezoidal integration. The integral, I_n, is examined every sample period to see if it is greater than a positive threshold, TH_n. If $I_n > TH_n$, then an output pulse occurs and the patient is given a bolus injection of insulin, $u_n = D_o \delta(t - nT)$. If $I_n < 0$, the integrator is also reset, but no output pulse is given; this prevents integrator "wind-up." System error is defined as $E_n = G_n - R_n$, where G_n is the plasma glucose reading at the nth sampling instant, and R_n is the corresponding value of the set point. A *three-point, central difference algorithm* is used to calculate the discrete derivatives of E and G, DE_n and DG_n, respectively. T is the sampling period.

$$DE_n = (1/2T)\left[E_n - E_{n-2}\right] \tag{7.30A}$$

$$DG_n = (1/2T)\left[G_n - G_{n-2}\right] \tag{7.30B}$$

The novel aspect of the controller lies in the use of RA_n, a radial measure of nonequilibrium:

$$RA_n = \sqrt{DE_n^2 + E_n^2}$$

RA_n is used to raise the IPFM threshold, effectively reducing the closed-loop system gain and response speed. That is,

$$TH_n = P_o + R_1 RA_n \tag{7.31}$$

RA_n is also used to weight the derivative term in the feedback path. (The system, in spite of using IPFM, basically uses a proportional plus derivative feedback strategy and hence is a type 0 closed-loop system, albeit a nonlinear one.) W_n drives the IPFM integrator and is given by:

$$W_n = E_n + DG_n \frac{R_3}{1 + RA_n/R_2} \tag{7.32}$$

where R_1, R_2, and R_3 are constants. As G_n approaches R in the steady state and $RA_n \to 0$, $W_n \to E_n + R_3 DG_n$. R_3 is chosen to make the system well damped. A low RA_n also lowers the IPFM threshold, effectively raising system dc loop gain and giving it good disturbance response. If RA_n is large, the derivative gain of W_n is small, making the closed-loop system underdamped. Northrop and Woodruff tested their controller with their plant model with a 1:8 range in insulin sensitivity, A_2, and found satisfactory performance restoring normoglycemia following simulated IVGTTs and OGTTs.

7.6.3 Type I Controllers for Artificial Beta Cells

In order to obtain zero average steady-state error between the set point and [pG], the artificial beta cell should be a type 1 controller. That is, the loop gain of the system, albeit nonlinear, should have a pole at the origin. This requires either a proportional-plus-integral (PI) or proportional-plus-integral-plus derivative (PID) compensator in the controller. We have examined the feasibility of using a PI "front end" to drive an IPFM bolus injector for insulin. The Simnon subroutine used to simulate the PI/IPFM controller is

```
"              CONTROLLER  (PI)
"
e  = pG - SP                    " System error. SP = setpoint.
dv = K2*e                       " v is error integral
u  = K3*(e + v)                 " P + I
ru = IF u < 0 THEN 0 ELSE u     " ru is nonneg. controller output.
"
```

Regulation of Blood Glucose

```
"              IPFM INJECTOR
"
dw = ru - z                      " Integrator
p = if w > phi then 1 else 0     " Threshold
s = DELAY(p, tau)                " Pulse generator
x = p - s
y = if x > 0 then x else 0       " Pulse rectification
z = y*phi/tau                    " Integrator reset feedback.
eI = Do*y/tau                    " Insulin bolus, Do units.
```

The PI/IPFM controller integrates the system error, e = [pG] − SP, and adds the integrator output to the error to form the PI output, u. In terms of a continuous, linear system transfer function, the PI operation is

$$\frac{U}{E}(s) = \frac{K_3(s + K_2)}{s} \qquad (7.33)$$

To make the IPFM bolus injector, u is rectified to form ru, which is passed to the unity gain IPFM integrator. If the IPFM integrator output, w, exceeds the pulse-generating threshold, phi, $p \to 1$. $s = p$ delayed one computation cycle, tau = 0.05 min in this case. A pulse, x, is formed by subtracting s from p. x is rectified to eliminate negative pulses. The rectified pulse is made a triangular pulse with base = 2 tau and height 1 by Simnon. Thus it has an area = tau. To reset the IPFM integrator we must feed back a pulse of area phi; thus z = y*phi/tau has area = phi. The bolus injection occurs when $p \to 1$. Thus the insulin injection, eI = $y*D_o$/tau, can be considered to be an impulse function with an area of D_o insulin units. That is, $eI \cong D_o\, \delta(t)$. eI is the control input to the ordinary differential equation for bI. From the blood insulin compartment, insulin propagates to the interstitial fluid and portal insulin compartments.

To test the effectiveness of the PI/IPFM controller in glucoregulation, we use the model for normal glucose regulation of Section 7.3, less its "pancreas." (The PI/IPFM controller replaces the pancreas.) The revised Simnon program is

```
CONTINUOUS SYSTEM GLUCOSE6    " Creation date 8/26/97. Mod.11/05/97.
" Type I diabetic model of 70 kg human. 14 States. With AEP.
STATE g nisG iscG lG gn gln bI ifI pI v "w
DER  dg dnisG discG dlG dgn dgln dbI difI dpI dv "dw
TIME t
"
" GLUCOSE PLANT (Because of IPFM insulin input, Must use Euler
"                 integration with delT = tau.)
"
dg = -dgu - dgnis - dgisc + nhgb + dGin    " dg is rate of glucose
"                                            mass input, g/min.
"               dGin is dietary, gluconeogenesis, and gtt sources.
"
" Utilization Uptake by Non-Insulin-Sensitive Cells.
"
```

```
dgnis = Kdnis*(rpG - cnisG)         " dgnis is rate of pG uptake by
"                                     non-insulin sens. cells, g/min.
dnisG = -nisG*Kmnis + dgnis         " nisG is mass of G inside NIS
"                                     cells.
cnisG = nisG/Vnis                   " Conc. Glucose in NIS cells, g/l
"
" Utilization Uptake by Insulin-Sensitive Cells.
"
discG = -iscG*Kmisc + dGisc         " iscG is total mass of glucose
"                                     inside insulin-sensitive cells.
dGisc = Hi*Kdisc*(rpG - ciscG)      " rate of mass diffusion into
"                                     iscs, g/min. from blood.
ciscG = iscG/Visc                   " Conc. isc Glucose, g/l
"
"           OTHER EQUATIONS:
"
Hi = K6 + (K7*Ki*bI)/(1 + Ki*bI)    " Hill function modelling effect
"                                     of insulin on diffusion constant.
R = rpG*Hi
"                                   " R function.
pG = g/Vol                          " Plasma glucose conc., g/l.
"
rpG = IF pG < 0 THEN 0 ELSE pG
"
dpG = dg/Vol                        " Rate of change of plasma
"                                     glucose conc.
dgu = IF pG < 1.8 then 0 ELSE Kk*(-1.8 + pG)   " Kidney spillover
"                                                 nonlinearity
"
"           NHGB EQUATIONS:
"
NHGB = Hi*Kld*(clG -pG)             " NHGB in g/min.
"
dlG = -NHGB - Ks*Hi*clG + fgn*Hgln  " lG is hepatic glucose mass.
"
dgn = Ks*Hi*clG - fgn*Hgln          " Glycogen stored in liver,
"                                     gms.
clG = lG/Vl                         " Conc. glucose in liver.
"
Hgln = bo + bm*Kgl4*gln/(1 + gln*Kgl4)   " Hill fctn for glucagon
"                                            action
fgn = IF gn < 0 THEN 0 ELSE K16     " Glycogen switch.
"
"           GLUCAGON EQUATIONS:
"
dgln = - h*gln + Kgl3*fgln          " Glucagon release from pancreatic
"                                     alpha cells.
fgln = Kgl1/(1 + Kgl2*R)            " glucagon release rate inverse to R
"                                     parameter.
"
"           3 INSULIN COMPARTMENTS: (K8, K9, K10, K11, KLpi, KLifi)
"
dbI = - (K10 + K11)*bI + K9*ifI + K8*pI + dUxin + eI    " Blood insulin
```

Regulation of Blood Glucose

```
difI = - (KLifi + K9)*ifI + K11*bI      " Interstitial fluid
"                                         compartment.
dpI = - (KLpi + K8)*pI + K10*bI         " Portal compartment
"
"              CONTROLLER (PI)
"
"e = pG - SP                            " System error
"dv = K2*e                              " v is error integral
"u = K3*(e + v)                         " P + I
"ru = IF u < 0 THEN 0 ELSE u            " u is controller output.
"eI = ru
"
"              CONTROLLER (PID)
"
e = pG - SP
dv = K3*e
u = K1*e + v + K2*dpG
ru = IF u < 0 THEN 0 ELSE u
eI = ru
"
"              IPFM INJECTOR
"
"dw = ru - z                            " Integrator
"p = if w > phi then 1 else 0           " Threshold
"s = DELAY(p, tau)                       " Pulse generator
"x = p - s
"y = if x > 0 then x else 0             " Pulse rectification
"z = y*phi/tau                          " Integrator reset feedback.
"eI = Do*y/tau                          " Insulin bolus, area Do.
"
"              OTHER EXOGENOUS INSULIN INPUT:
"
dUxin = IF t > tins THEN Iin*EXP(-(t - tins)) ELSE 0
"
"              GLUCOSE IVGTT:
"
dGin = Gin1 + Gin2 + Gneo
Gin1 = IF t > tgluc THEN Gin ELSE 0
Gin2 = IF t > (tgluc + 5) THEN -Gin ELSE 0
"
" CONSTANTS
"
tins:800
tgluc:400
SP:0.9        " g/liter.
Do:1
tau:0.05
phi:1
beta:0.1
Kf:.333
Ksi:1.7
K1:0.2
K2:.01
K3:.1
```

```
K4:.001
K5:0.333
K6:.05
K7:15
K8:.7
K9:.7
K10:.4
K11:.75
KLpi:.2
KLifi:.2
K14:.3
K15:2.
K16:12
gneo:.5
Gin:30     " g/min
Iin:5
Kdnis:.25
Kdisc:3
Kmnis:.4
Kmisc:.05
Vnis:1
Visc:50
Vl:5
Ki:0.05
Kk:5.5
Ks:3
a:.025
h:.1
bo:.05
bm:1
Kld:10
Kgl1:.25
Kgl2:50
Kgl3:3
Kgl4:.2
Kc:1
Vol:75
Kin:.2
point8:.8
zero:0
point9:.9
"
" INIT
g:150     " grams gives pG(0) = 2 g/l.
gn:1000
gln:.3
"
END
```

Figure 7.18 illustrates the closed-loop system's behavior for turn-on ([pG](0) = 2 g/l), followed by an IVGTT at $t = 400$ min. Notice how quickly the [pG] converges on the set point, 0.9 g/l. The glucagon concentration rises, and the [pG] slowly

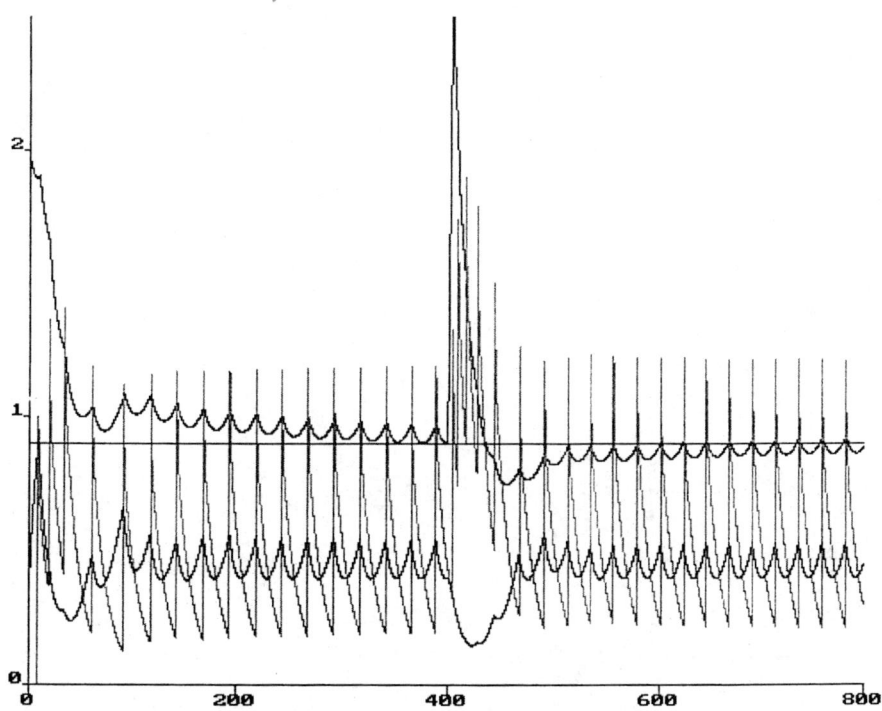

FIGURE 7.18 Simulated turn-on behavior of a type 1, continuous, PI/IPFM glucose regulator using the glucose plant of Figure 7.5 described in Section 7.3. Upper bold trace = [PG], middle light trace = [bI], bottom bold trace = [GLN]. Parameters: $D_o = 1$, $K_2 = 0.0075$, $K_3 = 0.1$, $K_{10} = 0.2$, $K_{11} = 0.1$. A simulated IVGTT was given at $t = 400$ min; 30 g/min was given for 5 min. Vertical axis in grams per liter of glucose. Note ripple on [PG].

recovers. In the steady state, the artificial endocrine pancreas gives a steady rate of injections to compensate for NHGB > 0 and gluconeogenesis. Note the pulsatile nature of the system's responses to bolus injections of insulin given by the controller.

Let us assume the same type 1 diabetic model system receives its insulin from a *continuous infusion pump*, rather than from an IPFM bolus injector. This simulation is shown in Figure 7.19. In this run, slightly different PI controller parameters were used: $K2 = 0.01$, $K3 = 0.75$, $Ks = 3$.

We also tried an "optimized" PID controller, the Simnon subroutine for which is given below:

```
"            CONTROLLER  (PID)
"
e  = pG - SP                      " System error
dv = K3*e                         " v is error integral
u  = K1*e + v + K2*dpG            " P+I+D
ru = IF u < 0 THEN 0 ELSE u       " u is controller output.
```

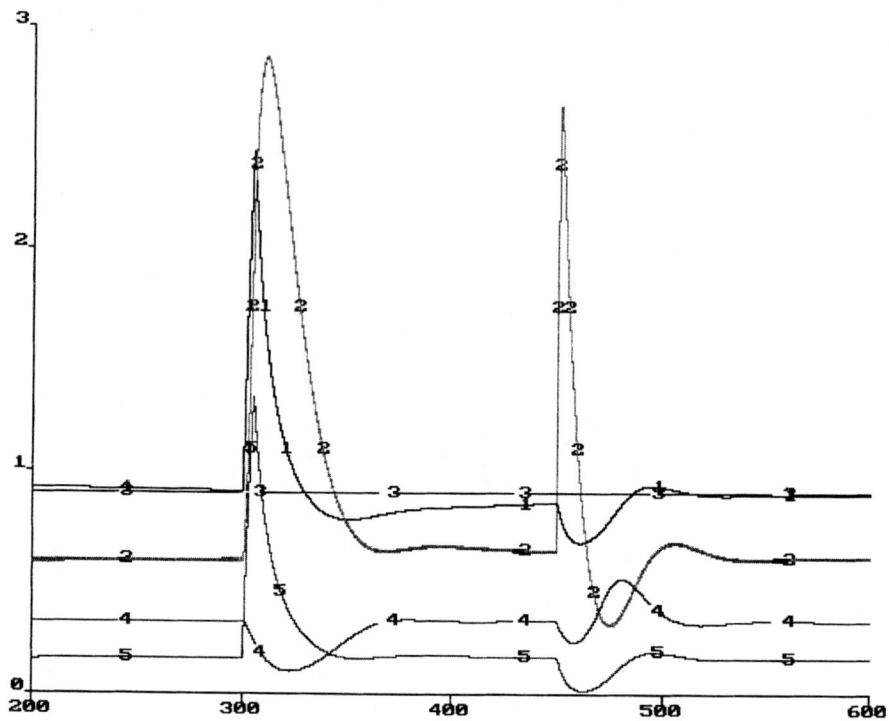

FIGURE 7.19 Simulated behavior of the same plant as in Figure 7.18. A continuous PI controller was used driving a continuous insulin infusion pump. Vertical axis: grams per liter of [PG] only. Horizontal axis: minutes. Trace 1 = [PG], trace 2 = [bI], trace 3 = SP, trace 4 = [GLN], trace 5 = $r u$. An IVGTT of 30 g/min for 5 min was given at $t = 300$; an IV bolus injection of insulin was given at $t = 450$, producing an interesting transient in the glucagon concentration [GLN] and causing the [PG] to dip and then return to the SP. Euler integration was used with $dt = 0.0005$. Controller gains: $K2 = 0.01$, $K3 = 0.75$.

In this case, the system's responses to the IVGTT and IV insulin bolus were very similar to Figure 7.19. The same plant parameters were used as for the PI controller, except $K1 = 0.2$, $K2 = 0.1$, and $K3 = 2.\text{E-}3$. There is not much difference between system performance for the PI vs. PID controllers. The performance of both controllers is dependent on plant parameters, underscoring the need for a self-tuning controller or initial off-line plant parameter identification to facilitate controller tuning.

Note that the PI/IPFM, PI, and PID controllers investigated above were simulated as continuous rather than discrete systems. In practice, the controller would be a discrete system, using periodically taken, digitized, plasma glucose samples. All features of the continuous controllers, of course, can be implemented by their discrete counterparts.

7.7 CHAPTER SUMMARY

Because of the prevalence and medical importance of type 1 diabetes mellitus, the endogenous control system for plasma glucose concentration has been thoroughly studied and is well understood. Section 7.1 introduced the important molecules in the glucoregulatory system: D-glucose, glucagon, and the hormones insulin and glucagon. Section 7.2 described the endogenous (normal) glucoregulatory system. This system is seen to be a high-order nonlinear type 0 (in a feedback control sense) and multiple input/multiple output. Principal sources of glucose were shown to include diet, liver glycogen, and gluconeogenesis. Sinks of glucose include insulin-dependent cells, noninsulin-dependent cells, loss through the kidneys, and storage in the liver. Pancreatic α and β cells secrete the hormones glucagon and insulin, respectively, in response to the plasma glucose concentration and to other hormones.

Sections 7.3 and 7.4 described two dynamic mathematical models of normal endogenous glucoregulation. A simplified dynamic model by the author was given in Section 7.3. It was simulated using Simnon and results were illustrated with an intravenous glucose tolerance test and bolus doses of insulin. In Section 7.3, the more complex glucoregulatory model of Cobelli and Mari was presented.

Section 7.5 described the differences between normal glucoregulation and type I and type II diabetes mellitus. Section 7.6 introduced the concept of exogenous control of plasma glucose in type I diabetics. Essential to any artificial beta cell is a long-term-stable, implantable glucose sensor or a reliable, accurate, noninvasive means of measuring plasma glucose concentration. Sadly, at this writing, there are no accurate implantable or noninvasive glucose sensors that retain their long-term calibration. Many enzymatic electrochemical glucose sensors that require direct contact with blood or extracellular fluid are accurate but, because of degradation of the enzyme and/or membranes in the body environment, lose calibration and eventually fail. Extracorporeal use of these sensors in a clinical setting has proved successful because they can easily be recalibrated and replaced.

Section 7.6.2 described early attempts at exogenous plasma glucose control. Some early, nonlinear controllers were seen to have been designed with little attention to the dynamics of the glucoregulation plant. The early controllers were type 0. Section 7.6.3 introduced a PI type 1 controller for an artificial endocrine pancreas, based on the IPFM artificial endocrine pancreas design by Woodruff and Northrop.[144] This PI/IPFM design was modeled, and the results of the simulations were presented. Note that in spite of the IPFM bolus-induced ripple on [pG], the steady-state average error tends to zero following a simulated IVGTT.

Future designs of exogenous [pG] controllers will probably include on-line plant identification algorithms and be self-tuning type 1 systems.

PROBLEMS

7.1 Listed below is a nine-state Simnon program to emulate the exogenous control of the glucose dynamics in a type 1 diabetic patient by insulin injection. v and w are controller states. There is a single, central glucose compartment containing glucose mass g. Losses from this central compartment include inputs to insulin-sensitive cells, to noninsulin-sensitive cells, and loss through the kidneys. Inputs are net hepatic glucose balance (NHGB > 0 when liver releases glucose stored as glycogen), from diet and IVGTTs and from gluconeogenesis. The pancreas is assumed to make no insulin, but does make glucagon. The program includes a continuous PI controller whose output determines the IV insulin infusion rate, $u(t)$. There is also an IPFM insulin bolus injector.

    A. Find the appropriate PI controller parameters, $K4$ and $K5$, that will enable the system to reach the set point (SP) = 1 g/l in minimum time without going below 0.8 g/l for any IVGTT. The IVGTT is given at tgluc = 200, so the system will be in steady state following turn-on. Demonstrate the effectiveness of your design by plotting SP, point8, pG, gn, NHGB, bI, and dGin over 500 min. *Note*: The IPFM injector must be disabled in the program by preceding each line of its code by a ". Also, the IPFM integrator state w and derivative dw must be removed by preceding them in the program by quotes, (").

    B. In order to implant the controller in the body, a reliable internal glucose sensor has been developed, and an IPFM insulin bolus injector pump will be used. The input to the IPFM is the continuous *numerical* output of the PI controller, ru. Find the $K2$, $K3$, and Do values that will allow the system to meet the specification in part A. Note the ripple on bI and pG. Give plots.

```
CONTINUOUS SYSTEM GLUCOSE2  " Creation date 11/01/94. Mod. 8/14/98.
" This is a model of glucose utilization and nhgb mediated by
" intrinsic glucagon (gl) and exogenous insulin. Includes nhgb and
" kidneys. dgin includes dietary, gluconeogenesis, and ivgtt
" sources of blood glucose (bG). Program must use Euler integration
" when simulating IPFM; Make ΔT of Euler = tau of IPFM.
"
STATE g cnsG iscG lG gn gl bI v w
DER  dg dcnsG discG dlG dgn dgl dbI dv dw
TIME t
"
" GLUCOSE PLANT
"
dg = -dgu - dgcns - dgisc + nhgb + dgin     " dg is rate of glucose
"                                             mass input.
pG = g/Vol                                  " Plasma & ECF glucose conc., g/l.
"
dgcns = Kdcns*(bG - ccnsG)                  " dgcns = rate of bG uptake by
"                                             non-insulin sens. cells, g/min.
```

Regulation of Blood Glucose

```
dcnsG = -cnsG*Kmcns + dgcns         " cnsG is mass of g inside CNS
"                                     cells.
ccnsG = cnsG/Vcns                   " Conc. cns Glucose, g/l
"
discG = -iscG*Kmisc + dGisc         " iscG is total mass of glucose
"                                     inside insulin-sensitive cells.
dGisc = fi*Kdisc*(bG - ciscG)       " rate of mass diffusion into
"                                     iscs, g/min. from blood.
ciscG = iscG/Visc                   " Conc. isc Glucose, g/l
"
fi = 0.25 + (5*Ki*bI)/(1 + Ki*bI)   " Hill function modelling effect of
"                                     insulin on diffusion constant.
r = pG*fi                           " r parameter.
"
dgu = IF pG< 1.8 then 0 ELSE Kk*((-1.8 + pG)   " Kidney spillover
"                                                nonlinearity
" NHGB EQUATIONS:
nhgb = - dlG                        " NHGB in g/min.
dlG = fi*Kld*(pG -clG) - Ks*fi*clG + rgn*hgl   " lG is hepatic
                                                 glucose mass.
dgn = Ks*fi*clG - rgn*hgl           " Glycogen stored in
"                                     liver,gms.
clG = lG/Vl                         " Conc. glucose in
                                      liver.
hgl = bo + bm*Kgl4*gl/(1 + gl*Kgl4) " Hill fctn for
                                      glucagon action
rgn = IF gn < 0 THEN 0 ELSE gn      " Rectification of
                                      glycogen
"
" GLUCAGON EQUATIONS:
dgl = -h*gl + Kgl3*fgl    " Glucagon release from pancreatic alpha cells.
fgl = Kgl1/(1 + Kgl2*r)   " glucagon release rate inverse to r parameter.
"
" INSULIN DYNAMICS
dbI = -j*bI + Kin*Ib    " Insulin input lag (single TC). Use ru input
"                         if continuous PI controller used alone.

" IVGTT
dgin = gin1 + gin2 + gneo
gin1 = if t > tgluc then Gin else 0    " Glucose input pulse, Gin
"                                        g/min.
gin2 = if t > (tgluc + 5)then -Gin else 0
"
" CONTROLLER (PI)
e = bG - SP
dv = K4*e
u = K5*(e + v)
ru = IF u < 0 THEN 0 ELSE u
"
" IPFM BOLUS INJECTOR
dw = ru - z                    " Unity-gain IPFM integrator output w.
p = IF w > phi THEN 1 ELSE 0   " Pulse generation when w > threshold.
s = DELAY(p, tau)
x = p - s
```

```
        y = IF x > 0 THEN x ELSE 0      " Positive output pulse w/area tau.
        z = y*phi/tau                    " Pulse z resets integrator
        Ib = Do*y/tau                    " Insulin bolus of area Do units.
        "
        " CONSTANTS
        SP:1.00
        tgluc:200
        K4:___
        K5:___
        Do:___
        tau:0.05    " set ΔT of Euler integrator = tau.
        phi:1
        gneo:1.
        Gin:49
        Inin:200
        Kdcns:1.
        Kdisc:2.
        Kmcns:.4
        Kmisc:.05
        Vcns:1
        Visc:50
        Vl:1
        Ki:1
        Kk:2
        Ks:1.
        h:.1
        bo:.25
        bm:5
        Kld:0.1
        Kgl3:.5
        Kgl4:10
        Kgl1:2
        Kgl2:5
        Kc:1
        Vol:75
        j:.05
        Kin:.2
        lower:.8
        "
        " INIT
        g:150
        "
        END
```

7.2 The hormone *leptin*, which is secreted by adipose cells, appears to be important in controlling an animal's appetite for food and hence its calorie input and body fat content. Make a detailed systems block diagram of the form of Figure 7.5 to show how leptin acts in the regulation of body fat. (Hint: Search the World Wide Web for current data on leptin from recently published research papers.) Note that leptin interacts with the secretion rate of insulin, neuropeptide Y, etc. How does a large protein molecule like leptin cross the blood–cerebrospinal fluid barrier?

7.3 Use the Simnon program "GLUCOSE6" in the text to illustrate the effect of glycogen exhaustion on the behavior of the closed-loop PID/IPFM glucoregulator. Glycogen exhaustion can be modeled by reducing the initial condition on glycogen (gn) from 1000 to 250 units. Give the IVGTT at $t = 400$. Plot [pG], [bI], [gln], [gn]/100. Why does the demand for insulin drop sharply after glycogen exhaustion? Use parameters: Do = 1.5, tau = 0.05, phi = 1, beta = 0.1, Kf = 0.333, Ksi = 1.7, K1 = 0.2, k2 = 0.01, k3 = 0.1, k4 = 0.001, k5 = 0.333, k6 = 0.05, k7 = 15, k8 = 0.7, k9 = 0.7, k10 = 0.2, k11 = 0.2, k14 = 0.3, k15 = 2, k16 = 12, KLpi = 0.2, KLifi = 0.2, gneo = 0.5, Gin = 30, Iin = 0, Kdnis = 0.25, Kdisc = 3, Kmnis = 0.4, Kmisc = 0.05, Vnis = 1, Visc = 50, V1 = 5, Ki = 0.05, Kk = 5.5, Ks = 1.75, a = 0.025, h = 0.1, bo = 0.05, bm = 1, Kld = 10, Kg11 = 0.25, Kg12 = 50, Kg13 = 3, Kg14 = 0.2, Kc = 1, Vol = 75, Kin = 0.2. INIT: g = 150, gn = 250, gln = 0.3, others 0.

7.4 Certain body fluids other than blood have been used to estimate blood (plasma) glucose concentration. Glucose in the urine is a sign that a diabetic's [pG] is too high. The aqueous humor of the eye is another extracellular fluid whose [G] follows [pG]. Optical noncontact means of measuring the [G] in aqueous humor have been proposed. Consider the concentration of glucose in saliva. Specifically, is there any, and does it follow [pG]?

7.5 Guyton[59] states that over 90% of the carbohydrates utilized by the body are used to form adenosine triphosphate (ATP) in the cells.
A. Describe some major uses of ATP in the body.
B. Describe the biosynthesis of ATP in *glycolysis*.
C. Describe the biosynthesis of ATP in the *citric acid cycle*.
D. Describe the biosynthesis of ATP by *oxidative phosphorylation*.
E. In what cellular structures is most ATP made?

7.6 Describe the effect of thyroid hormone on oxidative metabolism and the biosynthesis of ATP.

7.7 Describe the influence of insulin on the secretion rate of glucagon and the influence of glucagon on the secretion rate of insulin (cf. Chapter 49 in West[141]).

7.8 Among its other effects on plasma glucose concentration, the pancreatic hormone glucagon stimulates hepatic gluconeogenesis. Describe the principal biochemical substrates and pathways used in gluconeogenesis. How is the Krebs cycle involved?

7.9 Persons with untreated diabetes mellitus often suffer from blood acidosis. Describe the biochemical processes that produce acidosis.

7.10 Make a list of the various conditions and diseases that can occur as the result of poorly treated, severe type 1 diabetes mellitus.

8 Control of Mean Arterial Pressure by Sodium Nitroprusside Injection

8.0 INTRODUCTION

The human arterial blood pressure waveform [ABP(*t*)] varies with the site of measurement, e.g., the ascending aorta, the brachial artery, etc. (see Figure 8.1). In general, ABP(*t*) becomes lower in amplitude and more rounded as the measurement site becomes more distal to the heart. Since nearly all clinical, *noninvasive* measurements of systolic and diastolic blood pressures are made by sphygmomanometer cuff and stethoscope (or microphone) on the upper arm's *brachial artery*, we will assume in this chapter, unless otherwise noted, that brachial artery pressure is used in defining mean arterial pressure (MAP).

The MAP can be defined quantitatively for our purposes in this chapter as the blood pressure one would observe if ABP(*t*) were thoroughly low-pass filtered to give a dc output. Another way of quantifying MAP is to treat the ABP(*t*) pulse pressure waveform as a triangle, riding on a constant diastolic pressure. T is the period of the heartbeat. Thus MAP is approximately given by:

$$\text{MAP}(t) = \sum_{k=1}^{\infty} \{[P_s(k) + P_d(k)]/2\} \{U(t - t_k) - U(t - t_{k+1})\} \quad (8.1)$$

Note that MAP can be computed for every cardiac cycle in which the systolic and diastolic blood pressures are measured. Because the period of the heart beat varies normally (unless the patient is taking cardioregulatory drugs or is wearing a pacemaker), the MAP calculated on a beat-by-beat basis will be available as a rectangular waveform with an uneven period. Mathematically, the MAP output can be written as:

$$\text{MAP}(t) = \sum_{k=1}^{\infty} \{[P_s(k) + P_d(k)]/2\} \{U(t - t_k) - U(t - t_{k+1})\} \quad (8.2)$$

Here, $\{[P_s(k) + P_d(k)]/2\}$ is the MAP calculated from the systolic (peak) pressure and diastolic (minimum) pressure in the *k*th cardiac cycle which ends at t_k when

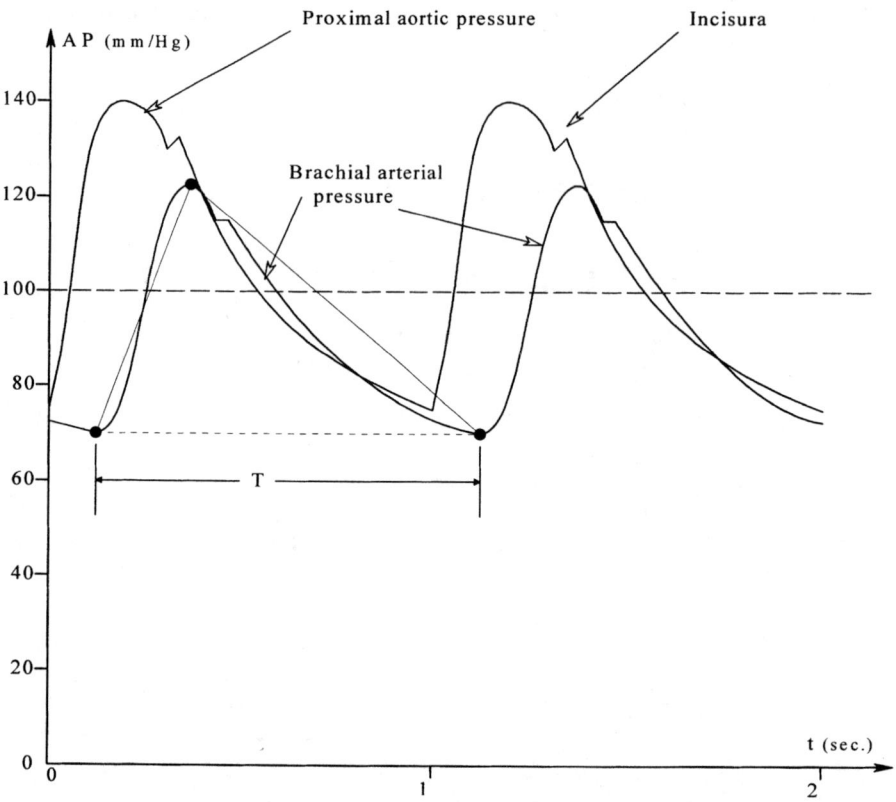

FIGURE 8.1 Typical pressure waveforms from the aorta and the brachial artery. Note that as the pressure waveform propagates along the arterial tree, it becomes delayed and attenuated with respect to the aortic pressure. The triangular approximation used in calculating the MAP estimate by Equation 8.1 is shown on one cycle of the the brachial arterial pressure waveform.

$P_d(k)$ is reached. $U(t - t_k)$ is a unit step function that begins at $t = t_k$ and is zero for $t < t_k$. $U(t - t_{k+1})$ is a unit step function which begins at the end of the $(k + 1)$th cardiac cycle. MAP(t) is thus seen to be a continuous waveform with discontinuities at the times $\{t_k\}$. (MAP$_k$ is not necessarily equal to MAP$_{k+1}$, etc.)

Many factors can affect the long-term set point for MAP. These include blood volume, the function of the kidneys as regulated by the hormones vasopressin and aldosterone, factors that modulate cardiac output and peripheral (circulatory system) resistance, and osmotic pressure of the blood (typically set by Na$^+$ ions and glucose). In this chapter, we are primarily concerned with elevated MAP which typically follows open chest surgery, such as cardiac bypass surgery, lung surgery, or any surgery that involves cardiac bypass pumping or slowing or stopping the heart. In the case of cardiac bypass surgery, it is desirable to keep postsurgical MAP down to minimize bleeding from sutures.

One possible reason postsurgical MAP becomes elevated is that the kidneys experience low blood flow during surgery, which may last a number of hours, which stimulates them to secrete renin. Renin leads to the formation of angiotensin I and angiotensin II in the blood. Angiotensin II is a potent systemic vasoconstrictor, raising peripheral circulatory resistance and hence MAP. Angiotensin II also acts directly on the kidneys to cause them to retain Na^+ and water. It also stimulates the secretion of aldosterone by the adrenal glands, which acts on the kidney tubules to increase salt and water retention.[59] Increased water and salt retention raises blood volume, which, in turn, raises cardiac output and hence MAP.

A short-term fix for elevated MAP is to decrease systemic peripheral vascular resistance. A drug that has been widely used for this purpose is sodium nitroprusside (SNP). In the following sections, the pharmacology and pharmacokinetics of SNP will be discussed, and several controllers (fixed and self-tuning) will be described for the administration of intravenous (IV) SNP in order to reduce and manage postoperative MAP.

Control of the SNP/MAP plant is challenging because it contains a destabilizing dead time or delay that is of the same order of magnitude as its dominant time constant. Also, there is wide variation of plant parameters among individuals, and certain parameters may drift with time.

8.1 THE PHARMACOLOGY AND PHARMACOKINETICS OF SNP

SNP is a rapidly acting vasodilator drug whose effect is of relatively short duration. Because of its short-term dynamics, SNP is ideally suited to external automatic control of MAP. SNP is a ferrous hydrated pentacyano compound with the formula $Na_2Fe(CN)_5NO-2H_2O$. Its hypotensive effect was first described in 1897 by Davidsohn, but it was not used therapeutically until 1929.[2] SNP acts to dilate arterioles and lower peripheral circulatory resistance in general, thus lowering MAP. The nitroso group (NO^-) interferes with the excitation–contraction coupling in blood vessel smooth muscle by inhibiting both the influx and the intracellular activation of calcium ions.[58] The typical IV dose of SNP is 0.5 to 8 µg/min/kg body weight from a 0.1 g/l SNP solution in 5% dextrose plus water.[137]

After an IV bolus injection of SNP, MAP begins to fall in about 40 to 90 sec; the peak decrease in MAP occurs from 90 to 120 sec. MAP then slowly rises to the untreated level over about 5 min from the injection time.

Long-term use of SNP is not without hazard. The initial metabolic product from SNP is cyanogen (CN^-), which is rapidly converted to thiocyanate ions (SCN^-) by a liver enzyme. Thiocyanate ions are cleared through the kidneys and may be oxidized back to CN^- by an erythrocyte enzyme, thiocyanate oxidase. Long, heavy use of SNP can result in symptoms of cyanide poisoning and hypothyroidism induced by the high levels of SCN^- ions in the blood. Ahearn and Grim[2] recommend that if the blood concentration of SCN^- exceeds 120 mg/l, SNP treatment should be discontinued.

In order to develop controller algorithms for the automatic injection of SNP to lower postoperative MAP, a quantitative description is needed of the dynamics relating MAP reduction to the rate of SNP infusion. That is, we need $\Delta MAP/\dot{Q}_{SNP}$. Stoelting[131] gave data for MAP reduction due to an IV bolus injection of SNP that can be modeled closely by a transfer function with two real poles and one zero (Woodruff and Northrop, unpublished). Stoelting's direct impulse response measurement did not show the twin peaks shown by Sheppard and Sayers'[123] pseudo-random binary noise (PRBN) transfer function identification method.

In the PRBN method, an SNP infusion pump is driven by a pseudorandom binary (0,1) noise signal so that it is either on or off for random times. This 0,1 input results in a mean infusion rate value (there can be no negative SNP infusion). The pseudorandom signal is generated digitally and approximates a random telegraph signal in which the random, 0,1 sequence repeats itself after a very large number of transitions. If the repeat period exceeds the system analysis time, then the PRBN sequence can be considered truly random, in which the probability of k transitions ($0 \rightarrow 1$ or $1 \rightarrow 0$) in a time interval T is given by the Poisson distribution:

$$P(k,T) = \frac{(\lambda T)^k}{k!} \exp(-\lambda T) \tag{8.3}$$

where λ is the average number of transitions per unit time. The autocorrelation function of the random, 0,1 waveform can be shown to be[34]

$$R_{xx}(\tau) = 1/4\left[1 + \exp(-2\lambda|\tau|)\right] \tag{8.4}$$

The autopower spectrum of the random 0,1 waveform is the Fourier transform of the autocorrelation function above:

$$\Phi_{xx}(\omega) = (\pi/2)\delta(\omega) + \frac{\lambda}{4\lambda^2 + \omega^2} \tag{8.5}$$

To identify the $\Delta MAP/\dot{Q}_{SNP}$ transfer function, MAP(t) is calculated and is cross-correlated with the pseudorandom 0,1 input signal driving the infusion pump. The frequency response of the desired transfer function can be shown to be given by:[89]

$$\mathbf{H}(j\omega) = \frac{\Delta MAP}{\dot{Q}_{SNP}}(j\omega) = \frac{\Phi_{yx}(\omega)}{\Phi_{xx}(\omega)} \tag{8.6}$$

where $\Phi_{yx}(\omega)$ is the cross-power spectral density (Fourier transform of the cross-correlation function). $\Phi_{yx}(\omega)$ is complex. $\Phi_{xx}(\omega)$ is the autopower spectrum of the PRBN input. It is a positive, real, even function of ω. The delta function is ignored in computing the SNP plant's transfer function. Slate et al. (1979) have given plots of typical $h(t)$ (plant impulse response) which they describe by the approximate frequency response function:

$$\Delta\text{MAP}/\dot{Q}_{\text{SNP}}(j\omega) \cong \frac{-K\exp(-T_i j\omega)\left[1 + \alpha\exp(-T_c j\omega)\right]}{(j\omega\tau + 1)} \quad (8.7)$$

Slate's model is now widely used by workers developing controllers for SNP-induced MAP reduction. The mean parameters and their ranges are shown in Table 8.1.[126] Twenty-six patients were studied experimentally with the PRBN cross-correlation method. A standard SNP in saline solution of 200 µg/ml was used.

It is noted that not all patients exhibited the second recirculation peak in their SNP transfer functions. Thus, in some individuals, $\alpha = 0$ or was <0.4. Also note the large range in the sensitivity parameter, K, a 9/0.25 = 36:1 spread. Such a large uncertainty in the plant gain necessitates a preliminary off-line plant parameter identification in order to tune the controller or the use of an on-line self-tuning controller for SNP infusion. Figure 8.2 illustrates a typical MAP/SNP weighting function (impulse response) obtained by Slate[126] which is approximated by the inverse Laplace transform of the transfer function above (when $j\omega$ is replaced by the complex variable, s). Note that ΔMAP is negative, because SNP reduces blood pressure. In modeling the effect of SNP on MAP, we assume that the MAP reduction is a linear phenomenon. That is,

$$\text{MAP}(t) = \text{MAP}_o + \Delta\text{MAP}(t) \quad (8.8)$$

MAP_o is the high MAP that would occur without SNP intervention. MAP_o itself can be made a function of time to reflect increasing cardiac output or an increasing titer of the vasoconstrictor angiotensin II. Measurement noise can also be added to MAP_o when modeling to test controller robustness.

TABLE 8.1
Parameters and Their Ranges Given by Slate[126]

Symbol	Parameter	Units	Mean	Minimum	Maximum
$K(1 + \alpha)$	Steady-state gain	mmHg/(ml/hr)	1.0	0.25	9.0
K	Sensitivity	mmHg/(ml/hr)	1.0	0.25	9.0
α	Recirculation fraction	none	—	0.0	0.4
T_i	Initial xport lag	seconds	30	20	60
T_c	Recirculation lag	seconds	45	30	75
τ	Time constant	seconds	40	30	60

Note: A 200-µg/ml SNP solution is assumed for a 70-kg patient in assigning K values.

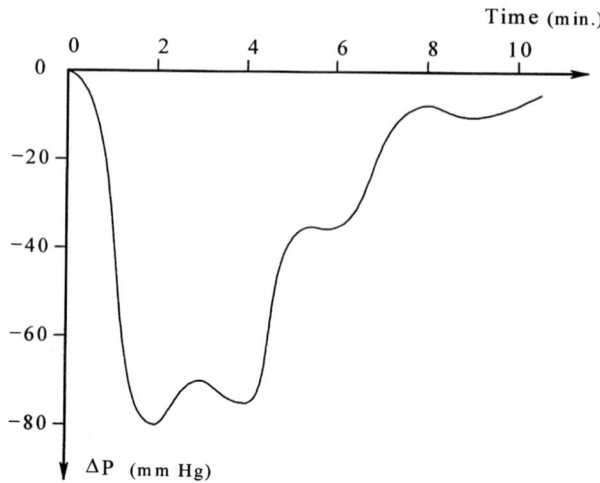

FIGURE 8.2 Impulse response of MAP to IV SNP infusion obtained by the PRBN method (drawn after a figure in Slate[126]). Ripples, presumably due to recirculation of unmixed SNP in the blood, are not seen with every patient. See text for description of the PRBN method.

8.2 MAP/SNP CONTROLLERS USING PROPORTIONAL-PLUS-INTEGRAL PROCESSING AND SMITH DELAY COMPENSATION

In Section 3.4 of this text, the general problems of controlling plants with appreciable delays or transport lags were discussed. The MAP/SNP plant is a classic example of such a difficult system to control. Of great importance in designing any kind of MAP/SNP controller is that there should be no excessive undershoot in MAP at turn-on and certainly no limit cycle oscillations for any plant parameter set in the broad physiological range expected. A type 1 control system, with one controller integrator in the loop gain, is desirable in order to realize zero steady-state error between MAP and the set point. Thus, proportional-plus-integral (PI) or proportional-plus-integral-plus-derivative (PID) controller architectures are required, although they are generally not suitable for plants with large delays. They will work, however, if they are carefully tuned to a plant with known parameters. A better approach is to use a Smith delay compensator (SDC) architecture,[128] followed by the PI or PID block. The SDC is a model reference control strategy that effectively cancels the system instability resulting from the plant's transport lag. It, too, requires tuning. (The SDC was introduced in Section 3.4 of this text.)

There is an extensive history of workers who have developed controller designs for the infusion of SNP to reduce postoperative MAP. A comprehensive review paper by Isaka and Sebald[63] summarizes most of these designs, some of which are discussed next.

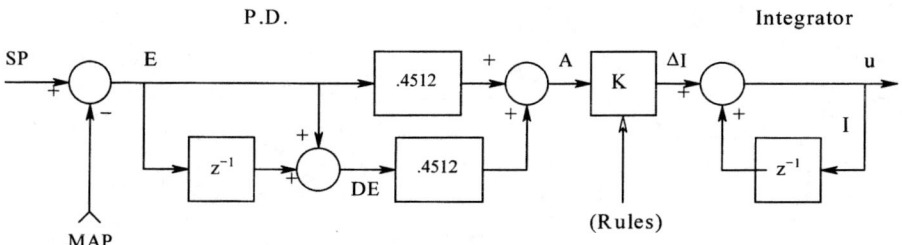

FIGURE 8.3 Nonlinear, discrete, PI controller used by Sheppard et al. to reduce MAP by IV infusion of SNP. The controller is described in Section 8.2 in the text.

Sheppard[121] described a discrete, nonlinear, PI controller for SNP infusion that was tested at the University of Alabama hospital. Figure 8.3 illustrates a block diagram for the discrete controller used by Sheppard. Note that the first block is a proportional-plus-derivative (PD) operation using first-difference derivative estimation. The PD output is multiplied by a variable gain block and then integrated by simple rectangular integration. The overall result is a PI controller with a fixed zero. This is a rather awkward way to realize a PI operation. A more direct means that avoids the necessity of computing the derivative is to simply integrate the error (with bounds) and add the error integral to the error with scaling. The gain K was scheduled by the following rules:[63]

1. If MAP < (SP + 5) and MAP > SP, then $K = -0.5$.
2. If MAP < SP and MAP > (SP − 5), then $K = -1$.
3. If MAP ≥ (SP + 5), then $K = -1$ and $\Delta I = \Delta I - 2$.
4. If MAP < (SP − 5), then $K = -2$.
5. If MAP < (SP − 5) and $\Delta I > 0$, then $\Delta I = 0$ ($K = 0$).
6. If MAP ≥ SP and $\Delta I > 7$, then $\Delta I = 7$.

Here, SP is the set point in millimeters of mercury, and ΔI is the output of the PD + gains blocks (input to the integrator). The sampling period was $T = 1$ min. Sheppard's controller was evaluated clinically in the ICU on 49 postcardiac surgery patients. Its performance was compared with a "skilled ICU nurse" regulating the SNP infusion on 37 patients. As in the case of John Henry, the machine won the contest, giving control within ±5 mmHg of the SP 94% of the time, while the nurse achieved 52% success.[35]

Other workers cited by Isaka and Sebald[63] who have used the nonlinear PI or PID controller architectures successfully are Smolen et al.,[129] Petre et al.,[113] and Reid and Kenny.[118] Colvin and Kenny[29] modified Sheppard's PI controller to simultaneously inject two vasodilator drugs, SNP and glyceryl trinitrate (nitroglycerine). IV nitroglycerine, like SNP, is fast-acting and rapidly inactivated. Nitroglycerine relaxes primarily venous smooth muscle, while SNP acts on both venous and arterial beds.[61]

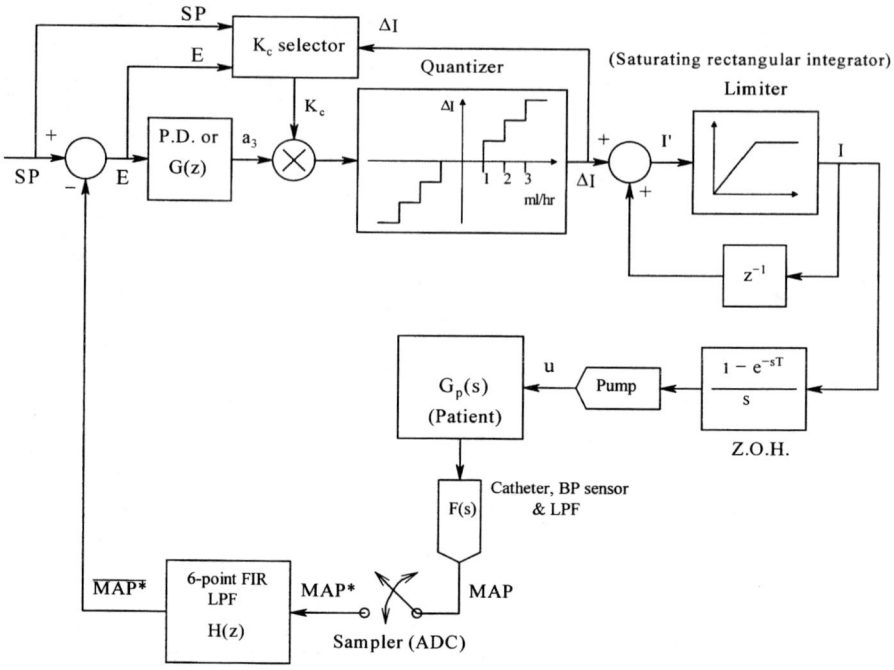

FIGURE 8.4 Block diagram of the modified nonlinear, discrete, PI controller used by Slate and Sheppard[127] to control MAP by SNP infusion. See text for description.

It is instructive to examine in detail the MAP/SNP controller described by Slate and Sheppard,[127] because it represents a fusion of the earlier work of Sheppard and Slate's Ph.D. thesis.[126] Figure 8.4 illustrates the block diagram of the Slate and Sheppard system. MAP was measured using a saline-filled, intra-arterial cannula connected to a blood pressure sensor and signal conditioning electronics. The arterial pressure, AP(t), was first low-pass filtered by a single-pole, analog, low-pass filter with a 4-sec time constant. The MAP was estimated by sampling the arterial pressure signal at 1-sec intervals and filtering it with a six-point FIR low-pass filter. The digital output of the FIR filter was sampled every 60 sec to generate the system error from the set point. That is, $E = SP - MAP$. The Slate/Sheppard controller used a nonlinear PI architecture in which a PD block is given rule-determined constants. The detail of the nonlinear, rule-dependent, PD block is shown in Figure 8.5. The output of the PD block is passed through a variable controller gain block, K_c, then through a quantizer whose output is 0, ±1, ±2, ±3, ... ml/hr. (Details of the PD block's rules and the quantizer can be found in Chapter 5 of Slate's Ph.D. dissertation.) The quantizer output, $\Delta i(kT)$, is then integrated by nonlinear rectangular integration with a 60-sec sample period. The integrator output, $i(kT)$, can be shown to have a PI format acting on E. $i(kT)$ is bounded by zero and an upper limit set by the maximum recommended dose rate for SNP, 10 µg/(kg body weight × min). $i(kT)$ is then passed to a zero-order hold which drives the SNP infusion pump. The authors claimed that over 1700 postoperative patients have been treated with this controller.

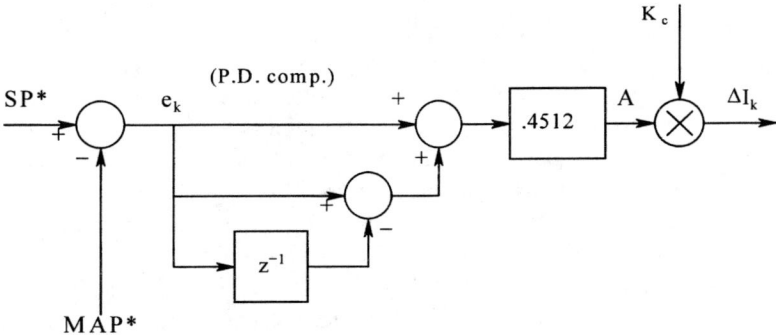

FIGURE 8.5 Details of the PD portion of the Slate and Sheppard controller. The gain, K_c, is selected by rules so $K_c = f(\text{SP}, E, \Delta I)$. AK_c is then input to a quantizer. Details on the rules for K_c and the quantizer are found in Slate's Ph.D. dissertation.[126]

Slate and Sheppard go on to describe an improved, complex, multimode controller for SNP infusion, based on Slate's Ph.D. research. In this system, low values of e and de/dt invoke a maintenance-mode controller similar to the PI system described above. For large, transient changes in set point (resulting in large e and de/dt), an SDC is used in the loop along with another version of a PI controller using parameter switching by rule. In addition, the authors use a recursive, least squares method of estimating the patient's dc gain sensitivity to SNP on-line. This estimate is used to "tune" certain controller parameters on-line for optimum performance in both the transient-mode and maintenance-mode PI controllers. Figure 8.6 illustrates the parameter-switching architecture of the PD section of the transient controller.

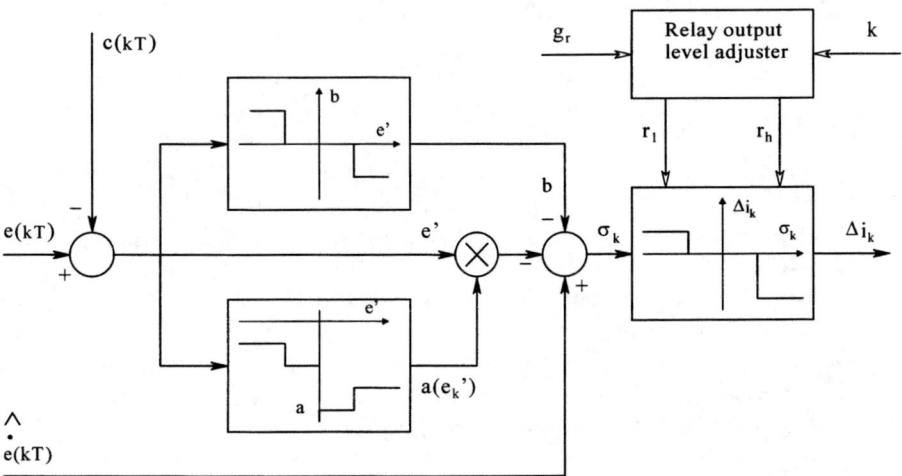

FIGURE 8.6 The unusual nonlinear PD system with three-level quantizer represents the "transient-mode" PD portion of the Slate and Sheppard MAP controller; it is based on Slate's Ph.D. work. See text for description.

Note that dc gain *is not* the most critical plant parameter in tuning an SDC-based controller. As we saw in Section 3.4, the plant's delay time is far more critical in determining overall system performance than is its dc gain or time constant. Also, it appears that the generation of a PD block and then integrating it to get an overall PI operation is awkward; direct generation of the PI operation with on-line adjustment of the zero's position and the gain is more efficient.

Using a modeling approach, a fixed-parameter, discrete, optimum controller for the MAP/SNP plant was developed by Behbehani and Cross.[11] They used the integrating self-tuning (IST) strategy developed by Tuffs and Clarke.[139] The IST controller is a general, adaptive control strategy that is effective for plants with transport lags. The control law minimizes system error *and* the total amount of SNP infused. Behbehani and Cross used the by-now familiar Sheppard and Slate MAP/SNP plant with two delays and one real pole. In developing the control algorithm, they assumed the plant model order to be 2 and to have a single delay, d. d was estimated off-line using the PRBN/cross-correlation method introduced by Sheppard. The combined dynamics of the infusion pump, the patient response, and the MAP measuring system was assumed to be modeled as a controlled, autoregressive, integrated, moving average process. Results of modeling the complex IST controller's performance under conditions of noise, and stepwise changes in the plant's primary delay and dc gain demonstrated its simulated performance to be adequate, but slow in response under most conditions. The system did not handle large step increases in the simulated plant delay well, either producing oscillations or a large undershoot in MAP to less than 40 mmHg.

8.3 ADAPTIVE AND SELF-TUNING CONTROLLERS FOR MAP

Isaka and Sebald[63] discuss the use of *adaptive controllers* in regulating MAP. They note that adaptive controllers may be categorized as (1) *self-tuning regulators*, (2) *model-reference adaptive controllers* (MRACs), and (3) *multiple-model adaptive controllers* (MMACs). Isaka and Sebald cite many examples of the adaptive control of MAP; the interested reader should consult their comprehensive review paper for details. A few representative MAP/SNP adaptive controllers will be considered here.

The SDC is easily adapted to the MRAC architecture. Shah[119] and Shah et al.[120] described an MAP/SNP controller design using an SDC MRAC with an integral pulse-frequency modulation (IPFM) bolus SNP injector. The SDC was tuned by replacing the Smith model's parameters with the on-line estimates of the plant's dc gain, time constant, and delay time. Shah used an infinite impulse response least mean squares (LMS) algorithm to estimate the plant's gain and time constant on-line. The delay time was estimated on-line using a modified version of the method described by Etter and Stearns.[39] Shah demonstrated the effectiveness of his MRAC by simulation. A supervisor was not used in Shah's modeling study. Shah used the simplified (no recirculation delay) plant model of Sheppard and Slate with transfer function:

$$G_p(s) = \frac{K_p e^{-s\delta_p}}{s\tau_p + 1} \tag{8.9}$$

Because Shah's adaptive algorithm is a discrete process with sample period T, the plant's transfer function needs to be written in discrete (z-transform) form:

$$G_p(z) = \frac{K_p z^{-d_p}}{1 - b_p z^{-1}} = \frac{\Delta P}{U}(z) \tag{8.10}$$

where $d_p = \text{INT}[d_p/T]$, and $b_p = e^{-T/\tau_p}$. (Note that d_p, d, D, k, etc. are integers.) The Smith model's discrete transfer function is similarly structured:

$$G_m(z) = \frac{\hat{K}_m z^{-\hat{d}_m}}{1 - b_m z^{-1}} = \frac{\Delta P_m}{U}(z) \tag{8.11}$$

where $\hat{K}_m$ is the *estimate* of plant gain used as the model gain, the discrete model's delay is $\hat{d}_m = \text{INT}(\hat{\delta}_m/T)$, $\hat{\delta}_m$ is the estimate of the plant's delay, $b_m = e^{-T/\hat{\tau}_m}$, and $\hat{\tau}_m$ is the estimate of the plant's time constant.

Figure 8.7 shows the block diagram of Shah's SDC/IPFM controller. Note that it is a type 0 controller. The IPFM bolus injector and the SDC model reference are discrete processes because they are implemented on a microcomputer. The MAP/

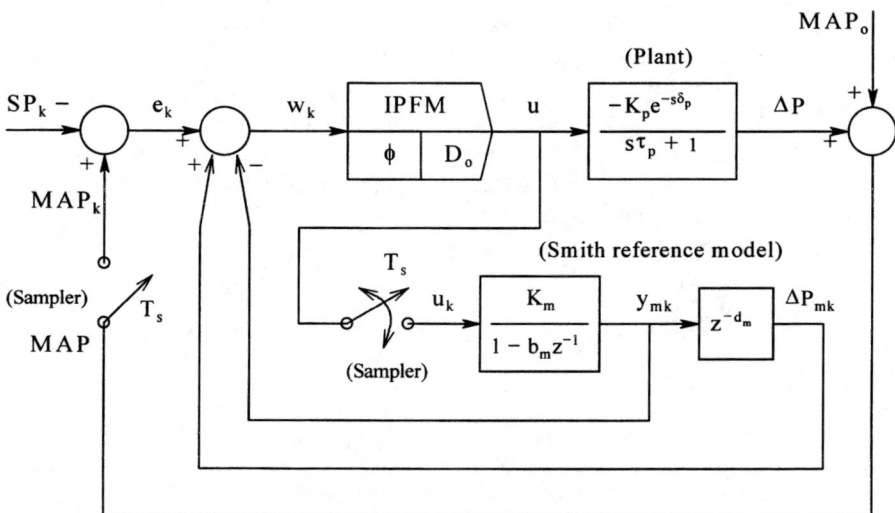

FIGURE 8.7 Block diagram of the SDC/IPFM controller used by Shah et al.[120] to control MAP by bolus injections of SNP. See text for description.

SNP plant is modeled as a continuous (analog) system with impulse inputs from the IPFM generator of area D_o at sampling instants if the integrated IPFM input, $se(k)$, exceeds the firing threshold, ϕ.

To estimate the plant's gain and time constant, Shah chose the infinite impulse response LMS algorithm; it is computationally inexpensive, requiring little memory and it is easily programmed.[120] Also, the LMS algorithm has a relatively slow convergence rate for parameter estimates; this is not a problem in a pharmacokinetic context because plant parameters change slowly. The LMS algorithm is robust in the presence of noise, as well. The parameter estimates from the parameter vector, $\hat{b}_m(k) = e^{-T/\hat{\tau}_p(k)}$, and $\hat{K}_m$ can be computed from the following algorithm:

$$\text{MAP}(t) \text{ is sampled,} \rightarrow \text{MAP}(k)$$

$$\Delta P(k) = \text{MAP}_o - \text{MAP}(k)$$

$$\varepsilon(k) = \Delta P(k) - \hat{\theta}^T(k)X(k)$$

$$\hat{\theta}(k+1) = \hat{\theta}(k) + 2\mu\varepsilon(k)X(k)$$

$$k = k + 1$$

Here, k is the kth sample ($t = kT$), $\hat{\theta}^T(k)$ is the kth parameter estimate vector ($\hat{\theta}^T(k) = [\hat{K}_m(k)\ \hat{b}_m(k)]$), μ is the convergence gain constant, typically 0.05 to 0.1, and $\varepsilon(k)$ is the prediction error which is minimized. Shah observed that the state vector, $X(k) = [D_o u_m(k - \hat{d}_m)\ \Delta P(k-1)]^T$, presents a problem because it is a function of $\hat{d}_m$. $\Delta P(k-1)$ is easily found from the previous sample of MAP. However, the delayed input term, $D_o u_m(k - \hat{d}_m)$, is a function of the delay estimate, $\hat{d}_m$. Even though the times at which the IPFM controller generates the plant input are known, the uncertainty in the delay estimate causes uncertainty in $X(k)$ and $\varepsilon(k)$. Thus the gain and pole estimation is intimately coupled with the delay estimation. To circumvent this problem, Shah used the facts that the IPFM controller output is either 0 or D_o at time $t = (k - \hat{d}_m)T$, and when an output impulse does occur, it causes (after the plant delay) a positive jump in ΔP of height $D_o K_p/\tau_p$ mmHg. Shah detected these jumps by a simple difference operation, $D(k)$. By detecting the jumps, he modified the LMS parameter estimator to either include the impulse input in the state vector, $X(k)$, or replace it with a zero. Thus the parameter estimation algorithm for $\hat{K}_m$ and $\hat{b}_m$ can be rewritten:

$$\text{MAP}(t) \text{ is sampled,} \rightarrow \text{MAP}(k)$$

$$\Delta P(k) = \text{MAP}_o - \text{MAP}(k)$$

$$D(k) = \Delta P(k) - \Delta P(k-1)$$

$$\text{IF } D(k) > \Psi,\ X(k) = [D_o\ \Delta P(k-1)]^T$$

$$\text{ELSE } \mathbf{X}(k) = [0 \, \Delta P(k-1)]^T$$

$$\varepsilon(k) = \Delta P(k) - \theta^T(k)\mathbf{X}(k)$$

$$\hat{\theta}(k+1) = \hat{\theta}(k) + 2\mu\varepsilon(k)\mathbf{X}(k)$$

$$k = k+1$$

Note that Ψ is a threshold set to detect a discontinuity in ΔP denoting an impulse input to the plant δ_p sec previously.

Shah adapted the delay estimation algorithm devised by Etter and Stearns[39] (E&S) to find the plant's delay and pass this value to the SDC reference model. In the E&S framework, we assume two nearly identical signal paths; one is delayed by the delay we wish to estimate, D, and the other by a delay varied in the estimation process, d. Figure 8.8 illustrates this system. An error, $e(k) = s(k-D) - s(k-d)$, is generated and squared. When we take the expected value of the squared error, we find:

$$E[e^2(k)] = 2R_{ss}(0) - 2R_{ss}(DT - dT) \tag{8.12}$$

where R_{ss} is the continuous autocorrelation of the signal s and T is the sampling period. If the autocorrelation function is assumed unimodal with a single maximum over the range of allowed values for the variable delay, d, then the mean squared error is unimodal with a single minimum. In the estimation of D, $\hat{d}$ will be treated like a continuous variable in the iterations, but the integer nearest to $\hat{d}$ will be designated as d.

The classical steepest descent algorithm used to find the D estimate has the form:

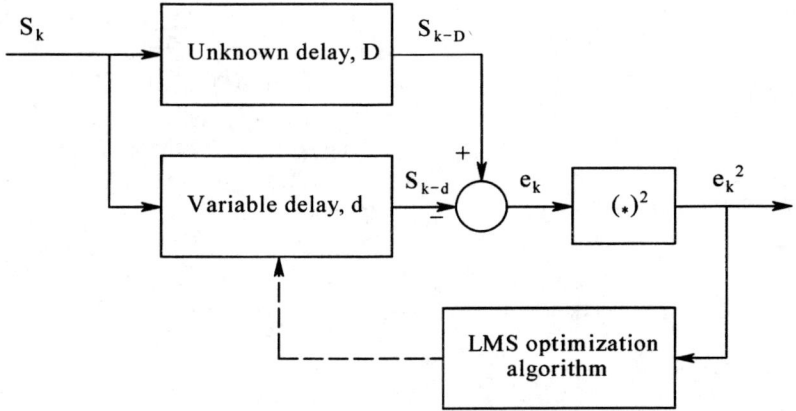

FIGURE 8.8 Block diagram of the delay estimation algorithm used by Shah[119] to tune the SDC model's delay on-line.

$$\hat{d}(k+1) = \hat{d}(k) - \mu \nabla(k) \tag{8.13}$$

where μ is the convergence constant and $\nabla(k)$ is the gradient of the mean squared error with respect to $\hat{d}$.

$$\nabla(k) = \frac{\partial E\left[e^2(k)\right]}{\partial d} = -2\frac{R_{ss}\left(DT - \hat{d}T\right)}{\partial d} \tag{8.14}$$

Through some arcane algebraic approximation and manipulation, Shah was able to write:

$$\nabla(k) \cong e(k)\left[s\left(k - \hat{d} - 1\right) - s\left(k - \hat{d} + 1\right)\right] \tag{8.15}$$

Thus the final generic, working version of the E&S algorithm is:

$$e(k) = s(k - D) - s(k - d)$$

$$\hat{d}(k+1) = \hat{d}(k) + \mu e(k)\left[s(k - d - 1) - s(k - d + 1)\right]$$

$$d(k) = \text{int}\left[\hat{d}(k)\right]$$

$$k = k + 1$$

where int[.] denotes rounding the argument to the nearest integer. To apply the above E&S algorithm, we need the delayed signal, $s(k - D)$, and its delay-free version, $s(k)$. (d is introduced into $s(k)$ in the E&S algorithm.) The delayed signal in the self-tuning MAP/SNP control system is $\Delta P(k) = MAP_o - MAP(k)$. ($\Delta P(t)$ is delayed δ_p sec; $D = \text{int}[\delta_p/T]$.) However, there is no specific delay-free signal to use to generate $e(k)$. We can approximate the delay-free signal by using the nondelayed output of the of the SDC model and delaying it some $\hat{d}_m$. If the SDC model is accurate enough, then the cross-correlation between $\Delta P(k)$ and $\Delta P_m(k)$ will be unimodal; hence the E&S algorithm will converge to a good estimate of $d(k) = \text{int}[\delta_p/T] = D$. The accuracy of parameter estimation is set by the minimum acceptable values for $|\varepsilon(k)|$ and $|e(k)|$. If $|e(k)|$ is above the threshold, Φ, then the value of $\hat{d}(k + 1)$ is maintained at $\hat{d}(k)$, its previous value.

Shah's overall adaptive SDC/IPFM controller is given by:

$$MAP(t) \text{ is sampled, } \rightarrow MAP(k)$$

$$\Delta P(k) = MAP_o - MAP(k)$$

$$D(k) = \Delta P(k) - \Delta P(k - 1)$$

$$\text{IF } D(k) > \Psi, \mathbf{X}(k) = \left[D_o \, \Delta P(k - 1)\right]^T$$

ELSE $\mathbf{X}(k) = \begin{bmatrix} 0 & \Delta P(k-1) \end{bmatrix}^T$

$\varepsilon(k) = \Delta P(k) - \theta^T(k)\mathbf{X}(k)$

$\hat{\theta}(k+1) = \hat{\theta}(k) + 2\mu\varepsilon(k)\mathbf{X}(k)$

IF $|\varepsilon(k)| < \Lambda$ CONTINUE, ELSE LOOP

$e(k) = \Delta P(k) - \Delta P_m(k - \hat{d})$

$\hat{d}(k+1) = \hat{d}(k) + \mu\varepsilon(k)\left[\Delta P_m(k - \hat{d} - 1) - \Delta P_m(k - \hat{d} + 1)\right]$

$d(k) = \text{int}\left[\hat{d}(k)\right]$

$k = k + 1$

$d(k)$ is the SDC model's delay, and the estimates of plant gain and time constant in $\hat{\theta}(k)$ are used in the SDC model for the $(k + 1)$th iteration.

Shah's simulation used multirate sampling. The plants were modeled using a $T = 1$-sec sample period so the delays could be modeled to the nearest 1 sec. The adaptive algorithms and the digital IPFM controller were modeled using a 3-sec sampling period. The simulations illustrated good convergence of the two-parameter estimators. However, for high plant gains, there was generally an unacceptable MAP undershoot at turn-on. Once the SDC model's parameters had settled, the response to a step change in set point showed no overshoot. Shah suggested, but did not implement, a supervisor to reduce IPFM controller gain at turn-on to prevent MAP undershoot transients. He also recommended that future investigation of his adaptive MAP/SNP controller use the recursive least squares algorithm instead of the LMS method. As we shall see below, by coupling the SDC architecture to a PI controller block and then to a drug infuser or IPFM bolus injector, it is possible to mitigate the turn-on MAP undershoot under certain conditions. It is prudent to first identify a patient's SNP (impulse) bolus response off-line and transfer the initial parameter estimates to the SDC model before closing the loop and starting the controller and parameter estimation.

A more complex approach to self-tuning control than that used by Shah was described by Martin et al.[92] They used a seven-model, MMAC using a Smith compensator and a PI element driving the SNP infusion pump. Martin et al. used a more complex MAP/SNP plant model than Slate and Sheppard, given by:

$$\frac{\Delta P}{U}(s) = \frac{e^{-sT} G(\tau_3 s + 1)}{\left[(\tau_3 s + 1)(\tau_2 s + 1 - \alpha)(\tau_1 s + 1)\right]} \tag{8.16}$$

where T = infusion delay = 50 sec, α = fraction of SNP recirculated = 0.5, G = plant dc gain = 0.25 to 9.0 mmHg/(ml/hr), τ_1 = time constant of SNP action = 50 sec,

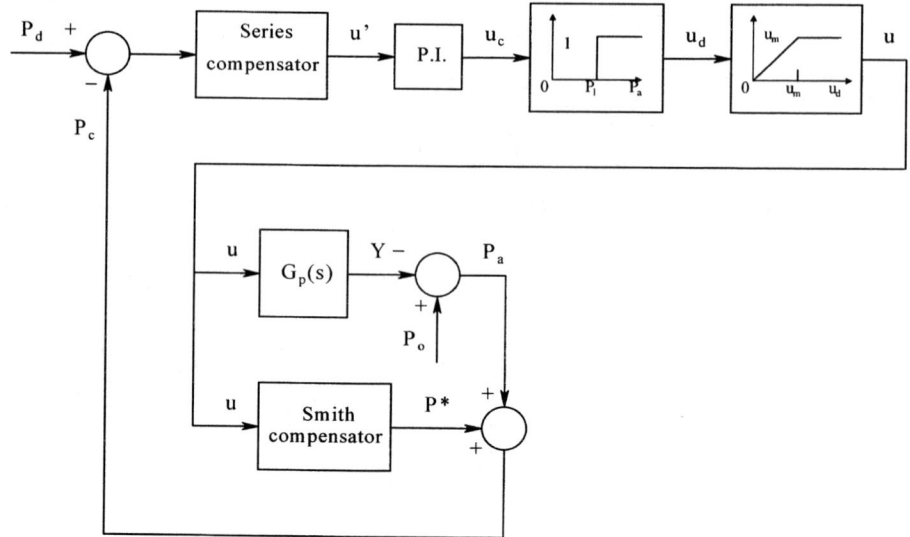

FIGURE 8.9 Block diagram of the general SDC/PI controller architecture used by Martin et al. The outputs of *seven* SDC plant models with different parameters are weighted and added together to optimize system performance by Martin's MMAC tuning algorithm.[91,92] See text.

τ_2 = time constant for flow through pulmonary circulation = 10 sec, and τ_3 = time constant for flow through systemic circulation = 30 sec.

MMAC is predicated upon the assumption that the plant can be represented by one of a finite number, N, of plant models. For each plant model, there will be a controller of the form shown in Figure 8.9. How each of the N = 7 models participates in plant control is computed by an adaptive algorithm. The input to the PI block, u', is given by:

$$u'(t) = \sum_{j=1}^{7} W_j(t) y_j(t) \qquad (8.17)$$

where $y_j(t)$ is the output from the *j*th Smith model, and $W_j(t)$ is the weighting factor computed for the jth model. The interested reader should consult Figure 1 in Martin et al.[92] for details on how the W_j are calculated. Because it takes time at turn-on for a dominant W_j to converge, Martin et al. used a conservative strategy to prevent MAP undershoot. The seven models were arranged in order of increasing model dc gain, G_m. Thus model 1 had G_m = 0.33 mmHg/(ml/hr), all the way to G_m = 9.03 for model 7. A high-gain model results in a lower SNP infusion rate. At turn-on, the compensator bank and the PI block were bypassed, and model 6 with G_m = 6.06 was selected as the default model. The infusion rate set by model 6 lasts for 2 min or until the MAP has fallen by 10 mmHg and W_j converges. If the MAP has not dropped 10 mmHg in the 2-min period, the infusion rate increases, based on model 4 which is selected with $G_m = 2.10$. If in another 2 min MAP does not drop 10 mmHg, model 2 is selected. Once a $W_j > 0.75$, the system is switched to its operating mode with the PI block included. The delay time of each Smith model in the model bank is initially set to 50 sec (for safe initial control). Then at every set point change greater than 20 mmHg the plant delay is reestimated. Martin et al. noted that a large set point change is necessary to accurately estimate the delay. Plant delay was estimated by fitting an exponential curve to sequential MAP measurements as MAP drops 3 to 10 mmHg. The time at which this curve would give zero drop is then subtracted from the time of the set point change to give the delay time.

To establish the effectiveness of their MMAC design, Martin et al. did a series of simulations in which they examined system performance under various conditions: G and T were varied from 0.25 to 9 mmHg/(ml/hr) and 10 to 50 sec, respectively, with other plant parameters held constant. They also varied plant α, τ_1, and τ_2, with other parameters constant, and tested system responses for nonstationary plant gain, G, and nonstationary delay, T. The MMAC algorithm compensated for $G(t)$ and $T(t)$ by switching models and W_js.

Martin et al. went on to modify their MMAC controller by introducing a *"supervisor"* which had three roles: (1) it set infusion rate limits, (2) it detected unphysiological changes in MAP (e.g., as caused by pressure sensor failure or a kink in the pressure catheter), and (3) it responded to potentially dangerous changes in MAP. The supervisor sampled MAP and certain system states every 2 sec, which was also the MMAC sample period. Their system was tested successfully on dogs and on 19 human cardiac surgery patients.[94]

The SDC has been used by a number of workers in their MAP controller designs. Accurate representation of the plant's parameters in the Smith reference model is necessary to obtain accurate, robust control. A model MAP/SNP control system devised by the author, which uses an SDC plus PI controller coupled to an IPFM drug bolus injector, is described below. A block diagram of the system is shown in Figure 8.10. The Simnon™ simulation program is

```
CONTINUOUS SYSTEM ipfmMAP1      " 10/19/94, 9/04/97
STATE v cr x1 x2 noise          " Use Algor EULER with delT = 0.002
DER dv dcr dx1 dx2 dnoise       " e.g., SIMU 0 10 0.002
TIME t                          " Time units minutes.
"
" 1st ORDER PLANT FOR MAP/SNP:
"
dx1 = u -ap*x1
Dp = Kp*DELAY(x1, Tp)           " Dp is MAP reduction in mm Hg.
"
" SMITH REFERENCE MODEL:
"
dx2 =Km*u - am*x2
Dm = DELAY(x2, Tm)
"
e = SP - MAP
se = -e - x2 + Dm               " SMITH SUMMER. Fixed SP.
```

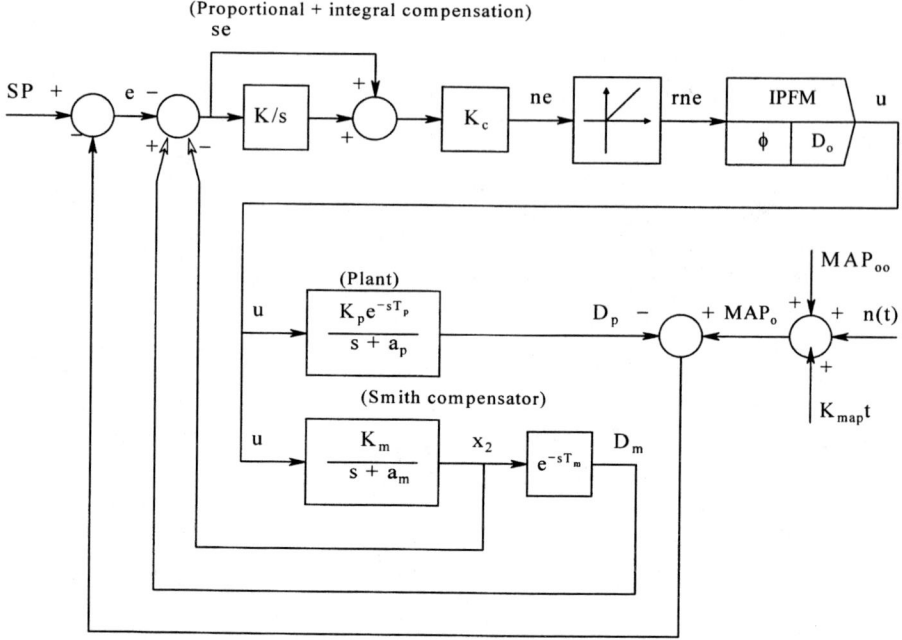

FIGURE 8.10 Block diagram of the SDC/PI/IPFM MAP control system proposed by the author. See text for description of the system.

```
MAP = MAPo - Dp              " MAPo is untreated MAP. Fixed MAPo.
MAPm = MAPo - Dm             " MAP predicted by reference model
"
MAPo = MAPoo + Kmap*t + noise  " Add BW-limited noise to MAP drift.
dnoise = -wo*noise + SD*NORM(t)  " BW limiting ODE for noise.
"
dcr = k*se                   " PI CONTROLLER; k is integrator gain.
ne = (cr + se)*Kc            " ne is PI controller output to IPFM.
rne = if ne > 0 then ne else 0
"
dv = rne - z                 " IPFM GENERATOR macro.
w = if v > phi then 1 else 0 " integrator threshold
s = DELAY(w, tau)            " pulse former
x = w - s
y = if x > 0 then x else 0
z = y*phi/tau                " IPFM reset
u = Do*y/tau                 " PLANT INPUT BOLUS
"
" PARAMETERS:
SD:100
MAPoo:185
wo:0.5
Kmap:1.5
ap:1.33       "r/min.
Kp:10
```

```
Tp:0.5       "minutes
am:1.33
Km:10
Tm:0.5
k:2.0
Kc:5.0
SP:100       "mm Hg set point
phi:15
tau:0.002
Do:2
upper:115
lower:85
"
END
```

Figure 8.11 illustrates the turn-on transient for a perfectly tuned system, i.e. the SDC model parameters equal the plant parameters. There is a 1:20 range in plant (and model) dc gain. Two MAP(t) curves are shown, one for $K_p = K_m = 0.5$ and the other for $K_p = K_m = 10$. At the high gain, the individual bolus injections from the IPFM controller cause a 20-mmHg peak-to-peak ripple on the MAP. See Figure 8.12 for details of the ripple; note the time scale. The ripple is not serious, as it is centered on the set point, in spite of the upward slope in MAP_o. Recall that MAP_o is the MAP of the *untreated* patient. In the simulation, MAP_o increases from about 185 to 250 mmHg in 60 min. This increase is not physiologically realistic, but it illustrates system robustness. High-frequency Gaussian noise is also added to $MAP_o(t)$, representing measurement noise and rapid fluctuations in MAP_o due to fast blood pressure reflexes.

The need for SDC reference model tuning is illustrated in Figure 8.13. Here, the SDC model gain is kept at $K_m = 5$, while the plant's dc gain, K_p, is made 2, 5, and 8. For $K_p = 8$, we see a dangerous undershoot in MAP. When $K_p = 2$, the system is sluggish; it takes 5 min to reach the set point. Note that in this figure and in all other simulations, the PI block forces zero average error in spite of MAP_o being noisy and ramping up.

It is also interesting to see how the SDC/PI controller will perform with a conventional, continuous SNP infusion. The Simnon simulation program is shown below:

```
CONTINUOUS SYSTEM SNPCONT1    " 10/07/94, 9/12/97
STATE x1 x2 td cr noise       " Continuous SNP infusion.
DER dx1 dx2 dtd dcr dnoise
TIME t
"
"   1st ORDER MAP/SNP PLANT
dx1 = u -ap*x1
Dp = Kp*DELAY(x1, Tp)
MAP = MAPo - Dp
"
"   SMITH REFERENCE MODEL
dx2 =Km*u  -  am*x2
```

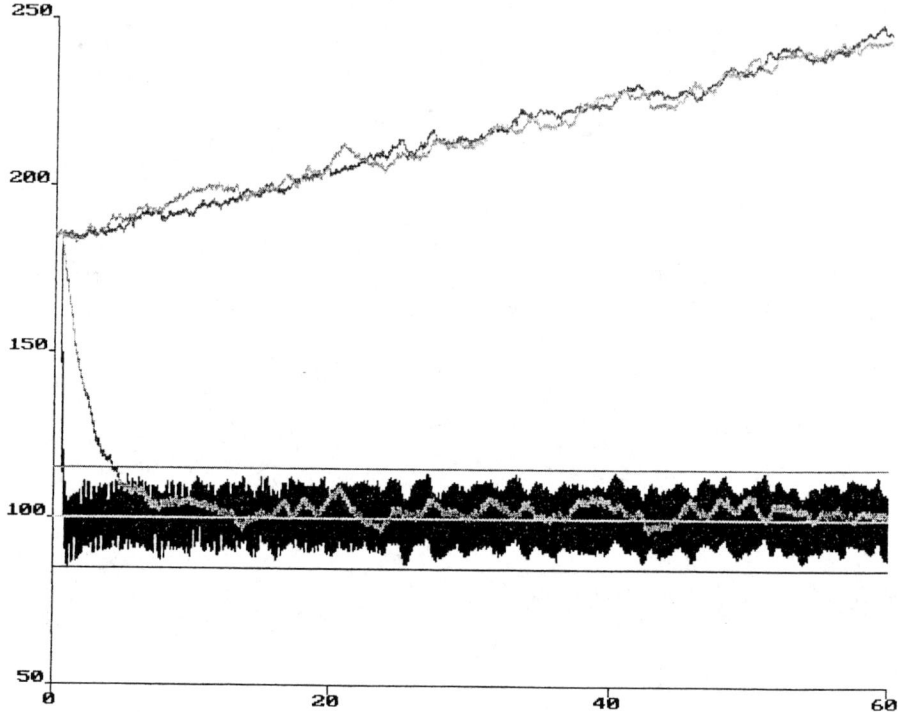

FIGURE 8.11 Simnon simulation of turn-on transients of the MAP/SNP control system of Figure 8.10. Bandwidth-limited Gaussian noise and a ramp were added to the untreated MAP, MAP_o, which is shown in the upper two traces. Time is in minutes; vertical axis in millimeters of mercury. Controller parameters were $K = 2$, $K_c = 5$, $D_o = 2$. The MAP was computed for $K_p = K_m = 0.5$ and $K_p = K_m = 10$ (a 1:20 range) with an SP = 100 mmHg. In both cases, in spite of the noise, MAP runs between the 85- to 115-mmHg safe range. In the high-gain case, the MAP(t) approaches the SP much faster than the low-gain system, and it shows a higher peak-to-peak ripple due to the bolus injections.

```
        Dm = DELAY(x2, Tm)
"
"       SMITH SUMMER
        e = MAP - SP            " System error.
        se = e - x2 + Dm        " Smith error
"
"       PI COMPENSATOR.
        dcr = k*se              " PI Compensator input is Smith error.
        ne = (cr + se)*kc       " ne is PI comp. output
        u = Do*ne               " SNP infusion, ml/min.
"
        dTd = u                 " Total SNP used = Td.
        cd = Td/10              " Scaled cumulative dose of SNP.
"
"       INPUTS:
        MAPo = MAPoo + Kmap*t + noise    " Add BW-limited noise to MAP drift.
```

Control of Mean Arterial Pressure by Sodium Nitroprusside Injection 315

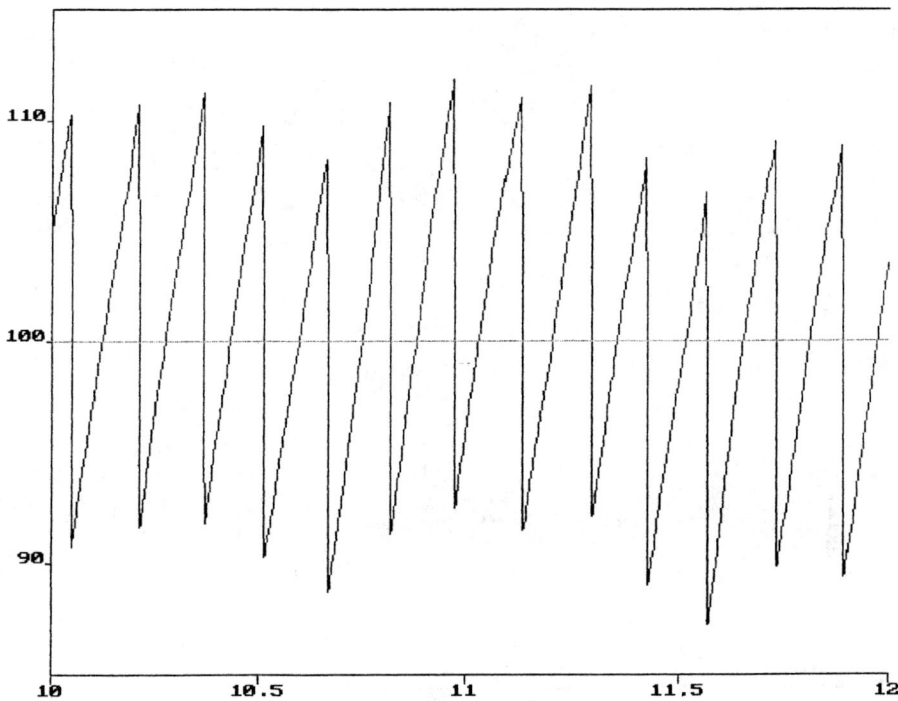

FIGURE 8.12 Expanded time and pressure scale plot of the steady-state MAP(t) for the high-gain case of Figure 8.11. Note sawtooth waveform which is the result of inputting impulses into a first-order plant (with delay). Irregularity in the waveform is due to noise added to MAP$_o$. $K_m = k_p = 10$, $K_c = 10$, $D_o = 2$, $K = 2$.

```
dnoise = -wo*noise + SD*NORM(t)    " BW limiting ODE for noise.
"
SP = 100
MAPoo:185
Kmap:1.
wo:.5
SD:100
Do:.5
ap:1.33
am:1.33
Kp:1
Km:1
Tp:0.5
Tm:0.5
k:2
kc:5
zero:0
upper:115
lower:85
"
END
```

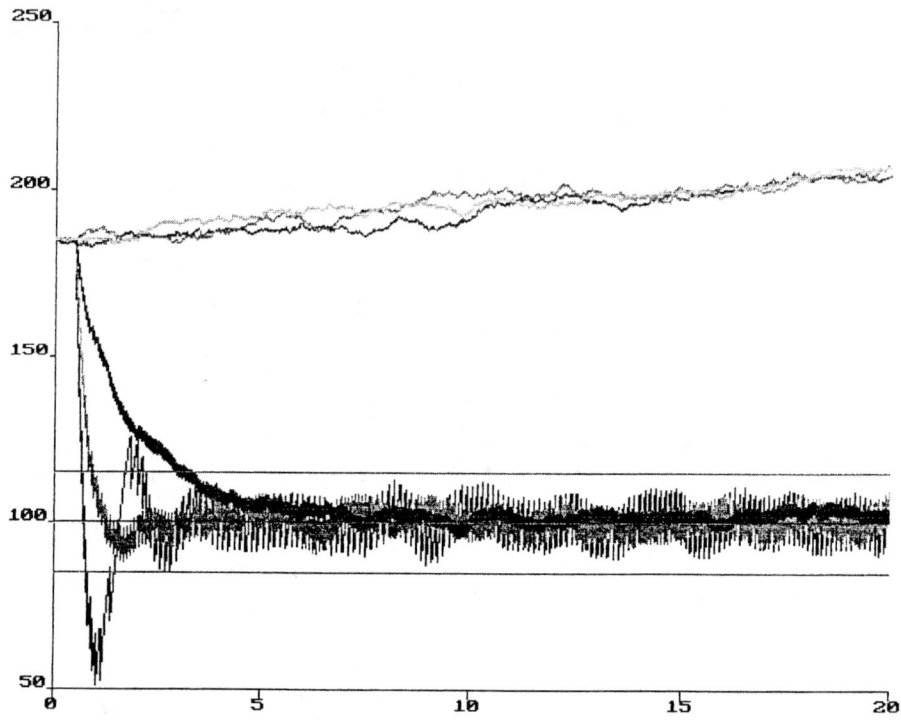

FIGURE 8.13 Simulated turn-on transients in MAP for the SDC/PI/IPFM controller when the SDC is detuned by varying the plant's dc gain. The three upper traces are MAP_o + noise. Controller parameters for all three cases: $D_o = 2$, $K = 2$, $K_m = 5$, $K_c = 5$, $a_m = a_p = 1.33$ r/min, $T_m = T_p = 0.5$ min. Slow MAP transient with low ripple for $K_p = 2$. Fast MAP transient with no overshoot and medium ripple for $K_p = 5$ (tuned). MAP transient with dangerous overshoot and large ripple for $K_p = 8$.

Figure 8.14 illustrates the turn-on transient for the continuous-infusion SDC/PI-controlled MAP/SNP system. The fast MAP(t) is for $K_m = K_p = 10$; the slower MAP trace is for $K_m = K_p = 0.25$ mmHg/(ml/min). The top traces are $MAP_o(t)$. In Figure 8.15, we see the effect of detuning the SDC gain. All runs are done with $K_m = 5$ mmHg/(ml/min). The MAP trace with severe undershoot is for $K_p = 10$, the fast MAP trace is for $K_p = 5$ (tuned), the next slowest is for $K_p = 1$, and there is a sluggish response for $K_p = 0.25$. Note that the extremes of detuning do not result in a useful system; it is too slow for $K_p = 0.25$, and the undershoot kills the "patient" when $K_p = 10$ is used in the simulation.

Other aspects of detuning the SDC portion of the controller are shown in Figures 8.16 and 8.17. Figure 8.16 shows the differences in the MAP turn-on transient for a low-gain system ($K_p = K_m = 1$) when the SDC reference model delay is detuned ±20%. The system is still safe to use, in spite of the undershoot for $T_m = 0.4$ min. Figure 8.17 shows that a 20% detuning of T_m results in dangerous MAP undershoots. Thus model delay parameter estimation is made more critical for a high-gain system.

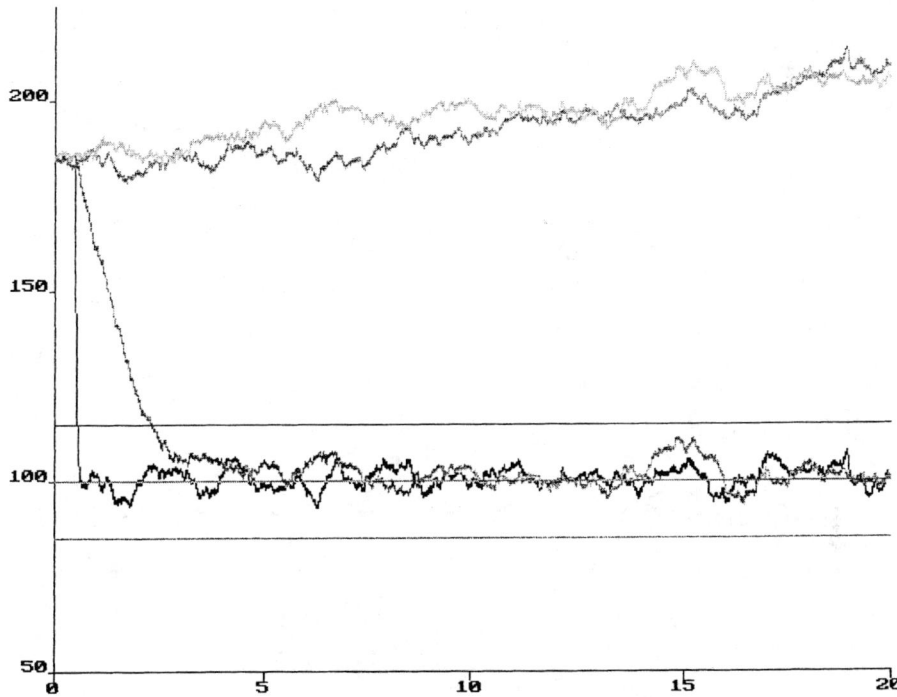

FIGURE 8.14 MAP(t) turn-on transients for the continuous SNP infuser controlled from a *tuned* PI/SDC controller. Common controller parameters: $K = 2$, $K_c = 5$, $a_m = a_p = 1.33$ r/min, $T_m = T_p = 0.5$ min, $D_o = 0.5$. The slower transient is for $K_m = K_p = 0.25$, and the quick MAP is for $K_m = K_p = 10$, a 1:40 range. Noise has been added to the increasing MAP$_o$.

The two simulations above of an SDC/PI controller for the MAP/SNP plant illustrate that this model reference architecture works well as long as the Smith reference model is tuned to the plant. Simple tuning requires several initial off-line bolus injections of SNP to estimate initial values for plant gain, time constant, and, most importantly, the delay. These parameters are put into the Smith model. Subsequent parameter updates can be done on-line by standard estimation techniques, such as used by Shah.[119,120]

A single MRAC for the MAP/SNP plant was described by Pajunen et al.[110] In a simulation study, they used the Slate and Sheppard plant with two time delays and single real pole. A 15-sec sampling period is used in their discrete plant model and controller. Their discrete plant is

$$\frac{\Delta P}{U} = \frac{q^{-d}(b_o + b_m q^{-m})}{1 - a_1 q^{-1}} \tag{8.18}$$

where q^{-1} is the unit delay operator (15 sec in this system), and the parameters have the extreme ranges and nominal values shown in Table 8.2.

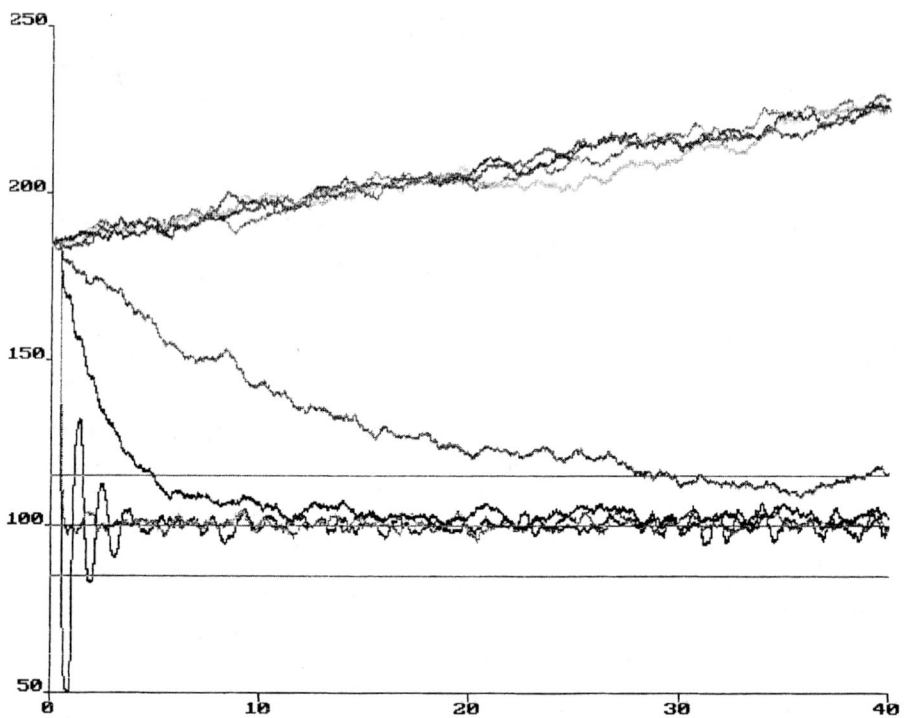

FIGURE 8.15 Simulated turn-on transients in MAP for the SDC/PI controller when the SDC is detuned by varying the plant's dc gain. The four upper traces are MAP_o + noise. Controller parameters for all four cases: $D_o = 0.5$, $K = 2$, $K_m = 5$, $K_c = 5$, $a_m = a_p = 1.33$ r/min, $T_m = T_p = 0.5$ min. Slowest $MAP(t)$ transient for $K_p = 0.25$. Next fastest MAP for $K_p = 1$. The fast, zero-overshoot, tuned response is for $K_p = 5$. Note that $K_p = 10$ gives a serious transient undershoot in MAP.

The delay times were estimated *off-line* (under open-loop conditions) using the pseudorandom binary signal (PRBS) method of Sheppard and Slate. This method also gave plant dc gain and time constant estimates for initial tuning of the reference model. Under closed-loop conditions, the delays were assumed to be constant, and the controller's parameters were adapted using an extended least squares (ELS) algorithm with covariance resetting following the stochastic MRAC algorithm described by Goodwin and Sin.[56] The control input $U(k)$ was generated from:

$$U(k) = \begin{Bmatrix} \Delta P_F{}^*(k+d) - \hat{k}(k)\Delta P_s(k) - \hat{\beta}_1(k)U(k-1) \\ -\ldots - \hat{\beta}_{d+m-1}(k)U(k-d-m+1) + \hat{c}_1(k)y_F(k+d-1) \\ +\ldots + \hat{c}_r(k)y_F(k+d-r) \end{Bmatrix} \Big/ \hat{\beta}_o(k) \qquad (8.19)$$

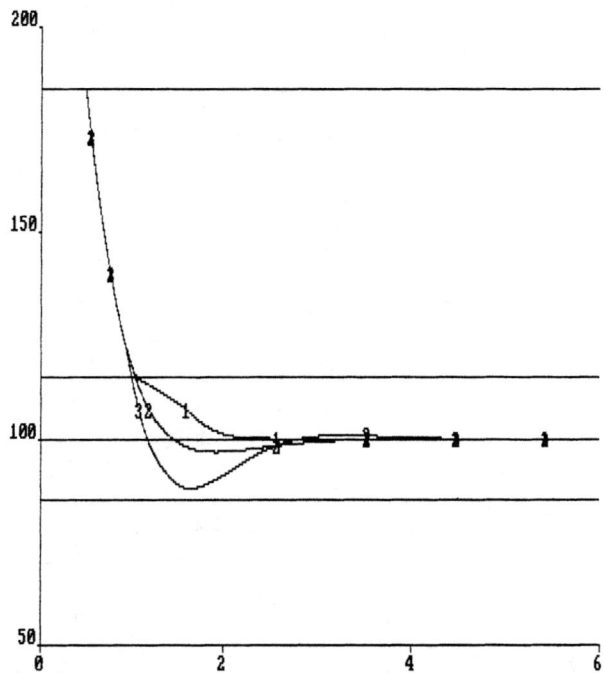

FIGURE 8.16 Simulated MAP turn-on transients when the SDC/PI controller is detuned by changing $T_m \pm 20\%$. In this case, $D_o = 0.5$, $K = 2$, $K_m = K_p = 1$ (low gain), $K_c = 5$, $a_m = a_p = 1.33$ r/min, $T_p = 0.5$ min. Trace 1, $T_m = 0.6$ min; trace 2, $T_m = 0.5$ (tuned); trace 3, $T_m = 0.4$. SP = 100 mmHg, MAP_o = 185 mmHg. No noise or ramp added to MAP_o.

where $\hat{k}(k)$, all $\hat{\beta}_j(k)$, and all $\hat{c}_j(k)$, are calculated by the MRAC algorithm. $y_F(k + d)$ is the d step-ahead adaptive prediction filtered through the the polynomial $E(q^{-1}) = 1 - eq^{-1}$. k stands for the kth sampling instant. $\Delta P_F^*(k)$ is the kth value of the filtered reference model output. That is, $\Delta P_F^*(q^{-1}) = \Delta P_F(q^{-1}) E(q^{-1})$. $e(k)$ is a computed parameter; $0 \le e \le 0.95$. The reader should consult the paper by Pajunen et al.[110] for the details of the adaptive algorithm.

The model studies showed that the MRAC system of Pajunen et al. was robust against noise and plant parameter changes (gain and time constant). The system had rapid settling times with modest overshoot. Plant delay was assumed to be estimated off-line; however, and no attempt was made to estimate it on-line. Clearly, on-line estimation of plant delay is critical in the design of a universally robust, adaptive, MAP/SNP controller.

8.4 CHAPTER SUMMARY

This chapter described exogenous physiological control systems in which the vasodilator drug SNP is injected in order to lower postoperative mean arterial blood

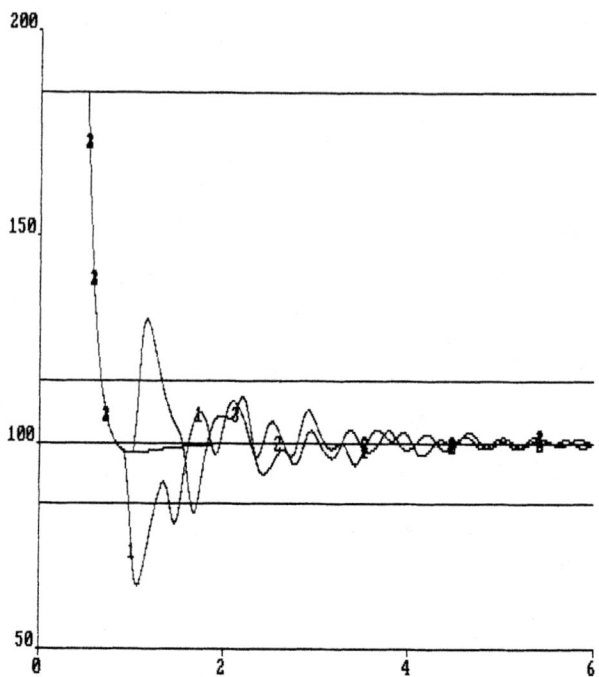

FIGURE 8.17 Simulated MAP turn-on transients when the SDC/PI controller is detuned by changing $T_m \pm 20\%$. In this case, $D_o = 0.5$, $K = 2$, $K_m = K_p = 4$ (higher gain), $K_c = 5$, $a_m = a_p = 1.33$ r/min, $T_p = 0.5$ min. Trace 1, $T_m = 0.6$ min; trace 2, $T_m = 0.5$ (tuned); trace 3, $T_m = 0.4$. SP = 100 mmHg, $MAP_o = 185$ mmHg. No noise or ramp added to MAP_o. Note that with the higher gain, the system is more sensitive to detuning T_m. Both $T_m = 0.4$ and $T_m = 0.6$ produce unacceptable settling transients for an MAP regulator.

pressure. SNP is a rapidly acting vasodilator which is quickly deactivated metabolically, making it ideal for the design of rapid responding systems. This compartmental pharmacokinetic/physiological plant is challenging to control because (1) there is an effective dead time of the same order as the dominant real pole, (2) plant

TABLE 8.2
Range of Parameters Used in the MAP/SNP Plant Model of Pajunen et al.[110] (Sample Period 15 sec)

Parameter	Minimum	Maximum	Nominal
b_o	0.053	3.564	0.187
b_m	0	1.418	0.075
a_1	0.606	0.779	0.741
d	2	5	3
m	2	5	3

parameters vary widely among individuals, and (3) plant parameters can "drift" in time. A controller for MAP must not exhibit undershoot at turn-on (an effective step input), nor must it show damped oscillations. The controller must also be robust against noise in the measured controlled variable, MAP.

Several MAP regulators using SNP which have been proven clinically effective on human patients and several theoretical MAP controllers, including the self-tuning SDC-based system designed by Shah[119] were described.

PROBLEMS

8.1 A closed-loop drug injection system is to be designed to continuously infuse the vasodilator SNP into a postoperative patient to reduce MAP following cardiac bypass surgery. The simplified plant model of Slate et al. is used:

$$\frac{\Delta P}{U}(s) = \frac{K_p e^{-s\delta_p}}{s + a_p}$$

The MAP is given by MAP = MAP_o − ΔP. MAP_o is the untreated MAP. System error is e = SP − MAP, where SP is the set point. An SDC is used to cancel the unstabilizing influence of the plant's delay time, δ_p. To give the system zero steady-state error for a constant SP, a PI controller is used to make the system type 1. Bandwidth-limited Gaussian noise associated with measurement of MAP is added into the model. Parameter SD sets the standard deviation of the noise source. The rectified output of the PI controller, rne, is the SNP infusion rate, u, in milligrams per minute. The worst-case turn-on transient in MAP to a fixed SP = 100 mmHg is to be designed so MAP

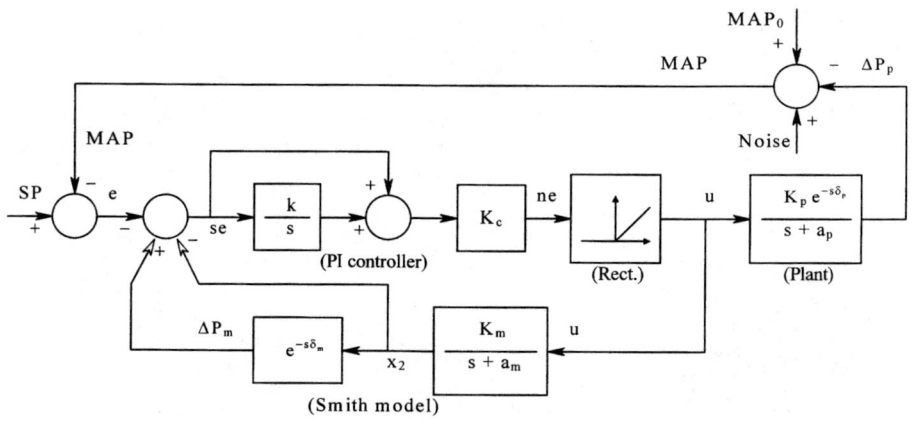

PROBLEM 8.1

reaches the SP in minimum time with a single undershoot no more than 85 mmHg. A block diagram of the system is shown in the figure. A Simnon program, "MAPpisdc," is also given below.

A. Simulate the MAP reduction controller turn-on transient. Plot SP, MAP, u, MAP$_o$, lower. Determine values for K_i and K_c that satisfy the criteria above. Examine tuned system response with and without (SD = 0) noise.

B. With the noise off, try detuning each of the SDC model's parameters, K_m, a_m, and T_m, by ±15% in turn. Plot what happens.

```
CONTINUOUS SYSTEM MAPpisdc      " 3/04/98
STATE cr x1 x2 noise            " Use Algor EULER with delT = 0.001
DER dcr dx1 dx2 dnoise          " e.g., SIMU 0 6 0.001 -mark
TIME t                          " Time units minutes.
"
" 1st ORDER PLANT FOR MAP/SNP:
dx1 = u -ap*x1
Dp = Kp*DELAY(x1, Tp)           " Dp is MAP reduction in mm Hg.
"
" SMITH REFERENCE MODEL:
Dm = DELAY(x2, Tm)
dx2 =Km*u - am*x2
"
e = SP - MAP
se = -e - x2 + Dm               " SMITH SUMMER. Fixed SP.
MAP = MAPo - Dp                 " MAPo is untreated MAP. Fixed MAPo.
MAPm = MAPo - Dm                " MAP predicted by reference model
"
" BW-LIMITED NOISE
MAPo = MAPoo + Kmap*t + noise   " Add BW-limited noise to MAP drift.
dnoise = -wo*noise + SD*NORM(t) " BW limiting ODE for noise.
"
" PI CONTROLLER
dcr = Ki*se                     " Ki is integrator gain.
ne = (cr + se)*Kc               " ne is PI controller output to IPFM.
rne = if ne > 0 then ne else 0
u = rne                         " Plant input infusion rate.
"
pr = IF t > 2 THEN 50*(t - 2) ELSE 0
"
" PARAMETERS:
SP:100
SD:100
MAPoo:200       " Untreated MAP
wo:0.5          " Radiar/sec BF of Gaussian noise.
Kmap:0
ap:1.33         "r/min.
Kp:10
Tp:0.5          "minutes
am:1.33
Km:10
Tm:0.5
Ki:
```

```
Kc:
upper:115
lower:85
"
```
END

8.2 SNP appears to be the drug of choice for closed-loop postoperative MAP reduction and control. Describe the physiological mechanism of action of SNP. Prolonged use of SNP infusion can lead to several toxic effects; describe them. (See Reference 61 or the pharmacological literature on SNP.[2,58,137])

8.3 The drug diazoxide (DO), given IV, provides an alternative mechanism to reduce postoperative MAP. Discuss the pharmacological mechanisms of action of DO. Formulate a linear transfer function approximating the dynamics of DO. That is, find a transfer function for MAP/DO, where DO is the rate of IV infusion of DO. (See Reference 61 and Johnson and Kapur. 1972. The influences of rate of injection upon the effects of diazoxide, *Am. J. Med. Sci.*, 263: 481-487; also see Velasco et al. 1976. A new technique for safe and effective control of hypertension with intravenous diazoxide, *Curr. Ther. Res.*, 19(2): 185–188.)

8.4 Compare the closed-loop control of MAP with DO with that by SNP. Use the controller model of Problem 1 above.

8.5 One strategy for the *long-term* reduction of elevated MAP is to block the renin → angiotensin I → angiotensin II biochemical pathway. What drugs are available to do this? How effective are they? Are there "side effects"?

8.6 Describe the physiological mechanisms whereby angiotensin II raises MAP.

8.7 Describe the medical risks of untreated severe hypertension.

8.8 Modify the program of Problem 1 to accommodate the two-delay plant model for SNP given by Slate:[126]

$$\Delta P = \frac{K e^{-T_i s} \left(1 + \alpha e^{-T_c s}\right)}{s\tau + 1} U(s)$$

where $K = 0.72$, $T_i = 30$ sec, $T_c = 45$ sec, $\alpha = 0.2$ and, $\tau = 40$ sec for the "average" patient.[62] Modify the SDC reference model to include the second delay. Note that the plant transfer function in Problem 1 is in Laplace format, and the transfer function above is in time constant form. Repeat parts A and B of Problem 1 for this plant.

8.9 In animal studies of MAP reduction by SNP infusion, it has been common practice to induce hypertension by first infusing norepinephrine (cf. Koivo, A.J., 1980. Automatic continuous-time blood pressure control in dogs by means of hypotensive drug injection, *IEEE Trans. Biomed. Eng.*, 27(10): 574–581). Describe the physiological effects of norepinephrine on the circulatory system. What are its pharmacokinetics? Estimate how long it takes for MAP to return to its normal, untreated baseline value following an IV bolus of norepinephrine. Does norepinephrine have to be continually infused to maintain a constant level of hypertension?

8.10 The drug Arfonad (trimethaphan camsylate) is used to relieve *chronic hypertension*. Arfonad is classified as a ganglionic blocking drug. Arfonad is generally given as an IV drip.
  A. What ganglia are blocked?
  B. By what mechanisms does Arfonad reduce MAP?
  C. What untoward side effects can result from use of Arfonad?

9 Control of Postoperative Pain by Self-Administered Opioids

9.0 INTRODUCTION

In this chapter, the contributions that control engineering is making to the relief of chronic, postoperative pain are examined. First, the neurophysiology and psychology of pain will be described, followed by the endogenous and exogenous mitigation of pain by endorphins, enkephalins, dynorphins, opioids, opiates, and other analgesic substances.

Relief of chronic pain in a clinical setting can be carried out by one of two means: (1) open-loop administration of an analgesic drug governed by a "safe dose" strategy and patient complaints and (2) closed-loop patient-controlled analgesia (PCA), in which the patient can literally dose himself or herself as the need arises by pressing a demand button. PCA systems generally have a dosing control algorithm between the patient and the infusion pump for the analgesic drug. Such a controller has two functions: (1) it acts as a supervisor, preventing overdose by calculating the maximum safe drug level in the body based on the known pharmacokinetics of the drug and the drug infusion history, and (2) it decides the actual dose rate, based on patient demand, the drug infusion history, the known pharmacokinetics of the drug, and a model for the patient's behavioral response to pain. The last two factors are imperfectly known. It is important to stress that individual patient responses to pain can vary widely (stoics vs. whiners), and PCA design must take this fact into account.

Research on PCA systems has been slow because of the "fuzzy" nature of the plant (pain/opiate) and because such systems use controlled substances which are generally considered toxic and addictive, especially if given in a nonpain context.

Sections 9.4 and 9.5 examine the clinically verified, stochastic PCA system devised by Reasbeck[116] and Jacobs et al.[66–68] and the integral pulse-frequency modulation (IPFM) PCA system devised by Liu and Northrop[82] and verified through model studies. The future design of PCA systems is discussed.

9.1 THE NEUROPHYSIOLOGICAL AND PHARMACOLOGICAL BASIS FOR PAIN

The experience of pain (perceived pain) can be subdivided into two components: *neurological pain* and *psychological pain*. The anatomy and pharmacology of neurological pain have been well described in the literature. Psychological pain forms the behavioral reaction to neurological pain. It determines how an individual reacts to pain and to changes in the pain experience. In terms of PCA, it determines the rate of analgesic drug demand button pressing.

9.1.1 SOURCES OF PAIN

There are many physical events which damage or threaten to damage body tissues that produce pain of one kind or another. Such noxious events produce pain by stimulating the specialized endings of pain neurons. Common experience tells us that many pain receptors in the skin are sensitive to extremes of heat and cold. Pain receptors also respond to certain noxious chemical stimuli, which include acids, proteolytic enzymes (as in spider venom), histamine, bradykinin (a small polypeptide split from α-2-globulins), acetylcholine, serotonin, and potassium ions.[59]

In addition to thermal and chemical stimuli, certain pain receptors respond to mechanical stimuli such as excess pressure, cuts, tearing of tissues, bone damage, etc. The presence of certain chemical pain stimulants not only stimulates chemical pain receptor neurons but also makes thermo- and mechano-pain sensors more sensitive by reducing their thresholds for stimulation. A well-known example of this hypersensitization occurs following tissue damage by severe sunburn or thermal burns.[59]

Tissue ischemia also causes pain. Any prolonged interruption of blood flow to an organ causes tissue damage, the release of lactic acid from cells as a result of anaerobic metabolism, and the production of bradykinin, proteolytic enzymes, etc., all of which are painful. In addition to other well-known symptoms, a coronary heart attack causes ischemia in the ventricular cardiac muscle and direct and referred pain. Coronary referred pain is often experienced in the left arm and the neck and jaw.

Bacterial and viral infection of tissues causes a complex of tissue changes called *inflammation*.[59] Inflammation involves (1) increased blood flow in the volume of inflamed tissue; (2) increased capillary permeability, resulting in local edema of the inflamed volume and various high-molecular-weight proteins leaking into the volume along with immune system cells; (3) clots form in the inflamed volume due to excessive amounts of fibrinogen and other proteins; (4) immune system T-cells, B-cells, and macrophages in the volume, release a complex of chemical messengers such as interleukins as they fight the infection. Tissue cells swell in response to bradykinin, histamine, serotonin, prostaglandins, and products of the immune system's complement system and release potassium ions. These biochemical products of infection stimulate chemical pain receptors, producing chronic pain. A commonly experienced infection pain is the sore throat accompanying colds or the flu.

Bone damage from injury, cancer, or bone surgery can cause intense, chronic pain. It is this deep, chronic pain which is effectively addressed by PCA. The work of Reasbeck[116] and Jacobs et al.[68] with PCA was directed toward patients who had undergone total hip replacement surgery, a procedure in which bone from the pelvis and head of the femur is removed and an artificial hip joint is installed by gluing and screwing into the remaining bone. Fortunately, as healing progresses, the pain stimulus is naturally reduced.

9.1.2 THE NEUROANATOMY OF PAIN

As implied above, pain receptor neurons are specialized as to stimulus category, responding to chemical, mechanical, and thermal stimuli. Pain receptors are distributed throughout the body. They are found in such diverse locations as the corneas and inside the cranial vault, as well as in all organs and the periosteum of bones. Pain receptor neurons have free endings. That is, they have no specialized terminal organelles that act as transducers. The pain stimulus acts directly on the pain neurons' terminal membranes. Pain receptors do not in general adapt to prolonged, constant stimuli, as do other types of sensory receptors; i.e., the firing frequency of pain neurons remains constant with the constant stimulus, or may even increase (the condition of *hyperalgesia*).

Pain receptors pass their signals to the spinal cord through two types of nerve fibers: The $A\delta$ fast fibers with conduction velocities of 6 to 30 m/sec carry fast-sharp pain signals. The slow-chronic pain signals travel on C fibers at velocities from 0.5 to 2 m/sec. On entering the spinal cord from the dorsal spinal roots, the $A\delta$ and C fibers ascend or descend one to three spinal segments in the tract of Lissauer. They then synapse on neurons in the dorsal horns. Pain signals from the fast fibers in some cases participate in a local, spinal-cord-level, motor withdrawal reflex. For example, sudden application of heat to a foot causes reflex flexion of the leg and withdrawal of the foot.

After synapsing with dorsal horn neurons, pain signals travel up to the brain through two pathways: the *neospinothalamic tract* and the *paleospinothalamic tract*. The neospinothalamic tract carries signals from the fast $A\delta$ fibers sensing mostly mechanical and thermal pain. The paleospinothalamic tract carries information from the slow C fibers as well as some $A\delta$ signals. Signals from these two tracts are distributed in the brain in an anatomically complex manner. The interested reader can pursue the details of these neuroanatomic pathways in Kitchell et al.[72] and in Chapter 48 of Guyton.[59] For example, *most* fibers in the neospinothalamic tract terminate in the ventrobasal complex of the thalamus, and other thalamic areas, and from the thalamus are projected to other basal areas and the somatic sensory cortex. The slow, chronic pain signals of the paleospinothalamic tract largely terminate in the brain stem; in the reticular nuclei of the medulla, pons; and mesencephalon; in the tectal area of the mesencephalon; and in the region of the aqueduct of Sylvius.

A very interesting aspect of the ascending pain systems is that their "gains" are under neural feedback control from the central nervous system (CNS). This is one reason why individual reactions to pain can vary tremendously. The pain control system consists of efferent nerve fibers originating in the raphe magnus nucleus in the lower pons–upper medulla area. These fibers travel down the spinal cord's dorsolateral columns to the appropriate segments' pain inhibitory complex in the dorsal horns, where they synapse with the dorsal horn pain neurons. The neurotransmitter stimulating the descending fibers originating in the raphe magnus nucleus is mostly enkephalin. The descending fibers secrete the neurotransmitter serotonin (5HT) to activate ventral horn pain control interneurons. These interneurons secrete the neurotransmitter enkephalin, which causes presynaptic inhibition of the primary Aδ and C pain neurons. This inhibition can last for many minutes or even hours,[59]

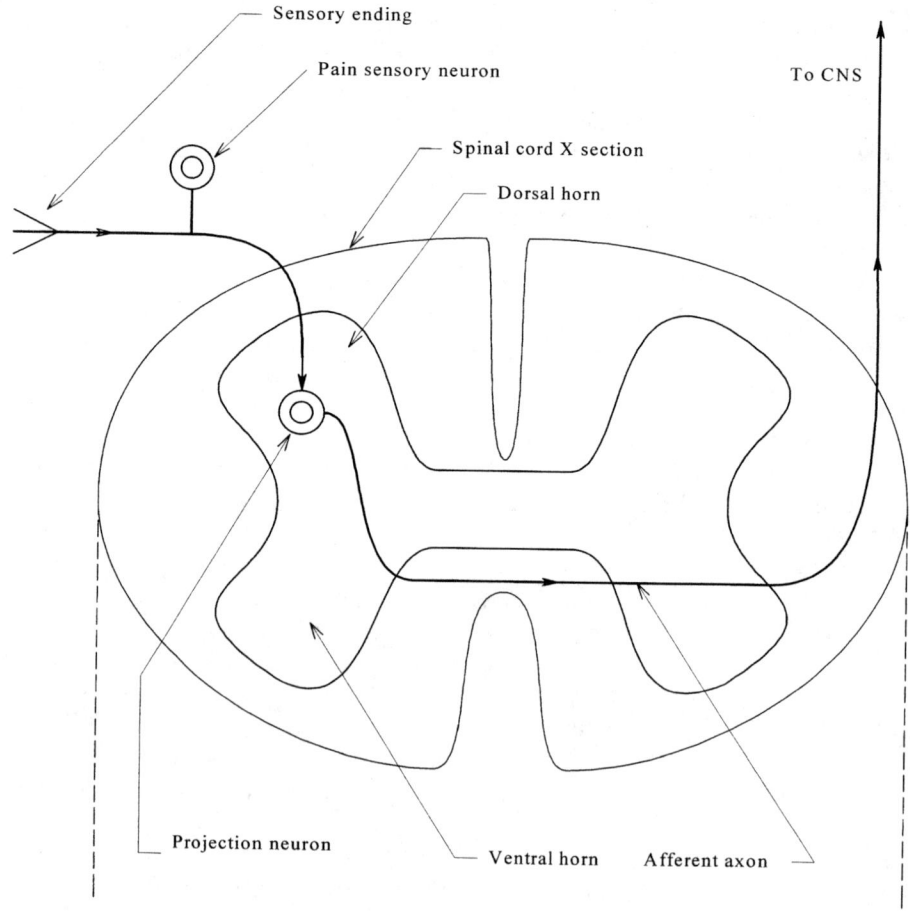

FIGURE 9.1A Direct afferent pathway for pain stimulus to the CNS.

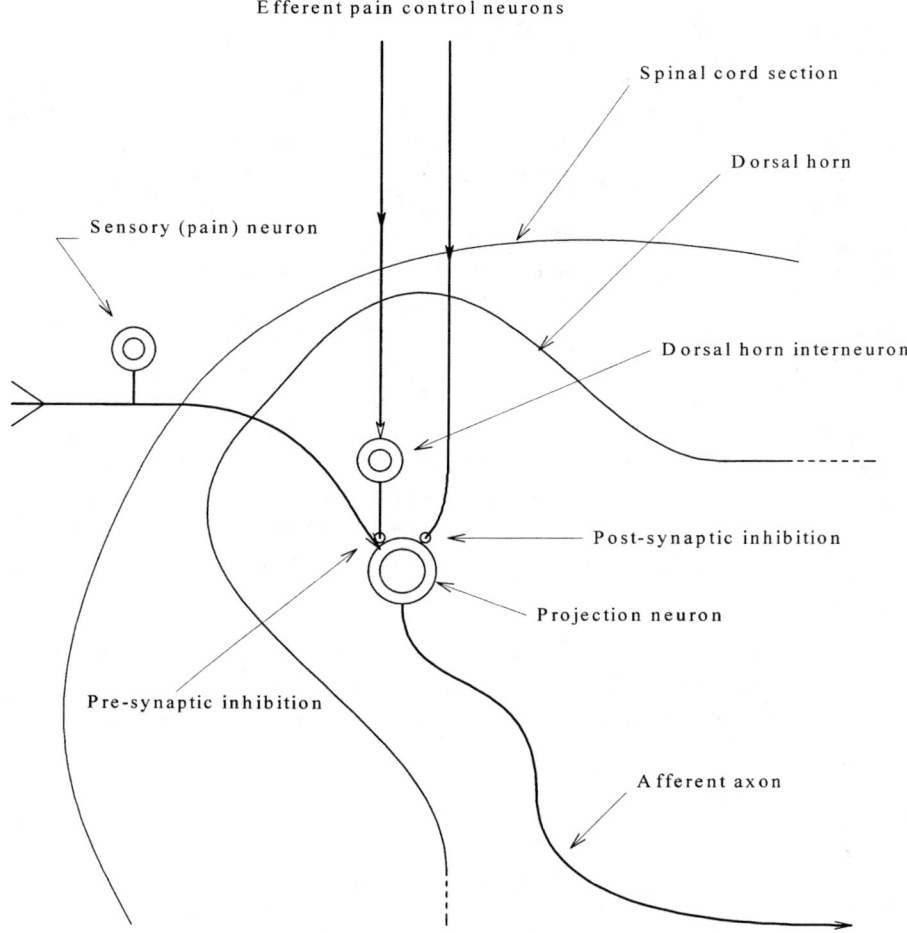

FIGURE 9.1B Schematic of efferent pain control. Both presynaptic inhibition by interneurons and direct postsynaptic inhibition of projection neurons can occur.

blocking primary pain signals at their entry point to the spinal cord and some local pain reflexes. Figures 9.1A and B show schematically the neural "wiring" associated with pain transmission to the CNS and endogenous, efferent pain control. The pain control system no doubt also acts in the CNS at the various relay points for pain information; however, few details are known of these actions.

Not addressed here are the analgesic effects of massage, acupressure, acupuncture, and transcutaneous electric nerve stimulation (TENS). TENS is an electric form of acupuncture. It is often effective in mitigating chronic back pain. These analgesic procedures do not work for everyone or for every cause of pain. That they do work at all is a challenge in understanding the dynamics of the endogenous neurophysiological control of pain.

9.1.3 THE PHARMACOLOGY OF PAIN

Analgesic substances may be subdivided into three classes: (1) exogenous natural substances including morphine and opiates related to morphine, (2) exogenous synthetic analgesic substances (opioids), and (3) endogenous neurotransmitters. The natural opiates and analgesics synthesized from them will be considered first.

The analgesic effects of the juice of the poppy (*Papaver somaniferum*) were known before the third century B.C. The concentrated juice of this poppy is a brown, gummy substance known as opium. Opium contains at least 20 distinct alkaloids (opiates) that have significant pharmacological action. The first opiate isolated was morphine in 1806; codeine was discovered in 1832, and papaverine was described in 1848. These three natural opiates have clinical usefulness as analgesics. Morphine, codeine, and thebaine belong to a class of opiate molecule called *phenanthrenes*.[61] The morphine molecule is shown in Figure 9.2. The natural opiates, papaverine and noscapine belong to the *benzylisoquinolines*. Many semisynthetic opiate derivatives are made by simple chemical modification of terminal groups on the morphine or thebaine molecules. For example, heroin (*diacetylmorphine*) is made from morphine by acetylation of the 3 and 6 positions (see Figure 9.2). (Goodman and Gilman[61] give a table in which 14 opiates related to morphine are described.)

Another class of synthetic analgesic opioids is based on *meperidine*, an opioid molecule that has the *phenylpiperidine* structure. Meperidine and its derivatives are chemically quite dissimilar from morphine and its derivatives, yet they have some

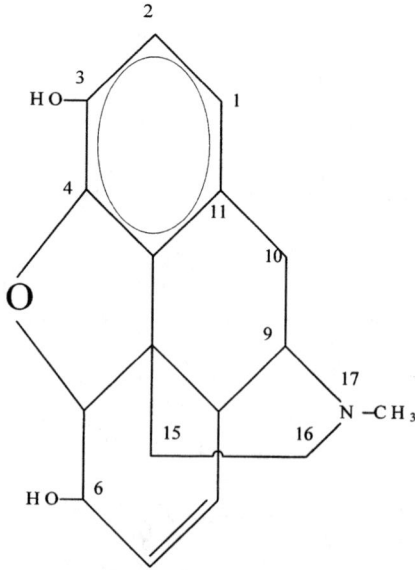

FIGURE 9.2 The morphine molecule.

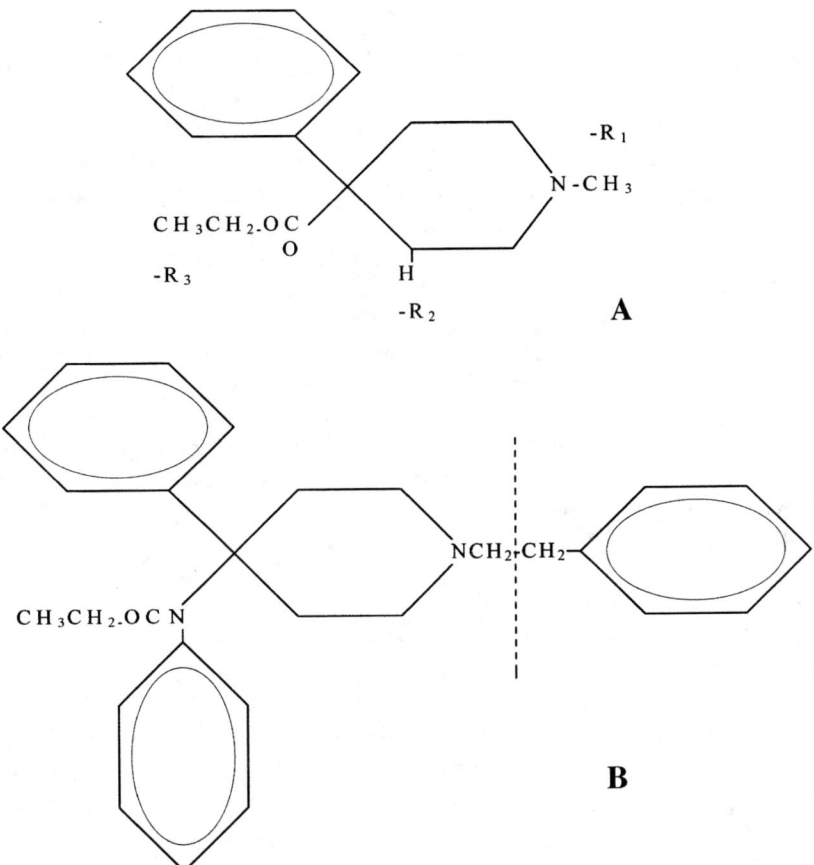

FIGURE 9.3 (A) The opioid meperidine. (B) The opioid fentanyl.

of their pharmacological effects. Figure 9.3 shows the meperidine and fentanyl molecules. Fentanyl is about 80 times as potent as morphine. When fentanyl is administered intravenous (IV), it has a half-life in the CNS described by a three-compartment, mammillary pharmacokinetic model.[116] The brain tissue level of fentanyl can be given by:[83]

$$\frac{X_o}{U}(s) = e^{-s\delta_p} G(s) = e^{-s\delta_p} \sum_{k=1}^{3} \frac{K_k}{(sT_k + 1)} \qquad (9.1)$$

where U is the input rate of fentanyl, X_o is the concentration of fentanyl in the CNS, $\delta_p \cong 1$ min (transport delay from blood central compartment into the CNS), T_k is the time constant of the kth partial fraction term, and K_k is the dc gain of the kth partial fraction term.

The *Physicians' Desk Reference*[10] lists the terminal elimination half-life for IV fentanyl citrate as 219 min (3.65 hr). Peak analgesic effect is noted in several minutes for an IV bolus. A "typical" analgesic IV dose of fentanyl citrate (Sublimaze®) is around 2 to 5 µg/kg.

The endogenous opioid peptides can be classified into three families: (1) *enkephalins,* (2) *endorphins,* and (3) *dynorphins.* Each family originates from a genetically unique polypeptide precursor: *proenkephalin, proopiomelanocortin,* and *prodynorphin,* respectively. These precursor polypeptides are found in blood and various tissue.[61] Each precursor contains in it various protein subunits of certain hormones and opioid peptides. For example, proopiomelanocortin contains the amino acid sequences for adrenocorticotropin (ACTH), γ-melanocyte-stimulating hormone (γ–MSH), and β-lipotropin (β-LP). Within the 91-amino-acid sequence of β-lipotropin are found the sequ (ences of β-endorphin and β-melanocyte-stimulating hormone. Curiously, a subunit of β-endorphin is met-enkephalin. However, met-enkephalin is not split from β-endorphin, but rather is formed from cleavage of the precursor protein, proenkephalin. Leu-enkephalin is formed by cleavage of both proenkephalin and prodynorphin. Prodynorphin can be cleaved to form two dynorphin A molecules, dynorphin B and α- and β-neodynorphin. The physical distribution of the precursor proteins and their cleavage opioid peptides is fairly specific and localized in the CNS and other peripheral nervous structures. The following quotations are from Goodman and Gilman:[61] "For example, in areas of the medulla that are involved in the modulation of pain, the prodynorphin peptides tend to be localized in neurons that are ventral to those containing peptides from proenkephalin." "The peptides from proenkephalin are also found in the adrenal medulla and in the nerve plexuses and exocrine glands of the stomach and intestine[s]." "While the endogenous opioid peptides appear to function as neurotransmitters, modulators of neurotransmission, or neurohormones, their role in physiological processes is not completely understood."

With the wide and specific distribution of endogenous opioids in the CNS and peripheral nervous system comes the necessary fact that there must also be a multiplicity of opioid receptors on these nerves. In the CNS there is evidence for at least four major types of opioid receptors, classified as µ, κ, δ, and σ. Undoubtedly, there are subclasses of these four classes. µ and κ receptors are associated with analgesia; however, not all receptors are associated with analgesia. Dysphoria and psychomimetic effects are associated with σ receptors in the limbic system of the CNS, and δ receptors may be involved with affective behavior.[61] The situation is made more complex because of cross-talk between exogenous opioids caused by nonunique affinities to a given class of receptor. This crosstalk is illustrated by the fact that certain exogenous opioids can have agonist (Ag), partial agonist (pAg), or antagonist (Ant) action at certain receptor types. Table 9.1, adapted from Goodman and Gilman,[61] illustrates some of these properties. Note that naloxone is the treatment of choice for opioid poisoning (as with morphine or heroin); it has no direct respiratory depressant action.

In summary, the neuropharmacology, neuroanatomy, and neurophysiology of pain and exogenous and endogenous pain control are very complex and are just

Control of Postoperative Pain by Self-Administered Opioids

TABLE 9.1
Action of Certain Exogenous Opioid Drugs at Endogenous Receptors

	Receptor Type		
Compound	μ	κ	σ
Morphine	Ag	Ag	—
Naloxone	Ant	Ant	—
Pentazocine	Ant	Ag	Ag
Butorphanol	—	Ag	Ag
Nalbuphine	Ant	pAg	Ag
Nalorphine	Ant	pAg	Ag
Buprenorphine	pAg	—	—
Propiram	pAg	—	—

beginning to be understood at the molecular level. Many exogenous analgesic drugs carry severe side effects, including headache, respiratory suppression, loss of gut motility, addiction, etc. The ideal analgesic drug would not be addictive and would only relieve pain.

9.2 THE PCA SYSTEM OF REASBECK AND JACOBS

In this section, the pioneering PCA system developed by Reasbeck and Jacobs et al. is examined. Much of the material follows Reasbeck's Ph.D. dissertation,[116] in which he described the pharmacokinetics of the opioid fentanyl, a speculative model for pain, button-pressing behavior of patients in pain, and several controllers that convert button pressing into opioid analgesic drug infusion rate.

9.2.1 THE PHARMACOKINETICS OF IV FENTANYL

As discussed in the preceding section, fentanyl is an analgesic based on the phenyl-piperidine molecule. Goodman and Gilman[61] tell us that fentanyl is about 80 times as potent as morphine for analgesia, acting primarily as an agonist at the endogenous μ receptors on pain system neurons. Fentanyl citrate (Sublimaze) is used for IV injections. In Chapter 3 of his dissertation, Reasbeck derives a three-term partial fraction form for fentanyl pharmacokinetics. He assumes a three-compartment mammillary model[55] in which the central compartment is the blood and the other two compartments are unspecified. The CNS level of fentanyl is related to the blood level by a simple delay term. The model is shown in Figure 9.4. The compartmental pharmacokinetic equations are

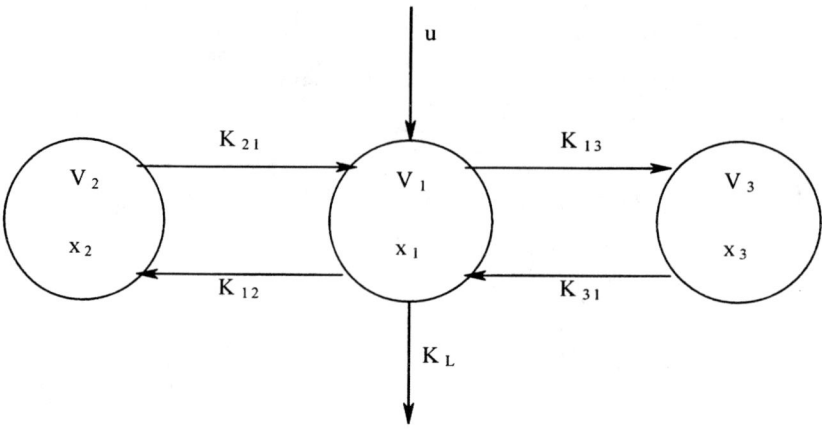

FIGURE 9.4 The three-compartment pharmacokinetic (3CPK) model for fentanyl.

$$\dot{x}_1 = -K_L x_1 - K_{12}(x_1 - x_2) - K_{13}(x_1 - x_3) + u/V_1 \qquad (9.2A)$$

$$\dot{x}_2 = -K_{21}(x_2 - x_1) \qquad (9.2B)$$

$$\dot{x}_3 = -K_{31}(x_3 - x_1) \qquad (9.2C)$$

$$\dot{x}_0 = x_1(t - \delta_p) \qquad (9.2D)$$

where x_j is the concentration in the jth compartment, K_L is the loss constant from the central compartment, V_1 is the volume of the central compartment, u is the drug mass input rate in micrograms per minute, x_0 is the CNS fentanyl concentration in micrograms per liter, and the K_{jk} are exchange rate constants. Standard techniques can be used to find the transfer function relating x_1 to u. In the rational polynomial form:

$$\frac{X_1}{U} = K_{SS} \frac{1 + a_1 s + a_2 s^2}{1 + b_1 s + b_2 s^2 + b_3 s^3} = G_p(s) \quad (\mu g/l)/(\mu g/\min) \qquad (9.3)$$

Note that this transfer function is in unfactored, time constant form and of order $N - M = 1$. If the denominator is factored, partial fraction expansion yields, in time constant form:

$$\frac{X_1}{U}(s) = G(s) = \sum_{k=1}^{3} \frac{K_k}{(sT_k + 1)} = G_p(s) \qquad (9.4)$$

TABLE 9.2
Pharmacokinetic Parameters of Five Opioid Drugs, Based on the Three-Compartment Mammillary Model[68]

Opioid	K_1 (min/l)	K_2 (min/l)	K_3 (min/l)	T_1 (min)	T_2 (min)	T_3 (min)
Morphine	0.133	0.188	0.5	1.92	18.9	244
Fentanyl	0.087	0.072	0.908	2.33	18.2	313
Buprenorphine	0.413	0.227	0.536	1.87	15.4	162
Diamorphine	0.771	0.134	0.37	5.81	13.3	165
Methadone	0.24	0.816	4.771	3	350	1300

The nominal values of the three K_k and T_k terms are given in Table 9.2. Figures 9.5A and B show the unit impulse response of $G_p(s)$ delayed 1 min to represent the diffusion time across the blood–brain barrier. Note that fentanyl in the CNS compartment has a half-life of about 4 min in this model, but an extremely long "tail" of residual concentration. Figure 9.6 shows the response of x_0 to four unit impulses spaced 3 min apart. Note that in the models for Figures 9.5 and 9.6, mixing of the IV bolus in the blood volume is assumed to be instantaneous.

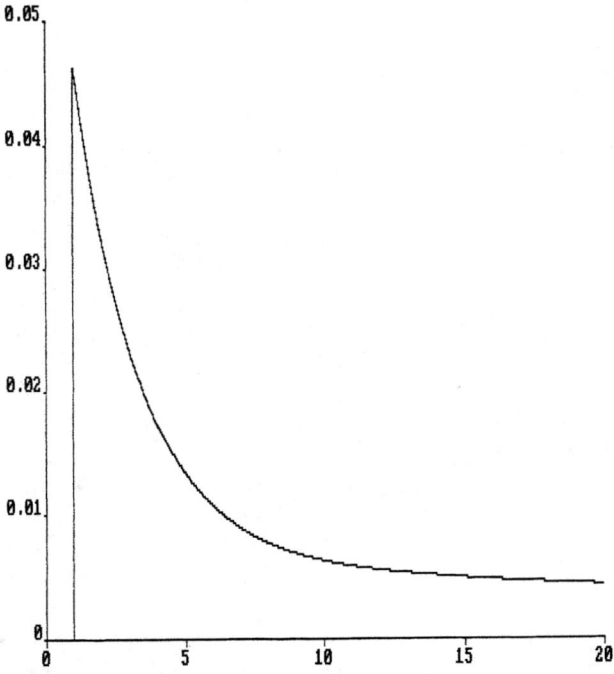

FIGURE 9.5A Simulated response of the delayed central (blood) compartment to a unit impulse (IV bolus) of fentanyl.

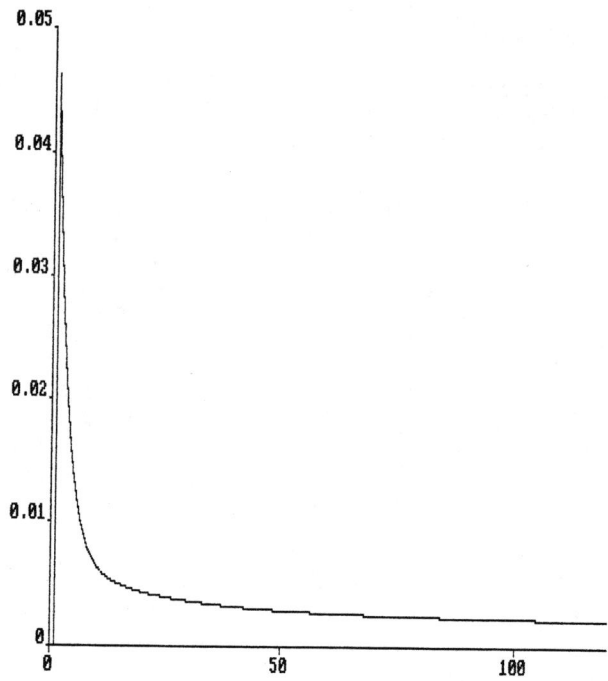

FIGURE 9.5B Simulated response of the delayed central (blood) compartment to a unit impulse (IV bolus) of fentanyl, using a longer time scale to show the "long tail."

Inspection of Figures 9.5 and 9.6 suggests that the pharmacokinetics of fentanyl are such that repeated demand-button presses cause the concentration to peak rapidly, giving relief, and then decaying rapidly. Such a rapid decay argues for a dual-mode of injection (e.g., IV and intramuscular). Because the pharmacokinetics of drug release from intramuscular injection are far slower than the IV route, CNS concentration would build up rapidly and decay more slowly following analgesia demand-button presses.

9.2.2 THE BUTTON-PRESSING MODEL OF REASBECK AND JACOBS

Figure 9.7 shows the block diagram used by Reasbeck[116] to describe human button-pressing behavior in demand analgesia. Note that the system is highly nonlinear and has two inputs (pain and opioid drug input rate) and one output, the button-pressing rate. The patient button-pressing rate was expressed as the number of button presses in consecutive, 1-min sample periods. Thus, at the end of every minute, the system output is the *integer* number of button presses occurring in that minute. Inspection of Figure 9.7 shows that the linear, three-compartment pharmacokinetic model has an opioid drug input rate of u μg/min. The blood concentration is x_1 μg/l, and the

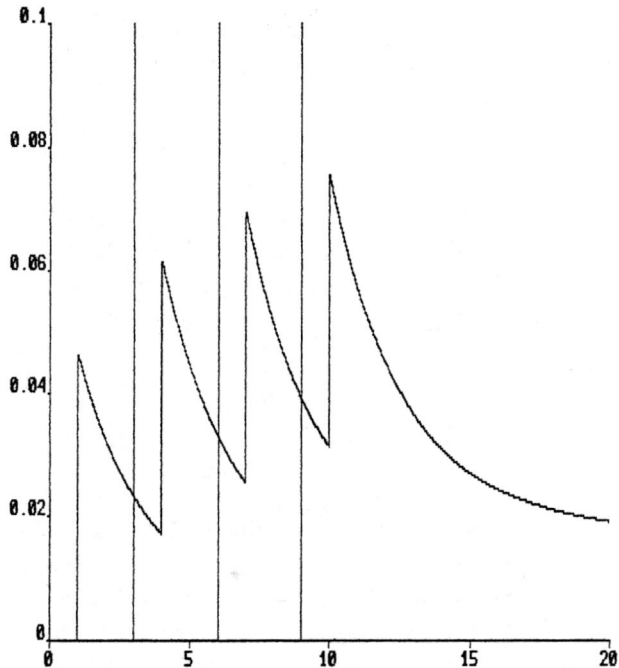

FIGURE 9.6 Simulated response of the delayed central (blood) compartment to four unit impulses of fentanyl. Note the delay.

CNS concentration, x_0, is x_1 delayed by 1 min. "Comfort," C, is defined as the product $x_0 x_5$. x_5 is called the "relief." The comfort C is subtracted from the "discomfort", D, to yield the "pain," P. That is, $P = D - C$.

Reasbeck defines the units of pain, comfort, and discomfort as the "pang." A pain has the units of one pang if it produces (on the average) one button press per

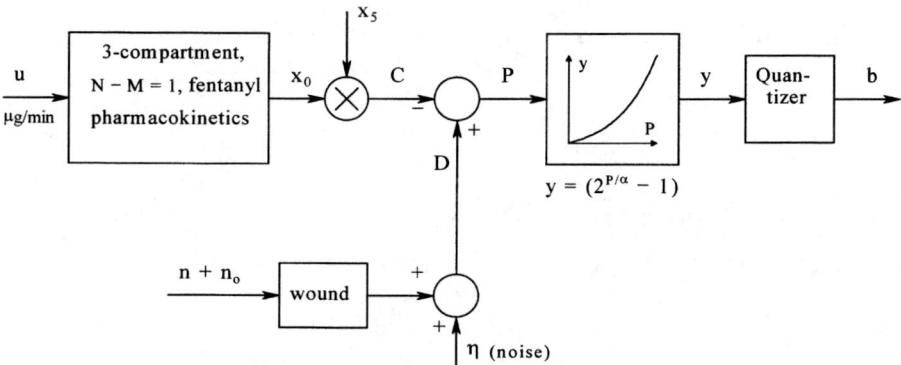

FIGURE 9.7 The PCA patient's button-pressing behavior model in response to pain given by Reasbeck.[116] See text for description.

second. To model the patient's psychomotor response to the pain, Reasbeck assumed that the button-pressing rate, y, is an exponential function of P. That is,

$$y = 2^{P/\alpha} - 1, \quad P \geq 0$$
$$y = 0, \quad P < 0 \quad (9.5)$$

When $P = 1$ pang, we want $y = 1$ button press per second. Thus $1 + 1 = 2^{1/\alpha}$, so $\alpha = 1$. Because the total number of button presses in a minute can only be zero or a positive integer, the exponential nonlinearity is followed by an integer operation:

$$b = \text{INT}(y) \quad (9.6)$$

Thus, if $P = 5$ pangs, $y = 2^{5/1} - 1 = 31$, which is an integer, so $b = 31$. In a later (1985) version of the button-pressing model, Jacobs et al.[68] eliminate the exponential nonlinearity and quantizer by a unity gain block.

The discomfort, D, in the Reasbeck model is derived from three additive pain sources: a decaying exponential term with an initial value $x_4(0)$ and time constant $T_4 = 480$ min (8 hr), a white noise source that drives a $1/(1 + T_4 s)$ low-pass filter, and a second white noise source, $\eta(t)$. The variance of the noise source driving the filter and the $h(t)$ noise source are assumed to decay exponentially from their initial values with time constant T_4, representing wound healing. The two noise sources are used to simulate the random, unexpected variations in experienced pain that are the result of such events as coughing, moving in bed, attention from the nurses, etc. All model noise sources decay with time, representing healing and endogenous analgesia. Thus $D = x_4(t) + x_4(0)e^{-t/T_4} + \eta(t)$.

Data from real patients were used to validate the model. The six compartmental pharmacokinetic constants, T_4, α, the noise variances, x_5, and $x_4(0)$ must be estimated. Table 5 in Reasbeck[116] gives the following values for "life-like" simulation:

Initial nominal wound pain $x_4(0) = 50$ centipangs (cp).
Initial $\sigma_n(0)$ of wound driving noise = 2.5 cp.
Initial $\sigma_n(0)$ of $\eta(t)$ noise = 25 cp.
Nominal discomfort decay time constant $T_4 = 900$ min.
Nominal relief constant $x_5 = 100$ cp/μg/l of fentanyl.

9.2.3 THE DRUG INJECTION CONTROLLER

Two major contributions of Reasbeck's Ph.D. work were in the development of a nonlinear patient-pain-model parameter estimation algorithm and the design of a nonlinear stochastic controller. The nonlinear controller used separate algorithms to estimate the five states of the controlled process $(x_1 \ldots x_5)$ and to implement control action. An on-line implementation of Bayes' rule generated the conditional mean and variance of the predictable component z of pain P $(z = x_4 - x_1 x_5)$. These conditional statistics are weighted to give a pseudomeasurement[68] y' of z which is used to drive an extended Kalman filter (EKF) which handles the nonlinearity due to uncertainty about x_5, and which generates estimates of the five states **x**. Good, but not optimum, estimation of the states was observed.

The stochastic control law was designed to make the comfort C greater than the predictable component of discomfort W by an amount proportional to the magnitude of the unpredictable component η of the discomfort. The one-step-ahead algorithm used was

$$kv_n(i+1) = \tilde{c}(i+1) - \tilde{w}(i+1) = \tilde{z}(i+1) \qquad (9.7)$$

k is a tuned, positive constant, typically 2.[68] The controller output is given by:

$$u(i) = \left\{ \begin{array}{l} kv_n(i+1) + a_4\tilde{x}_4 - a_{11}(\tilde{x}_1\tilde{x}_5 + \sigma_{15}) \\ - a_{22}(\tilde{x}_2\tilde{x}_5 + \sigma_{25}) - a_{13}(\tilde{x}_3\tilde{x}_5 + \sigma_{35}) \end{array} \right\} \Big/ \tilde{b}_1 x_5 \qquad (9.8)$$

Tildes signify estimates of states from the EKF. The index (i) has been omitted from the state estimates and σs for compactness. The a_{1k}, $k = 1, 2, 3$, and a_4 are from the **A** matrix in the state equation:

$$\mathbf{x}(i+1) = \mathbf{A}\mathbf{x}(i) + \mathbf{B}u(i) + \xi(i) \qquad (9.9A)$$

$$\mathbf{A} = \begin{vmatrix} a_{11} & a_{12} & a_{13} & 0 & 0 \\ a_{21} & a_{22} & a_{23} & 0 & 0 \\ a_{31} & a_{32} & a_{33} & 0 & 0 \\ 0 & 0 & 0 & a_4 & 0 \\ 0 & 0 & 0 & 0 & 1 \end{vmatrix} \qquad (9.9B)$$

$$\mathbf{b} = \begin{bmatrix} b_1 & b_2 & b_3 & 0 & 0 \end{bmatrix}^T \qquad (9.9C)$$

$$\xi(i) = \begin{bmatrix} 0 & 0 & 0 & n(i) & 0 \end{bmatrix}^T \qquad (9.9D)$$

The elements in **A** and **b** depend on the pharmacokinetic model K and T parameters. $n(i)$ in $\xi(i)$ is the decaying noise which drives w.

Jacobs et al.[68] performed clinical comparisons of three classes of PCA controller: (1) simple *proportional control* in which an IV bolus of opioid was injected for every button press, (2) the *separated stochastic controller* described by Jacobs et al., and (3) a hybrid combination of proportional control plus state estimation and separated stochastic control. Proportional control is used for a short initial period, while the Bayes/EKF system estimates the five states. After 10 button presses or 2 hr, whichever is sooner, the estimate of the relief gain, $\tilde{x}_5$, is frozen, and thereafter separated stochastic control is used on the first four states. All controllers had supervisors to prevent overdosing with fentanyl. $u(i)$ was set to zero if in any 1 min more than 50 μg was injected, in any 5 min more than 100 μg was injected, or in any 1 hr more than 200 μg was injected.

In evaluating the results of the PCA controller trials, we can see intuitively that few button presses indicate little need for analgesia, implying that either the patient did not experience severe discomfort or the controller maintained adequate blood levels of the drug, or both. Also, an effective controller used a minimum amount of opioid at the same time it minimized the number of button presses. Summary data presented by Jacobs et al.[68] showed that their so-called hybrid controller and the separated stochastic controller met these two criteria in an extensive modeling study. Both systems outperformed the purely proportional controller and a constant infusion of 40 μg/hr. In the latter case, button presses were very large, because mean tissue drug level was only about three fourths the levels reached in the active control schemes. (Button presses did not alter the infusion rate.) It should be noted that a *bolus* dose of 40 μg/kg body weight is considered a *high dose* of fentanyl (a total of 2.8 mg for a 70-kg patient).[10] An infusion of 40 μg over 1 hr is equivalent to 0.667 μg/min, a light pharmacological dose of fentanyl.

We simulated the pharmacokinetics of this dose using the values in Table 9.1, as given by Reasbeck.[116] Figure 9.8 illustrates the $x_1(t)$ response of the model to this

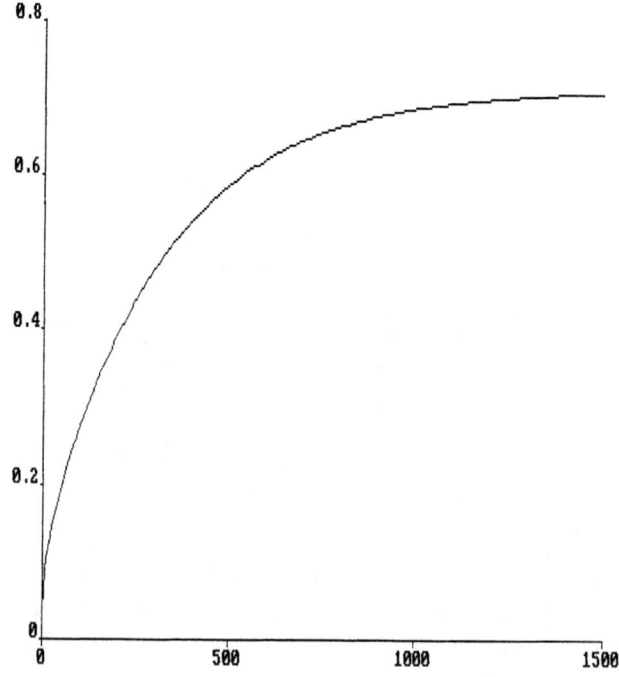

FIGURE 9.8 Simulated blood level of fentanyl in micrograms per liter given a continuous infusion of 40 μg/hr. Time scale in minutes. The fentanyl three-compartment pharmacokinetic model parameters used are in Table 9.2.

infusion rate. Note that steady-state fentanyl concentration in the central compartment is not reached until over 24 hr, a good reason for a patient's initial discomfort and excess button pressing. After 1500 min, $x_1 = 0.7066$ μg/l; Jacobs et al. reported a maximum tissue level of 0.76 μg/l. Clearly, a bolus injection approach is more effective in delivering analgesia. Choice of an opioid (other than fentanyl) that has a shorter third-compartment time constant would possibly give better pharmacokinetics for PCA. One such drug is buprenorphine (Buprenex®) with $T_3 = 162$ min vs. $T_3 = 313$ min for fentanyl. Buprenophine is an opiate analgesic derived from thebaine; it is about 30 times as effective as morphine sulfate in pain relief.[10]

9.3 THE PCA MODEL OF LIU AND NORTHROP

Liu and Northrop[83] designed a flexible, robust, deterministic PCA system. They wished to avoid the problem of the time the estimation system must take in accurately estimating parameters required for stochastic control. The estimation problem is made more severe by the nonstationarity of the patient system; i.e., as the wound heals, the deterministic and random components of discomfort decay to zero. The first major difference from the PCA system of Reasbeck and Jacobs is in the patient's button-pressing model, described in the next section.

9.3.1 THE PATIENT PAIN MODEL OF LIU AND NORTHROP

Figure 9.9 illustrates a block diagram of the "plant"; its inputs are the input rate of analgesic opioid, u, and the discomfort D produced by the wound. Its output is the number of button presses per sampling period. The fentanyl pharmacokinetic model used by Reasbeck[116] and Jacobs et al.[68] was used in the Liu and Northrop system for consistency. It will not be repeated here. The CNS level of fentanyl, x_0, is treated differently, however. Instead of subtracting relief from discomfort, a *multiplicative model* was used in which the neurological pain, P, is the *product* of the discomfort, D, and the fraction of *unbound* fentanyl receptor sites on pain control neurons (probably μ receptors[61]).

A steady-state, chemical kinetic model for pain relief is developed as follows: We assume the CNS neurons responsible for analgesia have membrane receptors that combine selectively with opioid analgesic molecules on a 1:1 basis. The binding of a single opioid molecule with a free receptor triggers (unknown) intracellular events which lead to analgesia. The binding can be described by a Michaelis–Menton-type chemical reaction:

$$O + R \underset{k_2}{\overset{k_1}{\rightleftharpoons}} O * R \xrightarrow{k_3} \overline{O} + R \quad (9.10)$$

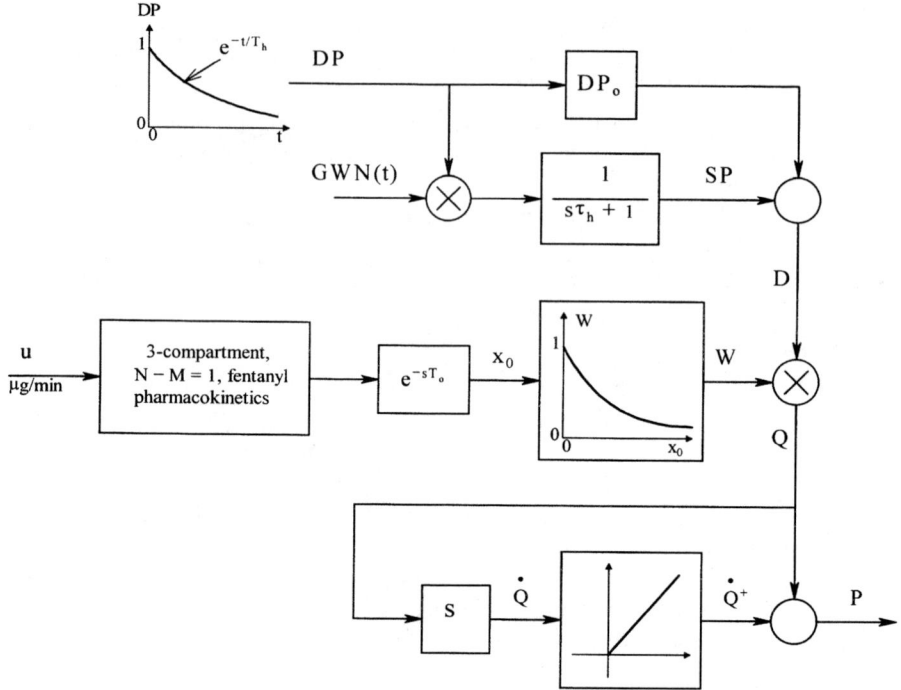

FIGURE 9.9 The patient pain model of Liu and Northrop.[83] (All summations are +.) See text for description.

where O is an opioid molecule, R is a free receptor molecule, $O * R$ is the active, bound complex, and $\overline{O}$ is an inactivated opioid molecule; k_1, k_2, and k_3 are reaction rate constants. The mass-action kinetics governing this reaction are

$$\dot{r} = -k_1 x_0 r + b(k_2 + k_3) \tag{9.11}$$

where r is the running density of R, and b is the running density of bound receptor sites. In the steady state, $\dot{r} = 0$, and we find that:

$$b_{SS} = \frac{k_1 x_0 r_{SS}}{k_2 + k_3} \tag{9.12}$$

Because $b_{SS} + r_{SS} = 1$, Equation 9.12 yields:

$$r_{SS} = W = \frac{k_2 + k_3}{k_2 + k_3 + k_1 x_0} = \frac{1}{1 + k_1 x_0/(k_2 + k_3)} \tag{9.13}$$

Control of Postoperative Pain by Self-Administered Opioids

W is the fraction of unbound receptors on the CNS' pain control neurons. Note that as the CNS concentration of opioid x_0 increases, W decreases hyperbolically. At very high x_0, nearly all receptors are bound and active, giving maximum analgesia. Thus the pain generated is assumed to be the product, WD. Liu and Northrop state: "Consideration of fentanyl pharmacology leads us to assume that $W = 0.05$ at a 'working' fentanyl concentration of $x_0 = 1$ µg/liter (95% of opioid receptors occupied = 95% analgesia. This assumption gives us:

$$W = \frac{1}{1 + 19x_0} \text{''} \qquad (9.14)$$

To model patient button-pressing behavior, we note that *neurological pain, Q,* was modeled by the multiplicative relation:

$$Q(t) = D(t)W \qquad (9.15)$$

The scaled, rectified derivative of the neurological pain is called the *psychological pain*. The *perceived pain, P,* is the sum of Q and the psychological pain, as shown in Figure 9.9.

The actual rate of button pressing is modeled by a relaxation pulse-frequency modulation (RPFM) process.[99] This process is illustrated in Figure 9.10. RPFM was chosen because it allows "forgetting" the pain input with time constant, T_R. An output "event" (in this case, a button press) is produced when the output of the RPFM low-pass filter exceeds its trigger threshold, *TH*. The output event also resets the low-pass filter output, $z(t)$, to zero. Note that the input to the RPFM process is always a nonnegative signal in this model. There is no problem with "negative pain," which is possible in the Reasbeck[116] model for patient behavior. The RPFM threshold can be considered to be part of the pharmacological response to pain. *TH* is assumed to rise with increasing CNS opioid concentration, x_0. That is,

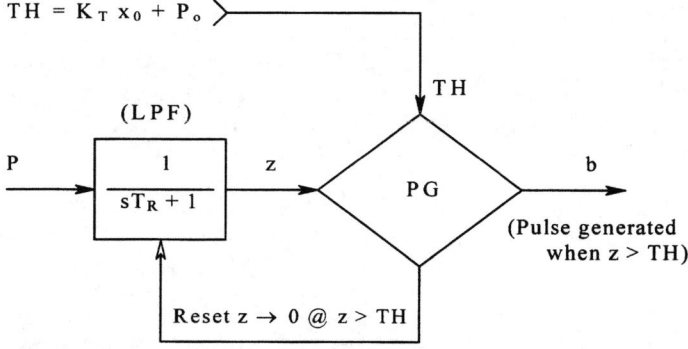

FIGURE 9.10 RPFM button-pressing model of Liu and Northrop. See text for discussion.

$$TH = K_T x_0 + P_0 \tag{9.16}$$

The generation of pain in the Liu and Northrop system is based on the model of Reasbeck.[116] There are two components to the discomfort, $D(t)$: (1) an exponentially decaying deterministic pain, DP, and (2) a nonstationary random pain modeled by passing Gaussian white noise (GWN) with an exponentially decaying standard deviation through a low-pass filter: Mathematically, the discomfort can be written as:

$$D(t) = DP_0 e^{-t/T_h} + \left(e^{-t/T_h}\right) \otimes \left[e^{-t/T_h}\right] GWN(t) \tag{9.17}$$

where T_h is the healing time constant, $\otimes$ denotes the operation of real convolution, and e^{-t/T_h} is the impulse response of the low-pass filter.

9.3.2 MODEL VALIDATION BY SIMULATION

Parameters of the model introduced in the previous section were adjusted to mimic the demand analgesia behavior of actual patients given by Reasbeck.[116] In validating the model, certain parameters were taken as fixed: $T_R = 0.15822$ min, $P_0 = 0.1$ pang (normalized, fixed threshold of RPFM process), $T_h = 480$ min (nominal wound healing time), and $\sigma_n(0) = 30.0$ pang (initial nominal standard deviation of Gaussian noise driving pain [input to low-pass filter]). Parameters that were varied in validating the model were $2.0 \leq K_Q \leq 10.0$ (psychological pain rate constant), $0.1 \leq K_T \leq 1.0$ (RPFM threshold sensitivity to CNS fentanyl concentration), and $0.5 \leq x_0(0) \leq 2.0$ µg/l (initial CNS fentanyl concentration when patient wakes up). Three patient sensitivities were used in the simulations, using the parameters $[KT, K_Q, x_0(0)]$ described in Table 9.3.

9.3.3 IPFM/SMITH DELAY COMPENSATOR CONTROLLER DESIGN

Because the button presses generated by either real or simulated patients represent a nonstationary, noisy point process, Liu and Northrop chose a simple, discrete, IPFM controller to convert the button-pressing point process to bolus drug injections, u.

$$u(t) = \sum_{k=1}^{\infty} D_o \delta(t - T_k) \tag{9.18}$$

The discrete IPFM controller is shown in Figure 9.11. If, for example, 2 button presses occur within a given sample period, T_{k-1}, the input to the IPFM rectangular integrator will be a "2" at the end of that sampling period. If the integrator output, z, reaches TH_2, the fixed threshold of the IPFM process, a single bolus injection is

TABLE 9.3
Simulated Patient Data

Patients	Parameters			
	K_Q	K_T	$x_0(0)$	BP
Min	2.0	1.0	1.97	23
Ave	6.0	0.5	1.02	56
Max	10.0	0.1	0.52	120

Note: BP = total button presses in 8 hr after waking.

given at T_k. If $z < TH_2$ at T_k, the integrator continues to accumulate the number of button presses in each successive sampling period until $z > TH_2$. At the time a bolus injection is given, the IPFM integrator is reset to zero output, and the process repeats itself. It should be noted that because of the resetting of the integrator when output pulses are generated, the IPFM pulse generator does not constitute a pure integration in the system's loop gain; thus the closed-loop system is type 0 and average discomfort will never go to zero. However, discomfort can be made to be very small by

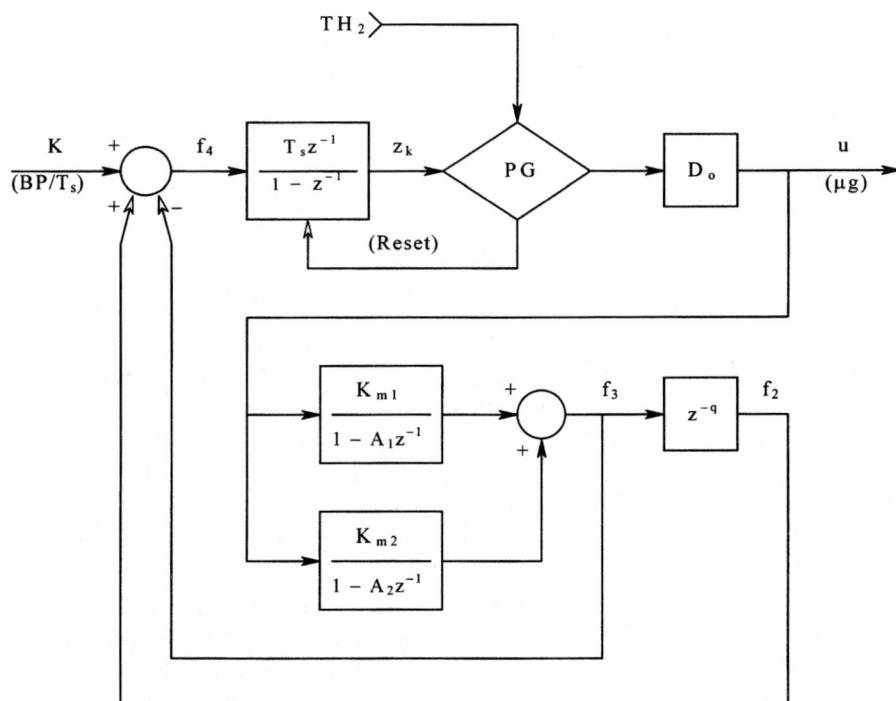

FIGURE 9.11 Discrete SDC/IPFM controller for PCA modeled by Liu and Northrop.

raising the dc loop gain (D_o/TH_2) so that there will be little motivation for button pressing.

In examining the modeled pharmacokinetic response of CNS fentanyl concentration to bolus injections (cf. Figure 9.6), it is apparent that the fall time of the concentration x_0 to one half its peak value is of the same order of magnitude as the transport leg between x_1 and x_0 (1 min). Thus at high proportional controller gains, the closed-loop PCA system using a pure IPFM controller will tend to be unstable. To circumvent this problem, a Smith delay compensator[128,146] (SDC) was used in the controller. (The SDC is discussed in this text in Sections 3.4 and 8.2.) Although the plant pharmacokinetics use a three-compartment model, only the two terms with the short (dominant) time constants were used in the design of the SDC. The complete controller shown in Figure 9.11 has constants given by:

$$K_{m1} = K_1/T_1$$
$$K_{m2} = K_2/T_2$$
$$A_1 = e^{-T_s/T_1}$$
$$A_2 = e^{-T_s/T_2}$$

The SDC reference model's delay is integer multiples of the sampling period, i.e., z^{-n}. $n = 1$ when $T_s = 1$ min, and $n = 2$ for $T_s = 0.5$ min.

9.3.4 RESULTS OF SIMULATIONS

Liu and Northrop investigated the performance of four different controllers using the patient pain model described above. **PAIN1** was a conventional proportional controller which injected IV 2 µg of fentanyl for every button press. This model was used to tune the patient pain model's variable parameters to match actual patient data from Reasbeck.[116] **PAIN2** used the IPFM/SDC controller shown in Figure 9.11 with $T_s = 1$ min, $D_o = 4$ µg, and $TH_2 = 0.05$ for the IPFM pulse generator. The **PAIN3** controller was the same as PAIN2 except $T_s = 0.5$ min. **PAIN4** was a hybrid system in which the PAIN1 proportional algorithm was used for the first hour after the simulated patient awoke, and the PAIN2 controller was used for the next 7 hr. Some of the results are summarized in Table 9.4.

Liu and Northrop also examined the effects of their PCA systems for "Min" and "Max" patient sensitivities to pain and varied the pharmacokinetic time constants of fentanyl over a 0.5 to 2.0 range. In almost every case, the simple proportional controller (PAIN1) required more button presses, and gave a larger INTP figure than the other IPFM/SDC configurations. Multiple runs were done on each type of patient with each type of controller in order to obtain performance statistics. In all cases, the PAIN3 controller with $T_s = 0.5$ min gave superior performance in terms of minimizing total demand and INTP. Curiously, the PAIN3 controller also gave a slightly

TABLE 9.4
System Performance Summary for the "Average" Patient

	PAIN1	PAIN2	PAIN3	PAIN4
Total demand	56	17	16	36
Total dosage	112	128	124	116
INTP	80	74	75	77
Max x_0 over 8 hr	0.54	0.59	0.51	0.52

Note: Total demand = total button presses in 8 hr, total dosage in micrograms in 8 hr, INTP = integral of perceived pain (pang * minutes) over 8 hr, Max x_0 = peak x_0 in micrograms per liter over 8 hr.[83]

lower mean peak x_0 than the other controllers. Figures 9.12A, B, and C illustrate the time history of button presses, doses injected, and x_0 over an 8-hr period for a representative, average, simulated patient given PCA with the PAIN3 IPFM/SDC controller.

9.4 CHAPTER SUMMARY

PCA using frequent, small IV bolus doses of analgesic drug has been shown to be an effective alternative to the use of large, widely scheduled bolus doses of opioid. The large doses do not take into consideration the patient's sensitivities to pain and

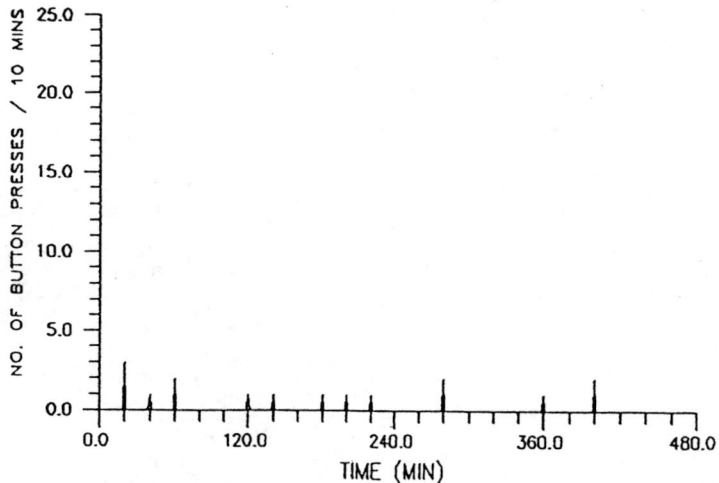

FIGURE 9.12A Graph of simulated "average" patient's button presses over an 8-hr period while under PCA by the system "PAIN 3" of Liu and Northrop. (From Liu, F-Y. and Northrop, R.B. 1990. *IEEE Trans. Biomed. Eng.*, 37(12): 1147–1158. ©1990 IEEE. With permission.)

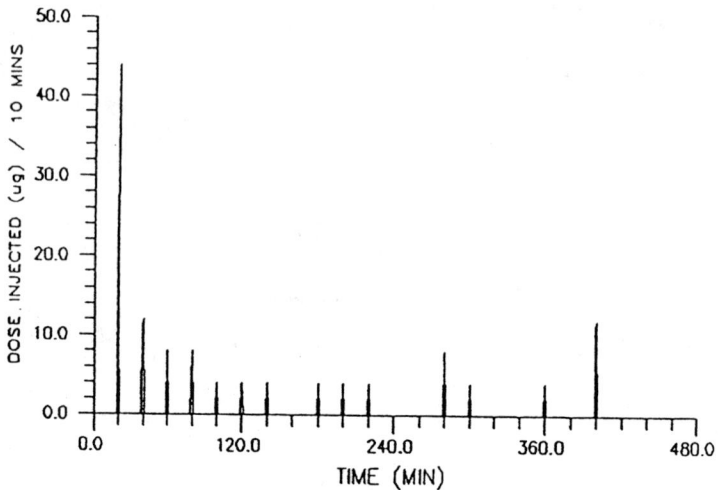

FIGURE 9.12B Time history of fentanyl injected for the "ave" patient by the PCA system of Liu and Northrop. Sample period T_s was 0.5 min. (From Liu, F-Y. and Northrop, R.B. 1990. *IEEE Trans. Biomed. Eng.*, 37(12): 1147–1158. ©1990 IEEE. With permission.)

the drug; the dose is generally based on the patient's weight. The pharmacokinetics of the drug typically produce intervals of time preceding the next scheduled dose where the CNS drug concentration has fallen below a concentration for effective analgesia. Also, peak drug concentration can easily exceed what is required at the

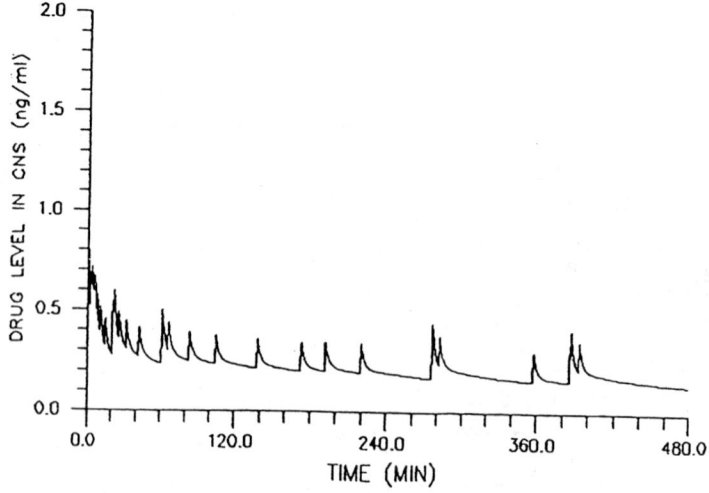

FIGURE 9.12C Simulated CNS concentration of fentanyl determined by the controller of Liu and Northrop. (From Liu, F-Y. and Northrop, R.B. 1990. *IEEE Trans. Biomed. Eng.*, 37(12): 1147–1158. ©1990 IEEE. With permission.)

time for effective analgesia. By using small bolus doses of opioid given on the demand of the patient experiencing the pain, much more effective analgesia can be realized. Both experienced pain and cumulative drug dose are generally lower for PCA.

Two different approaches to PCA were examined. The first was the pioneering work of Reasbeck[116] and Jacobs et al.[66–68] (Jacobs was Reasbeck's thesis adviser), and the second was the modeling study of Liu[82] and Liu and Northrop.[83] The approach of Reasbeck and Jacobs was to use a simple model for experienced pain and a complex, separated stochastic controller in which system parameters were estimated and used to "tune" the drug injection algorithm. Reasbeck and Jacobs did both modeling studies and clinical trials of their PCA system with good results.

The Liu and Northrop PCA system was also based on an *a priori* knowledge of fentanyl pharmacokinetics and a novel, multiplicative model for experienced pain in which a basic, steady-state, pharmacokinetic model for bound opioid molecule receptor sites acted as a "gain control" for neurological pain. The perceived pain was assumed to be the sum of neurological pain plus a psychological component which consisted of the weighted, positive derivative of neurological pain. This model was felt to be more realistic in generating patient button-pressing behavior than the simple summation of pain and "comfort" used by Reasbeck and Jacobs.

PCA controller design is made difficult because of the wide range of patient responses to pain and individual variability in response to the analgesic drug. Wound healing causes the deterministic and random pain inputs to the system to decrease in time. Such nonstationary behavior makes the design of a self-tuning controller for PCA difficult. Optimal control in which the pain (rate of button pressing) is minimized while minimizing the concentration of opioid in the blood is an approach that should be investigated. Also, a continuous infusion form for $u(t)$ that is updated every sampling period might be tried.

PROBLEMS

9.1 Discuss and compare the attempts of physiological psychologists to quantify the experience of pain.

9.2 *Audio analgesia* has been used in various forms for hundreds of years. In the first part of the 19th century, before chemical anesthetics were discovered, naval surgeons performed amputations while a drum was beat near the patient's head. This noise was found to mitigate the immediate pain of the procedure. More recently, audio analgesia has been investigated scientifically (Licklider, J.C.R. 1961. On psychophysiological models, in *Sensory Communication,* W.A. Rosenblith, Ed., MIT Press, Cambridge, MA, chap. 3) for various dental procedures involving drilling. Develop a theory of how audio analgesia may mitigate severe pain. Use the World Wide Web (and also see Gardner, W.J., Licklider, J.C.R., and Weisz, A.Z. 1960. Suppression of pain by sound, *Science,* 132: 32–33).

9.3 The opioid drugs naloxone and naltrexone are often administered to heroin addicts to prevent death by overdose and as part of a cure for addiction. Give the chemical structure of heroin (diacetyl morphine), naloxone, and naltrexone. Describe how naloxone and naltrexone can act to block the symptoms of heroin and to relieve addiction.

9.4 Suppose a patient under PCA using fentanyl becomes suicidal and decides to end it all in a flurry of button pressing in an attempt to receive a fatal overdose. *Design* a supervisor to interface between the IPFM injector and the patient that will prevent this event. Settings to the supervisor include patient weight, the compartmental pharmacokinetic model for the drug, the bolus dose D_o, and the maximum allowable steady-state drug level. Not only should the supervisor limit the dose, $u(t)$, but it should sound an alarm if an "event" occurs.

9.5 It is common knowledge that mammals sense pain and respond to painful stimuli. See what you can discover about pain sensors in other chordates (fish, reptiles, amphibians). Do insects, crabs, lobsters, and other invertebrate animals sense pain? Note that aversive behavior to a noxious stimulus is not necessarily the result of "pain."

9.6 The pain model of Liu and Northrop will be used to simulate the PCA button-pressing behavior of a model patient experiencing postoperative pain. A Simnon™ *analog* dynamic model will be used to simulate fentanyl pharmacokinetics, the generation and relief of pain, and RPFM for the patient's button-pressing behavior. The same program will also simulate a simple IPFM controller whose input is button presses (treated as impulses) and whose output u is a fentanyl bolus injection if the threshold, phi2, is exceeded by the integrator output. The analog patient model has eight states: three for the fentanyl compartmental pharmacokinetic plant ($x1$, $x2$, $x3$), one for the button-pressing RPFM system ($v1$), two for the noise low-pass filter (fn, ffn), and two for the pain derivative estimation filter ($y1$, $y2$). The IPFM controller has one state for the integrator ($v2$). The threshold for the RPFM model for button pressing is a linear function of the CNS fentanyl concentration, $x0$. The program follows:

```
CONTINUOUS SYSTEM PAIN1  " 9/27/98 Liu & Northrop analog pain model.
STATE x1 x2 x3 v1 v2 fn ffn y1 y2    " Use Euler integration w/ tau = .001
DER  dx1 dx2 dx3 dv1 dv2 dfn dffn dy1 dy2
TIME t
"
"    PK STATE EQUATIONS
"
dx1 = -x1/T1 + u*K1/T1
"
dx2 = -x2/T2 + u*K2/T2
"
dx3 = -x3/T3 + u*K3/T3
"
```

```
"   AUX. EQUATIONS:
xt = x1 + x2 + x3
x0 = DELAY(xt, Td)            " CNS fentanyl conc.
W  = 1/(1 + 19*x0)            " Fraction of unbound pain receptors
dfn = -fn/Tn + SD*NORM(t)/Tn  " Low-pass filter for GWN.
dffn = -ffn/Tn2 + fn/Tn2      " 2nd LPF for noise.
D = DP*exp(-t/Th)*(1 + ffn)   " Pain input to pain neurons.
Q = W*D                       " Neurological pain.
"
dQ  = 1.1E4*(y1 + y2)         " Deriv. estimator for Q, valid from zero
dy1 = -100*y1 - 10*Q          " to 100 r/min.
dy2 = -110*y2 + 11*Q
"
rdQ = IF dQ > 0 THEN dQ ELSE 0  " Patient responds only to increasing
"                                 pain.
PerP = Q + Kp*rdQ             " Perceived pain reacted to.
PHI  = Kt*x0 + P0             " RPFM variable threshold.
"
dv1 = -c1*v1 - z1 + c1*PerP   " RPFM button-pressing model LPF.
z1  = p1*phi/tau              " z1 Resets RPFM LPF to 0.
w1  = IF v1 > phi then 1 ELSE 0
s1  = DELAY(w1, tau)
r1  = w1 - s1
p1  = IF r1 > 0 THEN r1 ELSE 0
bp  = p1/tau                  " Button press bp due to pain experience.
"
"   IPFM INJECTOR:
dv2 = Kc*bp - z2              " Integrator of bps.
z2  = p2*phi2/tau             " z2 resets IPFM integrator to 0.
w2  = IF v2 > phi2 THEN 1 ELSE 0
s2  = DELAY(w2, tau)
r2  = w2 - s2
p2  = IF r2 > 0 then r2 else 0
inj = p2/tau
u   = p2*Do/tau               " Bolus injection of opioid, Do micrograms.
"
"   CONSTANTS
T1:2.33       " min.
K1:0.087      " min/l.
T2:18.2       " min.
K2:0.072      " min/l.
T3:313        " min.
K3:0.908      " min/l.
Kp:5          " Pain rate constant.
SD:3          " Noise SD.
Td:1          " CNS PK Delay time in minutes.
Th:600        " Wound healing TC in min.
Tn:7          " Noise LPF TC
Tn2:14        " 2nd noise LPF TC.
tau:0.001         " Delta t for RPFM pulse gen.
DP:10         " Pain input constant
c1:.2         " c1 = 1/TR for RPFM, radians/min.
Kt:1          " Threshold sensitivity
```

```
P0:1.3       " Fixed threshold for RPFM
Do:10        " Fentanyl bolus dose.
phi2:1       " IPFM drug injector.
Kc:1         " IPFM integrator gain.
"
```
END

Simulate the system for the parameters given over 24 hr (1440 min); observe D, $PerP$, $x0$, $p1$, and $p2$. Manipulate the parameters Kt, $P0$, Kc, Do, and $phi2$ to see if you can minimize the integral of $x0$ *and* the integral of $PerP$ over 24 hr.

9.7 Acupuncture with needles and TENS (Transcutaneous electric nerve stimulation) devices can often mitigate chronic pain associated with wound healing, cancer, arthritis, infection, bone injury, and spinal disc damage. Explain, in terms of the neuroanatomy and neurophysiology of pain, how such pain relief may work.

9.8 Untreated chronic pain can cause physiological stress. Stress in turn produces profound hormonal changes in the body affecting mainly the secretion of adrenocorticotropin, cortisol, and antidiuretic hormone (ADH). Describe these changes and their physiological effects. (See, for example, Chapter 77 in Guyton's *Textbook of Medical Physiology,* 9th ed., 1996, Saunders.)

9.9 Use of opioid drugs such as heroin to "get high" often leads to addiction. However, not all drug users become addicts. Some individuals can quit their habit "cold turkey" with minimum physiological and mental side effects and never return to their use. Develop a theory why some individuals can avoid addiction, while others cannot.

9.10 Investigate the efficacy of using *hypnosis* to mitigate chronic pain.

10 The Human Immune System Seen from a Biomedical Engineering Viewpoint

10.0 INTRODUCTION

In the opinion of the author, the human immune system is, compared to the central nervous system (CNS), the second most complex physiological system in the body. In order to manipulate its components by exogenous means to better fight diseases, we must attempt to understand the vast network of complex, causal interactions among its various components. These interactions are, in general, of a feedback as well as a feedforward nature. The immune system's components are self-regulating; i.e., the system is homeostatic. The components include specialized immune cells, the immunoregulatory *autacoids* they secrete, and the antibodies manufactured. The immune system responds to biochemical signals (autacoids) from immune system cells at the sites of infection, from the CNS, and from other tissues. Certain cellular components of the immune system are mobile and can move throughout the body's circulatory and lymphatic systems. Some of the immune system cells end up being fixed in certain tissues and organs.

The function of the immune system is to seek out and destroy invading bacteria, fungi, molds, viruses, protozoa, spirochetes, parasites, etc. Immune surveillance extends under certain conditions to the detection of cancer cells and their destruction. Immune system macrophages (monocytes) also act to clean up necrotic body tissues damaged by injury or disease. As we shall see, the immune system uses several mechanisms in its mission. These include *humoral* (molecular) *factors,* including antibodies (Ab) and complement, and *direct cellular attack,* including phagocytosis and lysis. Macrophages, natural killer (NK) cells, and cytotoxic T-cells (CTL) participate in the direct cellular attack of hostile invaders. The immune system also can respond to substances to which it should not react. These responses include allergy and autoimmunity. Certain foods, drugs, etc. taken in through the oral route can trigger a variety of adverse immune responses including nausea, headaches, hives, asthma, and anaphylaxis. Allergens such as dust, pollen, mold spores, etc. can also be inhaled, giving rise to rhinitis and asthma. For reasons poorly understood, the immune system can develop a sensitivity to certain normal cells or molecules in the body, creating autoimmune diseases.

This chapter describes the cellular components of the immune system, the major chemical signals it uses (*immunocytokines,* or immunoregulatory autacoids), how these components work and interact, and how they might be manipulated by exogenous means to better fight diseases and cancer (immunotherapy). It will be seen that immunocytokines including lymphokines, interleukins, interferons, prostaglandins, etc. form a complex hormonal regulatory network. Many immunocytokines have multiple effects on multiple target cells (called *pleiotropy*), adding to the difficulty of formulating meaningful mathematical models of the immune system. The problem in describing the role of immunocytokines is especially enigmatic because, in general, we have a poor understanding of the biochemical control mechanisms governing their synthesis and the molecular mechanisms whereby they exert their effects after binding to appropriate (or in some cases inappropriate) membrane receptor molecules.

The *cells* of the immune system are described first, starting with *macrophages*. The various types of *T-lymphocyces* (T-cells, so called because of their embryological origin in the thymus gland) are considered next. *NK cells* follow T-cells. NK cells come from a separate stem cell line; they have basically the same role as CTLs but lack antigen specificity possessed by CTLs and B-cells. *B-cells* (so called because of their embryological origin in the bursa of Fabricus of birds) and their role in manufacturing antibodies are then described. Finally, an attempt is made to make sense of the complex network of signaling substances, the immunocytokines, and their receptors on immune system cells.

Many of the immune system cells have aliases or alternate nomenclature. Early descriptions of immune system cells were based on how they were identified by various stains when prepared for viewing using light microscopy. More recent nomenclature relies on cell function and the specific identifying proteins found on their surfaces.

As mentioned above, a major reason to create effective, validated, mathematical modes of the immune system is to be able to predict immune system behavior in order to evaluate potential, exogenous, adoptive immunotherapies without the initial need to use animal or human subjects. In such therapies, suitable inputs of one or more immunocytokines, specific monoclonal immune cells raised *in vitro*, monoclonal antibodies, etc. might be used to stimulate or strengthen the immune system to fight a specific infection or cancer or to inhibit the immune system's attack on certain body tissues in autoimmune disease. Often, an exogenous adoptive immunotherapy tested *in vitro* produces counterintuitive, paradoxical results, or poor or equivocal results, *in vivo*. Such disappointing results are not unexpected, however, when dealing with an enormous, tightly coupled, nonlinear system in which initial conditions and past history can affect outcomes.

In subsequent sections of this chapter, the reader will be impressed with the enormous number of acronyms used to describe immune system functions. Although guilty of using acronyms, engineers are greatly outclassed in this sometimes frustrating literary peccadillo by the immunologists and biochemists who report on immune system function. An effort will be made to identify immune system acronyms, *pro re nata,* to minimize confusion.

Many excellent, detailed, illustrated references to the immune system and how it works can be found on the World Wide Web. These sites include lecture notes from various medical schools in the U.S. and abroad and commercial Web sites such as the on-line, venerable, *Merck Manual.*[153] A great part of the background material reviewed in this chapter came from Web sites (see Web listings in the Bibliography and References). (Caution: Web sites are ephemera; i.e., they are transitory. When a textbook goes out of print, it often can be found in stable form in a library. When a Web site is closed down, the material is *gone.*)

10.1 IMMUNE SYSTEM CELLS

The human immune system has four major classes of leukocytes, which carry out its mission of protecting the body from infection and exogenous foreign substances (inhaled or eaten): *macrophages, T-lymphocytes, B-lymphocytes,* and *NK cells.* Each of these four classes of cell has several subclasses which are based on the level of maturity, proteins found on the cell membrane, the cells' function, and their location in the body. Other cells such as *mast cells* and *granulocytes* also contribute to the inflammatory immune response, and *megakaryocytes* in the bone marrow make *platelets.* Platelets are 2 to 4 µm in diameter. They are not true cells in that they lack nuclei. However, they have membranes with receptors and contain mitochondria and RNA. Platelets also contain enzyme systems that allow them to synthesize prostaglandins, a class of immunocytokine.

Certain immune system cells can be described as *amplifiers,* responding to foreign proteins, etc. by activating cells of like kind and other immune system components. Other cells can be described as *effectors*; they carry out offensive actions. Still other cells are *modulators*; they regulate the reactivity of the sensor and effector cells. The amplifiers, which include macrophages and B-cells, perform *antigen presentation,* which has the function of causing the proliferation of helper T-cells, CTLs, and NK cells. Certain messenger immunocytokines are secreted by both the antigen-presenting cells and the receiver cells that cause the receiver cells to reproduce or clone themselves. Each daughter B-cell and CTL retains the specificity to bind with the presented antigen. The presented antigen is generally a piece of a coat protein of an invading bacterium, virus, etc. It is by this amplification process that a specific immune response to an invading organism is strengthened.

The normal, total leukocyte density in the human adult is 5000 to 10,000 cells per microliter. About 5.3% of these are monocytes, 30% are lymphocytes, 0.4% are basophils, 2.3% are eosinophils, and 62% are neutrophils. In general, when there is an acute viral infection (e.g., flu), the total leukocyte count is *decreased.* On the other hand, when a person suffers from an acute bacterial infection, the total white cell count is *increased.* Other specific diseases produce changes in the ratios of specifically stained leukocytes. For example, infection by the trichinosis parasite causes a signatory increase in eosinophils, where their ratio rises from about 2.3%

of the total leukocyte count to about 40 to 60%. Other white blood cell ratios stay the same, and the total white blood cell count increases about 30%.[28]

In the following subsections, the immune system's cells are examined in greater detail.

10.1.1 MACROPHAGES

Immature macrophages are called monocytes. They generally circulate freely in the blood and lymphatic system. As monocytes mature, they become more sessile, i.e., they assume fixed locations in tissues and organs.

The immunological functions of macrophages are broad. They are involved in all stages of the immune response. First, they provide a rapid, "frontline," nonspecific, cellular, immune defense against invading bacteria, fungi, and parasites. Activated macrophages can engulf and internalize such invaders. Internal enzymes break down the membranes and proteins of such phagocytosed bacteria, etc. into smaller molecular subunits which can then be used to activate helper T-cells in a complex process known as *antigen presentation*. Macrophages are the central effector and regulatory cells of the inflammatory response. They can secrete more than 100 different effector autacoid substances.[152] Some of these substances, such as hydrogen peroxide, lysozyme, neutral proteases, nitric oxide, etc., kill or damage the hostile target cells as well as normal cells (collateral damage) and cause inflammation. Other substances secreted also act to induce inflammation and then to repair tissues destroyed by infection and inflammation. In summary, like all immune system cells, macrophages are incredibly complex biochemical machines. All of their products and their actions are the subject of internal and external regulatory control by the molecular messengers of the immune system network.

Macrophages (Mϕs) are found throughout the body. They can migrate from the circulatory system and become resident in various tissues. In the liver, they are called *Kupfer cells*; in the brain, they are *microglia*; in the kidneys, they are *mesangial cells*; in connective tissue and skin, they are *histiocytes*; in the lungs, they are *alveolar macrophages*; in the lymph nodes and spleen, there are *fixed* and *free macrophages*; etc.

Macrophages are continually produced from bone marrow progenitor cells. Their rate of production is modulated by certain immunocytokines, notably *monocyte colony-stimulating factor* (MCSF) and *granulocyte–monocyte colony-stimulating factor* (GMCSF). (See Section 10.3 for identification of some immunocytokine abbreviations and their sources and functions.) Those Mϕs which are induced by granulocyte colony-stimulating factor (GCSF) are larger and have a higher phagocytic capacity than Mϕs induced by GMCSF, which are more cytotoxic against certain tumors, express more major histocompatibility complex (MHC) class II antigen molecules on their surfaces, more efficiently kill the bacteria *Listeria monocytogenes*, and secrete more prostaglandin E_2.[152] The very different structures and signal transduction mechanisms of the receptors for MCSF and GMCSF appear to be evidence for different differentiation pathways and thus two different macrophage populations. Immunological studies of the coat proteins of macrophages found in different immunological scenarios (dermatitis, gingivitis, osteoarthritis, tissue graft rejection, tumors, etc.) provide further evidence for functional subtypes of macrophages. Even so, current practice is to classify all macrophages by their level of *activation*.

For modeling purposes, Mϕs have been arbitrarily divided into three groups: *inactivated, primed,* and *fully activated*. In the *inactivated* condition, Mϕs can either be circulating or fixed in tissues. They have low oxygen consumption, little or no monokine (immunocytokine) secretion, and a low level of MHC class II gene expression. (MHC class II is an Mϕ membrane protein necessary for antigen presentation to helper T-cells.) Mature, *inactivated* Mϕs do have phagocytic activity, however, and can respond to chemotaxic signals and can proliferate in response to signals. Inactivated Mϕs are primed by gamma interferon (IFNγ) secreted from stimulated helper T-lymphocytes.

Primed Mϕs have increased oxygen consumption (increased metabolism) and they exhibit enhanced MHC class II expression. Other immunocytokines such as IFNα, IFNβ, interleukin-3 (IL3), MCSF, GMCSF, and tumor necrosis factor α (TNFα) can also prime Mϕs for selected functions. Primed Mϕs can respond to secondary signals to become *fully activated*. In this stage, they cannot proliferate and they have high O_2 consumption. They do maximal secretion of substances for cell killing and inflammation. On the other hand, they show decreased MHC class II protein production, as well as reduced antigen presentation.

Note that there is no sharp, metamorphic distinction between primed and active Mϕs. Among the members of a relatively isolated population of macrophages, such as in the spleen, at any moment one can find a spectrum of Mϕ development levels and capabilities. Such diversity adds zest (and confusion) to the challenge of modeling Mϕ actions.

In summary, Mϕs are seen to have a key role in activating the immune response. They phagocytose and destroy invading bacteria, viruses, parasites, fungi, protozoa, etc. After internal proteolysis, the Mϕs present fragments of the phagocytosed invader's protein coat to activate NK and helper T-cells. These cells respond to a specific protein fragment or epitope bound to the Mϕ's MHC coat protein, and certain immunocytokines are released, which causes the specific responding cells to reproduce themselves clonally. Other immune cells do antigen presentation, but the Mϕs are seen to have a key role in activation of helper T-cells. Mϕs also secrete substances toxic to invading organisms.

10.1.2 T-Lymphocytes: Helper T-Cells, Cytotoxic T-Cells, Suppressor T-Cells

T-lymphocytes originate in the bone marrow from pluripotent hemopoietic stem cells. Their rate of production is under control of the immune system. They are called T-cells because (1) they are "preprocessed" in the thymus gland (hence T-) and (2) they are mostly found in lymphoid tissue, including lymph nodes, the spleen, and submucosal areas of the gastrointestinal tract. The "processing" of immature

T-cells, or thymocytes, in the thymus includes selective deletion of any thymocytes that have a T-cell receptor (TCR) coat protein with an affinity for normal self-antigens. Such selective deletion prevents autoimmunity if portions of normal cell proteins are accidentally presented to T-cells by macrophages, etc.

T-lymphocytes are responsible for *cell-mediated immunity*, which is explained below. The three subsets of T-lymphocytes are characterized by their function in the immune system and further described by the proteins found on their cell surfaces. With the exception of cytotoxic T-cells (CTLs), T-lymphocytes do not directly attack invading organisms. Instead, they serve as activators (amplifiers) or suppressors of the immune response through the immunocytokines that they secrete in response to *antigen (Ag) presentation*. Helper T-lymphocytes are generally given the acronym Th, suppressor T-lymphocytes are Ts, and cytotoxic T-lymphocytes are CTL.

All T-lymphocytes have unique external TCR proteins which may bind to a presented antigenic *epitope* if affinity is high. In addition to the TCR proteins, there is a five-molecule protein complex, CD3, associated with each TCR. The CD3 complex can sense when the TCR has bound to an epitope, and then it initiates a complex sequence of intracellular events collectively called *T-cell activation*. In addition, Th have the CD4 protein complex next to their TCR proteins. CD4 is a coreceptor for MHC class II molecules. CTLs carry the CD8 protein complex next to their TCRs. Ts may also carry CD8 protein. CD8 is a coreceptor for MHC class I molecules found on *all* body cells that do not bear MHC II molecules.

Helper T-cell activation is a complex biochemical process whereby biochemical synthetic machinery in the Th cell reacts to structural changes in the intracellular portion of the CD3 molecular complex brought about by successful Ag presentation by an Mϕ. The Mϕ secretes IL1, which binds to receptors on the activated Th (ATh) cell's surface. Under the combined influence of IL1 and activated CD3, two subsets of ATh cells secrete certain immunocytokines. The ATh1 cells secrete IL2, IL3, and IFNγ; ATh2 cells secrete IL3, IL4, IL5, IL6, TNFα, and GMCSF. These immunocytokines have profound effects on other cellular components of the immune system, which will be described in Section 10.3.

Antigen presentation by infected cells that have MHC class I + Ag on their surfaces to CTLs is illustrated in Figure 10.1. When the CTL's TCR has affinity to the presented Ag, and the CD8 and CD3 molecules react, immunocytokines are released by the CTL and it is stimulated to undergo clonal proliferation. That is, it reproduces itself with the same complex, TCR that has affinity for the antigenic epitope that activated it through presentation. These CTL clones now can "recognize" and bind to other somatic cells infected by the same virus or parasite and carrying the presented Ag on their surfaces. (The invader generally leaves some of its coat proteins on the cell's surface.) Some CTLs that bind to infected somatic cells release a protein called *perforin*, which literally bores nonrepairable holes in the target cell's membrane. Ions and water pass through the holes, causing the target cell to rupture from osmotic lysis. Another means of cell killing is thought to be by the CTL inducing *apoptosis* in the target cell. Apoptosis is an internally mediated self-destruction mechanism that causes the cell and its contents to literally disintegrate

The Human Immune System Seen from a Biomedical Engineering Viewpoint

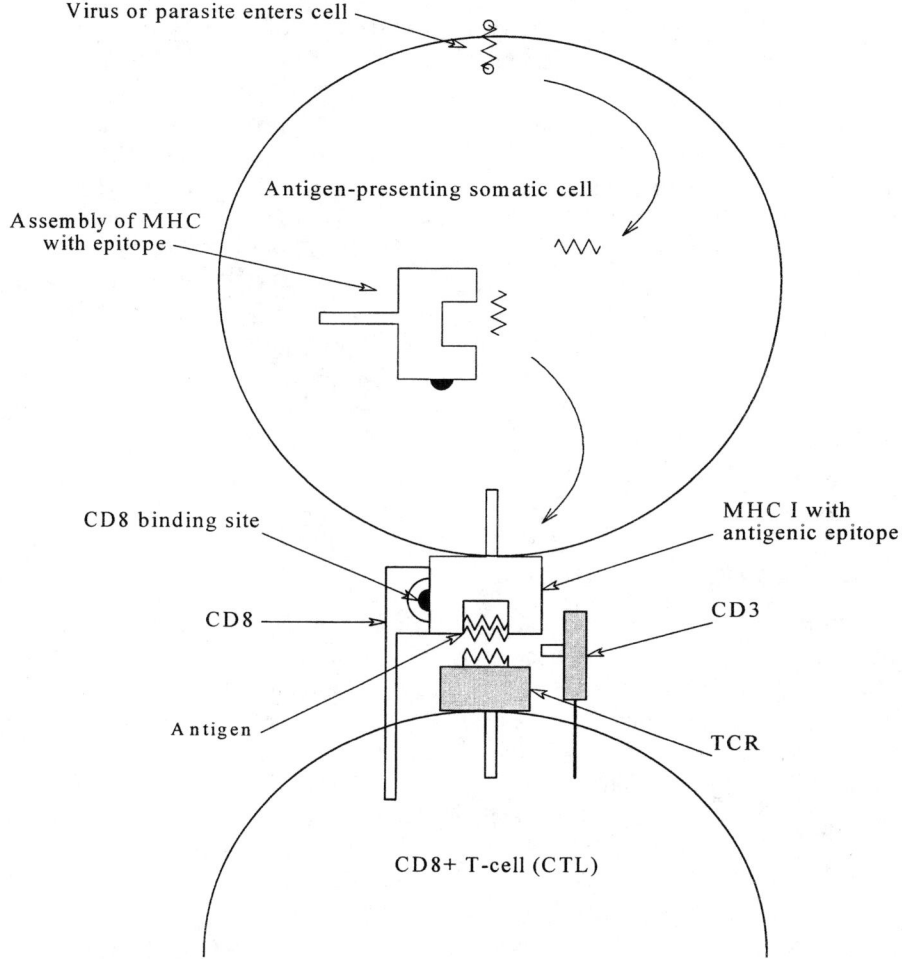

FIGURE 10.1 Schematic of the antigen presentation process by an MHC Type I somatic cell to a CD8+ T-cell (cytotoxic T-leukocyte or CTL). See text for description.

(see Section 10.4). By killing the infected target cell, the CTL prevents the internal virus or parasite from proliferating. Once the infected cell has ruptured, other components of the immune system (macrophages, antibodies, B-cells) attack the now externalized virions or parasites. One might wonder why there appears to be two separate methods for CTLs to kill infected cells. We speculate that redundancy ensures success and that apoptosis, operating from within the cell, causes viral nucleic acids to disintegrate, inactivating them.

Suppressor T-cells (Ts) are a controversial third class of T-lymphocyte. There is some evidence for their existence based on *in vitro* studies. However, it has not been possible to establish stable clones of Ts. It is reasonable to expect that the immune system, a complex network of cells and their signaling substances (immunocytokines, including interleukins, prostaglandins, interferons, and tumor necrosis factors), has developed mechanisms to halt the proliferation of CTLs, NK cells, plasma B cells, and antibodies, once the invading pathogen has been vanquished. This suppression of immune system actions is necessary to conserve immune system resources. The Ts cells apparently carry CD8, CD3, CD11b, and TCRαβ surface proteins, while CTLs carry CD8, CD3, CD28, and TCRαβ.[153]

At UCLA, scientists have recently isolated a naturally occurring immunoregulatory factor which activates Ts (www.research.ucla.edu/tech/97-003a.htm). The active portion of this suppressor T-cell-activating-factor (SCAF) has been synthesized and has a peptide sequence of about 10 amino acids. This synthetic SCAF has been shown *in vivo* to suppress the development of alloantigen-specific immune response and prolong the survival of a heart allograft. The potential advantage of synthetic SCAF is enormous; it may be able to suppress immune system rejection of transplanted organs. By developing a specific monoclonal antibody to the native SCAF protein, it should also be possible to unsuppress immune system action.

10.1.3 B-CELLS AND ANTIBODIES

B-lymphocytes are so-called because of their embryological origin in the organ called the bursa of Fabricus, found in birds (not humans). In humans, B-cells grow from stem cells in the bone marrow and mature in lymphoid tissue of the fetal liver. They are then released into the circulatory system, where they are distributed more or less randomly throughout the body. In humans, there is normally about a total of 10^{12} B-cells.[154] This translates into a plasma density of about 10^7 B-cells per milliliter. Because B-cells are morphologically identical to T-cells, they are identified by their coat proteins. Normally, about 10 to 20% of circulating blood lymphocytes are B-cells.

B-cells are the key effectors in the humoral arm of the immune system. They have the capability of fighting infection by producing huge quantities of specific, freely circulating antibodies (Abs) with affinity to a particular antigen (Ag). The Abs bind to the Ags, inactivating them, or marking them for destruction by NK cells and macrophages.

Each cell in the population of unstimulated, mature B-cells has a unique, surface-bound IgM (Ig stands for immunoglobulin) antibody that has great affinity to bind with a specific Ag. Nature produces millions of different B-cells with IgM molecules, each with a very different affinity to some Ag. When a specific B-cell's IgM Ab (paratope) binds with a soluble Ag, such as diptheria toxoid, the bound Ag is engulfed by the B-cell by the process of *receptor-mediated endocytosis*. The Ag is now digested into fragments which are picked up by MHC II molecules and then displayed on the B-cell's surface. A helper T-cell (Th) with a specific, complementary TCR protein binds to the antigen-presenting B-cell, as shown schematically in Figure 10.2. As in the case of macrophages, the proteins CD4 and CD3 are involved with activating the Th, which secretes IL4, IL5, and IL6. The local release of these interleukins causes the B-cell presenting the antigen to reproduce clonally, copying

The Human Immune System Seen from a Biomedical Engineering Viewpoint 361

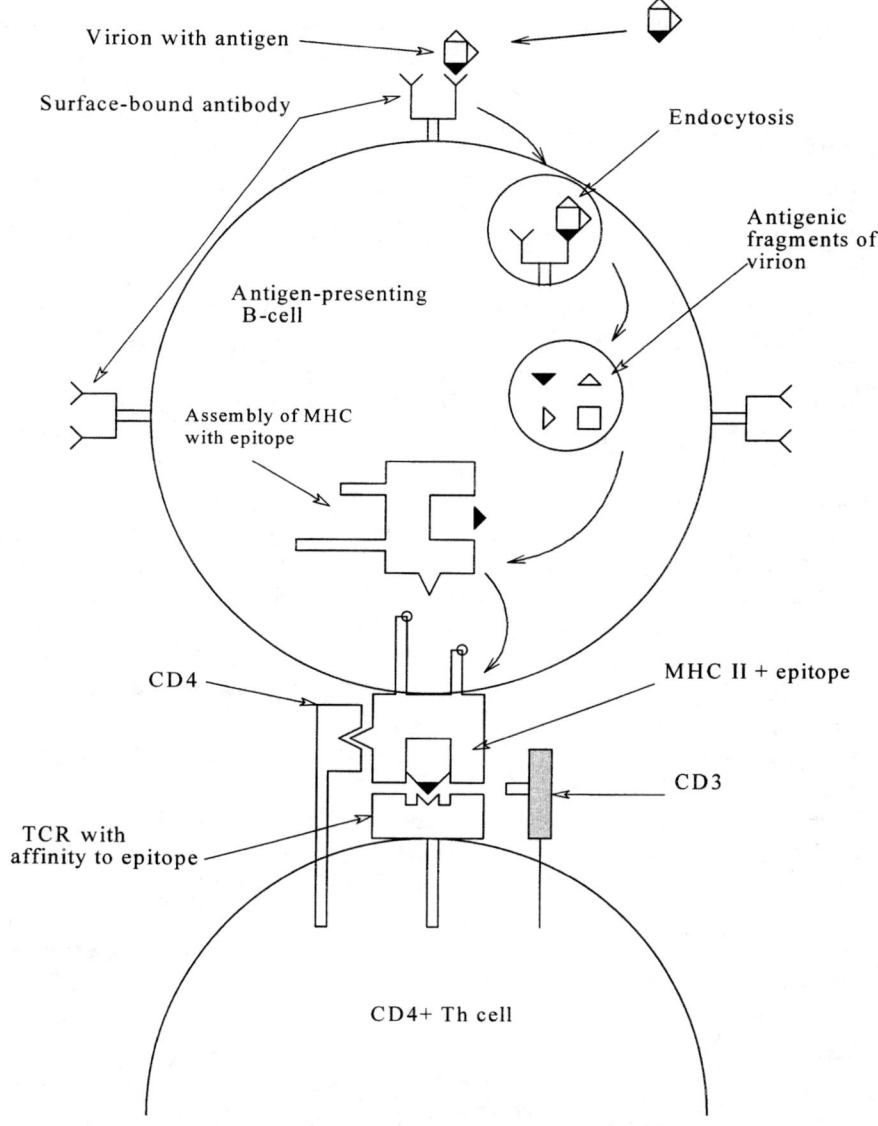

FIGURE 10.2 Schematic of antigen presentation by an MHC Type II B-cell to a CD4+ helper T-cell (Th). In order for the Th to be activated, the Th's TCR must have affinity to the presented antigen, and the CD4 and CD3 molecules must bind to specific sites on the MHC complex.

the specific IgM antibodies (paratopes) that bound to the antigen's epitope. The activated, mature B-cell clones (known as *plasma cells*) then release soluble antibodies, which can bind with free or cell surface-bound Ag. B-cell growth is also activated by IL1 from macrophages and by IL2. What makes this process more complex is that there are five different isotypes of a specific antibody.

The process of antigen presentation to Th by B-cells also leads to the production of special, clonally produced, circulating, inactive B- and T-cells that have long lives and specificity to the particular Ag. These cells are called *Memory B-Cells* (M-cells), and they evidently provide a rapid and strong humoral response if the pathogen with the Ag is reintroduced.

The antibodies produced by activated plasma B-cells have a unique molecular structure: They are "Y" shaped, with the Ag-binding domain (paratope) lying between the arms of the Y. Molecular diversity in encoding the binding domain structure gives the possibility of about 10^6 different Ab paratopes in the human at a given time. The stem of the Y is made from the paired ends of two "heavy protein chains" of over 400 amino acids (AAs) in length. The arms of the Y are also paired, the ends of the heavy chains each with a light chain of over 200 AA. (See Figure 10.3 for an antibody schematic.) As mentioned above, there are five classes of circulating antibodies: IgA, IgD, IgE, IgG, and IgM. The heavy chains that make up these Abs are called alpha, delta, epsilon, gamma, and mu, respectively. The light chains are called kappa or lambda. Genetically randomized, variable regions of the light and heavy chains form the paratope sites, one in each arm of the "Y". Some Abs, however, are formed from a fusion of as many as 10 light and 10 heavy chains and hence have 10 binding sites.[156] The protein structure of the stem region of the Y is relatively constant and contains a *constant Fc binding site* through which NK cells, macrophages, mast cells, etc. can recognize and bind to Abs. If the Ab has bound to a cell-surface Ag, NK cells can bind to the Fc site and then secrete perforin which kills the cell. Macrophages clean the mess up.

Abs can act in three different ways to fight invading pathogens: (1) They can directly bind to the invading pathogen, inactivating it by changing its gross structure. (2) They can activate the *complement system* which destroys the invader. Direct binding of Ab to Ag can lead to large clusters of Abs bound to many Ags; such a cluster is said to be an *agglutination*. Agglutinations can *precipitate*, becoming immobile, and prey for CTLs, NK cells, neutrophils, and macrophages. (3) The Fc regions of Abs bound to Ags on cell surfaces form attachment sites for NK cells, which kill the cells involved.

The complexity of the complement system is one reason the immune system is difficult to model. The complement system consists of about 20 soluble proteins; however, the reactions of the first 11 are best known. The complement system is activated by either the *classical pathway* or the *alternate pathway*. In the classical pathway, an Ab binds to an Ag, and a specific Fc site becomes uncovered. This Fc site can bind with the C1 molecule of the complement system, initiating a complex *cascade* of complement reactions, shown schematically in Figure 10.4. Some of the products of the classical cascade are immunocytokines that have activating effects on immune system cells. For example, the complement component, C3b, strongly stimulates phagocytosis by macrophages and neutrophils. They engulf the bacteria to which the Ab–Ag complexes are attached in a process called *opsonization*. Another important product of the complement reactions is C5b6789. This product

The Human Immune System Seen from a Biomedical Engineering Viewpoint

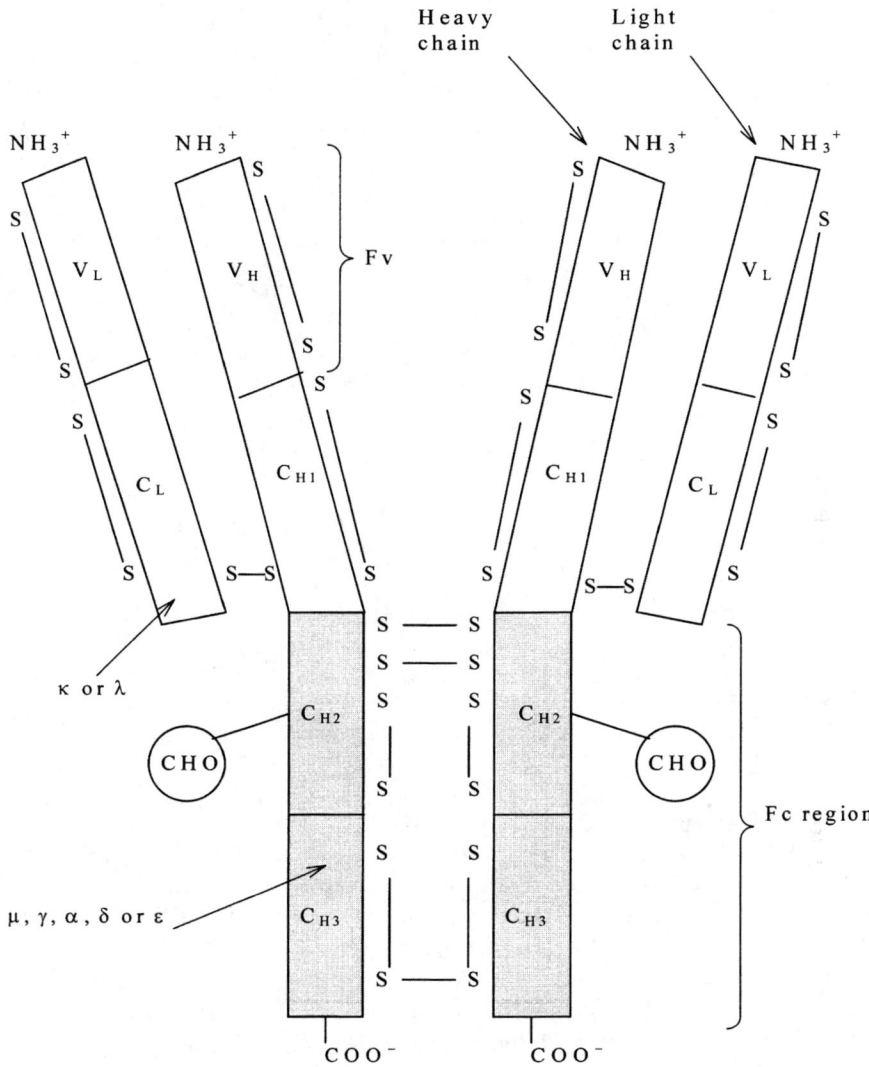

FIGURE 10.3 Schematic of a free (soluble) antibody molecule. See text for description.

does lysis, i.e., it ruptures cell membranes similar to perforin secreted by NK cells and CTLs.

10.1.4 Natural Killer Cells

NK cells represent a small fraction of the total leukocytes in the body. NK cells are unique because they can attack and kill cells in the body which lack normal MHC I cell surface proteins, such as cancer cells or cells infected by a virus. Thus they can

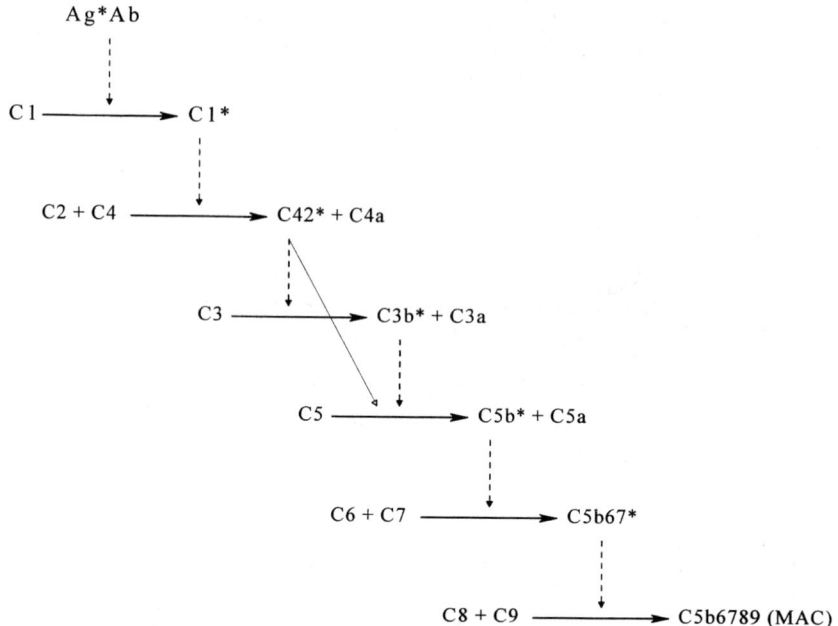

FIGURE 10.4 The complex cascade of complement reactions leading to the membrane attack complex (MAC) molecule, so-called C5b6789. MAC causes cell lysis. See Section 10.1.5 for a description of the complexities of the complement system.

form a fast, direct line of defense against cancer cells which bear altered MHC I proteins, in theory bypassing the need for antigen presentation. In a typical scenario, an NK cell uses an *activating receptor* on its surface to bind to a *critical, cell-surface glycoprotein* on a cell. If another receptor protein on the NK cell's surface binds with a normal MHC I protein on the cell surface, the NK cell is inhibited from secreting perforin or other cell killing autacoids. NK cells may have up to 11 variations of the MHC I binding protein, ensuring that all normal variations in MHC I will be bound.

As already described, the traditional mode of NK cell killing involves antibody-dependent cellular cytotoxicity (ADCC) in which an NK cell binds to a site in the Fc region of an IgG antibody bound to an Ag on an infected cell's surface. When the NK cell binds to the Fc region, it makes contact with the cell surface, then secretes perforin, etc. It is not known whether the MHC I-induced inhibition of perforin secretion is operative in ADCC by NK cells. Presumably, *both* Fc binding and cell surface contact are required for the release of perforin, etc. Otherwise, NK cells would bind to free Abs of any kind and secrete perforin at random, creating havoc for normal neighboring cells.

NK cells are stimulated by growth hormone, luteinizing hormone, IL2, IL12, IL15, and interferons α, β, and γ. Stimulated NK cells secrete IFNγ and TNFα. IL10 inhibits the production of IL2-induced IFNγ by NK cells.

10.1.5 MAST CELLS, PLATELETS, AND COMPLEMENT

Mast cells, platelets, and complement are important accessory immune effectors outside of the classical immune system structure (which includes B-cells, antibodies, T-cells, macrophages, and NK cells). All three contribute to the inflammatory response to infection and, when viewed along with the rest of the immune system, underscore the functional redundancy in the immune system. Mathematical models of immune system function generally ignore the behavior of mast cells, platelets, and complement because of their complexity or, at best, model inflammation as a function of the immunocytokines that induce it.

Mast cells are very important cells in allergic inflammatory reactions and, in particular, asthma. Mast cells are formed in tissues from undifferentiated precursor cells manufactured in the bone marrow and released into the blood. There are two types of mast cells; one is found in connective tissues and the other in mucosal sites. Connective tissue mast cells are found mostly in the skin, while mucosal mast cells reside in the gut and lungs. Both kinds of mast cells are sessile. They collect around blood vessels, nerves, and lymphatic vessels. They are concentrated around potential points of entry of pathogens and foreign microparticles (e.g., pollen, mold spores) into the body.

In humans, mucosal mast cells comprise a total of 1% of the cells located in the lungs and accessory tissues. About half of the mucosal mast cells are found in the intra-alveolar septa, with the other half located in the mucosa of the trachea, bronchi, and bronchioles. Mast cells contain cytoplasmic secretory granules of diverse morphology. Mast cells can release such substances as histamine, proteoglycans (heparin and chondroitin sulfates), leukotriene-C_4, proteolytic enzymes (tryptase and chymase), serotonin, IL4, IL5, IL6, TNFα, and prostaglandin-D_2. All of these substances contribute to the discomfort of an acute allergic reaction. Pain, edema (tissue swelling), inflammation, etc. are the result of mast cell stimulation. Stimulation of mast cells in the lungs can result in acute asthma, with bronchospasm (airway size reduction due to smooth muscle contraction), alveolar tissue swelling (vital capacity reduction), and thickened mucus that clogs small airways.

Mast cells are stimulated to release their granules by the binding of their high-affinity Fc receptors (FcRs) to the Fc region of IgE antibodies that, in turn, have high Fab affinities to allergen molecules such as pollens or danders. There can be as many as 5×10^5 FcRs per mast cell.[59] The actual trigger that initiates the release of mast cell products is the cross-linking of the bound IgE–Ag complexes attached to the mast cell's Fc receptors. Such cross-linking can occur if there are high enough concentrations of antigen and free IgE Abs and the Ag has multiple binding sites on it for the Abs. That mast cells secrete IL4, IL5, IL6, and TNFα means that other immune system cells such as eosinophils and T-cells are recruited into the inflammatory scenario, which is biochemically complex.

As far as is known, no one has incorporated mast cells into a mathematical model of inflammation dynamics; they are too complex, and little is known quantitatively about their behavior.

Platelets have two major functions in the body. One is to form clots to stop bleeding from wounds. Clot formation is a very complex biochemical process which will not be detailed here. The other function is to participate in inflammatory reactions of the immune system. There is normally from 1.5×10^5 to 4×10^5 platelets per cubic millimeter of blood. Platelets are made in the bone marrow by megakaryocytes and then enter the blood. Like red blood cells, they are cells without nuclei, ranging from 2 to 4 μm in diameter. Their resting shape is discoidal; however, when activated by appropriate stimuli, they become spheroidal and develop many protruding "tentacle-like" protrusions up to 5 μm in length. They have typical phospholipid bilayer plasma membranes with many embedded receptor molecules. Their cytoplasm contains a "skeleton" of microtubules, largely formed from actin and myosin, proteins normally associated with muscles. Also within platelets are mitochondria, several types of secretory storage granules, and an endoplasmic reticulum. The half-life of platelets is from 8 to 12 days. Most are destroyed by macrophages residing in the spleen.[59]

Platelet-activating factor (PAF) is synthesized by platelets, mast cells, macrophages, eosinophils, certain renal cells, and vascular endothelial cells. The PAF molecule is shown in Figure 10.5. PAF is a pharmacologically active autacoid with many diverse functions.[61] It causes vasodilation and it is over 1000 times more effective than histamine or bradykinin in promoting edema. PAF promotes platelet aggregation *in vitro* and *in vivo* and is a chemotactic factor for eosinophils, neutrophils, and monocytes, causing these leukocytes to aggregate at the source of its release. PAF also causes the contraction of smooth muscles in the gastrointestinal tract, the small airways, and the uterus. PAF certainly contributes to the edema which occurs when the immune system fights a bacterial infection introduced through the skin.

Complement is another remarkable, concatenated, biochemical system employed by the immune system. The complement system (C-system) consists of more than 20 large glycoproteins, most of which are soluble, some of which are bound to the surfaces of certain types of cells. When the C-system is activated, many of its component molecules are broken down, recombine with other complement molecules to form proteolytic enzymes which further break down other complement

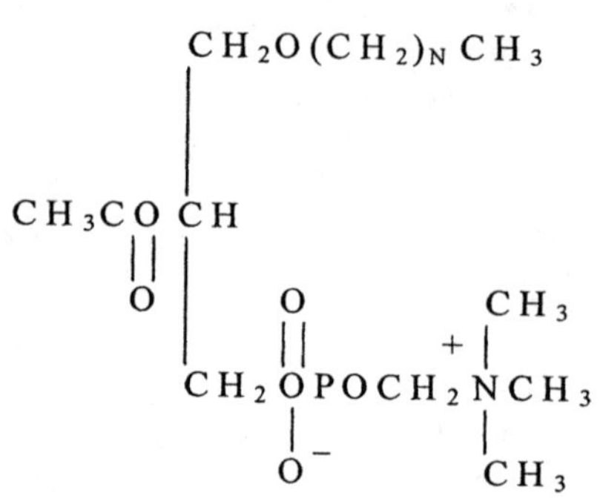

$$CH_3CO-CH \begin{array}{c} CH_2O(CH_2)_N CH_3 \\ | \\ | \\ CH_2OPOCH_2\overset{+}{N}CH_3 \\ | \quad \quad \quad | \\ O^- \quad \quad CH_3 \end{array}$$

FIGURE 10.5 Molecule of the autacoid PAF.

molecules, etc. Most of the complement glycoproteins are synthesized by liver cells; however, macrophages, fibroblasts, and epithelial cells may also produce complement components.[141]

The components of the C-system have four major roles in the immune system: (1) Certain C molecules (C3b and C4b) are *opsonins,* i.e., they opsonize or coat foreign objects bound to Abs. Once opsonized, the complex of foreign object, Abs, and opsonins can be phagocytosed by cells that have CR3 and CR4 opsonin receptors and then be lysed and destroyed. Portions of the foreign object can then be presented on an MHC molecule in "formal" antigen presentation. (2) Certain C molecules (C5a, C4a, and C3a) promote inflammation, increase vascular permeability, cause edema, recruit phagocytic cells (neutrophils and macrophages), and activate platelets which secrete PAF, histamine, etc. (3) The C-system can do lysis of cells, similar to NK or CTL cells. In this baroque scenario, C5b binds to the target cell surface, then to C6, C7, and C8. C7 and C8 undergo a conformational change which exposes hydrophobic domains which penetrate the lipid bilayer of the cell's membrane. The C5b678 complex catalyzes the polymerization of the final C9 component, which, like perforin, makes a permanent, 10-nm-diameter hole in the membrane, causing cell lysis from osmotic shock. The entire C5b6789 molecular assembly is known as the *membrane attack complex* (MAC). (4) The C-system participates in immune complex clearance. That is, the removal of groups of cross-linked Abs bound to Ags. This removal is accomplished by binding C3b and C4b covalently to the immune complex. CR1 receptors on erythrocytes (red blood cells) bind to the C3bC4bAbAg complex. Red cells carry these complexes to the spleen and liver, where they are destroyed by phagocytic cells.[157]

There are two pathways by which C-system activation is initiated or triggered: the *classical pathway,* which begins with AgAb complexes, and the *alternative pathway* in which activation begins at the pathogen surface. Glycoproteins in the classical pathway are prefixed with "C-". One of the reasons the C-system is difficult to understand is that many of its proteins are cleaved to become active component molecules. In the case of C1, it is formed from one unit of C1q, two units of C1r, and two units of C1s. C1q has six binding sites for IgFc. The binding affinity is low, so that at least two bound sites are required on C1q for C1r activation. This means that two adjacent IgGs or one distorted IgM is required. Thus random, unbound, soluble IgG or IgM will not activate complement; they have to be clumped around an antigen. When activated, C1r cleaves C1s to form the *C1s protease,* which in turn cleaves C4 $\rightarrow$ C4a + C4b, and C2 $\rightarrow$ C2a + C2b. C4b combines with C2a to form the *C3 convertase, C4bC2a,* which splits C3 $\rightarrow$ C3a + C3b. Now, at the risk of straining the reader's patience, we note that the complex C4aC2bC3b is formed and acts as *C5 convertase,* producing C5a and C5b. The complex C5b6789 is the MAC that causes cell lysis. Details of the steps in the classical and alternate pathways and their regulation are too complex to describe here; the interested reader should consult a text such as West.[141]

Complement is clearly important in the humoral response to infection, insofar as it works with antibodies to lyse cells and promotes inflammation. The fact that dormant complement molecules are always present in the blood means that the complement response needs little time to generate a local defense against an introduced pathogen. Protein fragments from lysed pathogens can be phagocytosed, and presented as antigens to Th cells by B-cells and macrophages, thus amplifying the classical responses of the immune system.

Because the activated complement proteins can cause extensive collateral damage to "good" cells, the inflammatory action of complement is tightly regulated in order to keep it localized. Regulatory actions are known to act at three points:[157] (1) A C1 inhibitor protein (C1INH) binds to free C1 and inhibits its spontaneous activation. C1INH is released upon activation by immune complexes (AgAb). It also limits the activation of C4 by inhibiting the C1r and C1s proteases. (2) Decay accelerating factor (DAF) found on cell surfaces, C4-binding protein (C4bp), and Factor H all act to speed the breakdown of C3 convertases, breaking the chain of reactions leading to MAC. (3) Two proteins found on cell surfaces, CD59 and Homologous Restriction Factor (HRF), inhibit the binding of C9 to C5b678 to form MAC. The factors controlling these inhibitors are not known.

Because of its complexity, the C-system has never been included in mathematical models of immune system function.

10.2 ANTIGEN PRESENTATION

Antigen presentation is a complex process whereby the immune system activates itself to fight specific invading microorganisms, including viruses, bacteria, parasites, molds, etc.

Antigens (Ag) are by definition molecules that elicit an immune response in the body. They are generally proteins and polysaccharides. Nonprotein molecules called *haptens* can cause an immune response as well. This response is generally thought to be due to the hapten reacting chemically with certain tissue molecules; the product of the reaction is the antigen. Antigens can be of *exogenous* or *endogenous* origin. Exogenous antigens can be *inhaled* proteins from animal dander, pollen, or mold spores. Exogenous antigens can also be *eaten*; nuts and shellfish are common foods to which people are allergic. Another route for internalization of antigens is *through the skin and mucus membranes*. These can include injected antigens such as used in allergy desensitization, pollen injected by a thorn stick, bacteria entering through a cut or abrasion, parasites that are injected by insects or that burrow into the skin, and poison ivy oil. Proteins from the cell walls of internalized bacteria, viruses, spirochetes, parasitic worms, etc. can activate immune responses.

Endogenous antigens include those which originate within the cells of the body. In the case of *autoimmunity,* a normal somatic cell protein is misidentified by some component of the immune system as a foreign invading protein. In other cases, cancer cells may have mutant proteins on their surfaces that the immune system can recognize and attack.

Antigen presentation is a crucial step in the amplification of specific immune responses. All antigen presentation involves the binding of an antigen molecule to a large protein molecule present on the antigen-presenting cell's surface. This molecule is called the *MHC* molecule or, equivalently, the human leukocyte antigen (HLA) molecule. Part of the MHC molecule projects through the antigen-presenting cell's plasma membrane into its interior. There are two major types of MHC molecule: Type I and Type II. Nearly all nucleated cells in the body carry MHC I proteins and can present antigens derived from foreign internal proteins. The *Type I MHC molecule* has three major parts: (1) a transmembrane protein which is exposed at the cell surface, the outermost portion of which is made from two alpha helices that form a groove between them; (2) a short peptide molecule attached in the groove between the alpha helices; and (3) a β_2 microglobulin molecule also attached to the alpha helices. To make matters more complex, humans have three subtypes of MHC I molecules, designated HLA-A, HLA-B, and HLA-C. The genes coding the structures of these molecules are inherited, so that if a person is heterozygous for HLA-A, HLA-B, and HLA-C, then *six* different MHC I proteins will be observed in such a person. MHC I proteins are found on all endogenous nucleated cell surfaces in the body, other than macrophages, neutrophils, and certain B lymphocytes

The MHC I processing pathway begins *inside* the MHC I cell, where peptides from a virus' coating or a parasite that has entered the cell are broken up by proteolytic enzymes from proteasomes in the cytosol. The peptide fragments, between 8 and 18 amino acids, are then transported into the lumen of the rough endoplasmic reticulum (RER). Special transporter molecules move the peptide fragments inside the RER lumen, where ribosomes form the MHC I molecules around them. The assembled MHC I–peptide complex is then transferred to the cell's plasma membrane by Golgi apparatus.[158] The MHC I–peptide complex is externalized through the plasma membrane; however, one transmembrane protein "root" projects back through the plasma membrane into the cell's interior. The three subunits in the externalized, active MHC I molecule are: (1) the "presented" antigenic peptide, (2) the β_2 microglobulin, and (3) the transmembrane polypeptide.

Antigen presentation from cells carrying active MHC I molecules is made to leukocytes that carry the CD8 surface glycoprotein, as well as receptor proteins for the MHC I complex. CD8+ leukocytes are generally CTL. The TCR protein for MHC I molecules and adjacent CD8 protein molecules project through the surface of the CTL. Antigen presentation is only possible if the TCR protein has an affinity for the presented antigenic peptide nestled in the MHC I cleft, *and* the CD8 protein on the CTL binds to the MHC I side site. (See Figure 10.1 for an illustration of this process.) Once the CD8 and TCR proteins have bound to the antigen-presenting cell's MHC I–peptide epitope complex, an internal, biochemical "message" is sent to the CTL, activating it to destroy the antigen-presenting cell. Destruction is necessary to prevent eclosion of reproduced virions or parasites from within the cell. The activated CTL either perforates the cell membrane of the infected cell with a special protein it secretes, or it sends chemical messengers to the infected cell that cause it to self-destruct by a process known as *apoptosis* (see Section 10.4).

The affinity match between the presented antigen and a TCR protein is accidental. That is, many billions of possible TCR configurations are randomly generated on CD8+ CTLs. The CTL receiving the message is also stimulated to divide, exactly reproducing a *clone* of CTLs with the TCR protein specific for the presented antigen. In this way, the immune system amplifies its ability to fight specific, internal viral or parasitic invaders.

A second kind of antigen presentation is done by cells which carry the *Type II MHC protein*. The Type II MHC protein is present on the surface of mature macrophages, and other phagocytic cells such as neutrophils and certain B-lymphocytes (B-cells). These cells can engulf, phagocytose, or endocytose an entire bacterium or virus, as well as fragments of cell debris. Once internalized, the foreign material is broken down enzymatically to smaller peptide fragments which are taken up by MHC II molecules, which in turn migrate to the cell surface with the presented peptide fragment antigen or epitope. When the antigen-presenting macrophage or neutrophil encounters a helper T-cell with a TCR with affinity to the presented antigen, and the Th's CD4 protein binds to a specific site on the MHC II molecule, signals are passed to the Th to activate it to secrete various immunocytokines that activate the cellular arm of the immune system. In particular, AThs secrete IL2, IL3, IL4, IL5, IL6, IL7, IL10, and IL13, also TNFβ, γIFN, and GMCSF. Antigen presentation by a macrophage is illustrated schematically in Figure 10.6.

The antigen-presenting B-cells have B-cell receptor (BCR) proteins bound on their outer surfaces. The BCRs are surface-bound IgG antibodies; they, too, have enormous variability in their affinities for epitopes. When a certain BCR has a strong affinity to an epitope, it binds to that epitope, and then is internalized where the Ag protein is broken down, and pieces of it are bound to MHC II proteins, which are then externalized on the B-cell. If a Th has a TCR protein with affinity to the peptide epitope nestled in the B-cell's MHC II protein, it binds to the antigen-presenting B-cell. A CD4 molecule on the Th must also bind to the side of the MHC II molecule to activate the Th, which secretes cytokines that cause the antigen-presenting B-cell to reproduce clonally with its BCRs that are specific for the Ag in question. This process is shown schematically in Figure 10.2. The clone of plasma cells grows identical BCRs with the affinity to the presented epitope. These BCRs are ultimately released as free antibodies. Since the original BCR bound to this epitope, the antibodies released are also specific for it and contribute to an amplified, humoral, immune defense.

10.3 IMMUNOCYTOKINES: THEIR SOURCES, EFFECTS, AND TARGETS

Immunocytokines are proteins and glycoproteins secreted mostly by immune system cells; however, as we shall see, under certain conditions other cells secrete substances that can also affect immune system cells. Collectively, these are called intercellular messengers *autacoids*, a term coined by Goodman and Gilman.[61] Autacoid is from the Greek *autos* = self and *akos* = medicinal agent or remedy. There are currently about 30 known immunocytokines (also known as cytokines). Immunocytokines include *interleukins* (17 are defined and more are being discovered every year), *interferons*, certain *prostaglandins, tumor necrosis factors*, various

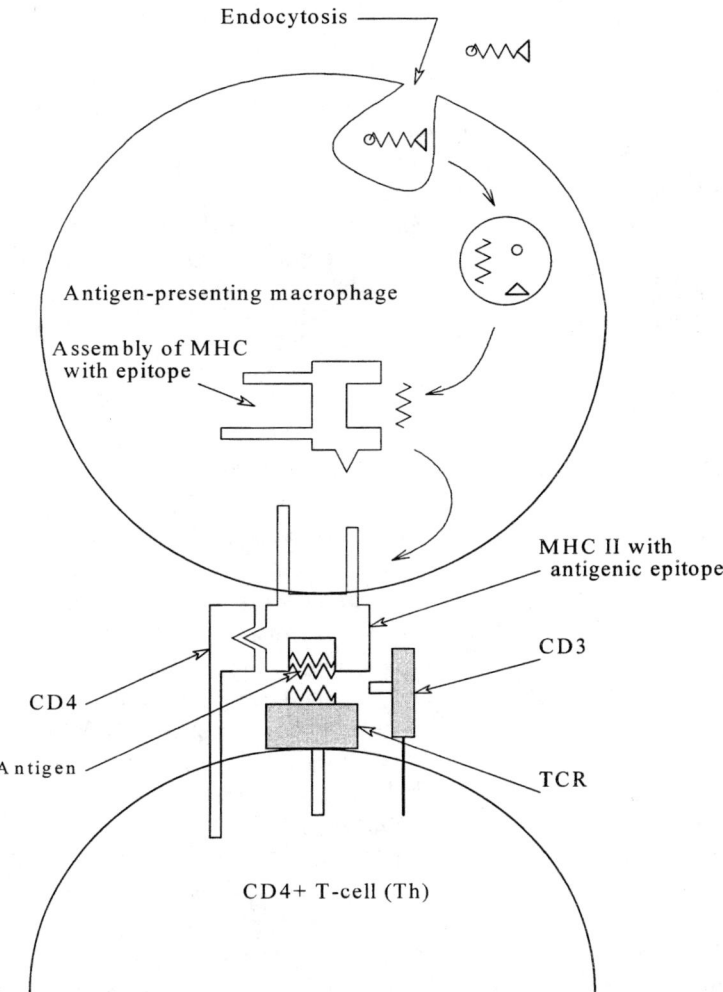

FIGURE 10.6 Schematic of antigen presentation to a CD4+ Th cell by an MHC Type II macrophage. Note that full Th activation requires not only that the TCR have affinity to the presented antigenic epitope, but also that CD4 and CD3 bind to sites on MHC II.

cell growth-stimulating factors, and certain *CD proteins.* Any one of these many and diverse substances can collide with an immune system cell where it can bind with a specific surface receptor protein (if such receptor exists). Once bound, a *molecular transduction mechanism* causes the cell to internally synthesize the same or other immunocytokines, more of the same or other membrane receptors for immunocytokines, proteins used in antigen presentation (TCRs, MHC, CD3, CD4, CD8, etc.), antibodies (if the cell is a B-lymphocyte), complement components, perforin, etc., or the binding can lead to the inhibition or switching off of an ongoing biochemical synthesis by the target cell. Individual immunocytokines can possess the property of *pleiotropism,* in which a cytokine can have different effects on different target cells that have receptors for it. Also, several cytokines can have the same biological function on the same or diverse target cells.

Presumably Ts (if they exist) synthesize immunocytokines that turn off or downregulate the cellular synthetic machinery that was activated in the initial stages of the immune response. Very important in incorporating immunocytokines into a quantitative, dynamic model of the immune network is a knowledge of the kinetics of their production, and their half-lives *in vivo*. Sadly, much of this information is nonexistent or is based on *in vitro* studies.

The immunocytokines are described in detail in the following subsections.

10.3.1 THE INTERLEUKINS

Interleukins (ILs) are high molecular weight glycoprotein autacoids secreted by leukocytes for purposes of signaling other leukocytes and somatic cells. The ILs have diverse functions such as cell attraction (chemotaxis), inducing target cells to secrete other ILs and other immune system autacoids and to manufacture more receptors for ILs. Most ILs are stimulatory in function; there are a few, however, that inhibit or downregulate certain immune system functions.

There are about 17 interleukins as of this writing. However, the ongoing research on the immune system reveals the identity of new cell surface proteins, receptors, and immunocytokines at a rate such that our knowledge base for the immune system is *never* in the steady-state; new things arise at an amazing rate. Thus our knowledge about the ILs and their origins and functions is continually changing. The descriptions below summarize the current knowledge.

IL1 is synthesized by activated macrophages that are presenting antigen to Th cells. There are *two* forms of IL1: IL1α and IL1β, with the same activity but not identical structures. IL1α is primarily bound to Mϕ surface membrane; IL1β circulates. Both forms bind to the same receptors. IL1 receptors are found on T- and B-cells, other macrophages, neutrophils, bone marrow cells, fat cells, bone osteoclasts, brain cells, cells of the adrenal gland, vascular endothelial cells, and smooth muscle cells. IL1 stimulates B-cell proliferation and T-cell production of certain other immunocytokines. High levels of IL1 in the blood can induce lethargy and fever, induce the wasting syndrome (*cachexia*), and trigger inflammation and tissue damage. Adrenal corticosteroids inhibit the actions of IL1; hence, prednisone is often used to reduce inflammation and swelling in certain diseases.

IL2 is also called *T-cell growth factor*. IL2 is made mostly by activated T-cells (ATh) and CTLs; it is a 15 kDa glycoprotein. IL2 synthesis requires a costimulus; in addition to Ag + MHC, IL1 or CD28/B7 is involved. Synthesis starts within hours of stimulation, peaks within 24 hr, and then falls off. Upon activation, T-cells make not only IL2 but also IL2 receptor (IL2R) proteins.

IL2 causes CD4+ T-cells (Th) to proliferate and causes CD8+ T-cells (CTL) to proliferate. It stimulates production of IFNγ by T-cells, which in turn stimulates macrophage production of IL1. IL2 directly stimulates Mϕs to secrete IL1 and

TNFα and to become cytotoxic. IL2 also stimulates the production of NK cells and lymphokine-activated killer (LAK) cells. Obviously, Th, CTLs, B-cells, and macrophages have IL2 receptors. If high levels of exogenous IL2 are given, the result is increased capillary permeability, leading to hypotension, ascites, generalized body edema, and pulmonary edema. Thus the use of high-dose IL2 as an immune system stimulator has been of questionable value.

IL3 is an early-acting hematopoietic growth factor, also known as multicolony-stimulating factor and mast cell growth factor because of its diverse effects in stimulating precursor immune cell growth. It also stimulates B-cell differentiation while it *inhibits* lymphokine-activated NK (LAK) cell activity. IL3 molecular weight ranges from 14 to 36 kDa. Mast cells under the influence of IL1, IL3, IL4, and IL9 secrete IL3. T-cells under the influence of ILs 1, 2, 4, 6, 7, 12, 15, and 16 and TGFβ secrete IL3. IL3 receptors (IL3R) are found on mast cells, B-cells, and some mature granulocytes; mature neutrophils lack IL3 receptors.

Exogenous IL3 given as an intravenous bolus to normal human volunteers produced a transient increase in neutrophil count which began within one-half hr, peaked at about 4 to 6 hr, and returned to the steady-state level after about 12 to 24 hr. IL3 injected subcutaneously daily for 2 weeks produced a gradual increase in lymphocytes, monocytes, neutrophils, and eosinophils; the eosinophils increasing by a factor of up to 50.[164]

IL4 is a 15- to 19-kDa, glycosylated protein dimer. Its production is highly regulated; it is made by CD4+ Th cells, CD8+ memory T-cells, mast cells, and basophils. It induces CD4+ Th cells to differentiate into Th2 cells while suppressing the development of Th1 cells. IL4 also acts as a growth factor for B-, T-, and mast cells; stimulates MHC II expression on B-cells (for B-cell antigen presentation); and promotes plasma cells to switch Ig production to IgE1 and IgE Abs. IL4 receptors are found on T-cells, B-cells, macrophages, fibroblasts, and endothelial cells.

IL5 has also been known as T-cell replacing factor, eosinophil differentiation factor, and B-cell growth factor II. It exists as a 45 kDa dimer linked by disulfide bonds. It is produced by CD4+ Th2 cells, NK cells, and mast cells. IL5 stimulates the proliferation of immature B-cells, causes the activation and differentiation of eosinophils, stimulates Ig class switching to IgA, and stimulates mast cells. It may be considered to be involved in the stimulation of the humoral inflammatory response.

IL6 is a ubiquitous cytokine that has a number of aliases, underscoring its high degree of activity in the immune system. IL6 is also known as IFN-β2, hybridoma/plasmacytoma growth factor, hepatocyte-stimulating factor, B-cell-stimulating factor-2 (BSF-2), and B-cell differentiation factor (BCDF). IL6 is a 212-amino-acid glycoprotein with a molecular weight of about 26 kDa. It is made by a wide variety of immune system cells: macrophages, mast cells, T-cells, fibroblasts, and neutrophils. IL6 receptors are found on stimulated B-cells, macrophages, hepatocytes, CD4+ and CD8+ Th cells, and fibroblasts. IL6 stimulates acute phase protein synthesis by the liver. It acts as a growth factor for B-cells and induces their final maturation into Ab-secreting plasma cells. It is involved in T-cell activation and differentiation and stimulates their production of IL2 and IL2 receptors. Interestingly, IL6 inhibits the production of TNF, limiting the acute inflammatory response. Glucocorticoids, IL4, and IL13 inhibit the production of IL6 in monocytes (macrophages).

IL7 is also known as T-cell growth factor. It is a glycoprotein with a molecular weight of about 25 kDa. It is made by cells in the bone marrow and by thymus stromal cells. IL7 stimulates the development of pre-B- and T-cells and early thymocytes.

IL8 is produced by activated T-cells, macrophages, and lymphocytes. It is a chemotactic agent (attractant) for macrophages, neutrophils, and basophils. The process of chemotaxis can attract the affected cells over a distance of about 100 μm to the site of inflammation and release of the chemotactic substance. IL8 also stimulates granulocytes to release granules and the production of leukotrienes (inflammatory agents).

IL9 is made by CD4+ Th2 cells as well as by some B lymphomas. Its production is dependent on Th2 cells receiving IL2, IL4, and IL10. IL9 *inhibits* the production of lymphokines by CD4+ Th cells producing IFNγ; it also stimulates the growth of CD8+ T-cells. It also promotes the production of immunoglobulins by B-cells and the proliferation of mast cells.

IL10 is also known as B-cell-derived T-cell growth factor or *cytokine synthesis inhibitory factor* (CSIF). It is an 18-kDa glycoprotein made by CD4+ T-cells, activated CD8+ T-cells, and activated B-cells. IL10 *inhibits* IFNγ production by activated T-cells; it also *inhibits* IL2-induced IFNγ production by NK cells. IL10 *reduces* antigen-specific T-cell (CTL) proliferation, and it *inhibits* IL4- and IFNγ-induced MHC II expression on macrophages. Since IL10 can be produced by Th2 cells and it *inhibits* Th1 cells by preventing them from secreting their cytokines, especially IFNγ, it has been called an "anticytokine." Clearly, IL10 would be a key cytokine to be secreted by a putative suppressor T-cell line. Interestingly, the Epstein–Barr retrovirus produces a homolog to human IL10; vIL10 is 84% homologous, and acts to suppress the immune system's natural response to Epstein–Barr virus infection. Neat.

IL11 is a 26-kDa cytokine produced by bone marrow stromal cells, macrophages, and some fibroblasts. It is a functional homolog to IL6 and mimics the activity of IL6. In addition, IL11 can stimulate Th-dependent B-cell immunoglobulin secretion, increase platelet production, and induce IL6 expression by CD4+ T-cells. IL11 is a hemopoietic growth factor (HGF), including increasing the rate of platelet production.

IL12 has also been called NK stimulatory factor (NKSF) and cytotoxic lymphocyte maturation factor (CLMF). It is a disulfide-linked, heterodimeric cytokine with two distinct subunits of 40 and 35 kDa each. IL12 is secreted by activated B-cells and antigen-presenting cells (macrophages and B-cells). Its production is inhibited by IL4 and IL10. Also, IL12's activating effect on Th1 cell development is inhibited by IL4, which stimulates Th2 cell development. IL12's activities include the increased rate of production of CTL, NK, and LAK cells; increased NK cell cytotoxicity, induction of IFNγ by NK cells and T-cells, and inhibition of IgE synthesis by B-cells.

IL13 is another cytokine primarily involved in regulating the humoral inflammatory response. IL13 is produced by activated Th cells (Ts cells?). It is a 12 to 17 kDa protein which acts on macrophages to *inhibit* the activation and release of their inflammatory cytokines, e.g., IL1β, TNFα, IL6, and IL8. IL13 enhances macrophage and B-cell differentiation and proliferation, increases CD23 expression, and induces IgG_4 and IgE class switching.[165]

IL14 is a 53-kDa protein product of T-cells. Like IL4, it induces B-cell proliferation; however, it also *inhibits* their differentiation into plasma B-cells, thus inhibiting the secretion of Igs.

IL15 is a protein that contains 114 amino acids with a molecular weight of approximately 15 kDa. It is produced by activated macrophages, epithelial cells, and fibroblasts. IL15 shares many biological properties with IL2 as an immune system activator, including the stimulation of NK, B-, and T-cells. IL15 also stimulates CTL and LAK cell activity as well as Ig production; it is also an attractant for T-cells. IL15 does not have any sequence homology with IL2, although the IL15 receptor (IL15R) protein shares two chains in common with the IL2R (β and γ), but has a unique α chain.

IL16 consists of four linked homotetrameric chains with a molecular weight of 56 kDa. IL16 was originally known as lymphotactin or lymphocyte chemoattractant factor. Evidence suggests that IL16 exists preformed in CD8+ T-cells and is released in response to histamine or serotonin (5HT). IL16 is also released by eosinophils. It is not made by CD4+ T-cells. CD4+ T-cells and eosinophils are strongly attracted by IL16; macrophages are less strongly attracted, and it has no taxic effect on neutrophils. All the cells that IL16 attracts exhibit CD4. In fact, CD4 is the receptor molecule for IL16. Thus the effects IL16 induces in its target cells are mediated through CD4 molecules. IL16 treatment of monocytes leads to increased MHC II expression, similar to, but independent from, that induced by IFNγ. IL16 has been shown to inhibit HIV replication in monkey and human CD4+ T-cells; the mechanism is unknown.

IL17 is one of the newest (1995) immunocytokines to be described. A 17.5 kDa protein, IL17 is the product of activated CD4+ T-cells, and its biological activities are diverse. IL17 can induce the release of IL6, IL8, G-CSF, and PGE_2. TNFα and IFNγ were found to act additively with IL17 on the release of IL6, while neither IL17 nor TNFα induced the release of GM-CSF; however, the combination of IL17 and TNFα was effective in releasing GM-CSF.

By the time you read this section, more interleukins will have been discovered, and more of their synergistic, pleiotropic actions in the immune network will have been clarified. Note that some interleukins act locally at the site of infection, while others circulate more generally in the body. To effectively signal, or carry *information* to immune system cells, all interleukins must eventually be broken down to inactive forms and their amino acids recycled metabolically. *In vivo* interleukin half-lives are generally unknown, but are necessary for formulating any sort of dynamic mathematical model of the immune system.

10.3.2 THE INTERFERONS

Three types of interferon are currently known: IFNα, IFNβ, and IFNγ. IFNα is thought to have 15 subtypes, which presumably act alike, so we can deal with generic IFNα.[160] IFNα is made by virus-infected monocytes and lymphocytes, IFNβ is from virus-infected fibroblasts, and activated T- and NK cells release IFNγ.

All three interferons have antiviral activity; they increase NK and CTL cell activity and increase MHC I expression on cells. Specifically, IFNα and IFNβ activate NK cells and CTLs, stimulate B-cell differentiation into plasma cells (which release soluble antibodies), induce MHC I protein kinase, and cause other internal biochemical events that inhibit viral messenger RNA translation, preventing viral multiplication in the target cell. IFNα is the first cytokine to be used effectively in clinical trials as an exogenous immunotherapy agent; it has been approved for the treatment of several types of human cancer. It has been demonstrated that IFNα has a direct antitumor action: It downregulates oncoproteins, induces tumor suppressor genes, has antagonistic effect against the action of growth factors, induces cell suicide (apoptosis), inhibits angiogenesis (growth of new blood vessels that nourish the tumor), and also increases MHC I production and tumor-associated antigens.[160]

IFNβ1a and IFNβ1b have both been successfully used to treat the autoimmune disease multiple sclerosis. Both interferons slow the progress of disability, reduce the rate of relapses, and reduce the severity and number of MRI-detected lesions. They do not cure multiple sclerosis, however.

IFNγ increases expression of MHC II molecules and receptors for the Fc region of antibodies (FcγR) on macrophages, increases the activity of neutrophils and NK cells, promotes T- and B-cell differentiation, and increases IL1 and IL2 synthesis. In addition, IFNγ increases type IgG_{2a} antibodies while suppressing IgE, IgG_1, IgG_{2b}, and IgG_3 synthesis. IFNγ also inhibits proliferation of Th2 cells, but not Th1 cells.

All exogenously administered interferons risk losing their effectiveness due to the production of host antibodies.[161] It is ironic that a potentially effective immunotherapy may be thwarted by the host's immune system.

10.3.3 TUMOR NECROSIS FACTORS

Two TNFs are known: TNFα and TNFβ. The principal effect of the TNFs is inflammation.

TNFα, or cachectin, is a 17-kDa soluble protein trimer. It is the product of activated macrophages, fibroblasts, mast cells, and some T- and NK cells. Peritoneal mast cells have preformed, reserve TNFα available for immediate release upon appropriate stimulation. TNFα can induce fever, either by stimulating the release of pyrogenic prostaglandins or by causing the release of IL1, which is also pyrogenic. TNFα plays a major positive role in fighting local infections. It induces the production of acute-phase proteins, mobilizes neutrophils, activates T- and B-cells, increases the release of antibodies and complement, increases the adhesion of platelets to blood vessel walls, and increases the extravasation of lymphocytes and macrophages (*diapedesis*) to fight the infection in intracellular space. These actions result in the phagocytosis of the pathogens, local vessel occlusion, and the drainage of cells, debris, and fluid into the lymphatic system. These actions, carried out with other relevant cytokines, lead to the removal of the infecting pathogen and eventual tissue repair. TNFα also can bind to TNF receptors on tumor cells and kill them. Some tumor cells shed their TNF receptors, which become soluble and bind to TNFα, inactivating it. Excess TNFα can also lead to the body creating antibodies to it, giving similar deactivation. The mechanism of infected cell killing by TNFα is not known.

High levels of TNFα cause pain, systemic edema, hyperproteinemia, and neutropenia. TNFα binds to TNF receptors on virus-infected cells, causing apoptosis and cell death. Acute infections can lead to the overproduction of TNFα, leading to fever, immune suppression, septic shock due to the loss of blood volume to extracellular space, fatigue, anorexia, and cachexia (wasting of the body tissues).

TNFβ, or **lymphotoxin**, is made by activated CD4+ and CD8+ T-cells. It binds to the same receptor sites as TNFα. It has similar properties to TNFα and induces apoptosis in many types of virally infected cells, tumor cells, and damaged cells. Certain endotoxins, such as that from staphylococcus bacteria, cause high production of TNFs, which contributes to the onset of toxic shock syndrome. Chronic high production of TNFs may be responsible for the cachexia observed in many chronic parasitic infections and some cancers.

10.3.4 PROSTAGLANDINS AND OTHER EICOSANOIDS

The prostaglandins are part of a family of autacoids that includes the leukotrienes, thromboxanes, and prostacyclin. Collectively, they are called *eicosanoids* because they are synthesized from 20-carbon fatty acids that are in turn derived from the 20-carbon *arachidonic acid* (see Figure 10.7). The metabolites of arachidonic acid are varied and have diverse pharmacological effects. Subtle changes in eicosanoid structure can produce dramatic alterations in their bioactivity. Receptors for eicosanoids are highly specific for affinity and effect.

Two major biosynthetic pathways exist for the products of arachidonic acid. The prostaglandins, prostacyclin, and thromboxanes are made through the *cyclooxygenase route*. The second major pathway of arachidonic acid metabolism begins with *lipoxygenase*. This pathway leads to the biosynthesis of various leukotrienes. These pathways are shown in Figure 10.8.

The discussion of the immune system's use of eicosanoids as effectors and signaling substances will be restricted to the consideration of those autacoids considered by the author as having a *direct effect* in the functioning of the immune system. First, it is relevant to remark that the eicosanoid immunocytokines are short-lived. They are chemically unstable and are rapidly broken down in the body (mostly in the pulmonary circulation) to inactive forms. Their half-lives are thus on the order of minutes, and they must be continually synthesized to have effective concentrations.

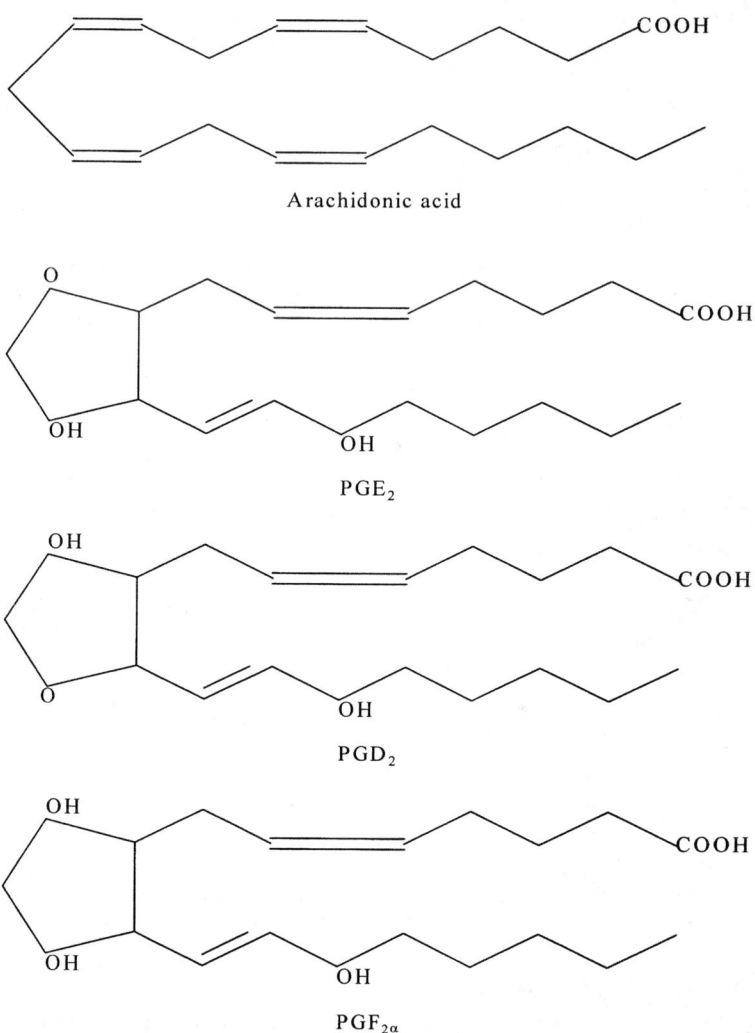

FIGURE 10.7 Some prostaglandins that are synthesized from the 20-carbon arachidonic acid molecule. These eicosanoids include PGE_2, PGD_2, and $PGF_{2\alpha}$.

PGE_2 is formed by macrophages and granulocytes; **PGD_2** comes from mast cells. Both of these PGs are formed through the cyclooxygenase pathway, followed by an arcane series of biochemical reactions. PGE_2 causes the relaxation of smooth muscle, both in blood vessels and in the lungs; thus it acts as a bronchodilator when given as an exogenous aerosol. PGE_2 and PGI_2 enhance edema formation, the infiltration of leukocytes into the site of infection/inflammation, and they potentiate the production of pain by bradykinin.[61]

Most importantly, PGE_2 also acts as an immune response suppressor substance. Both PGE_2 and PGD_2 decrease the release of proteases from stimulated

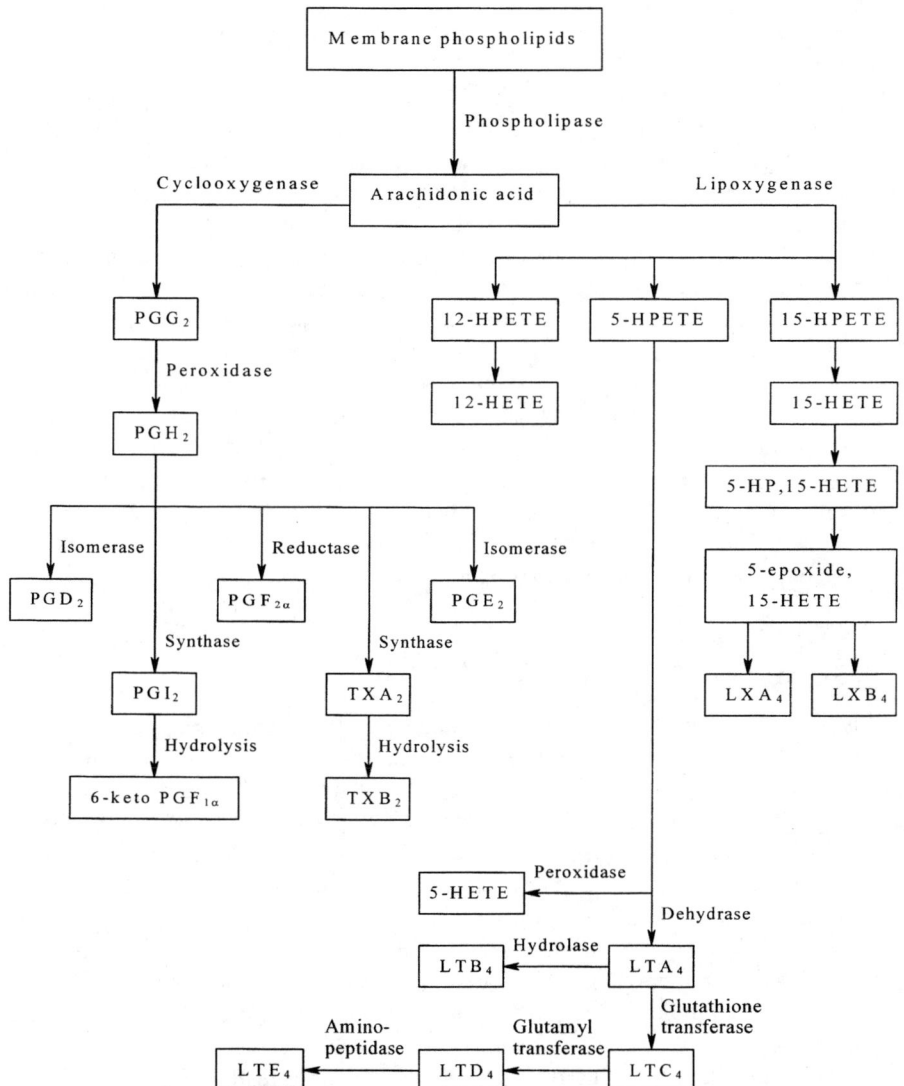

FIGURE 10.8 The *cyclooxygenase* and *lipoxygenase* pathways whereby arachidonic acid is converted to various prostaglandins, leukotrienes, and other 20-carbon autacoids. Important molecules: PG_ = prostaglandins, TXA_2 = thromboxane A_2, x-HPETE = x-hydroperoxy-eicosatetraenoic acid, x-PETE = x-monohydroxyeicosatetraenoic acid, LTX_4 = leukotriene X_4, LXA_4 = tetraene trihydroxy lipoxin A_4.

granulocytes, thus limiting the inflammatory reaction. In addition, PGE_2 inhibits histamine release from basophils and generally inhibits the release of activating immunocytokines from sensitized T-cells and B-cells. It also inhibits CTLs. PGE_2 simulates the release of the hormone ACTH from the anterior pituitary gland. ACTH in turn stimulates the formation of certain adrenocortical hormones, notably *cortisol*. Cortisol is a potent anti-inflammatory hormone that has the following effects on the inflammatory immune response:[59] (1) it lowers fever, mostly by suppressing the release if IL1 by macrophages; (2) it suppresses the reproduction of T- and B-cells, reducing the number of circulating Abs; (3) it reduces the chemotaxis of macrophages, NK cells, and CTLs into the inflamed area, thus reducing cell death and the release of inflammation-producing autacoids; (4) it decreases capillary permeability and hence edema; and (5) it stabilizes and strengthens lysosomal membranes so that when cells are ruptured, there is reduced spread of proteolytic enzymes, causing reduced collateral damage to adjoining cells.

It is interesting that aspirin (acetlysalicylate) acts as an antipyretic (fever reducer) and anti-inflationary substance. Aspirin inhibits the function of the cyclooxygenase enzymes responsible for the conversion of arachidonic acid to various prostaglandins. In particular, certain cytokines (IL1β, IL6, IFNα, IFNβ, and TNFα) induce PGE$_2$ production by the circumventicular organs near the hypothalamic area in the brain. PGE$_2$ triggers the hypothalamus to cause the body temperature to rise by readjusting the body's heat balance. Aspirin and other nonsteroidal anti-inflammatory drugs block the production of PGE$_2$ and hence lower fever.

10.3.5 Section Summary

In this section, the more important autacoids with strong signaling and/or effector actions in the immune network are summarized. Clearly, most of these immunocytokines are stimulators or activators that increase the degree of immune response at a number of levels. Very few immunocytokines have definite inhibitory actions in the immune network; these include IL10 and PGE$_2$. This apparent disparity between the number of activators and suppressors challenges our understanding of how the overall immune response works. All immunocytokines have finite lifetimes *in vivo*; they are broken down by enzymes so that their effects as immune system activators naturally decay in time. Thus if the stimulus (e.g., a bacterial endotoxin) for immunocytokine production disappears because of successful combat by the immune system, immunocytokine production decreases, and their titer is reduced enzymatically, leaving the immune network in a resting state.

Another possible natural scenario to explain why there are so few suppressor immunocytokines may lie in the concept of *anti-idiotypic antibodies*. Vigorous production of Abs for a particular antigen, X, leads to the demise of the pathogen associated with it. An antigen-presenting, phagocytic B-cell *by chance* has a surface antibody with an affinity to the Abx's X-binding site. That is, the particular unique IgG on the B-cell's surface has a site that looks like X to the Abx. The bound Abx is phagocytosed by the B-cell, broken down enzymatically, and then the X binding site is combined with MHC II and externalized to the B-cell's surface, where it is presented to a Th. The Th is stimulated by the process of antigen presentation to release immunocytokines that stimulate clonal expansion of the antigen-presenting B-cell with its particular IgG with affinity to the Abxs. Eventually, the B-cell daughters change to plasma cells and release soluble $\overline{\text{Abx}}$ with affinity to Abx.

These $\overline{A}$bxs combine with the binding sites on the Abxs, inactivating them, thus damping the humoral immune response, inflammation, etc. One can speculate that if there are enough anti-idiotypic $\overline{A}$bxs in circulation, anti-anti-idiotypic antibodies might be made to suppress the $\overline{A}$bxs.

In summary, the complexity of the immune network is imposed by the great number of immunocytokines and their often pleiotropic actions.

10.4 APOPTOSIS IN THE IMMUNE SYSTEM

Apoptosis is also called programmed cell death (PCD). Apoptosis comes from the Greek *apo-*, meaning away from or off, and *ptosis*, meaning to fall. The preferred pronunciation of apoptosis is "$\overline{a}$ 'pototis"; i.e., the second p is silent.

All nucleated body cells have built-in "machinery" which, upon receiving the appropriate biochemical signals, causes the cell to literally commit suicide (apoptosis). A cell undergoing apoptosis shrinks and pulls away from neighboring cells. Its innards appear to boil; blebs form on its surface, disappear, and reform. The cell's nucleus is broken down, and its DNA is broken up and randomized; little fragments of DNA are included in "apoptic bodies" when the cell breaks up. This debris is usually cleaned up by the usual phagocytic scavenger cells, such as macrophages.

Many animal cells undergo apoptosis as part of normal development, growth, and maturation. Some examples follow: (1) The lens of the eye, which forms during embryonic development, is formed from apoptic cells that die and leave the clear protein crystalline in layers, somewhat like the layers of an onion. (2) Cells composing the intestinal epithelial covering on the villi continuously arise at the bases of the villi and migrate toward the tips. This process takes several days. At the tips, the intestinal epithelial cells undergo apoptosis, slough off, and are replaced by new cells. (3) The shedding of the uterine lining during menstruation is accomplished by apoptosis. The signals for this PCD are apparently hormonal. (4) Skin cells (keratinocytes) arise in the deepest layers of the dermis and migrate toward the surface. After about 21 days, they undergo apoptosis. The resulting layers of dead cells are the epidermis, which is the body's first line of defense from a hostile environment or invading organisms. (5) Thymic preselection of T-cells to eliminate defective cells or those that have autoimmune potential is accomplished by apoptosis. (6) An intracellular "parity check mechanism" can cause virally infected cells or cells with damaged DNA (as by radiation) to undergo apoptosis. It is not unreasonable to expect some viruses to have evolved a biochemical means to inhibit this PCD from within the cells they invade.

The molecular mechanisms that control apoptosis are just beginning to be understood. The internal protein-cleaving enzymes that effect PCD are called *ICE-like proteases*. (They are called ICE-like because they structurally resemble IL1-converting enzyme [ICE]). The intracellular molecules used to control the ICE-like proteases that effect PCD belong to a large family of proteins called the *Bcl-2 family*. Identified so far are seven Bcl-2-type *apoptosis inhibitors*. There are three Bcl-2 *apoptosis promoters*: *Bad, Bax,* and *Bak*. Certain infecting viruses, once inside an infected cell, make Bcl-2-like apoptosis inhibitors. Five are currently known. Another intracellular substance, *nuclear factor kappa B* (NF-κB), is reported to be an inhibitor of apoptosis induced by TNF.[162,163]

Apoptosis of circulating T-cells is seen as a natural means of reducing the cellular immune response once a pathogen has been successfully fought. At least two mechanisms may be involved in triggering apoptosis in T-cells. Reduced concentration of IL2 as an infection is abated can lead to PCD of T-cells. Another cause of apoptosis may involve the transmembrane protein *Fas/CD95*. Some Fas is present on the membrane surface of resting T-cells. When T-cells are activated by antigen presentation, they produce more but initially nonfunctional Fas. They also produce a cell surface protein, *Fas ligand/CD95L* (FasL). After a few days, the Fas becomes active; now it can bind with FasL on itself or neighboring T-cells. The binding of Fas to FasL transduces messages to activate the ICE-like proteins, and apoptosis takes place. Thus Th and CTLs are given several days to fight the infection before they are programmed to die. Presumably low IL2 concentration expedites this process.

There is evidence that some cancers protect themselves from the immune system's TNF by turning off their PCD mechanism. This protection may be due to the continuous intracellular production of the protective substance, *nuclear factor kappa B* (NF-κB). In fact, one proposed approach to killing cancer cells is to inhibit their production of NF-κB so that TNF will be more effective.

One scenario for the depletion of CD4+ T-lymphocytes in HIV infection is anomalous or premature induction of apoptosis of Th caused by antibodies made against glycoprotein 120 (gp120, the external contact protein on the HIV virion's surface with which it binds to CD4+ T-cells). gp120 molecules can detach from the HIV's surface and are suspended in the blood of an HIV-infected person. These free gp120 molecules still retain their affinity to CD4 molecules on the surfaces of T-cells. The free gp120 binds with CD4. Antibodies to gp120 can also bind to another site on the gp120 that is bound to the CD4. If two adjacent CD4*gp120 sites bind with the two combining sites on the arms of the gp120 Ab, it is hypothesized that signals (1) are sent to the Th that induce the premature formation of the apoptosis-inducing protein, Fas, on its surface. If the Fas molecules encounter Fas ligand (FasL) on the surface of another mature Th, the first Th can be made to undergo premature PCD (2), even though it may not yet be actually infected with HIV. This putative process is illustrated in Figure 10.9.

10.5 THE ART OF MATHEMATICALLY MODELING SCENARIOS IN THE IMMUNE SYSTEM

10.5.1 INTRODUCTION

As is apparent from the foregoing sections of this chapter, the immune system has considerable complexity, second only to the CNS. The complexity arises not from

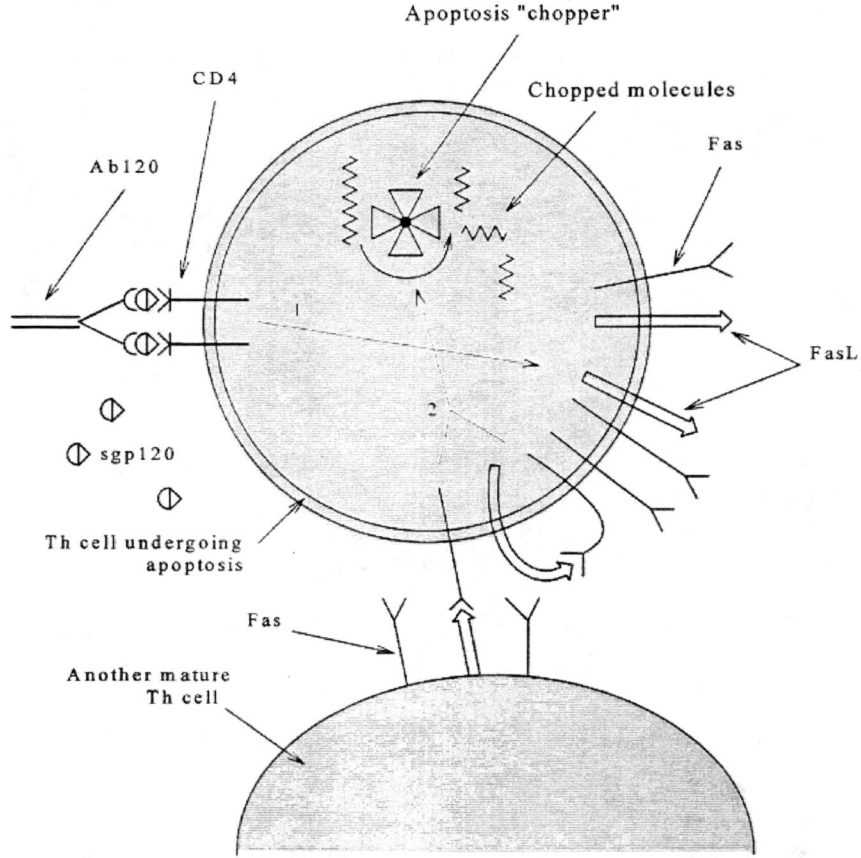

FIGURE 10.9 Schematic illustrating the hypothesis that apoptosis can be prematurely induced (1) in mature Th cells by antibodies binding to the HIV coat protein, gp120. The binding of the Th coat protein, Fas to FasL is thought to trigger the production and activation (2) of internal proteolytic enzymes (the "chopper") that destroy the cell from within (PCD). See text for description.

the finite number of cell types contributing to immune responses (macrophages, NK cells, cytotoxic T-cells, helper T-cells, B-cells, memory B-cells, other memory cells, mast cells, neutrophils, granulocytes, basophils, etc.) but from the bewildering number of immunocytokines (immune system *autacoids*), including the molecules of the complement system, molecules from platelets, and immunoglobulins (antibodies and anti-idiotypic antibodies). Complexity is also engendered from the *pleiotropy* of the immunocytokines. The effects they have on their target cells vary with the level of maturity and activation of the cells and the prior reception of other cytokine signals by these cells.

In order to accurately model the dynamics of immune system cell growth, development, control, action, and death, we need to know the rates at which immune system cells are normally produced in the bone marrow, the rate at which they can undergo clonal expansion under the influence of various immunocytokines, and the normal half-lives of the cells. We also need to know the dynamics governing the controlled intracellular manufacture of immunocytokines, the dynamics of their release, and their normal half-lives. Immunocytokines combine with receptors on cell surfaces at rates that are generally assumed to be governed by the laws of mass action. Cytokine molecules binding to receptors trigger the internal biochemical machinery of the cell to either manufacture or stop manufacturing biochemical products. How this signal transduction works is currently being studied. Every cell has a finite number of receptors for certain cytokines; hence a cytokine message's effect can saturate. That is, a cell can make a product at a maximum rate, given 100% saturation of its receptors by a cytokine. These saturation effects must also be incorporated in an accurate model.

Immune system events that are interesting to model, and which have been modeled, are (1) the immune system vs. a growing cancer, (2) the immune system vs. a viral or bacterial infection in a tissue, (3) how the HIV virus affects the immune system and leads to AIDS, and (4) autoimmune disease, etc. A good, validated, mathematical model should be able to predict how well a proposed drug or immunotherapy will work in killing the pathogen or restoring normal immune system homeostasis.

In formulating a simulation of an immune system event, we must make certain assumptions about where in the body it is assumed to take place. It is unsuitable to model the body as a whole because of the differences between various organs and tissues and the fact that most infections are localized in a particular organ or tissue. Also, we have seen that certain types of immune system cells (e.g., macrophages, mast cells) are sessile, i.e., they are fixed in some tissue or organ and therefore are not "well mixed", a requirement for mass-action kinetic assumptions. They would participate in an immune reaction only if the pathogen being fought were nearby.

For example, the site, or modeling scenario, can be in a "unit volume" surrounding a cancer beginning to grow in the lungs, or it can be in a unit volume of blood, in a lymph node, in a small volume of tissue surrounding an injury or in the pleural cavity, etc. In a model, the immune system cell states can be considered to be an absolute number in the defined volume or a density (number per cubic millimeter). Cytokines can be represented as a molecular density (molecules per cubic millimeter) or by weight; e.g., picograms per cubic millimeter, moles per liter, or "units" [an operational definition] per cubic millimeter. Once secreted, cytokines diffuse away from the source and are broken down enzymatically, contributing to their loss rate in the scenario volume.

An immune system model generally consists of a large set of coupled, nonlinear ordinary differential equations based on mass-action kinetics or the probability of interaction between receptors on cells and soluble immunocytokines. The model should include the effects of certain cytokines on cell growth, maturation, clonal expansion (if it occurs), and cell function.

There is a natural tendency in modeling a complex system to simplify it. Such simplification can follow the best of intentions. For example, suppose a relation between the advent of a certain cytokine and the induced manufacture of a second cytokine by a cell has been observed qualitatively *in vivo* or *in vitro*. An ordinary differential equation can be formulated to describe this process. One might argue that if we do not know the numbers for the rate constants, decay constants, etc., we should not include the putative relation in the model. However, if the product cytokine is known to be critical in other aspects of the model, the process is considered important to the overall model. Even though numbers for rate constants may not have been measured on living systems for the process, reasonable estimates of rate constants and other parameters can be made for the speculative ordinary differential equation for purposes of validation. If the ordinary differential equation and the trial numbers do not contribute to the accuracy of model validation, then the numbers can be revised and the model run again, etc. If no improvement is seen over a reasonable range of parameter values, then the ordinary differential equation is probably not relevant to the phenomenon being simulated and should be deleted from the model.

Model simplification may be made for a less pure motive; i.e., we do not want the complex system to confuse or bewilder our intended audience. In any case, such *reductionism* can rob the model of realistic functional detail. High-order, coupled, nonlinear systems often exhibit unexpected or counterintuitive behavior which will be missing in a reduced model, but which can be verified *in vivo*. Thus it should be a paramount goal of anyone seeking to model immune function to include as much validated detail as possible. The modeler is urged to eschew reductionism. Models are cheap; the computer does the work. They are an ideal platform on which to test hypotheses on immune system interactions.

10.5.2 THE IMMUNE SYSTEM VS. CANCER

There are many theories why normal human cells in organs and tissues become cancerous. The mutation from normal cell function to cancer (*oncogenesis*) is often correlated to family genetic predisposition (breast cancer, intestinal cancer), diet, environmental factors including the inhalation of tobacco smoke, excess sunlight (ultraviolet) on the skin, drinking water contaminated with chemical carcinogens, exposure to ionizing radiation (radon, X-rays), etc. The genetic machinery of *one cell* becomes mutated, either because its DNA is damaged, or because foreign DNA is inserted into its normal DNA by a viral infection. (In some types of cancer, viruses are known to be the oncogenic agent.) In any case, the cell produces altered surface proteins; its adhesion to other cells often decreases. Cancer cells' apoptosis mechanisms may also be disabled so they can reproduce clonally with altered DNA and proteins. Cancerous cells are generally characterized by rapid growth, even though the normal cells in the tissue in which they lie have slow, controlled growth. At first, the cancerous cells grow exponentially in time due to clonal "twinning." The time between divisions can be as short as 10s of hours in a fast-growing cancer. After the tumor reaches a certain size, its growth slows. Then autacoids secreted by the cancer cells cause blood vessels from the host to invade the mass of the tumor to provide nourishment. This vascularization is called *angiogenesis*. Once the tumor is vascularized, its growth rate again increases. A 1-g tumor may contain about 10^9 cells. If untreated, cancers kill their hosts by crowding out normal cells and disrupting organ function. Malignant cancers can *metastasize*, or spread to other sites in the body. Lung cancers, for example, can metastasize to the bone marrow and the brain.

As mentioned above, certain surface proteins on cancer cells are altered, providing, in theory, a means whereby the immune system can recognize their difference from normal cells and attack and destroy them. Obviously, in some cases, the immune system, for whatever reason, does not do this, and the cancer grows until it is diagnosed. The rate of cancer growth appears to be important in eliciting immune response. A rapidly growing cancer is generally more likely to excite the host's immune defenses than is a slowly growing tumor. One reason for this phenomenon may be the half-life of helper T-cells. If the half-life is too short, and the tumor grows slowly, the density of activated Th does not reach a critical threshold, and the tumor "sneaks through," avoiding immune rejection.[37]

An interesting phenomenon to contemplate is the *spontaneous regression* of tumors. Anecdotal, circumstantial evidence of spontaneous regression[78] leads us to speculate that the immune system, having once missed the cancer, can return in full force and destroy it, even though it has reached a mass large enough to permit diagnosis. This reawakening of immune control may accompany a simultaneous stimulation of the immune system by a viral infection.

Certain cancers are "inoperable," and chemotherapy may be unsuitable because of the patient's medical condition. Needless to say, using exogenous stimulation of the patient's immune system to fight cancer is a current active area of research. Certain aspects of immunotherapy for cancer can be modeled mathematically. One of the first, detailed mathematical models of the immune system vs. cancer was described by DeBoer et al.[36] Their model pitted the cellular arm of the immune system vs. cancer and was based on the experimental behavior of an ascitic lymphoma injected into the peritoneal cavity of mice.

DeBoer et al.'s 1985 model used *eight states*: the population densities of cytotoxic T-lymphocyte precursor cells (CTLP), cytotoxic T-lymphocytes (CTL), "normal" macrophages (MPH), activated cytotoxic macrophages (ANGRY), unprimed precursor helper T-lymphocytes (HTLP), activated helper T-lymphocytes (HTL), tumor cells (TUMOR), and debris (DEBRIS) from tumor cells killed by CTL and ANGRY. B-cells, antibodies, and NK cells were not considered in the DeBoer model. The only specific immunocytokine used in the DeBoer model was IL1 (see Section 10.3). In order to keep poorly estimated or unknown rate constants to a minimum, DeBoer et al. defined algebraic parameters which pool the effects of immunocytokines known to be active in the cellular arm of the immune system. Their model included antigen presentation by macrophages to helper T-cells (CD4+) in the parameter APC.

$$APC = \frac{(MPH + ANGRY)*DEBRIS}{KMD + DEBRIS}$$

The constant KMD = 10^7. The parameter FACTOR acts to activate macrophages to ANGRY and also stimulates clonal expansion of CTL and HTL (Th). FACTOR may have an analog in IL2 (see Section 10.3.1). IL2, however, is produced by CD4+ Th that have been activated by antigen presentation by macrophages. (It would make more sense to have FACTOR a function of APC*HTL rather than TUMOR*HTL.) The constant KMT = 50.

$$FACTOR = \frac{HTL*TUMOR}{KMT + TUMOR}$$

The role of INFLAM in the model is to increase the production of hematopoietic stem cells and cause their differentiation into macrophages, CTL, and Th. INFLAM is a saturating function of FACTOR. IL3 and IL12 are now known to promote the proliferation of stem cells and their differentiation and maturation. A number of other immunocytokines also promote hematopoiesis. DeBoer et al. give:

$$INFLAM = \frac{H*FACTOR}{KMF + FACTOR}$$

where H = 9 and KMF = 50. The De Boer model's state equations are given below in Simnon™ form:

$$dCTLP = I1*(1 + INFLAM) - A*CTLP*TUMOR - EL*CTLP \quad (1)$$

$$dCTL = A*CTLP*TUMOR - EL*CTL + \frac{R*CTL*FACTOR}{KMF + FACTOR} \quad (2)$$

$$dHTLP = I2*(1 + INFLAM) - A*HTLP*APC - EL*HTLP \quad (3)$$

$$dHTL = A*HTLP*APC - EL*HTL + \frac{R*HTL*FACTOR}{KMF + FACTOR} \quad (4)$$

$$dMPH = I3*(1 + INFLAM) - A*MPH*FACTOR - EM*MPH \quad (5)$$

$$dANGRY = A*MPH*FACTOR - DM*ANGRY \quad (6)$$

$$dTUMOR = \frac{R*TUMOR}{1 + TUMOR/KR} - \frac{KILL*(ANGRY + CTL)*TUMOR}{KMK + TUMOR} \quad (7)$$

$$dDEBRIS = \frac{KILL*(ANGRY + CTL)*TUMOR}{KMK + TUMOR} - ED*DEBRIS \quad (8)$$

The constants that DeBoer et al. initially used are A = 0.001, EL = 0.02/day, I1 = 0.5 or 10, I2 = 0.01 to 100, I3 = 1.25 × 10^5, EM = 0.05 per day, R = 1, KMF = 50, DM = 1 per day, KILL = 10, KR = 10^9, KMK = 10^5, ED = 2 per day, KMD = 10^7, KMT = 10^3, H = 9, KMF = 50.

DeBoer et al. state:

> We consider one compartment in which a tumor grows autonomously and in which tumor cells are killed upon contact with cytotoxic effector cells. Effector cells are generated upon local (i.e., within the compartment) activation of precursor cells. Cells do not recirculate: precursors immigrate into the compartment, leave (decay), or become effector cells: effector cells, on the other hand, only leave the compartment (or decay locally).

Simulations of the DeBoer et al. model equations were done by the author using Simnon. The equations are stiff; this is seen in the computational "noise" on some of the traces when using the default Runge–Kutta/Fehlberg order 4/5 integrator at low (1.E-3) error tolerance. As might be expected, this noise is much reduced when the error tolerance was lowered to 1.E-6. Figure 10.10 illustrates the time course of the DeBoer model's states when the scenario does not include enough aggressive immune behavior toward the growing cancer cells. The logarithms of states are plotted, since an active tumor can reach millions of cells in a month. Figure 10.11 illustrates an interesting, nonintuitive result of the DeBoer model. Here the tumor growth rate parameter is low (R = 0.9, instead of 1), and the kill parameter = 17, instead of 15. Note that the tumor regrows to a low cell count after apparently being destroyed. In Figure 10.12, the growth rate R = 1, kill = 17, and the tumor does not regrow.

In their papers, DeBoer et al. used their models to demonstrate the effect of antigenic modulation of the tumor cells. In this scenario, the tumor cells progressively lose surface antigen proteins as the tumor grows. This loss causes the tumor to appear less "visible" to the immune system, which may never reach a level of stimulation necessary to successfully fight the tumor.

Clapp et al.[22] reported on the results of an 18-state model of the immune system vs. cancer. Their model expanded the model of DeBoer et al. and included the then more recent knowledge about the action of certain immunocytokines. It included B-cells, antibodies, and NK cells as well as CTL, macrophages, and Th. It also used certain specific immunocytokines. This model has been slightly revised and is presented below. The states are

pNK	precursor natural killer cells
NK	natural killer cells
pTh	precursor T-helper cells
Th	T-helper cells
pTc	precursor cytotoxic T-cells (CTL)
aTc	first-stage activated CTL
Tc	fully activated cytotoxic T-cells
pMp	precursor macrophages

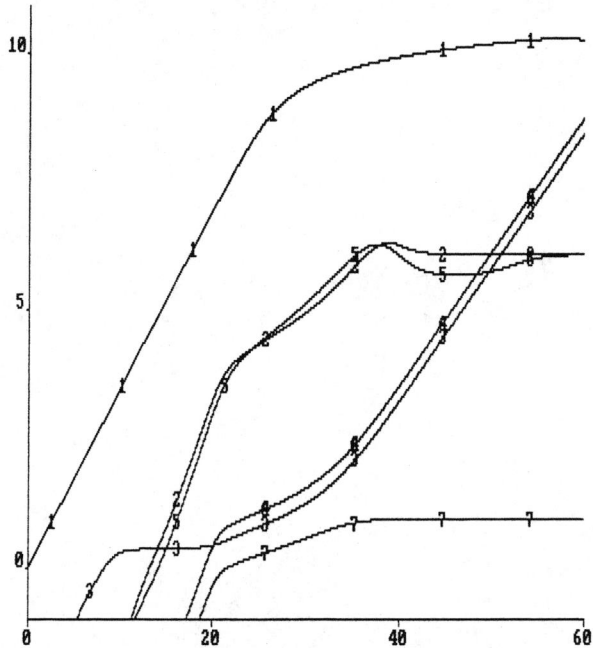

FIGURE 10.10 Simulated runaway growth of cancer using the DeBoer eight-state model for the immune system vs. cancer. Trace 1 = $\log_{10}$(cancer cell population density [pd]), trace 2 = log(ANGRY pd), trace 3 = log(CTL pd), trace 4 = log(HTL pd), trace 5 = log(APC pd), trace 6 = FACTOR, trace 7 = INFLAM. Parameters: R = 0.8, KILL = 10, I1 = 0.5, A = 1.E-3, EL = 0.2, KMF = 50, I2 = 1, I3 = 1.25E5, EM = 5.E-2, DM = 1, KR = 1.E9, KMK = 1.E5, ED = 2, KMD = 1.E7, KMT = 1.E3, H = 9. Vertical axis in $\log_{10}$ units unless noted; horizontal axis in days.

aMp	first-stage activated macrophages
Mp	fully active macrophages
Lys	cellular debris
CA	cancer/tumor
IL2	interleukin-2
nB	nonactivated B-cells
aB	antigen-activated B-cells
pB	plasma B-cells specific for CA
AbCA	antibodies (free and cellular) for CA
mBCA	memory cells for CA

```
************* STATE EQUATIONS (18) ***************
dpNK =(1 + inflam)*mf1 - mat1*rpNK + ep1*rpNK - rpNK/tpNK
dNK = mat1*rpNK - rNK/tNK
dpTh = (1 + inflam)*mf3 - apf*rpTh - rpTh/tpTh
dTh = rpTh*apf + ep2p*rTh - rTh/tTh
dpTc = (1 + inflam)*mf5 - rpTc*apf - rpTc/tpTc
daTc = rpTc*apf - raTc*(gIFN + rIL2)/(1 + PGE2) - raTc/taTc
```

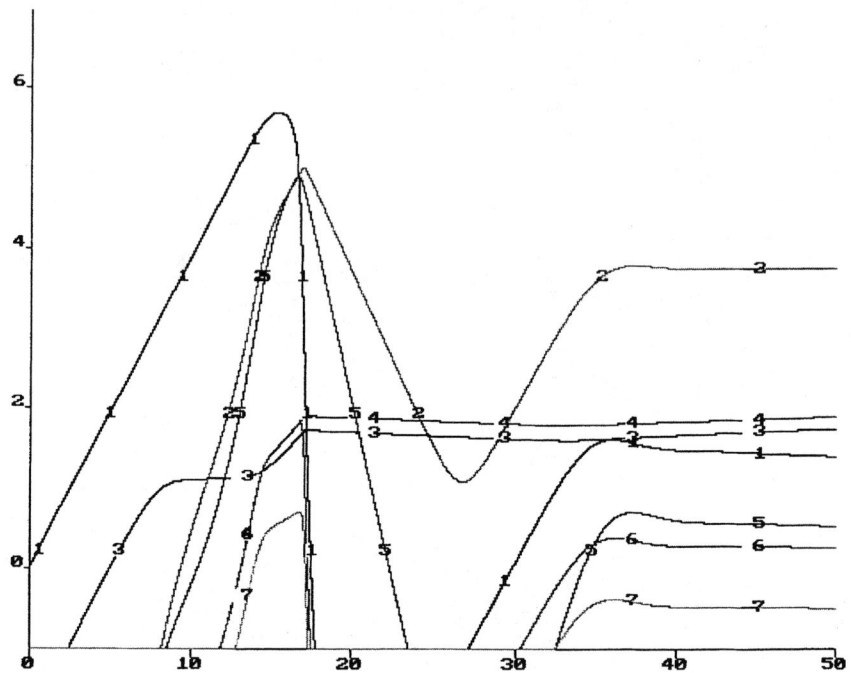

FIGURE 10.11 Simulated regression of cancer, followed by low-level regrowth. De Boer model. Same parameters as in Figure 10.10 except R = 0.4, KILL = 17, and KR = 1.E7. Trace 1 = log(TUMOR pd), trace 2 = log(ANGRY pd), trace 3 = log(CTL pd), trace 4 = log(HTL pd), trace 5 = log(APC pd), trace 6 = FACTOR, trace 7 = INFLAM.

```
dTc = raTc*(gIFN +rIL2)/(1 +PGE2) + rTc*IL2*c6/(sf10 + rIL2) + addTc
addTc = dmTc - rTc/tTc
dpMp = (1 + inflam/(1 + PGE2))*mf6 - gIFN*rpMp - rpMp/tpMp
daMp = gIFN*rpMp - raMp*rTh*rCA*c7/(sf11 + rCA*raMp*rTh) + addaMp
addaMp = -raMp/taMp
dMp = raMp*rTh*rCA*c7/(sf11 + rCA*raMp*rTh) - rMp/tMp
dLys = rCA*(rNK*co1 + (rMp + c5*raMp)*co2 + rTc*co3)*c14 - rLys/tLys
dCA = gf*rCA/(1 + rCA/grs) - nvmp*rCA + addCA
addCA = -rCA*(rNK*rAbCa*co1 + raMp*co2 + rTc*co3)*c14
dIL2 = IL2in - 1f4*rIL2 + rCA*rTh*c13/(sf26 + rCA*rTh)
dnB = (1 + inflam)*mf7 - c18*rnB - c19*rCA*rnB*fi12
daB = c19*rCA*rnB*fi12 - c20*raB + (c21-c22)*raB*BCFca*factor - raB*c27
dpB = c22*factor*BCFca*raB - c23*rpB + c29*rmBCA*BCFca
dmBCA = c27*raB - c28*rmBCA - c29*rmBCA*BCFca
dAbCA = c24*rpB*rCA/(sf32 + rpB) - c25*rAbCA - c26*rAbCA*rCA
```

(Note that c26 governs the antigenicity of cancer surface proteins.)
The **immunocytokines** and their functions are

gIFN = rTh*IL1/(sf9 + rTh) + rNK/(sf23   Gamma interferon
+ rNK) + gIFNin

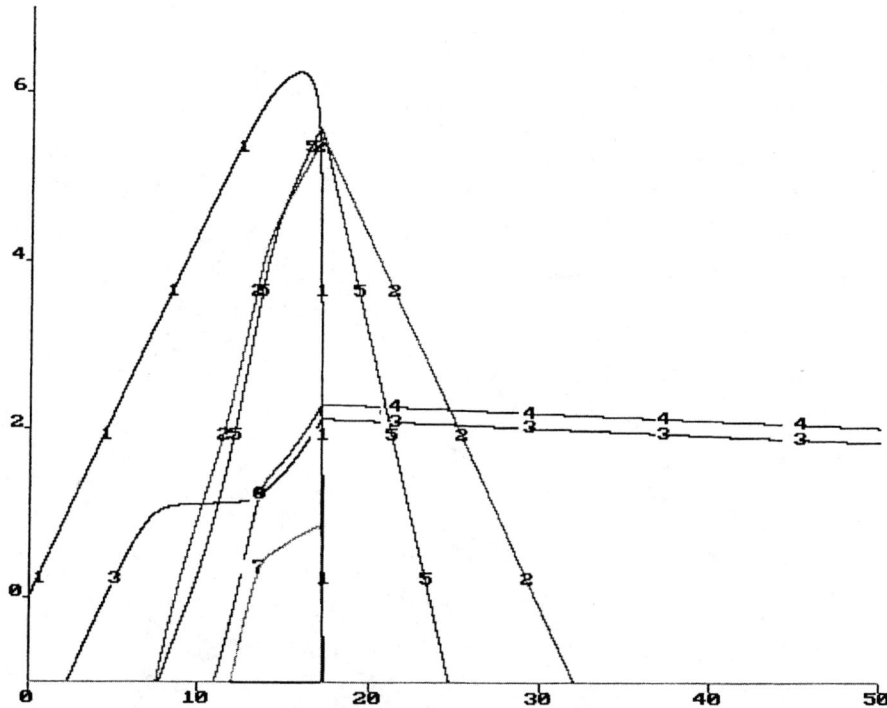

FIGURE 10.12 Simulated regression of TUMOR, with no regrowth. DeBoer model. Same parameters as in Figure 10.10 except KILL = 17, R = 1.0. Trace 1 = log(TUMOR pd), trace 2 = log(ANGRY pd), trace 3 = log(CTL pd), trace 4 = log(HTL pd), trace 5 = log(APC pd), trace 6 = FACTOR, trace 7 = INFLAM.

IL1 = c16*(raMp + rMp)*rTh*rCA/ ((sf19 + rTh)*(sf20 + rCA)) + IL1in	Interleulin-1 from Mp
HPF = c17*factor/(sf27 + factor)	Hematopoeitic factors
fil2 = c30*rIL2/(sf29 + rIL2) + 1	Hill function of interleukin-2
BCFca = IL1*rTh*rCA*c31/(sf30 + rTh)	B-cell maturation factor
PGE2 = c8*rMp/(sf12 + rMp)	Prostaglandin E2
aIFN = c10*(rMp + raMp)/(sf17 + rMp + raMp) + rNK/(sf24 + rNK)	Alpha interferon
factor = rTh*IL1/(sf28 + IL1)	Immunocytokines causing B-cell maturation and inflammation
apf = agf*(raMp +rMp)*rLys*(gIFN/ (1 +PGE2))/(sf4 +(raMp +rMp)*rLys)	Antigen presentation factor to activate Th
cxf1 = aIFN*c11/(sf16 + aIFN) + IL2*c12/(sf18 + IL2)	

Other functions used in the state equations are

```
"   OTHER FUNTIONS:
mat1 = c15*(gIFN + IL2)/(sf1 + gIFN + IL2)
ep1  = c1*(gIFN/(sf2 + gIFN) + IL2)
rp1  = c2*PGE2/(sf3 + PGE2)
ep2p = (IL1/(1 + PGE2))/(sf9 + rTh)
co1  = cxf1/(sf13 + rNK*rCA)
co2  = cxf2/(sf14 + (raMp*c5 + rMp)*rCA)
co3  = cxf3/(sf15 + rTc*rCA)
```

The default constants for the revised Clapp et al. model were derived from the DeBoer papers, the immunology literature, and clever estimates. Note that tXYZ = time constants in days, sfx = saturation factor, txyzinj = time at which exogenous therapy given, dtx = duration of exogenous therapy, etc. The colons are the equivalent of "=" in Simnon notation.

```
"   CONSTANTS:
tpNK:2, tNK:.1, tpTh:15, tTh:5, tpTc:30, taTc:50, tTc:1, tpMp:40,
taMp:50, tMp:5, t2Mp:100, tLys:0.5, mf1:0.1, mf2:0, mf3:25, mf4:1,
mf5:20, mf6:200, mf7:8, agf:0.2, sf1:1e5, sf2:100, sf3:5000,
sf4:3500, sf5:50, sf6:1000, sf7:10000, sf8:1.0e5, sf9:200,
sf10:1000, sf11:2500, sf12:5000, sf13:500, sf14:500, sf15:2500,
sf16:2000, sf17:1000, sf18:1000, sf19:1000, sf20:1000, sf21:1000,
sf22:1000, sf23:200, sf24:600, sf25:2e3, sf26:1e3, sf27:15, sf28:5,
sf29:20, sf30:100, sf31:5e3, sf32:500, lf1:0.0001, lf2:0.0001,
lf3:0.0001, lf4:0.2, nvmp:0. 001, gf:1, grs:1e7, c1:5.E-2, c2:0,
c3:1, c4:3, c5:0.5, c6:10, c7:100, c8:5, c9:2, c10:10, c11:5,
c12:2, c13:2, c14:0.2, c15:1e-7, c16:0.1, c17:20, c18:0.1,
c19:5.E-2, c20:0.1, c21:0.01, c22:0.2, c23:1e-2, c24:4, c25:0.01,
c26:1.E-5, c27:0.1, c28:1e-4, c29:0.1, c30:9, c31:3.3E-5, cxf2:4,
cxf3:1000, dt1:7, dt2:14, dt3:7, dt5:1, til1inj:14, til2inj:14,
tgifinj:14, gifinj:0, il1inj:0, il2inj:0, mTcinj:0
```

The default initial conditions were

```
pNK:2e3, NK:200, pTh:15, Th:10, pTc:25, Tc:0, aTc:10, pMp:2500,
aMp:0, Mp:2, Lys:0
CA:1, IL2:0.5, nB:10. Other state ICs are 0.
```

The model of Clapp et al. is useful because it can be used to explore the effects of putative immunotherapies such as the infusion of IL2, γ-IFN, activated NK cells, or monoclonally grown, activated CTLs on model cancer growth. Figure 10.13 illustrates the response of the Clapp model to one cancer cell arising at time $t = 0$. The model tumor grows exponentially until it reaches a critical size at around day 20, when growth abruptly slows. No immune rejection is seen in the model over 45 days. In Figure 10.14, 15 units of IL2 are given per day as an infusion from day 14 to day 28. Here, the model tumor grow exponentially until about day 20, when the population begins to decay. The population crashes at around day 33; there is no tumor regrowth. Figure 10.15 summarizes the effect of IL2 infusions on cancer growth rate. Infusion rates of over 9 units/day cause the model cancer cell density

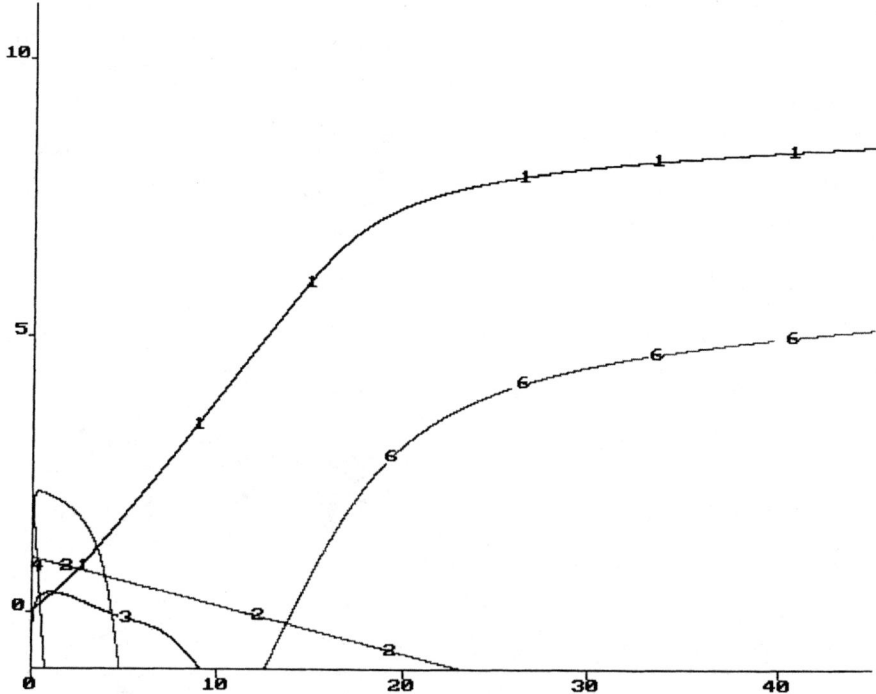

FIGURE 10.13 Simulation of a cancer growth that avoids immune rejection. Eighteen-state Clapp et al. model. Trace 1 = log(Ca), trace 2 = log(Th), trace 3 = log(CTL), trace 4 = log(NK), trace 5 = log(aMp), trace 6 = log(AbCa). See text for parameters.

to crash to zero inside the 0 to 45 daytime window, i.e., there is complete immune rejection of CA. The time at which the CA population density crashes is dose dependent.

Other adoptive immunotherapy scenarios can also be investigated with the model. For example, we consider the bolus infusion of *in vitro* grown, monoclonal CTLs specific for the cancer antigen, given at day 10. Figure 10.16 illustrates the model's response to 100 CTLs at $t = 10$. Note that there is a temporary regression of cancer cells until the population of CTL (Tc) dies out, and then the tumor regrows. In order to see if a critical number of CTL cells will cause a permanent regression, a series of escalating mTcinj values was given and the log(Ca) values plotted. In Figure 10.17, we see that regrowth appears to cease for mTcinj ≥ 400 cells for $t < 35$ days. On a 0 to 100 day scale, we see that Ca regrows at $t = 50$ for mTcinj = 400 cells. An mTcinj = 500 gives [Ca] = 7.E-3 at $t = 100$. mTcinj = 1000 cells gives [Ca] = 0 at $t = 100$.

These modeling adventures illustrate, or suggest, general trends in immunotherapies. They should be viewed skeptically because they are gross simplifications of what are very complex situations. Rate constants and half-lives are often estimates or even educated guesses. Some are taken from *in vitro* data, or mouse *in vivo*

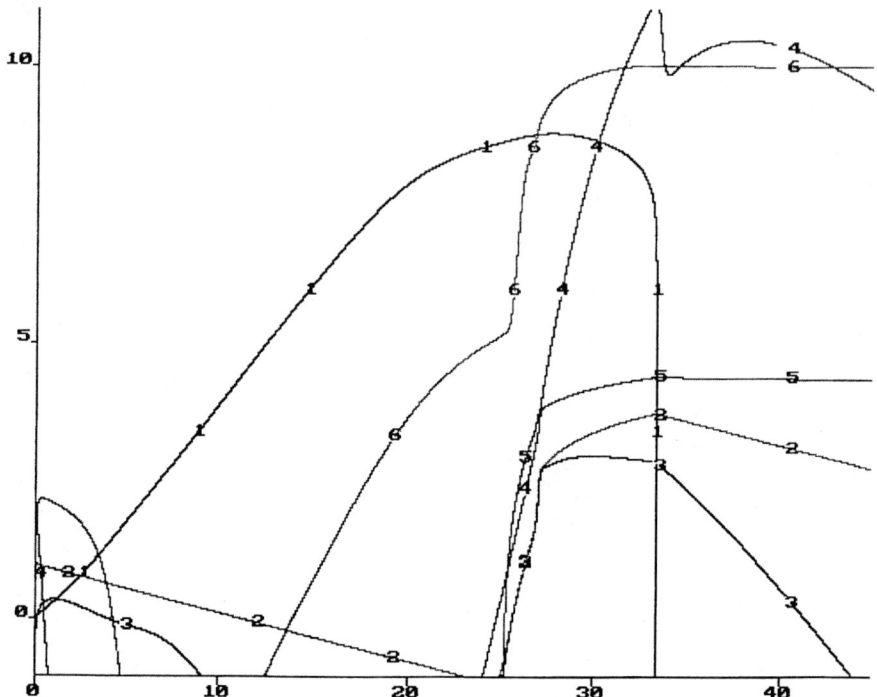

FIGURE 10.14 Simulation of the effect of an infusion of IL2 to the Clapp model of Figure 10.13. Fifteen units/day of IL2 given from day 14 to day 28. Trace 1 = log(Ca), trace 2 = log(Th), trace 3 = log(CTL), trace 4 = log(NK), trace 5 = log(aMp), trace 6 = log(AbCa). See text for parameters. Note pd of Ca cells crashes at day 33. Note general immune activation occurring around day 25.

data, and may be far from human numbers. Also, different kinds of tumors vary in their antigenicity and growth rates. Tumors with weak antigenicity may avoid strong attack by antibodies and CTLs, but still may fall prey to NK cells under circumstances where NK cells are attracted to the tumor cells and are not inhibited from killing them.

10.5.3 MODELS OF HIV INFECTION AND AIDS

The realization in 1983 that AIDS, the acquired immune deficiency syndrome, was the result of a virus that attacked specific cells in the human immune system has led to intense research on the physiology and molecular biology of the immune system and on the different varieties of the human immunodeficiency virus (HIV). HIV is a *retrovirus* in which the genetic information for its reproduction is coded in RNA base sequences, rather than DNA. HIV, like certain other retroviruses, infects cells in the human immune system. HIV is unique, however, in that it causes a gradual

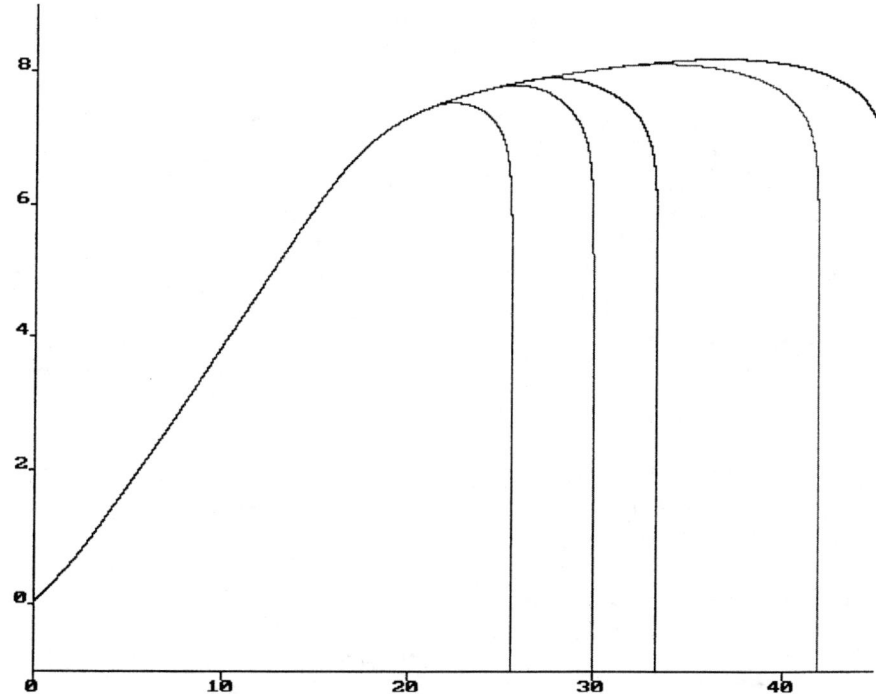

FIGURE 10.15 Same scenario as in Figure 10.14, except just log(Ca) plotted for various dose rates of IL2 infusion given from day 14 to 28. From left to right: IL2 rate is 40, 20, 15, 10, 9 units/day. No effect on Ca growth was seen for IL2 rate <9 units/day.

and generally fatal loss of effectiveness of the immune system's ability to fight infections. AIDS is characterized by a terminal illness from "opportunistic infections" such as *Pneumocystis carinii* pneumonia, Kaposi's sarcoma, aspergilliosis fungal infection (usually of the lungs or brain), tuberculosis, syphilis, etc.

It is worth remarking that other retroviruses attack immune system cells; for example, the feline leukemia virus (FeLV) and the human T-lymphotropic virus 1 (HTLV-1) both infect an immune system's T-cells, causing leukemia and immune suppression. The Visna virus of sheep attacks their monocytes/macrophages, causing progressively debilitated immune function and CNS problems. The Visna virus and human HIV are structurally similar, yet Visna does not attack sheep's T-cells.[111]

Significant in human HIV infection is the fact that HIV attacks, and is found within, both CD4+ helper T-cells and monocytes/macrophages and related cells in the CNS. In about 95% of the human population, HIV infection is characterized by a slow, progressive decline in the number of CD4+ T-cells (Th) over a period of 5 years or so. Eventually, the ability of the immune system to respond successfully to opportunistic infections is sufficiently impaired, so that the patient is killed by such an infection. Normal CD4+ Th density in the blood is about 800 per cubic millimeter. When the Th density has fallen to about 200 per cubic millimeter, the patient is

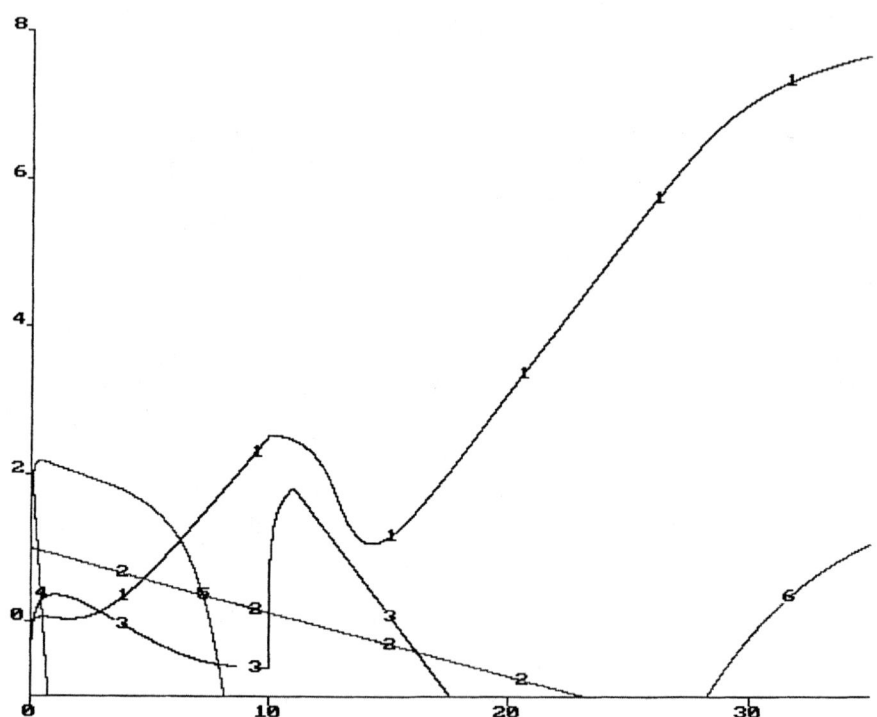

FIGURE 10.16 Simulation of the effect of another possible immunotherapy on the growth of Ca. One hundred, monoclonally grown CTL specific for the Ca were injected at $t = 10$ days. Note the jump in CTL pd and its decay. The Ca pd dips, then regrows. Trace 1 = log(Ca), trace 2 = log(Th), trace 3 = log(CTL), trace 4 = log(NK), trace 5 = log(aMp), trace 6 = log(AbCa).

considered to have AIDS. The cause of AIDS is certainly not as simple as a decreased count of Th cells, however. As will be discussed, there are many other factors that contribute to the loss of immune function due to HIV infection. Significantly, from 5 to 10% of the human population exposed to HIV maintains normal CD4+ Th cell counts and does not develop AIDS-related complex (ARC)/AIDS for over 10 years after infection, or the progression of the disease is very slow. The reason for this resistance has been recently discovered and, hopefully, may lead to effective immunization and treatment for HIV infection.

Figure 10.18 illustrates a schematic cross-sectional structure of the HIV virion. It is a very efficient, compact, self-replicating, molecular "nanomachine." (The reader might ask philosophically, "Is it alive?") An HIV virion is a 20-sided (icosohedron) about 100 nm in diameter. It has an outer membrane and a nucleocapsid, or core, containing two, identical, single strands of RNA of about 9000 base pairs which code its proteins and reproductive enzymes. Also inside of the nucleocapsid are the HIV's reproductive enzymes, including reverse transcriptase (RT = p66), protease (PR = p9), and integrase (IN = p32). Reverse transcriptase makes a

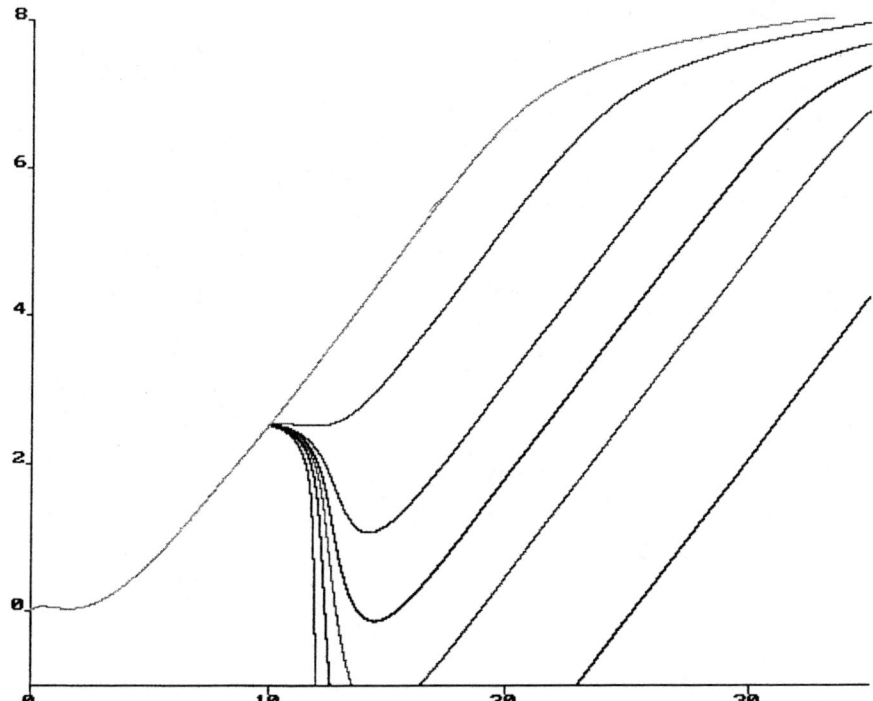

FIGURE 10.17 Simulated dose-dependency of Ca growth to exogenous, monoclonally grown CTLs specific for the Ca injected at $t = 10$ days. Just log(Ca) plotted. Top curve, no mCTL; second curve down, 20 mCTLs; third down, 100 mCTLs; fourth down, 150 mCTLs; fifth down, 200 mCTLs; sixth down, 300 mCTLs, seventh down; 400 mCTLs (the Ca does regrow for 400 at time $t = 50$). See text for details.

DNA copy of the virion's RNA. This DNA copy is inserted in the host cell's DNA by integrase. When activated, the host cell makes HIV RNA and proteins using the host cell's RNA polymerase II. New HIV virions are assembled within the host cell and then released in the process of *eclosion*. Portions of the host cell's plasma membrane, including some host cell surface proteins such as MHC I, become the eclosed virions' outer membrane.

On the outer surface of the HIV membrane are a dense array of about 72 to 80 glycoprotein 120 (gp120) molecules with which the virion makes contact with its host cells. gp120 is known to have a strong affinity for a binding site on the CD4 glycoprotein found on the surface of helper T-cells and macrophages. (Recall that CD4 is a critical cofactor in antigen presentation by certain B-cells and macrophages. It combines with a site on MHC II + Ag to trigger Th activation or B-cell activation.) The gp120 "cap" is attached to the virion by a "stem," glycoprotein 41 (gp41), that penetrates the virion's coat; the entire "lollypop" assembly is called gp160. The cytosol of the virion is filled with matrix protein p17.

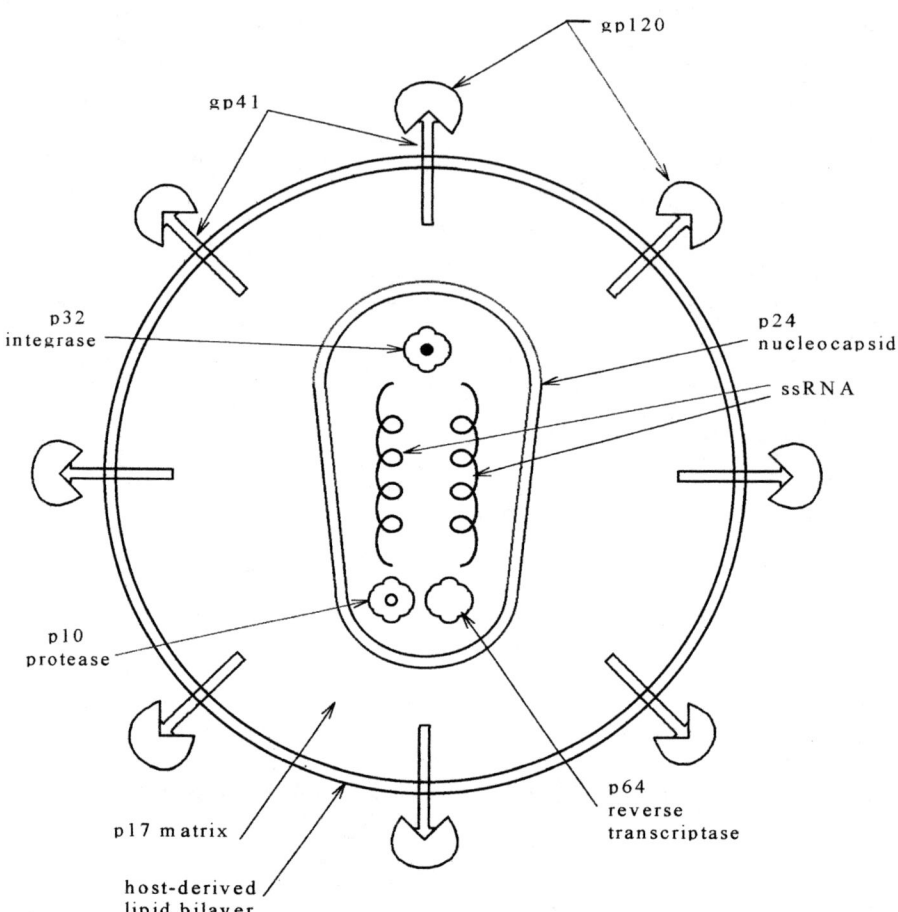

FIGURE 10.18 Cross-sectional schematic drawing of the important features of an HIV virion. The outer membrane forms an icosahedron about 100 nm in diameter, shown as a sphere.

The mechanisms of introduction of HIV into the body are well documented and will not be discussed here. However, we will examine the mechanisms by which HIV virions enter CD4+ Th and CD4+ macrophages and infect them. The first step in HIV replication is the binding of the virion to the cell it will infect. Until recently, it was believed that HIV binding to its host cell was due to the affinity of gp120 to CD4 surface proteins, found on macrophages and CD4+ Th cells. In 1995, it was reported that β-chemokines (*chemokines* are attractants of immune system cells secreted by other immune system cells) are major suppressors of HIV infection.[167] Subsequent studies showed that the excess β-chemokines combined with and thus saturated and blocked their receptors on Th and macrophages. Finally, it was discovered that HIV binding and entry required not only CD4, but also one of several chemokine receptors adjacent to the CD4 molecule. Apparently gp120, once bound to CD4, sends out a *v3 loop* which has affinity for an adjacent chemokine receptor. This second binding triggers the *inclosion* of the virion. The virion coat with its attached gp120 molecules and some free gp120 remain outside the infected cell, while the core with its RNA and enzymes enters through a hole in the plasma membrane, which reseals.

Once inside the cell, the virion's core capsule uncoats, and the RNA and enzymes are released into the infected cell's cytosol. The RNA is copied into viral DNA (vDNA) by the viral *reverse transcriptase*. The vDNA is then transported into the host cell's nucleus, where the viral enzyme *integrase* inserts the vDNA into the host's DNA, where it becomes an addendum to the host cell's genes. The vDNA remains dormant until some factor activates its reading and protein synthesis, which leads to the reassembly of viral enzymes, vRNA, a core coat, and the eventual eclosion of many new virions. Mass eclosion of virions kills the infected cell. HIV eclosion is not the only cause of death of HIV-infected cells, however.

Let us go back and examine the role and importance of the chemokine receptor cofactors for HIV binding and entry. Macrophages carry several chemokine receptors, some of which have affinity for part of the HIV's gp120 molecule. The macrophage CCR5 chemokine receptor normally binds with RANTES (which stands for *regulated on activation, normal T-cell expressed and secreted,* the ultimate acronym). Macrophage CCR5 also binds with an HIV1 gp120 site. Helper T-cells carry a chemokine receptor called CXCR4 which normally binds to stromal cell-derived factor 1 (SDF-1). The CXCR4 receptor also has an affinity to HIV1's gp120. An HIV virion is called M-tropic or T-tropic depending on whether the v3 loop of gp120 has an affinity for macrophage CCR5 or T-cell CXCR4, respectively. What controls this tropism switch is not known.

Back in the late 1980s, research was focused on confusing the HIV by injection of *soluble CD4 molecules* (sCD4) or CD4 molecules bound to erythrocyte membranes or other molecules. In theory, the excess sCD4 would bind to the free virion's gp120 molecules, preventing them from binding to CD4 on macrophages and T-cells. This strategy was largely ineffective *in vivo* because, in the writer's opinion, the sCD4 combined with its normal binding site on macrophage and B-cell MHC II, and presumably blocked normal antigen presentation by macrophages and B-cells, severely weakening the immune response. Free, soluble gp120 in the blood could combine with the sCD4, giving a null result. HIV virions already inside macrophages and CD4+ Th in the process of assembly could not be affected by sCD4.

The discovery of chemokine cofactors required for HIV binding and entry has opened up a whole new avenue of research for controlling HIV infection. The chemokine receptors are evidently *not* vital for the operation of the immune system, as is CD4, but are vital to HIV. Thus a new therapy based on selectively blocking the v3 loop from combining with its M- or T-chemokine receptors appears possible.

We note that there are a large number of chemokines, and not all their receptors have affinity to the v3 loop of gp120. There are about 16 known chemokines that attract one or more kinds of leukocytes. Chemokines generally attract immune system cells to sites of inflammation. Some chemokines, such as *lymphotactin,* attract only T-cells and NK cells.[168] Other chemokines, like RANTES, attract macrophages, eosinophils, basophils, T-cells, and NK cells, but not neutrophils. IL8 attracts neutrophils and T-cells, while complement C5a attracts neutrophils, macrophages, and eosinophils and no others. Such complexity appears to be the rule as we gain understanding of the immune system.

There are a number of theories why HIV infection is so difficult to control, let alone cure. One involves the large number of soluble gp120 molecules in the blood of an infected person. As we have seen, gp120 has affinity for CD4. If a free sgp120 molecule binds with an uninfected Th cell's CD4, antigen presentation can be blocked.[87] Another scenario involves antibodies the body has made to gp120, *Ab120*. If an Ab120 has affinity to the gp120 site that binds with CD4, it may also have affinity to the CD4 *receptor* on the MHC II molecule, blocking antigen presentation. Anti-idiotypic antibodies (AIAbs) may form against the Ab120s. These AIAbs may also have affinity for the CD4 site that binds to MHC II, again blocking normal antigen presentation. On the other hand, antigen presentation to HIV-infected Th appears to activate HIV virion production from these cells, leading to their death. If antigen presentation is blocked by soluble gp120, Ab120, or AIAb120, the rate of viral production will be slowed, but at the expense of reduced overall immune function. It has also been shown that HIV-infected macrophages can transmit HIV to Th during antigen presentation.[87] Thus the HIV infection process appears to be very complicated, including inherent negative feedback on HIV production, as well as positive feedback.

Still another scenario of immune system debilitation in HIV infection involves collateral damage of Th cells by NK cells. NK cells combine with the Fc region of Ab120s on the surfaces of HIV-infected and uninfected CD4+ T-cells that have had the misfortune to have soluble gp120 (sgp120) bind with their CD4. The NK cell secretes perforin, which lyses the marked infected *and* normal T-cells. Another theory involving sgp120 has each arm of one Ab120 binding to gp120 bound to two adjacent CD4 molecules. This binding may induce the cell to produce the Fas molecule. If one Fas molecule encounters the Fas ligand on itself or an adjacent cell, premature Th cell death by apoptosis may be induced. See Figure 10.9 for a schematic of this putative process. Because HIV infects key cells in the antigen presentation process using cell surface molecules used by the immune response, and the immune system responds to viral infection by killing infected cells, we see that the immune system is in fact attacking itself. HIV-infected macrophages pass on the HIV to uninfected Th in antigen presentation. Infected CD4+ T-cells are depleted because of viral eclosion and attack by NK and CTL cells. Ab120 and AIAb120 may jam antigen presentation, and premature apoptosis may be induced. Fewer activated Th cells mean lower concentrations of key immunocytokines and hence less immune system activation in response to opportunistic infections.

Another scenario which contributes to a weak immune defense against HIV involves the continual mutation of the noncritical structure of the gp120 molecule. This mutation rate can exceed the immune system's ability to mount an Ag-specific attack with CTLs and Ab120s. Thus the normal lags in clonal expansion of CTLs and plasma B-cells specific for flavor A of gp120 produce a response that is weakly effective against flavor B, etc.

Several workers have sought to describe the dynamics of HIV infection with mathematical models. As we have seen above, HIV infection is a very complex process, and not all scientists working on the problem of HIV and AIDS agree on why the immune system gradually loses competence. Obviously, there are many factors involved. One advantage of a good mathematical model is that one can explore possible new immunotherapies before one commits them to animal and then human studies. It is possible to introduce viral or bacterial "opportunistic" pathogens into the model at various times to test the model immune system's competence.

Overly simple mathematical models exploring the dynamics of HIV infection in different classes of immune system cells have appeared in the literature of mathematical biology over the past 10 years or so. Bailey et al.[9] described a simple five-state model for HIV infection of CD4+ T-cells. Their model was based on the results of experimental *in vitro* studies of HIV infection of CD4+ Th cells. The cells were cultured in the presence of IL2 for 50 to 60 days. The HIV-infected cells did not express virions or secrete extra IL2 unless they were stimulated with *phytohemagglutinin* (PHA). (PHA is a protein derived from beans that activates Th cells.) PHA-stimulated, infected CD4+ Th cells secreted IL2 for 2 days, eclosed virions transiently after 5 to 6 days, and then lysed and died (apoptosis?) after 7 to 10 days. Bailey et al. concluded that "...the critical series of events relevant to T4 cell depletion and disease progression appeared to be: infection, stimulation [as by Ag presentation], IL2 secretion, virus expression, and finally cell death."

A compartmental diagram of the Bailey model is shown in Figure 10.19. The five Bailey state equations are, in Simnon notation:

```
" STATE EQUATIONS:
"
" Tur = Uninfected resting T-cell population density.
dTur = -P*rTur -k10*rTur*rV +(k1 -k2)*rTur + k0*rTua
"
" Tir = Infected resting T-cell p.d.
dTir = -P*rTir + k10*rTur*rV +(k3 -k4)*rTir
"
" Tua = Uninfected, activated T-cell p.d.
dTua = P*(rTur + rTua) - k11*rTua*rV +(k5 - k6 - k0)*rTua
"
" Tia = Infected, activated T-cell p.d.
dTia = P*rTir + k11*rTua*rV + (k7 - k8)*rTia
"
" V = free HIV p.d.
dV = k12*k8*rTia - (k10*rTur + k11*rTua)*rV - k9*rV
```

The parametric input is

```
P = IF t > to THEN A*EXP(-(t - to)/tau) ELSE 0.
```

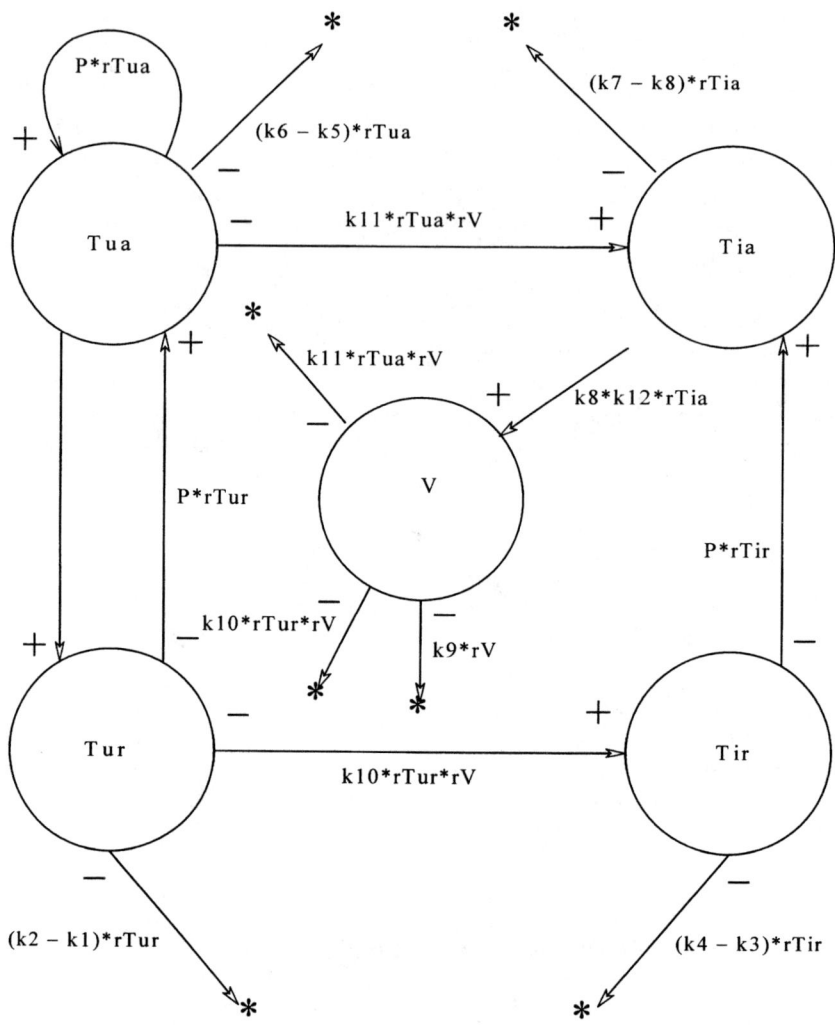

FIGURE 10.19 Five-compartment model of the HIV infection of CD4+ Th cells given by Bailey et al.[9] See the text for the ordinary differential equations.

The constants and initial conditions are

```
" CONSTANTS:
"
to:50, Poo:10, A:1, k00:0.01, k1:0.10, k2:0.05, k3:0.05, k4:0.05, k5:0.05
k6:0.1, k7:0.1, k8:0.25, k9:2.64, k10:5E-7, k11:1E-7, k12:75, TAU:5 days.
"
" INITIAL CONDITIONS:
Tur:1E6, V:1E6
```

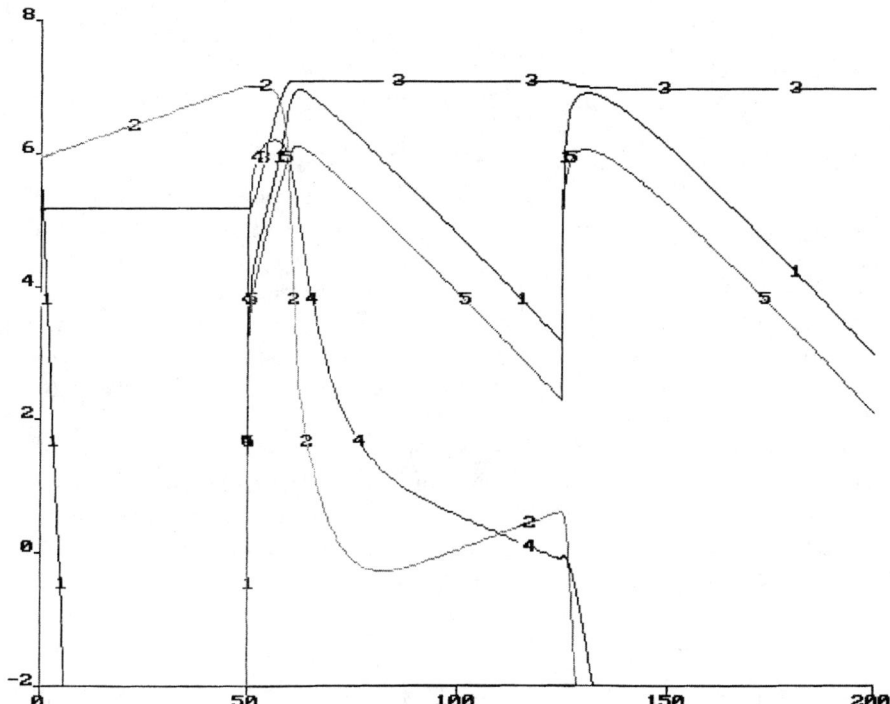

FIGURE 10.20 Simulation of the behavior of four types of CD4+ Th cells and free HIV virions in a test volume using the five-state model of Bailey et al.[9]. Boluses of PHA are given at $t = 50$ and again at $t = 125$ days. PHA is a mitogen for lymphocytes. Its effect is simulated by transiently increasing the parameter P in the model. Trace 1 = log(V), trace 2 = log(Tur), trace 3 = log(Tir), trace 4 = log(Tua), trace 5 = log(Tia). Vertical axis in log units; time axis in days. Note that initially, the HIV pd decays exponentially while the pd of Tur increases and there is a constant, steady-state pd of Tir. When the PHA is given, it stimulates the infected Th to make and eclose HIV virions; hence the abrupt increase in V and Tir. Tur and Tua decrease because they are infected and hence become Tir and Tia. Tia follows $V(t)$. At the second bolus of PHA, the pds of Tur and Tua plummet.

Figure 10.20 illustrates the behavior of the model when two simulated boluses of PHA are given. There are rapid eclosions of HIV virions which decay exponentially. Also, note that in the steady state following activation, there is a constant population of infected, resting Th which acts as a reservoir for future HIV eclosions.

Another simple four-state model of HIV infection of CD4+ Th cells was set forth by Layne et al.[77] This model's states were I = population density (pd) of successfully infecting HIV; V = pd of live, free virions; F = pd of free, soluble gp120 molecules; and C = pd of complexed gp120 molecules. They used their model to examine the dynamics of exogenous, soluble CD4 in neutralizing HIV. Both the Bailey and Layne models are good examples of reductionism. As we have seen, the immune system is so enormously complex that it defies mathematical description. To single out one component subsystem of a complex, tightly coupled, nonlinear system and to seek to draw meaningful conclusions about its isolated behavior is naïve.

A more complex 19-state mathematical model of HIV infection was presented by Reibnegger et al.[117] They included a self-replicating pathogen, HIV virions, and CD4+ Th cells with TCRs specific for pathogen Ag and also for HIV Ag. They also used monocytes/macrophages; CTLs specific for pathogen or HIV Ag, pathogen debris, and HIV debris; and nonstate functions of state variables for macrophages presenting Ag for pathogen (Mx) or HIV (My). They also used a nonstate function for "factor" (F), which is evidently pooled, stimulatory immunocytokines secreted by active Th following antigen presentation by Mx and My. Reibnegger et al. did not include HIV-infected macropages, B-cells, antibodies, NK cells, or specific immunocytokines in their model, a simplification which we will accept considering the detail in the rest of their model. Figures 10.21A, B, and C illustrate the 19 compartments (states) used by Reibnegger et al. and the transfer rates between them. The Simnon model for the 19-state model is given below. Note that any line preceded by a quotation mark is a comment (nonexecutable).

```
CONTINUOUS SYSTEM reibsys2
" Version of 6/14/88. Rev. 11/30/94, 5/30/98.
" Original 19-state Reibnegger model with R's numbers.
" See: 'Theoretical implications of cellular immune reactions against
" helper lymphocytes infected by an immune system retrovirus.'
" Reibnegger, G. et al; P.N.A.S. 84: 7270 - 7274, Oct. 1987.
"
TIME t " Weeks.
"
STATE x1 x2 x3 x4 x5 x6 x7 x8 x9 x10 x11 x12 x13 x14 x15 x16 x17 x18 x19
"
DER  dx1 dx2 dx3 dx4 dx5 dx6 dx7 dx8 dx9 dx10 dx11 dx12 dx13 dx14 dx15
DER  dx16 dx17 dx18 dx19
"
" STATE EQUATIONS:......................................................
"
" x1 = p.c. pathogen P
dx1 = r*rx1/(1 + rx1/kr) - k1*rx1*rx7/(kk + rx1) -k2*rx1*
rx3/(kk + rx1) + pin
"
" x2 = p.c. precursor CTL w/ specificity to P.
dx2 = ic*(1 + h*F)/(kf + F)) - A*rx2*rx1 - e1*rx2
"
" x3 = p.c. CTL w/ specificity to P.
dx3 = A*rx2*rx1 - e1*rx3 + p*F*rx3/(kf + F)
"
" x4 = p.c. of precursor Th w/ TCR specificity to P.
dx4 = ih*(1 + h*F)/(kf + F)) - A*rx4*Mx -e1*rx4 -a1*rx4*rx10/(ky + rx10)
"
" x5 = p.c. of Th w/ TCR specificity to P.
dx5 = A*rx4*Mx - e1*rx5 + p5*F*rx5/(kf + F) - k5*rx5*rx10/(ky + rx10)
"
```

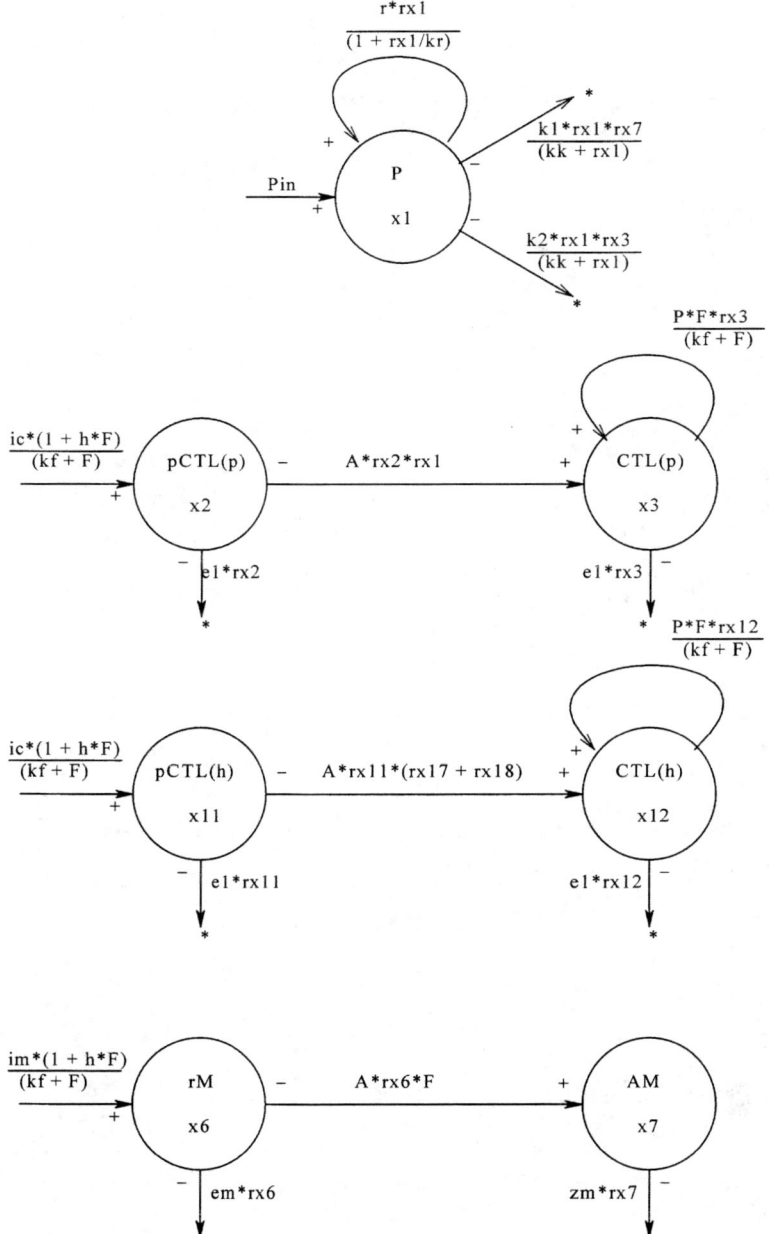

FIGURE 10.21A Compartments in the 19-state model of HIV infection of Reibnegger et al.,[117] *Reibsys2*. The self-feedback on states x1, x3, and x12 represents cytokine-induced, clonal expansion.

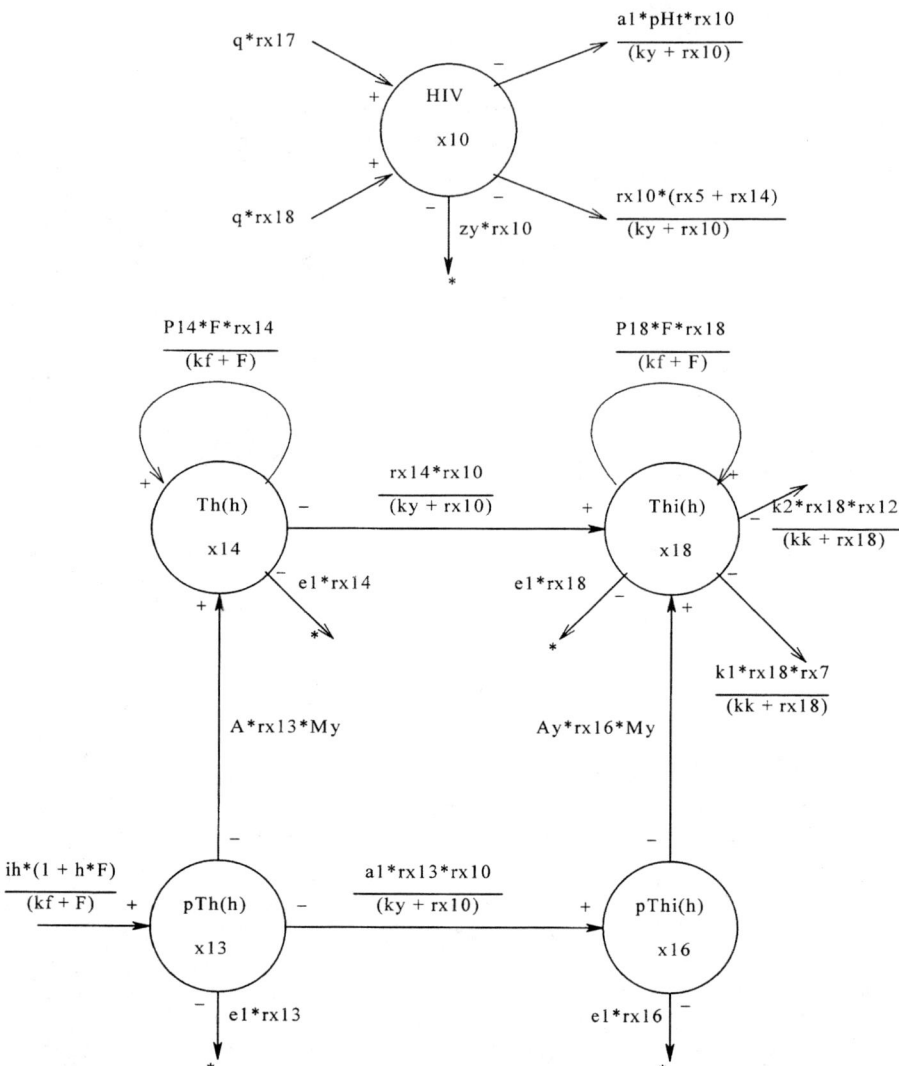

FIGURE 10.21B More compartments in the Reibnegger model, *Reibsys2*.

```
" x6 = p.c. of resting macrophages.
dx6 = im*(1 + h*F)/(kf + F)) - A7*rx6*F - em*rx6
"
" x7 = p.c. of cytotoxic (angry) macrophages.
dx7 = A7*rx6*f - zm*rx7              .
"
" x8 = p.c. of pathogen debris (Dp).
dx8 = k1*rx1*rx7/(kk + rx1) + k2*rx1*rx3/(kk + rx1) - ed*rx8
"
" x9 = p.c. of precursor Th w/ TCR specificity exclusive of pathogen.
```

The Human Immune System Seen from a Biomedical Engineering Viewpoint 407

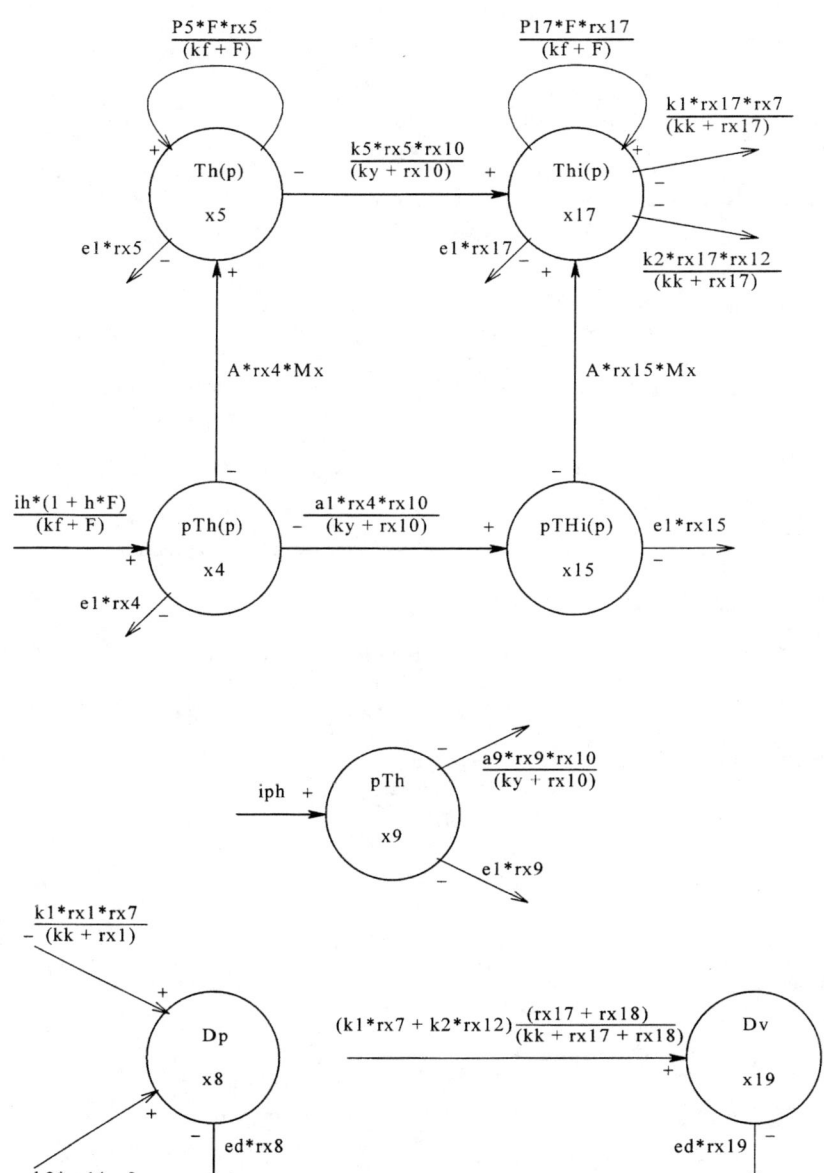

FIGURE 10.21C More compartments in the Reibnegger model, *Reibsys2*.

```
dx9 = iph - e1*rx9 - a9*rx9*rx10/(ky + rx10)
"
" x10 = p.c. of free HIV virions.
dx10 = q*(rx17 + rx18) - zy*rx10 - a1*pHt*rx10/(ky + rx10) + addx10
addx10 = -(rx5 + rx14)*rx10/(ky + rx10) + av1in + av2in
"
```

```
" x11 = p.c. of preCTL specific for HIV (gp120).
dx11 = ic*(1 + h*F)/(kf + F)) - A*rx11*(rx17 + rx18) - e1*rx11
"
" x12 = p.c. of CTL specific for HIV.
dx12 = A*rx11*(rx17 + rx18) - e1*rx12 + p*F*rx12/(kf + F)
"
" x13 = p.c. of preTh w/ TCR specific for HIV.
dx13 = ih*(1 +h*F)/(kf +F)) - Ay*rx13*My - e1*rx13 - a1*rx13*
rx10/(ky + rx10)
"
" x14 = p.c. of Th w/ TCR specific for HIV.
dx14 = A*rx13*My - e1*rx14 + p14*F*rx14/(kf + F) - rx14*rx10/(ky + rx10)
"
" x15 = p.c. of HIV-infected, preTh w/ TCR specific for pathogen P.
dx15 = a1*rx4*rx10/(ky + rx10) - A*rx15*Mx - e1*rx15
"
" x16 = p.c. of HIV-infected preTh w/ TCR specific for HIV.
dx16 = a1*rx13*rx10/(ky + rx10) - Ay*rx16*My - e1*rx16
"
" x17 = p.c. of HIV-infected Th w/ TCR specific for pathogen P.
dx17 = k5*rx5*rx10/(ky + rx10) + A*rx15*Mx + p17*F*rx17/(kf + F) + addx17
addx17 = - e1*rx17 - k1*rx17*rx7/(kk + rx17) - k2*rx17*rx12/(kk + rx17)
"
" x18 = p.c. of HIV-infected Th w/ TCR specific for HIV.
dx18 = rx14*rx10/(ky + rx10) + Ay*rx16*My + p18*F*rx18/(kf + F) + addx18
addx18 = - e1*rx18 - k1*rx18*rx7/(kk + rx18) - k2*rx18*rx12/(kk + rx18)
"
" x19 = p.c. of HIV debris (sgp120).
dx19 = ((rx17 + rx18)/(kk + rx17 + rx18))*(k1*rx7 + k2*rx12) - ed*rx19
"
" INPUTS ............
av1in = if t <t1avinj then 0 else if t<(t1avinj + dt) then vinj1 else 0
av2in = if t <t2avinj then 0 else if t<(t2avinj + dt) then vinj2 else 0
"
pin1 = if t<t1pinj then 0 else if t <(t1pinj + dt) then pinj1 else 0
pin2 = if t<t2pinj then 0 else if t <(t2pinj + dt) then pinj2 else 0
pin = pin1 + pin2
q = if t < t1avinj then 0 else qmax
r = if t < t1pinj then 0 else rv
"
" QUASI-SS VARIABLES & FUNCTIONS.................................
"
Mx = rx8*(rx6 + rx7)/(kd + rx8)      "APC's for pathogen.
My = rx19*(rx6 + rx7)/(kd + rx19)    "APC's for viral gp-120 protein.
"
" Lymphoid "factors" resulting from antigen presentation to Th.
F = (rx5 +rx17 *Mx/kf +Mx*(rx5 +rx17)) +(rx14 +rx18)*My/(kf +My*(rx14
+rx18))
"
pHt  = rx4 + rx9 + rx13                   " Total ininfected pre-Th.
Ht   = rx4 + rx5 + rx9 + rx13 + rx14      " Total uninfected CD4+Th
Hit  = rx18 + rx17                        " Total HIV infected Th.
TOTHt = rx4 + rx5 + rx9 + rx13 + rx14 + rx15 + rx16 + rx17 + rx18
```

```
Ratio = Hit/TOTHt          " Ratio of HIV-infrcted Th to Total Th.
"
" ...RECTIFICATION OF STATES...
rx1  = if x1  < 0 then 0 else x1
rx2  = if x2  < 0 then 0 else x2
rx3  = if x3  < 0 then 0 else x3
rx4  = if x4  < 0 then 0 else x4
rx5  = if x5  < 0 then 0 else x5
rx6  = if x6  < 0 then 0 else x6
rx7  = if x7  < 0 then 0 else x7
rx8  = if x8  < 0 then 0 else x8
rx9  = if x9  < 0 then 0 else x9
rx10 = if x10 < 0 then 0 else x10
rx11 = if x11 < 0 then 0 else x11
rx12 = if x12 < 0 then 0 else x12
rx13 = if x13 < 0 then 0 else x13
rx14 = if x14 < 0 then 0 else x14
rx15 = if x15 < 0 then 0 else x15
rx16 = if x16 < 0 then 0 else x16
rx17 = if x17 < 0 then 0 else x17
rx18 = if x18 < 0 then 0 else x18
rx19 = if x19 < 0 then 0 else x19
"

" LOGGING STATES & Variables ......
lx1  = if x1  < .01 then -2 else log(x1)
lx2  = if x2  < .01 then -2 else log(x2)
lx3  = if x3  < .01 then -2 else log(x3)
lx4  = if x4  < .01 then -2 else log(x4)
lx5  = if x5  < .01 then -2 else log(x5)
lx6  = if x6  < .01 then -2 else log(x6)
lx7  = if x7  < .01 then -2 else log(x7)
lx8  = if x8  < .01 then -2 else log(x8)
lx9  = if x9  < .01 then -2 else log(x9)
lx10 = if x10 < .01 then -2 else log(x10)
lx11 = if x11 < .01 then -2 else log(x11)
lx12 = if x12 < .01 then -2 else log(x12)
lx13 = if x13 < .01 then -2 else log(x13)
lx14 = if x14 < .01 then -2 else log(x14)
lx15 = if x15 < .01 then -2 else log(x15)
lx16 = if x16 < .01 then -2 else log(x16)
lx17 = if x17 < .01 then -2 else log(x17)
lx18 = if x18 < .01 then -2 else log(x18)
lx19 = if x19 < .01 then -2 else log(x19)
lMy = if My < .01 then -2 else log(My)
lMx = if Mx < .01 then -2 else log(Mx)
lf  = if f  < .01 then -2 else log(f)
lHt = if Ht < .01 then -2 else log(Ht)
lHit = if Hit < .01 then -2 else log(Hit)
lTOTHt = if TOTHt < .01 then -2 else log(TOTHt)
lRatio = if Ratio < .01 then -2 else log(Ratio)
"
" CONSTANTS:..................................................
```

```
a1:0.01         " conversion raye const., pThp to Thp.
A:0.001         " activition rate const.
Ay:0.1          " conversion rate const. pre to active Thvi
A7:9.E-4        " activation rate for macrophages.
a9:0.1          " viral loss constant for pH cells.
e1:0.02         " efflux of precursor T-cells.
ed:2.0          " efflux of debris, per day.
em:0.05         " efflux of macrophage, M.
h:9.0           " inflammation constant.
ic:0.25         " influx of the precursor T-cells into the compartment.
ih:5e-2         " influx of pHx.
im:1.25e5       " macrophage influx rate.
inbc:2.0        " influx rate of unactivated (naive) B-Cells.
iph:1e4         " influx of pH.
k1:10           " killing capacity constant.
k2:10           " killing capacity constant.
k5:1.2          " HIV infection r.c., Thp to Thpi
kd:1e7          " presentation saturation.
kf:65.          " lymphoid factor saturation.
kk:5e4          " killing saturation constant.
kt:1e3          " restimulation saturation.
ky:1e3          " saturation constant.
kr:1e7          " pathogen repro. saturation const.
p:1             " activated CTL prolif. factor.
p5:1            " activ. Th prolif. factor.
p14:2           " activ. Thv prolif factor.
p17:2           " activ. Thpi prolif. factor.
p18:4           " activ. Thv1 prolif. factor.
qmax:1e3        " IDV production constant, max.
rv:0.95         " pathogen replication rate.
dt:3            " time over which infusion given.
t1avinj:0       " time of first injection of HIV.
t2avinj:1000    " time of 2nd HIV injection.
t1pinj:75       " time of first introduction of pathogen.
t2pinj:300      " time of second pathogen introduction.
zm:1.414        " effector macrophage decay rate, per day.
zy:0.693        " decay constant.
vinj1:100       " First injection of HIV virus.
vinj2:100       " Second injection of HIV virus.
pinj1:1e5       " Injected pathogen cells; first injection.
pinj2:1e5       " Second injection of pathogen cells.
"                 For actual values used, see parameter file REIBPAR8.T below.
"
" INITIAL CONDITIONS ...............................................
" See parameter file REIBPAR8.T below.
"
END
```

A great deal of line space in this program is taken up with the rectification of states and logging of states and variables. Rectification is required because in the simulation of large nonlinear systems with poorly known rate constants, states can go negative under certain conditions of initial conditions and inputs. Negative states are meaningless when dealing with population densities and concentrations; hence rectification is used to preserve nonnegativity on the ordinary differential equations. Logging states allows us to use one set of axes to display all states, because in sets of stiff equations, which we have here, some states can be very large while others approach zero.

The program above follows the Reibnegger et al. model faithfully with the exception of the quasi-steady-state variable F for "lymphoid factors." In their original paper, Reibnegger et al. state: "These factors, F, are assumed to (i) be identical kinetically, (ii) be produced by helper T-cells Hx upon presentation of antigen X [X is P, the pathogen], (iii) be produced instantaneously, and (iv) decay rapidly. Thus the different factors [e.g., IL2, 3, 4, 5, 6, 7, 10, 13, γIFN, αIFN, GMCSF, TNFβ, etc.] are simply represented by *one* quasi-steady-state variable: F = H$_x$ * X/(k$_f$ + X)." Here, $H_x = x5$ and $X = x1$ above. Antigen presentation by macrophages, as we have seen, initially involves an MHC II molecule containing the Ag binding with a helper TCR specific for that antigen. Following the binding of the Th's CD4 to the MHC II molecule, and activation of CD3, the Th is activated and, among other things, releases the many "factors" comprising F. Reibnegger's expression for F implies, by mass-action kinetics, that the Th combines directly with pathogen. This is not the case. The Th combines with an antigen-presenting macrophage bearing the pathogen antigen. Reibnegger et al. give an algebraic expression for the population concentration (pc) of antigen-presenting macrophages presenting pathogen antigen, M_x. In their notation:

$$M_x = D_x(M + M^*)/(k_d + D_x)$$

Here, D_x is the pc of pathogen debris (after lysing of pathogen by macrophages and CTLs). $D_x = x8$. M = pc of resting macrophages = $x6$. M^* = pc of activated cytotoxic macrophages = $x7$.

A more realistic expression for F involving the presentation of pathogen and HIV antigen to appropriate Th cells is

$$F = (rx5 + rx17) * M_x / [k_f + M_x^* (rx5 + rx17)]$$
$$+ (rx14 + rx18) * M_y / [k_f + M_y^* (rx14 + rx18)]$$

Here, we assume that antigen-presenting macrophages with pathogen Ag combine with Ths with TCRs specific for pathogen Ag. Both normal Th and Th infected with HIV are considered. The factors so produced are given saturation by use of a Hill function. The second term mirrors the first term except it is HIV antigen (gp120) that is being presented to Ths with TCRs specific for HIV Ag. The immunocytokines produced by the antigen-presentation-activated Th are identical and independent of the Ag presented. In the model they stimulate Th and CTL cell clonal expansion and the input of precursor cells.

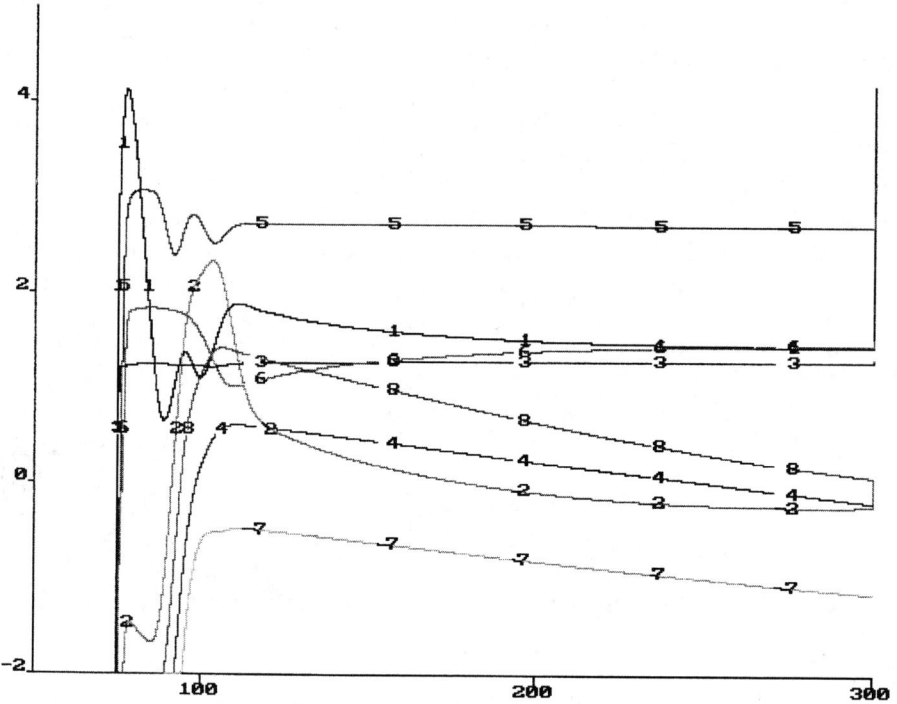

FIGURE 10.22A Simulation of HIV infection and pathogen growth using the *Riebsys2* program in the text with REIBPAR8 parameters. At $t = 0$ days, the model is "infected" with 100 HIV virions. At $t = 75$, 1.E5 pathogen pd is inoculated. The pathogen expands rapidly, then is quickly brought down to about 1.5 log units by immune system action, giving a chronic infection. Following the pathogen peak, there are peaks of the pd of HIV and Hit. Trace 1 = $x1$ = pathogen, trace 2 = $x10$ = HIV, trace 3 = $x3$ = CTL(p), trace 4 = $x12$ = CTL(v), trace 5 = $x7$ = AM, trace 6 = $x5$ = Th(p), trace 7 = $x14$ = Th(v), trace 8 = Hit.

Figure 10.22 illustrates how the model responds to an HIV input at $t = 0$, then a pathogen inoculation at $t = 75$ weeks, and a further infection by pathogen at week 300. Note that the pathogen, when first introduced, is fought to an innocuous, chronic pc. The parameters used in the simulation were

```
[reibsys2] " Reibpar8.t, 7/28/98
x1:0, x2:30, x3:0, x4:15, x5:0, x6:2.5E5, x7:0, x8:0, x9:5.E5,
x10:0, x11:30, x12:0, x13:15, x14:0, x15:0, x16:0, x17:0, x18:0,
x19:0.
kr:1.E6, k1:90, kk:6.E4, k2:90, ic:0.25, h:1, kf:50, A:5.E-3,
e1:0.02, p:1, ih:1.5, a1:0.2, ky:1000, p5:1, k5:1, im:6.E4,
A7:1.E-3, em:0.05, zm:1, ed:2, iph:1.E4, a9:0.01, zy:2, Ay:1.E-3,
p14:1, p17:1, p18:1, t1avinj:0, dt:2, vinj1:100, t2avinj:250,
vinj2:0, t1pinj:75, pinj1:3000, t2pinj:300, pinj2:5.E4, qmax:2500,
rv:0.8, kd:1.E7, inbc:2, kt:1000.
```

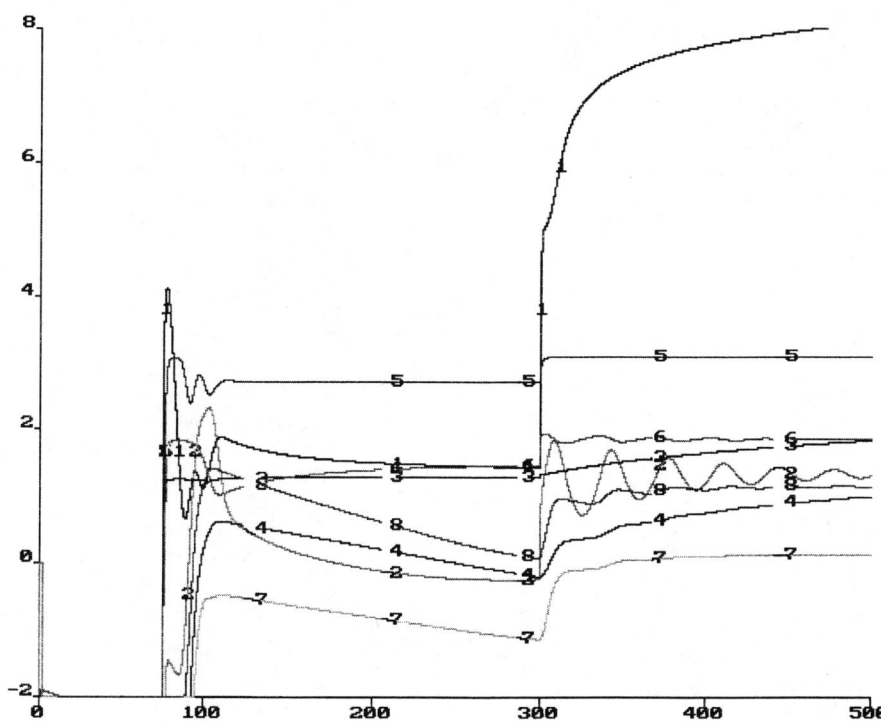

FIGURE 10.22B The same simulation as Figure 10.22A, except longer time scale used. A second inoculation of 1.E5 pathogen pd is given at $t = 300$. The pathogen rapidly grows and overcomes the "weakened" model immune system, leading to AIDS. Note the strong growth of the pd of free HIV and Hit in the AIDS condition for $t \geq 400$. Trace $1 = x1 =$ pathogen, trace $2 = x10 =$ HIV, trace $3 = x3 =$ CTL(p), trace $4 = x12 =$ CTL(v), trace $5 = x7 =$ AM, trace $6 = x5 =$ Th(p), trace $7 = x14 =$ Th(v), trace $8 =$ Hit.

As a result of immune system activation, HIV virions are released by infected Th cells. When the pathogen is introduced for the second time, the immune system is weakened, and it grows rapidly, in effect killing the model host. As in the immune system vs. cancer model of DeBoer et al., it is possible to introduce to the model exogenous immunotherapies, such as monoclonally grown CTLs specific for HIV antigen.

In order to explore more detailed immunotherapies, including specific exogenous immunocytokines, the author and certain graduate students in 1990 developed a more detailed mathematical model based on Reibnegger's approach, but including, in addition to the cells modeled by Reibnegger et al., a *viral pathogen* that infects certain somatic cells in the body (SCp), NK cells, B-cells specific for HIV Ag and viral pathogen Ag, plasma B-cells secreting Abs for pathogen and HIV, HIV infection of macrophages as well as CD4+ Th cells, states for IL2 and γIFN, and quasi-steady-state variables for IL1, antigen-presenting macrophages (not B-cells), and B-cell factors (e.g., IL6) causing Ab release. The author's model uses 32 states. While complex to set up and interpret, it allows more depth in simulations than do previous models. Like all of the models described in this section, it is an oversimplification of nature. Many rate constants are taken from the literature, and others are "reasonable estimates." The model allows us to investigate immunotherapies such as exogenous inputs of IL2, γIFN, AZT, monoclonal CTL specific for HIV, etc. Our HIV Simnon program, AIDS12, follows:

```
CONTINUOUS SYSTEM AIDS12       "v.0411prime
"
" Program to simulate IDV infection of the immune system.  Includes IDV
" infection of Th and Mp cells, B cells and antibodies to IDV and viral
" pathogen.  Mps and NK cells have Fc receptors.  32 States.  04/11/90.rev,
" 6/01/98
"
STATE  P  pCx  Cx  M  AM  Mi  AMi  pHx  Hx  pHxi  Hxi  pCy  Cy  pHy  Hy  pHyi
STATE  Hyi  Dv  Bx  By  PBx  PBy  AbX  AbY  pNK  NK  gIFN  Y  SCp  mBx  mBy  IL2
"
DER  dP  dpCx  dCx  dM  dAM  dMi  dAMi  dpHx  dHx  dpHxi  dHxi  dpCy  dCy  dpHy  dHy  dpHyi
DER  dHyi  dDv  dBx  dBy  dPBx  dPBy  dAbx  dAby  dpNK  dNK  dgIFN  dY  dSCp  dmBx  dmBy
DER  dIL2
"
TIME  t

"STATE EQUATIONS ........................
"
" P is viral pathogen.
dP =DELAY(rSCp*fab, 1)/(1 + rP/kps) - (k1*rAM + k2*rAMi)*rP - k3*rP*rAbX
+c0
c0 = Pin - k33*rP
"
" SCp are somatic cells infected by viral pathogen, P.
dSCp = k88*rP - k33*rSCp - k5*rSCp*rAbX*rNK - k4*rCx*rSCp - k1*rSCp*rAM
"
" pCx are precursor CTL for pathogen.
dpCx = ic*ff - k8*rpCx*rSCp*fIL2 - k9*rpCx
"
" Cx are activated CTL for pathogen.
dCx = k8*rpCx*rSCp*fIL2 - k9*rCx + pf*rCx*ff*fIL2
"
" M are resting macrophages.
dM = im*ff - k11*rM - k12*rM*fgIFN - k13*rM*satY
"
" AM are activated macrophages secreting IL1, etc.
dAM = k12*rM*fgIFN - k15*rAM - k13*rAM*satY
"
" Mi are resting macrophages infected with HIV.
dMi = k13*rM*satY - k17*rCy*rMi - k18*rMi - k31*rAbY*rMi*rNK -
k12*rMi*fgIFN
"
" AMi are activated macrophages infected with HIV.
dAMi = k12*rMi*fgIFN + k13*rAM*satY -k17*rCy*rAMi -k31*rAbY*rAMi*rNK + c2
C2 = -k19*rAMi
```

" Y is HIV virions.
dY = f20*satHi*fIL2+ f21*(satAMi +satAPi) + avin1 + avin2 + avin3 + avin4 +C3
C3 = avin5 - k34*rY -k25*rY*rAbY -k23*(rHx + rHy + rpHx +rpHy + rpH)*satY +C4
C4 = -k24*rY*sCD4
"
" pNK are precursor NK cells.
dpNK = ink*ff + k26*rpNK*fIL2 - k28*rpNK - k29*rpNK*fIL2
"
" NK are active NK cells.
dNK = k29*rpNK*fIL2 - k30*rNK
"
" pCy are precursor CTL specific for HIV Ag.
dpCy = ic*ff - k8*rpCy*totVI*fIL2 - k9*rpCy
"
" Cy are active CTL specific for HIV Ag.
dCy = k8*rpCy*totVI*fIL2 - k9*rCy + pf*rCy*ff*fIL2
"
" pHx are precursor Th with TCR for pathogen Ag.
dpHx = ih*ff -k37*rpHx*satY -k35*rpHx-k36*rpHx*(APCxi+APCx)/(1+rDv/k32)
"
" Hx are mature Th with TCR for pathogen Ag.
dHx = k36*rpHx*APCx/(1+rDv/k32) - k38*rHx - k37*rHx*satY + k39*rHx*fIL2
"
" pHxi are precursor CD4+ Th w/ TCR specific for patho. infected w/ HIV.
dpHxi = k37*rpHx*satY - k40*rpHxi - k41*rpHxi*rAbY*rNK
"
" Hxi are mature Th w/ TCR for pathogen Ag, infected with HIV.
dHxi = k39*rHxi*fIL2 + k37*rHx*satY - (k46*rCy + k45*(rAM + rAMi))*rHxi + c10
c10 = -k41*rHxi*rAbY*rNK + k36*rpHx*APCxi/(1 + rDv/k32) - k43*rHxi
"
" pHy are precursor CD4+ Th w/ TCR specific for HIV Ag.
dpHy = ih*ff - k36*rpHy*APCy/(1+rDv/k32) -k37*rpHy*satY -k35*rpHy + c16
c16 = -k36*rpHy*APCyi/(1 + rDv/k32)
"
" Hy are mature Th w/ TCR for HIV Ag.
dHy = k36*rpHy*APCy/(1+rDv/k32) + k39*rHy*fIL2 - k38*rHy - k37*rHy*satY
"
" pHyi are pre-Th w/ TCR for HIV Ag, infected w/ HIV.
dpHyi = k37*rHy*satY - k40*rpHyi - k41*rpHyi*rAbY*NK
"
" Hyi are mature Th w/ TCR for HIV Ag, infected w/ HIV.
dHyi = - k43*rHyi + k37*rHy*satY - k41*rHyi*rAbY*rNK + k39*rHyi*fIL2 + c13
c13 = -(k46*rCy + k45*(rAM + rAMi))*rHyi + k36*rpHy*APCyi/(1 + rDv/k32)
"
" Dv is HIV debris (e.g., gp120).
dDv = (rMi +APCxi +APCyi +rpHxi +rHxi +rpHyi +rHyi)*(k50*rCy+k51*rNK*AbY) +c14
c14 = - k52*rDV + k34*rY + Dvin
"

" gIFN is gamma interferon, units.
dgIFN = k82*satHx*(APCx + APCxi)/(k87 + APCx + APCxi) + gIFNin + c21
c21 = - k9*rgIFN + k82*satHy*(APCy + APCyi)/(k68 + APCy + APCyi)
"

" Interleukin 2
dIL2 = IL2in +k70*fIL1*rHx*(APCx +APCxi) + k72*fIL1*rHy*(APCy +APCyi)+
C22
C22 = - k96*rIL2
"

"..B-CELL SECTION ....
"

" By are inactive B-cells specific for HIV Ag.
dBy = k54*rnBy*BCF*satV + k58*rBy*BCFy*fIL2 - k56*rBy - k57*rBy*BCF -
k92*rBy
"

" Bx are inactive B-cells specific for pathogen Ag.
dBx = k55*rnBx*satP*BCF + k58*rBx*BCFx*fIL2 - k56*rBx - k61*rBx*BCF -
k92*rBx
"

" PBy are plasma B-cells making Ab for HIV Ag.
dPBy = k57*rBy*BCF - k59*rPBy + k91*rmBy*(fIL2 - 1)
"

" PBx are plasma B-cells making Ab for pathogen Ag.
dPBx = k61*rBx*BCF - k59*rPBx + k91*rmBx*(fIL2 - 1)
"

" AbX are Abs for pathogen Ag.
dAbX = k63*rPBx*BCF - k64*rAbX - k60*rAbX*SatP
"

" AbY are Abs for HIV Ag.
dAbY = k63*rPBy*BCF - k64*rAbY - k66*rAbY*satV + c15
c15 = - k67*(rHxi + rHyi + rpHxi + rpHyi + rMi + rAMi + APCyi + APCxi)*rAbY
"

" mBx are memory B-cells for pathogen Ag.
dmBx = k92*rBx - k62*rmBx - k91*rmBx*(fIL2 - 1)
"

" mBy are memory B-cells for HIV Ag.
dmBy = k92*rBy - k62*rmBy - k91*rmBy*(fIL2 - 1)
"

" QUASI-STEADY-STATE VARIABLES ....
"

rDx = rP*k49 + rSCp*(rAbX*rNK + rCx)*k89 + rSCp*k90   " Pathogen debris
"

APCx = rM*satP*fIL2*k76 +k77*rBx*fIL2*satP   " Antigen presenting Mp
"   for patho.
"

APCxi = rMi*satP*fIL2*(k78 +k79*rAbX)   " HIV-infected Antigen presenting
"   Mp for patho.
"

APCyi = rMi*satV*fIL2*(k78 +k79*rAbY)   " HIV-infected Antigen presenting
"   Mp for HIV.
"

APCy = rM*satV*fIL2*k76 +k77*rBy*fIL2*satV   " Antigen presenting Mp
"   for patho.
"

The Human Immune System Seen from a Biomedical Engineering Viewpoint

```
IL1   = k80*(rM + rMi)*rP/(1+ rP/kps) + k81*(rM + rMi)*satHx + c22
c22   = k81*(rM + rMi)*satHy + k80*(rM + rMi)*rY + IL1c
"
fIL1  = 1 + 9*IL1/(kIL1 + IL1)
"
fIL2  = (1 + 9*rIL2/(k10 + rIL2))
"
fgIFN = 1 + 9*rgIFN/(k27 + rgIFN)
"
BCFx  = 1 + kbcf*fIL1*rHx/(k16 + rHx*fIL1)
"
BCFy  = 1 + kbcf*fIL1*rHy/(k83 + rHy*fIL1)
"
BCF   = BCFx + BCFy
"
qv1   = k20max - dk20*AZT/(k85 + AZT)   " Functions to model reduced HIV
"                                          reproduction rate due to AZT.
qv2   = k21max - dk21*AZT/(k85 + AZT)
"
f20   = if t < tav1inj then 0 else qv1  " Variable rate constants to
"                                          model effect of AZT
f21   = if t < tav1inj then 0 else qv2
"
Thtot = (rHx + rHxi + rHy + rHyi)
"
f     = foo + k84*(rHx + rHy)*IL1/(kIL1 + IL1)
"
ff    = 1 + 9.0*f/(kf + f)
"
fab   = rv/(1 + ABI/k7)
"
satV  = (rY + rDv)/(1 + (rY + rDv)/k73)
"
satP  = (rP + rDx)/(1 + (rP + rDx)/k71)
"
satY  = rY/(k14 + rY)
"
satHx = rHx/(1 + rHx/k69)
"
satHy = rHy/(1 + rHy/k69)
"
satHi = (rHxi + rHyi)/(1 + (rHxi + rHyi)/k44)
"
satAMi = rAMi/(1 + rAMi/k32)
"
satAPi = (APCxi + APCyi)/(1 + (APCxi + APCyi)/k44)
"
Htot  = rpHx + rpHy + rHx + rHy + rpH
"
Hitot = rpHxi + rpHyi + rHxi + rHyi
"
"Hratio = Hitot/(Htot + Hitot)
"
"totH  = Htot + Hitot
```

```
"
"Miratio = (rMi + rAMi)/(rM + rAM + rMi + rAMi)
"
totVI = rMi + rAMi + APCXi + APCyi + rpHxi + rpHyi + rHxi + rHyi

" INPUTS ..................................
"
avin1 = if t < tav1inj then 0 else if t<(tav1inj + dt) then vinj1 else 0
avin2 = if t < tav2inj then 0 else if t<(tav2inj + dt) then vinj2 else 0
avin3 = if t < tav3inj then 0 else if t<(tav3inj + dt) then vinj3 else 0
avin4 = if t < tav4inj then 0 else if t<(tav4inj + dt) then vinj4 else 0
avin5 = if t < tav5inj then 0 else if t<(tav5inj + dt) then vinj5 else 0
"
Pin1 = if t < tp1inj then 0 else if t < (tp1inj + dt) then pinj1 else 0
Pin2 = if t < tp2inj then 0 else if t < (tp2inj + dt) then pinj2 else 0
Pin3 = if t < tp3inj then 0 else if t < (tp3inj + dt) then pinj3 else 0
Pin4 = if t < tp4inj then 0 else if t < (tp4inj + dt) then pinj4 else 0
Pin5 = if t < tp5inj then 0 else if t < (tp5inj + dt) then pinj5 else 0
Pin = Pin1 +Pin2 +Pin3 +Pin4 +Pin5
"
IL2in = if t<til2inj then 0 else if t <(til2inj + dt1) then IL2inj else 0
"
gIFNin = if t <tgifinj then 0 else if t <(tgifinj + dt2) then gIFinj else 0
"
sCD4 = if t<tcd4inj then 0 else if t<(tcd4inj + dt3) then sCD4inj else 0
"
AZTin1 = if t<taz1inj then 0 else if t<(taz1inj+dtaz) then AZin1 else 0
AZTin2 = if t<taz2inj then 0 else if t<(taz2inj+dtaz) then AZin2 else 0
AZTin3 = if t<taz3inj then 0 else if t<(taz3inj+dtaz) then AZin3 else 0
AZTin4 = if t<taz4inj then 0 else if t<(taz4inj+dtaz) then AZin4 else 0
AZT = AZTin1 + AZTin2 + AZTin3 + AZTin4
"
ABIin1 = if t<tab1inj then 0 else if t<(tab1inj +dtab) then ABin1 else 0
ABIin2 = if t<tab2inj then 0 else if t<(tab2inj +dtab) then ABin2 else 0
ABIin3 = if t<tab3inj then 0 else if t<(tab3inj +dtab) then ABin3 else 0
ABIin4 = if t<tab4inj then 0 else if t<(tab4inj +dtab) then ABin4 else 0
ABI = ABIin1 + ABIin2 + ABIin3 + ABIin4
"
Dvin = if t< tdvinj then 0 else if t<(tdvinj + dt6) then Dvinj else 0
"
" RECTIFICATION of STATES ................
"
" LOGGING STATES & VARS. ..................
"
" CONSTANTS ..............................
" See PAR727.T file below.
"
" INPUT CONSTANTS ........................
" See PAR727.T file
"
" INITIAL CONDITIONS > 0 .................
" See PAR727C.T file
"
END
```

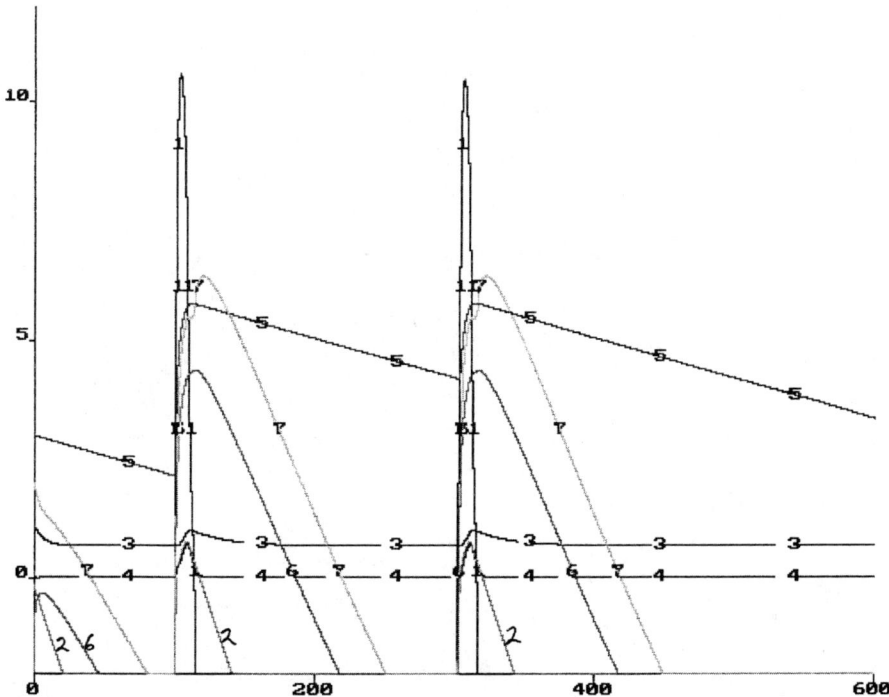

FIGURE 10.23 Simulated response of the 32-state model of the immune system, "AIDS12," to two inoculations of a viral pathogen when HIV not present. Vertical scale, log units; horizontal scale, time in days. Scales are the same in all simulation plots. Trace 1 = log(P), trace 2 = log(Cx), trace 3 = log(AM), trace 4 = log(NK), trace 5 = log(Hx), trace 6 = log(PBx), trace 7 = log(Abx). Initial pathogen inoculation = 1.E4, second inoculation = 100.

All the simulations described below were done using Euler integration with Δt = 0.001 over a time scale of 0 to 600 days. The vertical axis is calibrated in $\log_{10}(pd)$ of the states and variables.

Figure 10.23 illustrates the response of the system when the model immune system is free of HIV virions. Pathogen (P) is introduced as an injection P at $t = 100$ days. P initially grows rapidly, then is quickly defeated by the immune system. A second pathogen injection is given at $t = 300$ with the same effect. In Figure 10.24, an initial inoculation of pathogen is given at $t = 100$; the pathogen is quickly destroyed by the "normal" immune system model. An inoculation of 1000 HIV virions is given at $t = 200$. A second pathogen inoculation occurs at $t = 300$. However, this time the immune system is weakened by the HIV infection, and there is pathogen regrowth to a steady-state level, representing ARC/AIDS. Figure 10.25 illustrates what happens in the model when 1.E4 "units" of AZT are infused from $t = 250$ to $t = 450$. Up to $t = 250$, the system's response is the same as in Figure 10.25. Note that at $t = 250$, the pd of HIV begins decreasing; the free HIV pd [2] crashes at about $t = 390$, but the pd of total HIV-infected Th cells (Hitot) [8] remains

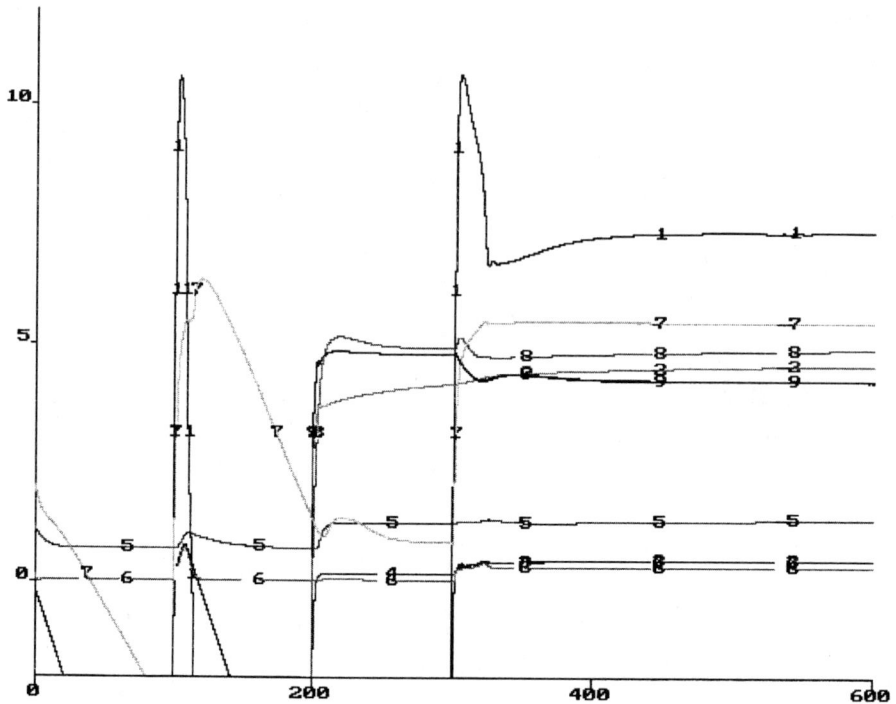

FIGURE 10.24 Response of the 32-state model when 10^3 HIV virions are inoculated at $t = 200$. Trace 1 = log(P), trace 2 = log(Y) [Y = HIV pd], trace 3 = log(Cx), trace 4 = log(Cy), trace 5 = log(AM), trace 6 = log(NK), trace 7 = log(AbX), trace 8 = log(PBy), trace 9 = log(Hitot). After second pathogen spike, P reaches a chronic, steady-state pd corresponding to ARC/AIDS.

high. Curiously, the pathogen pd regrows rapidly to a third peak at $t = 430$ as a result of the AZT infusion, and then abruptly decreases when AZT infusion is stopped at $t = 450$. This behavior is counterintuitive. Following the cessation of AZT infusion, the pds of pathogen and HIV again increase. Figure 10.26 illustrates the model's response to an infusion of IL2 of 1.E3 "units"/time from $t = 250$ to $t = 450$. Again, the normal immune model conquers the pathogen inoculated at $t = 100$. HIV is introduced at $t = 200$, as in the previous simulations. Now, a simulated infusion of IL2 of 1.E3 "units"/day is begun at $t = 250$ and continued until $t = 450$. Note that there is an increase in the slope of the log HIV concentration and a jump in the log pd of antibodies (AbX) for the viral pathogen when IL2 infusion is begun. At the second pathogen inoculation, its pd spikes and then remains flat at a log pd of about 6.3. When the IL2 infusion is stopped, the pathogen again grows rapidly. To investigate what happens if exogenous IL2 is given along with AZT, we examine Figure 10.27. Here, an IL2 infusion of 100 units/time and an AZT infusion of 250 units/time are given from $t = 250$ to $t = 450$. (These are the minimum simulated doses to cause the pathogen pd to $\rightarrow 0$.) After the IL2 and AZT infusions are stopped, the

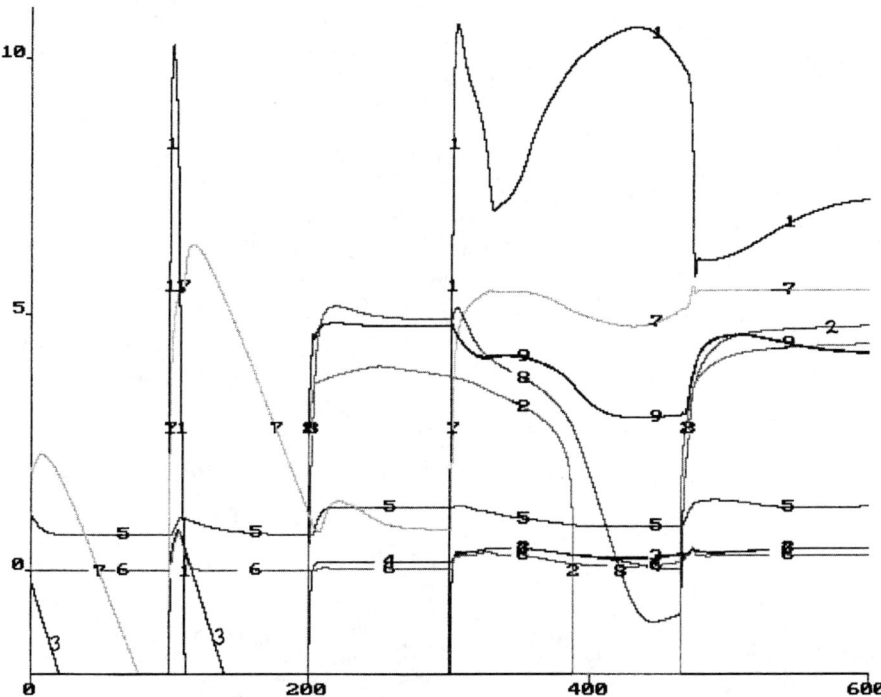

FIGURE 10.25 Response of the 32-state model to pathogen inoculation at $t = 100$, HIV infection at $t = 200$, and a second pathogen inoculation at $t = 300$. An infusion of 1.E4 "units"/day of AZT is begun at $t = 250$ and ends at $t = 450$. Trace 1 = log(P), trace 2 = log(Y) [Y = HIV pd], trace 3 = log(Cx), trace 4 = log(Cy), trace 5 = log(AM), trace 6 = log(NK), trace 7 = log(AbX), trace 8 = log(PBy), trace 9 = log(Hitot). Note bizarre behavior of pathogen during AZT infusion. Also note that log(HIV) < –2 at about $t = 390$. Free HIV increases once the AZT infusion is stopped.

pathogen tries to regrow, but the system has regained enough strength to cause the pathogen to $\to 0$ at $t = 570$. Stronger doses of AZT and IL2 cause the pathogen pd to crash at about $t = 450$. This simulation illustrates the possible merit of multimodal therapies to treat HIV infection.

The constant simulation parameters for the figures above are

```
[AIDS12]  "  PAR727C 7/27/98
P:0, pCx:0.9, Cx:0.7, M:2.45E4, AM:13.5, Mi:0, AMi:0, pHx:128,
Hx:6.2E5, pHxi:0, Hxi:0, pCy:1.65, Cy:0, pHy:2.47E5, Hy:0, pHyi:0,
Hyi:0, Dv:0, Bx:10, By:0, PBx:0, PBy:0, AbX:100, AbY:0, pNK:0.2,
NK:1.3, gIFN:1.66, Y:0, SCp:0, mBx:100, mBy:0, IL2:0, kps:1.E9,
k1:9.E-6, k2:5.E-5, k3:5.E-5, k33:0.01, k88:0.3, k5:9.E-6,
k4:9.E-6, ic:0.25, k8:3.E-5, k9:0.2, pf:10.E-5, im:1000, k11:0.05,
k12:2.E-4, k13:5.E-4, k15:1, k17:0.01, k18:0.05, k31:2.E-5, k19:1,
k34:4.E-3, k25:10.E-7, k23:10.E-7, rpH:1.E6, k24:0.2, ink:0.5,
k26:0.125, k28:0.3, k29:1, k30:0.5, ih:1.E4, k37:0.05, k35:0.05,
```

```
k36:4.E-3,  k32:1.E7,  k38:0.02,  k39:5.E-4,  k40:0.1,  k41:2.E-5,
k46:1,  k45:2,  k43:0.2,  k50:0.02,  k51:0.02,  k52:0.5,  k82:2.E-4,
k87:1.E5,  k68:5000,  k70:10.E-11,  k72:10.E-11,  k96:1,  k54:0.01,
rnBy:10,  k58:2.E-3,  k56:0.1,  k57:0.05,  k92:5.E-5,  k55:0.03,
rnBx:10,  k61:0.01,  k59:0.2,  k91:1.E-3,  k63:5,  k64:0.3,  k60:5.E-4,
k66:5.E-3,  k67:5.E-3,  k62:3.E-4,  k49:0.3,  k89:0.1,  k90:0.1,
k76:10.E-11,  k77:10.E-7,  k78:10.E-5,  k79:10.E-6,  k80:5.E-7,
k81:5.E-7,  IL1c:0.2,  kIL1:1.E5,  k10:10,  k27:10,  kbcf:9,  k16:3.E5,
k83:2.E5,  k20max:0.4,  dk20:0.498,  k85:10,  k21max:0.4,  dk21:0.498,
foo:0.5,  k84:10.E-5,  kf:20,  rv:250,  k7:10,  k73:1.E5,  k71:5000,
k14:300,  k69:5000,  k44:200,  tav1inj:0,  dt:4,  vinj1:0,  tav2inj:60,
vinj2:0,  tav3inj:120,  vinj3:0,  tav4inj:150,  vinj4:0,  tav5inj:200,
vinj5:1000,  tp1inj:0,  pinj1:0,  tp2inj:30,  pinj2:0,  tp3inj:60,
pinj3:0,  tp4inj:100,  pinj4:1.E4,  tp5inj:300,  pinj5:100,  til2inj:250,
dt1:200,  IL2inj:100,  tnd1inj:150,  dt2:240,  ND1inj:0,  tcd4inj:150,
dt3:180,  CD4inj:0,  taz1inj:250,  dtaz:200,  AZin1:250,  taz2inj:240,
```

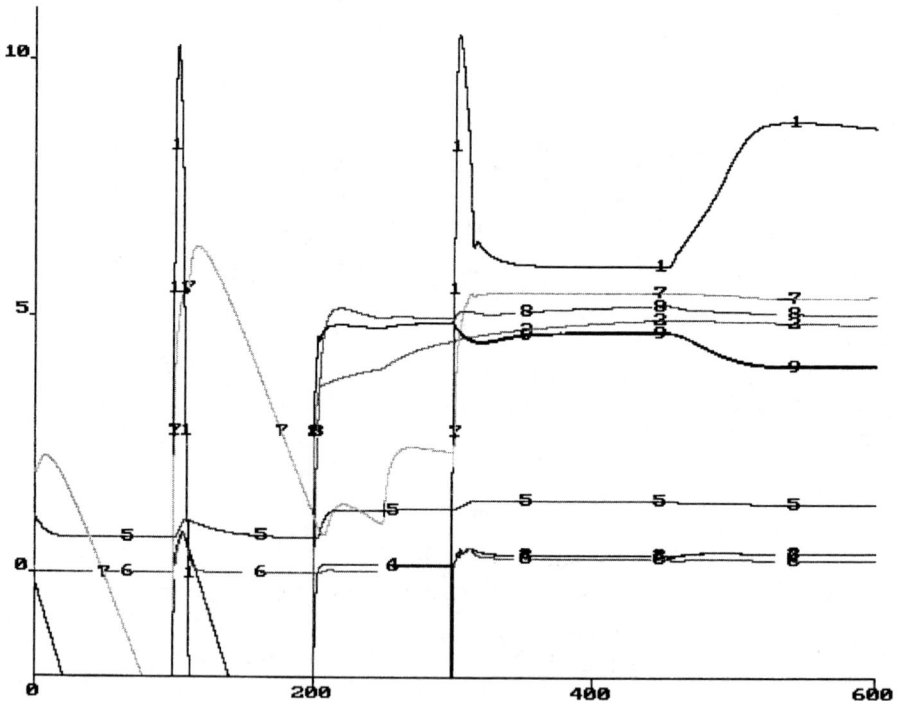

FIGURE 10.26 Response of the 32-state model to pathogen inoculation at $t = 100$, HIV infection at $t = 200$, and a second pathogen inoculation at $t = 300$. An infusion of IL2 of 1.E3 "units"/day is begun at day 250 and stopped at day 450. Trace 1 = log(P), trace 2 = log(Y) [Y = HIV pd], trace 3 = log(Cx), trace 4 = log(Cy), trace 5 = log(AM), trace 6 = log(NK), trace 7 = log(AbX), trace 8 = log(PBy), trace 9 = log(Hitot). Note the jump in the pd of AbX when the IL2 infusion is started. Also, IL2 causes the pd of HIV virions to increase at a higher rate because it stimulates viral reproduction in various infected Th cells. At the cessation of IL2 infusion, the pathogen pd increases dramatically (remember the log scale).

The Human Immune System Seen from a Biomedical Engineering Viewpoint 423

```
AZin2:0, taz3inj:300, AZin3:0, taz4inj:440, AZin4:0, tab1inj:300,
dtab:300, ABin1:0, tab2inj:240, ABin2:0. tab3inj:300, ABin3:0,
tab4inj:440, ABin4:0, tdvinj:140, dt6:3, Dvinj:0, kcd4:100,
inbc:150, k6:2, k22:0.1, k42:0.1, k47:0.05, k48:10.E-5, k53:0.1,
k65:2, k74:10.E-5, k75:10.E-10, k86:2.E-3, k93:10.E-5, k94:100.
k95:10.E-5
```

Other possible therapies for HIV can also be explored using the model. For example, exogenous inputs of γIFN, externally cultured and propagated CTL specific for HIV (Cy), and antiviral drug for pathogen can be tried. Also, various scenarios for sCD4 infusion can be investigated. As discussed previously, sCD4 will not only decrease the number of active gp120 sites for HIV attachment to CD4+ cells, thus reducing the rate of HIV infection, but it may also decrease the effective number of MHC II antigen-presenting cells by binding to MHC II and preventing antigen presentation. (These possible scenarios can be inserted in the appropriate ordinary differential equations and algebraic equations.)

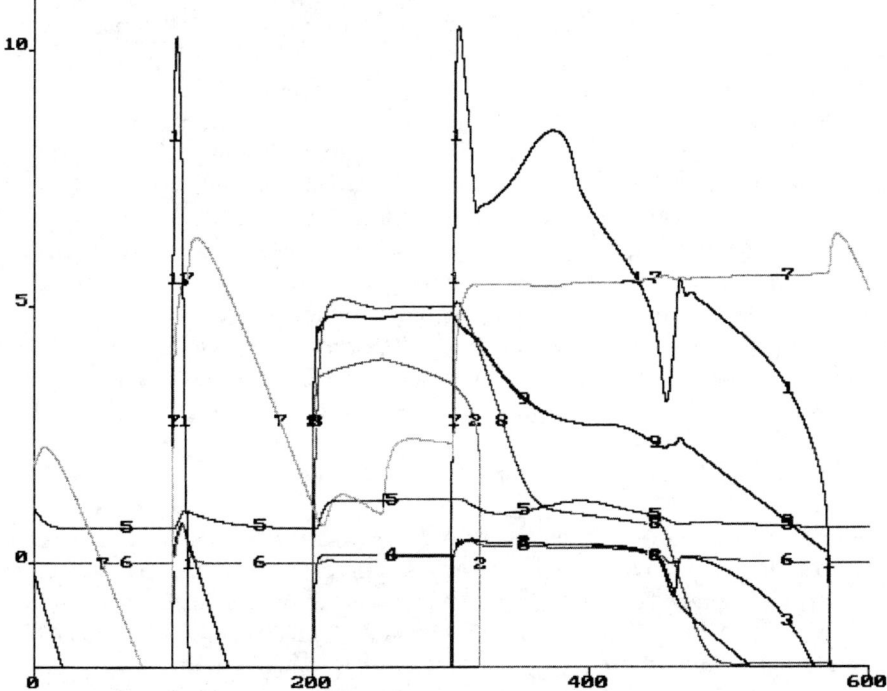

FIGURE 10.27 Response of the 32-state model to pathogen inoculation at $t = 100$, HIV infection at $t = 200$, and a second pathogen inoculation at $t = 300$. In this case, a multimodal therapy is explored. Both exogenous IL2 at 100 units/day and 250 units/day of AZT are infused at $t = 250$ to $t = 450$. Trace 1 = log(P), trace 2 = log(Y) [Y = HIV pd], trace 3 = log(Cx), trace 4 = log(Cy), trace 5 = log(AM), trace 6 = log(NK), trace 7 = log(AbX), trace 8 = log(PBy), trace 9 = log(Hitot). Note that the pathogen regrows at the cessation of infusions, but crashes eventually at about $t = 570$ as a result of this minimal dose therapy.

10.6 CHAPTER SUMMARY

In this chapter, the reader was introduced to the complexity of the human immune system, and some basic attempts to mathematically model immune system function were illustrated. The basic types of cells active in the immune system (monocytes/macrophages, T-lymphocytes, NK cells, B-cells, mast cells, and platelets) were described first. All immune cells, as the result of their growth and maturation and their responses to immunocytokines secreted by other immune system cells, can be further categorized by their stages of morphological and functional development. Such parsing of immune system cells is somewhat arbitrary and adds to the complexity of modeling the immune system.

Antigen presentation was described as a complex biochemical process whereby the immune system is activated and its response is amplified and gains molecular specificity for an antigen. Macrophages, B-cells, and virally infected somatic (nonimmune) cells are the major antigen presenters. Following antigen presentation by somatic cells, clones of cytotoxic T-cells with specificity for the particular antigen are produced. Antigen presented by a B-cell causes its clonal expansion and production of antibodies with specificity for the presented Ag. Antigen presentation is thus the major activation process for specific immune response.

Immune system autacoids (also known as immunocytokines), including the interleukins, interferons, tumor necrosis factors, prostaglandins, and chemokines, are secreted by immune system cells and diverse somatic cells. These signaling substances must bind to receptor proteins on their target cells. After binding, specific internal biochemical processes are initiated, leading to events such as cell growth, differentiation, clonal expansion, and production of receptor proteins and other immunocytokines. Autacoid receptors and other immune system cell surface molecules are generally classified as *cluster of differentiation* (CD) antigens. Over 182 CD molecules have been described.[169] For example, CD127 found on bone marrow lymphoid precursor cells, pro-B-cells, mature T-cells, and monocytes is the IL7 receptor; CD25 found on activated T-cells, B-cells, and monocytes is the receptor for IL2; and CD37 found on mature B- and T-cells and myeloid cells has unknown function.

One of the problems in formulating a mathematical model of an immune system scenario is immunocytokine *pleiotropy*; that is, a given cytokine may have different effects on different cells. Also, many immune system "battles" with pathogens are local; they are fought in a small volume around the point of pathogen introduction to the body. Thus there are local concentration gradients for secreted immunocytokines. The cells at the focus of the battle are subject to the highest concentrations; those at the periphery receive reduced stimulation. Such scenarios properly should involve spatiotemporal diffusion equations. The assumption of a well-mixed pool of molecules required for mass-action kinetics can seldom be made. A further complication in modeling detailed immune system dynamics results from the finite number of receptors on a cell's surface. It is necessary to invoke saturating

Hill functions to account for the finite numbers of receptors and rate-limiting steps in the biochemical machinery activated by the receptors binding with their specific autacoid.

To illustrate the art of immune system modeling, examples of the immune system vs. cancer, and the immune system vs. HIV were chosen. Both models from the literature and the author's own work were described. The models are generally nonlinear, high-order, and stiff. Stiff ordinary differential equations require special integration techniques to minimize errors and numerical artifacts.

The central dichotomy in modeling any sort of scenario in the immune system is that validated, lifelike behavior lies in the details of the model, which, in turn, requires a number of rate constants that increases faster than a linear function of the number of states. In fact, if k states or compartments ($k \geq 2$) can exchange mass bidirectionally *with each state* of the group of k, then there will be a *maximum* of N rate constants governing these mass exchanges. It can be shown that N is given by:

$$N = 2\left[2k - 3 + \sum_{m=3}^{k}(k-m)\right]$$

This relation reduces to the simpler quadratic form $N = k^2 - k$. N vs. k is plotted in Figure 10.28. In practice, there will be fewer than N rate constants for a k-state system because obviously not every compartment exchanges mass with every other compartment, and some mass transfers are one-way. Note that loss rates from compartments were not considered in finding N; only exchange rates were. Note that for $k = 32$, $N = 992$. Thus the burden of validating a system with a large (and more realistic) number of states rests on identifying or correctly estimating an oppressively large number of rate constants.

An important goal of developing validated, high-order models of the immune system vs. cancer and HIV is to be able to model proposed adoptive immunotherapies. The results of mathematical modeling scenarios can suggest *in vitro* and animal experiments to further pursue novel therapies. Data already exist on how the immune system responds to exogenous IL2 and γIFN in the case of cancer and soluble CD4 protein in the case of HIV infection. Since it is now known that HIV gp120 protein not only must bind to CD4 but also requires an adjacent chemokine receptor protein as well to effect inclosion into Th cells and macrophages, it should be possible to qualitatively model a therapy that artificially blocks the chemokine protein. Such blocking may inhibit chemotaxis of certain immune system cells, but this untoward effect may be far offset by the reduction of HIV infection. The ability of HIV to develop resistance to certain drugs such as AZT, 3TC, and Abacavir can also be modeled by adjusting appropriate rate constants in time.

Because of its ability to hide dormant in certain immune system cells that lie far from locally stimulatory cytokines, it is probable that no therapy will ever "cure"

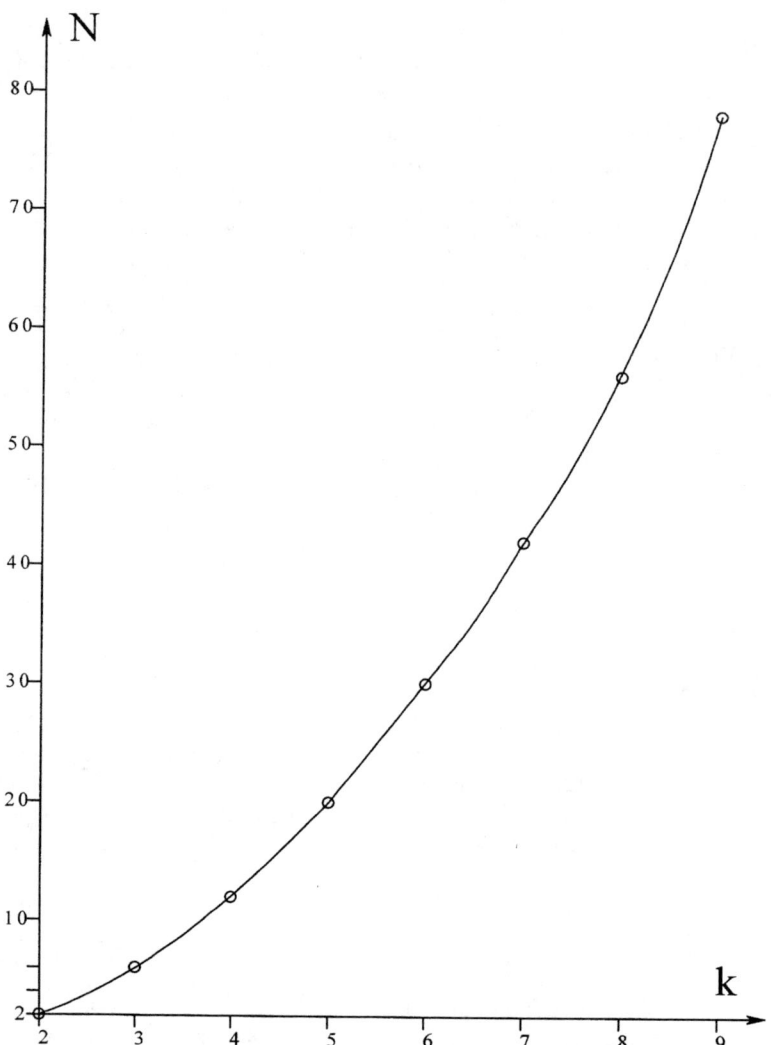

FIGURE 10.28 Graph of the maximum number of rate constants, N, for a k-compartment pharmacokinetic/mass-action system where every compartment can exchange mass bidirectionally with every other compartment. See text for discussion.

HIV infection by wiping out all HIV virions. Certain combinations of anti-HIV drugs appear to be able to reduce blood HIV titers by as much as a factor of 3.2×10^{-3} and restore the CD4+ Th cell count to near normal. The more vigorous immune system is then able to successfully fight the opportunistic infections that give rise to ARC/AIDS, and the patient can lead a near-normal life.

PROBLEMS

10.1 A six-state mathematical model of the effect of HIV infection on CD4+ T-cells was proposed by Hraba and Dolezal (1996) (www.cdc.gov/ncidod/EID/vol2no4/hraba.htm). The system is written in Simnon notation. Note that there are no specific cytokines in the model. pT4 = concentration of immature CD4+ T-cells, T4 = concentration of mature CD4+ T-cells, pT8 = concentration of immature CD8+ T-cells, T8 = concentration of mature CD8+ T-cells, HIV = concentration of free HIV, and CTH = concentration of cytotoxic T-cells specific for HIV coat protein.

$$dpT4 = \{I_T + k18*[(T4_0 - T4) + (T8_0 - T8)]\}/D - k1*pT4 - k2*HIV*CTH*pT4 \tag{1}$$

$$dT4 = k1*pT4 - k3*T4 - k4*HIV*CTH*T4 \tag{2}$$

$$dpT8 = \{I_T + k18*[(T4_0 - T4) + (T8_0 - T8)]\}/(1.5*D) - k5*pT8 \tag{3}$$

$$dT8 = k5*pT8 - (K6 - k7)*T8 \tag{4}$$

$$dHIV = HIV*(k8 - k8 - k10*CTH) \tag{5}$$

$$dCTH = k11*HIV*(k12*I_c + k13*CTH)*T4/T4_0 - (k14 - k15)*CTH \tag{6}$$

$$D = \begin{cases} 1 & \text{If } \ln(HIV/HIV_0) < k16 \ [\ln(x) = \text{natural log of } x] \\ k17*\ln(HIV/HIV_0) & \text{If } \ln(HIV/HIV_0) \geq k16 \end{cases} \tag{7}$$

Time is in days. I_T = input rate of T-cell concentration, per day = 1.0. I_c = input rate of CTH concentration = 0.2 per day. k1 = 0.2, k2 = 0, k3 = 0.01, k4 = 20, k5 = 0.2, k6 = 0.1, k7 = 0, k8 = 0.02, k9 = 0, k10 = 0.3, k11 = 1, k12 = 0.512, k13 = 0.7, k14 = 0.01, k15 = 0, k16 = 3, k17 = 3.5, k18 = 0.01. Initial conditions: $pT4_0 = 5$, $T4_0 = 100$, $pT8_0 = 3.33$, $T8_0 = 66.7$, $CTH_0 = 0$, $HIV_0 = 0.0005$.

(Note that in simulating the six ordinary differential equations above, some states may go negative. If this happens, physical reality has been violated [there are no negative concentrations of cells], and you will need to rectify the states used on the right-hand sides of the ordinary differential equations. For example, use rHIV = IF HIV > 0 THEN HIV ELSE 0, rT8 = IF T8 > 0 THEN T8 ELSE 0, etc.)

For plotting purposes, it may be necessary to use the $\log_{10}$ of the states if certain cell numbers increase dramatically while others keep a low value during a simulation. For example, plot lpT4 = IF pT4 < .001 THEN -3 ELSE log(pT4), etc.

The purpose of this simple model is to explore some various therapies against HIV infection of CD4+ T-cells.

A. AZT is given in a therapeutic dose starting at a time T_o for a duration of N days. You are to determine the best T_o and N to bring pT4 and T4 toward normal levels and to cause HIV to have minimum growth rate. AZT infusion is modeled by parametrically increasing k9 from 0 to 0.005 during the infusion.

B. Anti-CD8 antibodies (Ab8) grown *in vitro* are given to reduce the number of CD8+ T-cells. This therapy adds loss rate terms to Equations 3 and 4: –Kab*pT8*Ab8 and –Kab*T9*Ab8, respectively. See what happens. Try different Kab, Ab8 (assume constant, with T_o and N to optimize effect on reducing HIV and raising T4).

C. Explore how different initial conditions affect the simulations, e.g., try $HIV_0 = 0.001$, $pT4_0 = 20$, etc., one at a time.

10.2 The Simnon program below, "TUMOR.5," crudely describes the immune system events that take place in the vicinity of a rapidly growing tumor. If no external injections are given, the tumor will grow steadily until it kills its host, a rat. Inspection of the state and auxiliary equations reveals that it is possible to simulate bolus injections or infusions of IL1, IL2, γIFN, and monoclonally grown (*in vitro*) activated cytotoxic T-cells with specificity for the Ca (mTc).

Determine the minimum doses (if they exist) of the above substances and cells (each therapy taken one at a time) that will cause regression of the tumor. Observe plots of $\log_{10}$(CA, Th, Tc, NK, MP, AbCA, IL1, IL2, gIFN) for 0 to 150 weeks. You may alter the time and duration of the infusions, but do not alter the model's parameters. Note that the ordinary differential equations in this model are stiff and may give noisy results. Try the Dormand and Prince/Runge–Kutta integrator (DOPRI45R) or Euler integration with a small ΔT (here you will trade off run time for noise). Note that the Ca has a bistable behavior; it either dies out or grows. Curious.

CONTINUOUS SYSTEM TUMOR5
```
" This model represents the third version of an EE 320 Immune model
" modified 14/10/88; B-Cell subsystem added. RBN. Tinkered with 5/08/98.
"
" DEFINE STATE VARIABLES:  (18 States)
" pNK = precursor natural killer cells
" NK = natural killer cells
" pTh = precursor T-helper cells
" Th = T-helper cells
" pTc = precursor cytotoxic T-cells
" aTc = 1st stage activated CTL.
" Tc = fully activated cytotoxic T-cells
" pMp = precursor Macrophages
" aMp = 1st stage activated Macrophages
```

The Human Immune System Seen from a Biomedical Engineering Viewpoint

```
"  Mp   = fully active Macrophages
"  Lys  = cellular debris
"  CA   = Cancer/tumor.
"  IL2  = Interleukin-2
"  nB   = non-activated B-cells.
"  aB   = Antigen-activated B-cells.
"  pB   = Plasma B-cells specific for CA.
"  AbCA = Antibodies (free & cellular) for CA.
"  mBCA = Memory cells for CA.
"
TIME t
"
STATE pNK NK pTh Th pTc aTc Tc pMp aMp Mp Lys CA IL2 nB aB pB AbCA mBCA
"
DER dpNK dNK dpTh dTh dpTc daTc dTc dpMp daMp dMp dLys dCA dIL2 dnB daB
DER dpB dAbCA dmBCA
"
"************* STATE EQUATIONS **************
dpNK =(1 + inflam)*mf1 - mat1*rpNK + ep1*rpNK - rpNK/tpNK
dNK  = mf2 + mat1*rpNK - rp1*rNK - rNK/tNK
"
dpTh = (1 + inflam)*mf3 - apf*rpTh - rpTh/tpTh
dTh  = mf4 + rpTh*apf + ep2p*rTh - rTh/tTh
"
dpTc = (1 + inflam)*mf5 - rpTc*apf - rpTc/tpTc
daTc = rpTc*apf - raTc*(gIFN + rIL2)/(1 + PGE2) - raTc/taTc + rTc/tTc
dTc  = raTc*(gIFN +rIL2)/(1 +PGE2) + rTc*IL2*c6/(sf10 + rIL2) + addTc
addTc = dmTc - rTc/tTc
"
dpMp = (1 + inflam/(1 + PGE2))*mf6 - gIFN*rpMp - rpMp/tpMp
daMp = gIFN*rpMp - raMp*rTh*rCA*c7/(sf11 + rCA*raMp*rTh) + addaMp
addaMp = -raMp/taMp + rMp/t2Mp
dMp  = raMp*rTh*rCA*c7/(sf11 + rCA*raMp*rTh) - rMp/tMp - rMp/t2Mp
"
dLys = rCA*(rNK*co1 + (rMp + c5*raMp)*co2 + rTc*co3)*c14 - rLys/tLys
"
dCA  = gf*rCA/(1 + rCA/grs) - nvmp*rCA - c26*rCA*rAbCA + addCA
addCA = -rCA*(rNK*co1 + (rMp + c5*raMp)*co2 + rTc*co3)*c14
"
dIL2 = IL2in - lf4*rIL2 + rCA*rTh*c13/(sf26 + rCA*rTh)
"
dnB  = (1 + inflam)*mf7 - c18*rnB - c19*rCA*rnB*fil2
daB  = c19*rCA*rnB*fil2 - c20*raB + (c21-c22)*raB*BCFca*factor - raB*c27
dpB  = c22*factor*BCFca*raB - c23*rpB + c29*rmBCA*BCFca
dAbCA = c24*rpB*rCA/(sf32 + rpB) - c25*rAbCA - c26*rAbCA*rCA
dmBCA = c27*raB - c28*rmBCA - c29*rmBCA*BCFca
"
"                    AUXILLARY EQUATIONS:
"       CYTOKINES:
gIFN = rTh*IL1/(sf9 + rTh) + rNK/(sf23 + rNK) + gIFNin
IL1  = c16*(raMp + rMp)*rTh*rCA/((sf19 + rTh)*(sf20 + rCA)) + IL1in
inflam = c17*factor/(sf27 + factor)
fil2 = c30*rIL2/(sf29 + rIL2) + 1
```

```
BCFca  = IL1*rTh*rCA*c31/(sf30 + rTh)
PGE2   = c8*rMp/(sf12 + rMp)
aIFN   = c10*(rMp + raMp)/(sf17 + rMp + raMp) + rNK/(sf24 + rNK)
factor = rTh*IL1/(sf28 + IL1)
apf    = agf*(raMp +rMp)*rLys*(gIFN/(1 +PGE2))/(sf4 +(raMp +rMp)*rLys)
cxf1   = aIFN*c11/(sf16 + aIFN) + IL2*c12/(sf18 + IL2)
"
"     INPUTS:
IL2in  = if t<til2inj then 0 else if t<(til2inj+dt2) then il2inj else 0
gIFNin = if t<tgifinj then 0 else if t<(tgifinj+dt3) then gifinj else 0
IL1in  = if t<til1inj then 0 else if t<(til1inj+dt1) then il1inj else 0
dmTc   = if t < 10 then 0 else if t < (10 + dt5) then mTcinj else 0
"
"     OTHER:
mat1 = c15*(gIFN + IL2)/(sf1 + gIFN + IL2)
ep1  = c1*(gIFN/(sf2 + gIFN) + IL2)
rp1  = c2*PGE2/(sf3 + PGE2)
ep2p = (IL1/(1 + PGE2))/(sf9 + rTh)
co1  = cxf1/(sf13 + rNK*rCA)
co2  = cxf2/(sf14 + (raMp*c5 + rMp)*rCA)
co3  = cxf3/(sf15 + rTc*rCA)
"
"        RECTIFICATION of STATES
rpNK  = if pNK  < 0 then 0 else pNK
rNK   = if NK   < 0 then 0 else NK
rpTh  = if pTh  < 0 then 0 else pTh
rTh   = if Th   < 0 then 0 else Th
rpTc  = if pTc  < 0 then 0 else pTc
raTc  = if aTc  < 0 then 0 else aTc
rTc   = if Tc   < 0 then 0 else Tc
rpMp  = if pMp  < 0 then 0 else pMp
raMp  = if aMp  < 0 then 0 else aMp
rMp   = if Mp   < 0 then 0 else Mp
rLys  = if Lys  < 0 then 0 else Lys
rCA   = if CA   < 0 then 0 else CA
rIL2  = if IL2  < 0 then 0 else IL2
rnB   = if nB   < 0 then 0 else nB
raB   = if aB   < 0 then 0 else aB
rpB   = if pB   < 0 then 0 else pB
rAbCA = if AbCA < 0 then 0 else AbCA
rmBCA = if mBCA < 0 then 0 else mBCA
"
"        LOGGING
lpNK = if pNk < .1 then -1 else log(pNK)
lNK  = if NK  < .1 then -1 else log(NK)
lpTh = if pTh < .1 then -1 else log(pTh)
lTh  = if Th  < .1 then -1 else log(Th)
lpTc = if pTc < .1 then -1 else log(pTc)
laTc = if aTc < .1 then -1 else log(aTc)
lTc  = if Tc  < .1 then -1 else log(Tc)
lpMp = if pMp < .1 then -1 else log(pMp)
laMp = if aMp < .1 then -1 else log(aMp)
lMp  = if Mp  < .1 then -1 else log(Mp)
lLys = if Lys < .1 then -1 else log(Lys)
```

```
lCA = if CA < .1 then -1 else log(CA)
lIL2 = if IL2 < .1 then -1 else log(IL2)
lIL1 = if IL1 < .1 then -1 else log(IL1)
lgIFN = if gIFN < .1 then -1 else log(gIFN)
lnB = if nB < .1 then -1 else log(nB)
laB = if aB < .1 then -1 else log(aB)
lpB = if pB < .1 then -1 else log(pB)
lAbCA = if AbCA < .1 then -1 else log(AbCA)
lmBCA = if mBCA < .1 then -1 else log(mBCA)
"
"                   CONSTANTS
tpNK:2
tNK:1
tpTh:15
tTh:5
tpTc:30
taTc:50
tTc:1
tpMp:40
taMp:50
tMp:5
t2Mp:100
tLys:0.5
mf1:0.1
mf2:0
mf3:25
mf4:1
mf5:20
mf6:200
mf7:8
agf:0.2
sf1:50
sf2:100
sf3:5000
sf4:3500
sf5:50
sf6:1000
sf7:10000
sf8:1.0e5
sf9:200
sf10:1000
sf11:2500
sf12:5000
sf13:1000
sf14:1000
sf15:5000
sf16:2000
sf17:1000
sf18:1000
sf19:1000
sf20:1000
sf21:1000
sf22:1000
sf23:200
```

```
sf24:600
sf25:2e3
sf26:1e3
sf27:15
sf28:5
sf29:20
sf30:100
sf31:5e3
sf32:500
lf1:0.0001
lf2:0.0001
lf3:0.0001
lf4:0.2
nvmp:0.0001
gf:1
grs:2000
c1:5.E-2
c2:0
c3:1
c4:3
c5:0.5
c6:10
c7:100
c8:5
c9:2
c10:10
c11:5
c12:2
c13:2
c14:0.5
c15:0.5
c16:0.1
c17:20
c18:0.1
c19:5.E-2
c20:0.1
c21:0.01
c22:0.2
c23:1e-2
c24:4
c25:0.01
c26:1.E-5
c27:0.1
c28:1e-4
c29:0.1
c30:9
c31:3.3E-5
cxf2:3
cxf3:900
dt1:7
dt2:14
dt3:7
dt5:1
tillinj:14
```

```
til2inj:14
tgifinj:14
gifinj:0
illinj:0
il2inj:0
mTcinj:0
"
"    INITIAL CONDITIONS:
pNK:2e3
NK:200
pTh:15
Th:10
pTc:25
aTc:10
Tc:0
pMp:2500
aMp:250
Mp:2
Lys:10
CA:1.0
IL2:0.5
nB:10
"
END
```

10.3 To better appreciate the complexity of the immune system, on a large sheet of paper sketch representations of all types of immune system cells, including bone marrow. Also include the autacoids they secrete, and show the target cells that have receptors for those autacoids, whenever known.

10.4 Make a list of all the autoimmune diseases you can find. Discuss the treatments used to relieve the symptoms of various autoimmune diseases.

10.5 Discuss the role of the immune system in organ transplants. What is meant by tissue matching (donor to acceptor)?

10.6 Review the literature on suppressor T-cells. Do they exist, or is *active* suppression of immune system function a transient, autacoid-induced behavior of other leukocytes, such as helper T-cells?

10.7 In certain individuals, many common foods and injectable substances cause a severe, life-threatening, *anaphylactic* immune reaction.
  A. Describe the symptoms and physiology of anaphylaxis.
  B. What emergency treatment is usually given for anaphylaxis?
  C. List some of the common foods and injectable substances that more frequently induce anaphylaxis. (Note that in some cases, it may be the way a food is prepared or cooked that causes the allergic reaction.)

10.8 Discuss the immunological role of mother's milk for neonates.

10.9 The mind–immune system causality is just beginning to be understood. Persons under severe, prolonged, physical and mental stress appear to have reduced immune system function and may be more likely to develop cancer or opportunistic infections.

A. List some of the physical, environmental, and social factors that create stress.
B. Describe the hormonal events triggered by stress. Discuss how certain hormones may influence immunity. (See, for example, references on stress and the general adaptation syndrome; see also Chapter 77 in Guyton[59] and Berczi, I. 1986. *Pituitary Function and Immunity*, CRC Press, Boca Raton, FL.)

10.10 The endogenous opioid peptide β-endorphin is found in certain CNS neurons. While endorphins are normally associated with relief from pain, β-endorphin also affects the immune system. Describe the role of β-endorphin in analgesia and how β-endorphin affects the immune system.

Bibliography and References

1. Ackerman, E. 1962. *Biophysical Science,* Prentice-Hall, Englewood Cliffs, NJ.
2. Ahearn, D.J. and C.E. Grim. 1974. Treatment of malignant hypertension with sodium nitroprusside, *Arch. Intern. Med.,* 183: 187–191.
3. Albisser, A.M. 1979. Devices for the control of diabetes mellitus, *Proc. IEEE,* 67: 1308–1320.
4. Albisser, A.M. et al. 1974. An artificial endocrine pancreas, *Diabetes,* 23: 389–396.
5. Alvis, J.M., J.G. Reves, J.A. Spain, and L.C. Sheppard. 1985. Computer-assisted continuous infusion of the intravenous analgesic fentanyl during general anesthesia — an interactive system, *IEEE Trans. Biomed. Eng.,* 32(5): 323–329.
6. Angelo, E.J., Jr. 1969. *Electronics: BJTs, FETs, and Microcircuits,* McGraw-Hill, New York, Section 15-3.
7. Arnsparger, J.M., B.C. McInnis, J.R. Glover, Jr., and N. Normann. 1983. Adaptive control of blood pressure, *IEEE Trans. Biomed. Eng.,* 30(3): 168–176.
8. Åström, K.J. and B. Wittenmark. 1989. *Adaptive Control,* Addison-Wesley, Reading, MA.
9. Bailey. J.J., J.E. Fletcher, et al. 1992. A kinetic model of CD4+ lymphocytes with the human immunodeficiency virus (HIV), *BioSystems,* 26: 177–183.
10. Barnhart, E.R. 1988. *Physicians' Desk Reference,* 42nd ed., Medical Economics, Montvale, NJ. (www.medec.com)
11. Behbehani, K. and R.R. Cross. 1991. A controller for regulation of mean arterial blood pressure using optimum nitroprusside infusion rate, *IEEE Trans. Biomed. Eng.,* 38(6): 513–521.
12. Bellomo, G. et al. 1982. Optimal feedback glycaemia regulation in diabetics, *Med. Biol. Eng. Comput.,* 20: 329–335.
13. Beyer, J., G. Schulz, and U. Cordes. 1983. A glucose-controlled insulin infusion system (Biostator®) in pancreatectomized patients, in *Artificial Systems for Insulin Delivery,* P. Brunetti et al., Eds., Raven Press, New York, 535–542.
14. Bischoff, K.B., R.L. Dedrick, et al. 1971. Methotrexate pharmacokinetics, *J. Pharm. Sci.,* 60(8): 1128–1133.
15. Botz, C.K. 1976. An improved control algorithm for an artificial β-cell, *IEEE Trans. Biomed. Eng.,* 23(3): 252–255.
16. Broekhuyse, H.M., J.D. Nelson, B. Zinman, and A.M. Albisser. 1981. Comparison of algorithms for the closed-loop control of blood glucose using the artificial beta cell, *IEEE Trans. Biomed. Eng.,* 28(10): 678–686.
17. Browne, A.F. 1998. A New Approach to Monitoring Glucose Concentrations Based on Reflection of Polarized Light from a Liquid/Lens Interface, and Detection by an Improved, Closed-Loop Optical Polarimeter, Ph.D. dissertation, University of Connecticut, Storrs.
18. Candas, B. and J. Radziuk. 1994. An adaptive plasma glucose controller based on a nonlinear insulin/glucose model, *IEEE Trans. Biomed. Eng.,* 41(2): 116–124.

19. Carson, E.R., C. Cobelli, and L. Finkelstein. 1983. *The Mathematical Modeling of Metabolic and Endocrine Systems,* John Wiley & Sons, New York.
20. Chestnut, H. and R.W. Mayer. 1951. *Servomechanisms and Regulating System Design,* Vol. 1, Wiley, New York.
21. Christiansen, J.S. et al. 1979. Studies in order to optimize constants used in the algorithms of the Biostator GCIIS, in *Proc. Workshop Artificial Beta Cells Diabetes Res. Management,* Héviz, Hungary, September, D-11.
22. Clapp, K.P., R.B. Northrop, and Q. Li. 1988. The immune system versus cancer: a modelling study, in *Proc. 10th Annu. IEEE/EMBS Conf.,* New Orleans, November 4 to 7, 1023–1024.
23. Clemens, A.H. and R.W. Meyers. 1997. Blood Glucose Control Apparatus, U.S. Patent 4,055,175, October 25.
24. Clement, K., C. Vaisse, et al. 1998. A mutation in the human leptin receptor gene causes obesity and pituitary dysfunction, *Nature,* 392: 398.
25. Cobelli, C. and A. Mari. 1983. Validation of mathematical models of complex endocrine–metabolic systems. A case study on a model of glucose regulation, *Med. Biol. Eng. Comput.,* 21: 390–399.
26. Cobelli, C. and A. Mari. 1985. Control of diabetes with artificial systems for insulin delivery — algorithm independent limitations revealed by a modeling study, *IEEE Trans. Biomed. Eng.,* 32(10): 840–845.
27. Cobelli, C. and A. Ruggeri. 1983. Evaluation of portal/peripheral route and of algorithms for insulin delivery in the closed-loop control of glucose in diabetes — a modeling study, *IEEE Trans. Biomed. Eng.,* 30(2): 93–103.
28. Collins, R.D. 1968. *Illustrated Manual of Laboratory Diagnosis,* J.B. Lippincott, Philadelphia.
29. Colvin, J.R. and G.N.C. Kenny. 1989. Automatic control of arterial pressure after cardiac surgery, *Anaesthesia,* 44: 37–41.
30. Considine, R.V. et al. 1996. Serum immunoreactive-leptin concentrations in normal weight and obese humans, *New Engl. J. Med.,* 334: 292.
31. Coté, G.L., M.D. Fox, and R.B. Northrop. 1992. Noninvasive optical polarimetric glucose sensing using a true phase measurement technique, *IEEE Trans. Biomed. Eng.,* 39(7): 752–756.
32. Coté, G.L., M.D. Fox, and R.B. Northrop. 1993. Optical Glucose Sensor Apparatus and Method, U.S. Patent 5,209,231, May 11.
33. Cramp, D.G. and E.R. Carson. 1981. The dynamics of short-term blood glucose regulation, in *Carbohydrate Metabolism: Quantitative Physiology and Mathematical Modeling,* C. Cobelli and R.N. Bergman, Eds., Wiley, New York, 349–367.
34. Davenport, W.B., Jr. and W.L. Root. 1958. *An Introduction to the Theory of Random Signals and Noise,* McGraw-Hill, New York.
35. DeAsla, R.A., A.M. Benis, et al. 1985. Management of postcardiotomy hypertension by microcomputer-controlled administration of sodium nitroprusside, *J. Thorac. Cardiovasc. Surg.,* 89: 115–120.
36. DeBoer, R.J., P. Hogeweg, et al. 1985. Macrophage T lymphocyte interactions in the anti-tumor immune response: a mathematical model, *J. Immunol.,* 134(4): 2748–2758.
37. DeBoer, R.J., S. Michelson, and P. Hogeweg. 1986. Concomitant immunization by the fully antigenic counterparts prevents modulated tumor cells from escaping cellular immune elimination, *J. Immunol.,* 136(11): 4319–4327.
38. Dorf, R.C. 1967. *Modern Control Systems,* Addison-Wesley, Reading, MA.

39. Etter, D.M. and S.D. Stearns. 1981. Adaptive estimation of time delays in sampled data systems, *IEEE Trans. Acoust. Speech Signal Process.*, 29(3): 582–589.
40. Fadali, M.S. 1987. Continuous drug delivery system design using nonlinear decoupling: a tutorial, *IEEE Trans. Biomed. Eng.*, 34(8): 650–653.
41. Finkelstein, L. and E.R. Carson. 1979. *Mathematical Modeling of Dynamic Biological Systems*, Vol. 3, Medical Computing Series, Hill, D.W., Ed., Research Studies Press, Forest Grove, OR.
42. Fischer, U. et al. 1978. Derivation and experimental proof of a new algorithm for the artificial B-cell based on the individual analysis of the physiological insulin–glucose relationship, *Endokrinologie*, 71: 65–75.
43. Fischer, U. et al. 1980. Assessment of an algorithm for the artificial B-cell using the normal insulin–glucose relationship in diabetic dogs and men, *Diabetologia*, 18: 97–107.
44. Fisher, M.E. 1991. A semiclosed-loop algorithm for the control of blood glucose levels in diabetics, *IEEE Trans. Biomed. Eng.*, 38(1): 57–61.
45. Fisher, M.E. and K.L. Teo. 1989. Optimum insulin infusion resulting from a mathematical model of blood glucose dynamics, *IEEE Trans. Biomed. Eng.*, 36(4): 479–486.
46. Flügge-Lotz, I. 1968. *Discontinuous and Optimal Control*, McGraw-Hill, New York.
47. Flügge-Lotz, I. and W.S. Wunch. 1955. On a nonlinear transfer system, *J. Appl. Phys.*, 26(4): 484–488.
48. Flügge-Lotz, I., C.F. Taylor, and H.E. Lindberg. 1958. Investigation of a Nonlinear Control System, NACA Tech. Report 1391.
49. Franklin, G., J.D. Powell, and A. Emami-Naeini. 1986. *Feedback Control of Dynamic Systems*, Addison-Wesley, Reading, MA.
50. Freund, E. 1975. The structure of decoupled nonlinear systems, *Int. J. Control*, 21: 443–450.
51. Furler, S.M., E.W. Kraegen, R.H. Smallwood, and D.J. Chisholm. 1985. Blood glucose control by intermittent loop closure in the basal mode: computer simulation studies with a diabetic model, *Diabetes Care*, 8(6): 553–561.
52. Ganda, O.P. et al. 1978. Reproducibility and comparative analysis of repeated intravenous and oral glucose tolerance tests, *Diabetes*, 27(7): 715–725.
53. Gawthrop, P.J. 1987. *Continuous-Time, Self-Tuning Control*, Research Studies Press, Letchworth, Herts., England.
54. Gibaldi, M. 1971. *Introduction to Biopharmaceutics*, Lea & Febiger, Philadelphia.
55. Godfrey, K. 1983. *Compartmental Models and Their Application*, Academic Press, London.
56. Goodwin, G.C. and K.S. Sin. 1984. *Adaptive Filtering, Prediction and Control*, Prentice-Hall, Englewood Cliffs, NJ.
57. Graupe, D. 1975. *Identification of Systems*, Krieger Publishing, Malabar, FL.
58. Greiss, L., N.A.G. Tremblay, and D.W. Davies. 1976. The toxicity of sodium nitroprusside, *Can. Anaesth. Soc. J.*, 23(5): 480–485.
59. Guyton, A.C. 1991. *Textbook of Medical Physiology*, 8th ed., W.B. Saunders, Philadelphia.
60. Hang, C.C., C.H. Tan, and W.P. Chan. 1980. A performance study of control systems with dead time, *IEEE Trans. Ind. Electron. Control Instrum.*, 27(3): 234–241.
61. Hardman, J.G. and L.E. Limbird, Eds. 1996. *Goodman & Gilman's The Pharmacological Basis of Medical Practice*, 9th ed., McGraw-Hill, New York.

62. Hultquist, P.A. 1988. *Numerical Methods for Engineers and Computer Scientists*, Benjamin/Cummings, Menlo Park, CA.
63. Isaka, S. and A.V. Sebald. 1993. Control strategies for arterial blood pressure regulation, *IEEE Trans. Biomed. Eng.*, 40(4): 353–363.
64. Isidori, A. 1985. The matching of a prescribed linear input–output behavior in a nonlinear system, *IEEE Trans. Autom. Control*, 30(3): 258–265.
65. Izzo, J.L. 1975. Pharmacokinetics of insulin. B. Degradation of insulin, in *Handbook of Experimental Pharmacology*, Hasselblatt and Bruchhausen, Eds., Springer-Verlag, New York.
66. Jacobs, O.L.R., R.E.S. Bullingham, et al. 1981. Feedback control of post-operative pain, *IEE Conf. Publ.*, 194: 52–56.
67. Jacobs, O.L.R., M.P. Reasbeck, R.E.S. Bullingham, and H.J. McQuay. 1982. On-line estimation in the control of post-operative pain, in IFAC Identification and System Parameter Estimation Conf. Proc., Washington, D.C.
68. Jacobs, O.L.R. et al. 1985. Modelling, estimation and control in the relief of post-operative pain, *Automatica*, 21(4): 349–360.
69. James, H.J., N.B. Nichols, and R.S. Phillips. 1947. *Theory of Servomechanisms*, McGraw-Hill, New York.
70. Jones, R.W. 1973. *Principles of Biological Regulation*, Academic Press, New York.
71. Kienitz, K.H. and T. Yoneyama. 1993. A robust controller for insulin pumps based on H-infinity theory, *IEEE Trans. Biomed. Eng.*, 40(11): 1133–1137.
72. Kitchell, R.L. and H.R. Erickson. 1983. *Animal Pain: Perception and Alleviation*, American Physiological Society, Bethesda, MD.
73. Kochenburger, R.J. 1950. A frequency response method for analyzing and synthesizing contactor servomechanisms, *AIEE Trans.*, 69: 270–284.
74. Kraegen, E.W. and L. Lazarus. 1973. Feedback control of blood glucose, in *Regulation and Control in Physiological Systems*, A.S. Iberall and A.C. Guyton, Eds., Instrument Society of America, Pittsburgh, 470–474.
75. Kraegen, E.W. et al. 1977. Control of blood glucose in diabetics using an artificial pancreas, *Aust. N.Z. J. Med.*, 7: 280–286.
76. Kuo, B.C. 1982. *Automatic Control Systems*, 4th ed., Prentice-Hall, Englewood Cliffs, NJ.
77. Layne, S.P., J.L. Spouge, and M. Dembo. 1989. Quantifying the infectivity of human immunodeficiency virus, *Proc. Natl. Acad. Sci. U.S.A.*, 86: 4644–4648.
78. Lewison, E.F., Ed. 1976. Conference on Spontaneous Regression of Cancer, DHEW Publ. No. (NIH) 76-1038, National Cancer Institute Monograph 44.
79. Li, C-C. 1961. Integral Pulse Frequency Modulated Control Systems, Ph.D. dissertation, Northwestern University Evanston, IL.
80. Lindorff, D.P. 1965. *Theory of Sampled-Data Control Systems*, Wiley, New York.
81. Linkens, D.A. 1992. Adaptive and intelligent control of anesthesia, *IEEE Control Systems*, December, pp. 6–11.
82. Liu, F-Y. 1988. Model and Control of Postoperative Pain, M.S. dissertation, University of Connecticut, Storrs.
83. Liu, F-Y. and R.B. Northrop. 1990. A new approach to the modeling and control of post-operative pain, *IEEE Trans. Biomed. Eng.*, 37(12): 1147–1158.
84. Ljung, L. 1987. *System Identification: Theory for the User*, Prentice-Hall, Englewood Cliffs, NJ.
85. Ljung, L. and T. Söderström. 1983. *Theory and Practice of Recursive Identification*, MIT Press, Cambridge, MA.

86. Maciejowski, J.M. 1989. *Multivariable Feedback Design*, Addison-Wesley, Reading, MA.
87. Mann, D.L., S. Gartner, et al. 1990. HIV-1 transmission and function of virus-infected monocytes/macrophages, *J. Immunol.*, 144(6): 2152–2158.
88. March, W.F., B. Rabinovitch, and R.L. Adams. 1982. Noninvasive glucose monitoring of the aqueous humor of the eye. II. Animal studies and the scleral lens, *Diabetes Care*, 5(3): 259–265.
89. Marmarelis, P.Z. and V.Z. Marmarelis. 1978. *Analysis of Physiological Systems*, Plenum Press, New York.
90. Maron, S.H. and C.F. Prutton. 1958. *Principles of Physical Chemistry*, MacMillan, New York.
91. Marshall, T.G., N. Mekhiel, W.S. Jackman, K. Perlman, and A.M. Albisser. 1983. A new microprocessor-based insulin controller, *IEEE Trans. Biomed. Eng.*, 30(11): 689–695.
92. Martin, J.F., A.M. Schneider, and N.T. Smith. 1987. Multiple-model adaptive control of blood pressure using sodium nitroprusside, *IEEE Trans. Biomed. Eng.*, 34(8): 603–611.
93. Martin, J.F., A.M. Schneider, M.L. Quinn, and N.T. Smith. 1992. Improved safety and efficacy in adaptive control of arterial blood pressure through the use of a supervisor, *IEEE Trans. Biomed. Eng.*, 39(4): 381–388.
94. Martin, J.F., N.T. Smith, M.L. Quinn, and A.M. Schneider. 1992. Supervisory adaptive control of arterial pressure during cardiac surgery, *IEEE Trans. Biomed. Eng.*, 39(4): 389–393.
95. Mathews, C.K. and K.E. van Holde. 1990. *Biochemistry*, Benjamin/Cummings, Redwood City, CA, 357.
96. McMullen, J.K. and C.E. Tindall. 1983. A new pulsed insulin delivery pump, in *Artificial Systems for Insulin Delivery*, P. Brunetti et al., Eds., Raven Press, New York, 115–118.
97. Meier, R., J. Nieuwland, et al. 1992. Fuzzy logic control of blood pressure during anesthesia, *IEEE Control Systems*, December, pp. 12–17.
98. Mendelson, Y. et al. 1990. Blood glucose measurement by multiple attenuated total reflection and infrared absorption spectroscopy, *IEEE Trans. Biomed. Eng.*, 37(5): 458–465.
99. Meyer, A.U. 1961. Pulse Frequency Modulation and Its Effect in Feedback Systems, Ph.D. dissertation, University of Illinois.
100. Milsum, J.H. 1966. *Biological Control System Analysis*, McGraw-Hill, New York.
101. Miron, D.B. 1989. *Design of Feedback Control Systems*, Harcourt Brace Jovanovitch, San Diego.
102. Monjanel, S., J.P. Rigault, et al. 1979. High-dose methotrexate: preliminary evaluation of a pharmacokinetic approach, *Cancer Chemother. Pharmacol.*, 3: 189–196.
103. Montague, C.T. et al. 1997. Congenital leptin deficiency is associated with severe early-onset obesity in humans, *Nature*, 387: 903.
104. Northrop, R.B. 1990. *Analog Electronic Circuits: Analysis and Applications*, Addison-Wesley, Reading, MA.
105. Northrop, R.B. and E.A. Woodruff. 1986. Regulation of a physiological parameter or *in vivo* drug concentration by integral pulse frequency modulated bolus drug injections, *IEEE Trans. Biomed. Eng.*, 33(11): 1010–1020.

106. Northrop, R.B., X-Z Liu, and Q. Li. 1989. A modelling study of human immune deficiency disease, in *Proc. 15th Annu. Northeast Bioeng. Conf.*, S. Buus, Ed., IEEE Press, New York, 171–172.
107. Ogata, K. 1970. *Modern Control Engineering,* Prentice-Hall, Englewood Cliffs, NJ.
108. Ogata, K. 1990. *Modern Control Engineering,* 2nd ed., Prentice-Hall, Englewood Cliffs, NJ.
109. Owens, D.H. 1981. *Multivariable and Optimal Systems,* Academic Press, London.
110. Pajunen, G.A., M. Steinmetz, and R. Shankar. 1990. Model reference adaptive control with constraints for postoperative blood pressure management, *IEEE Trans. Biomed. Eng.*, 37(7): 679–687.
111. Pauza, C.D. 1988. Commentary: HIV persistence in monocytes leads to pathogenesis and AIDS, *Cell. Immunol.*, 112: 414–424.
112. Pavlidis, T. 1964. Analysis and Synthesis of Pulse Frequency Modulation Feedback Systems, Ph.D. dissertation, University of California, Berkeley.
113. Petre, J.H., D.M. Cosgrove, and F.G. Estefanous. 1983. Closed loop computerized control of sodium nitroprusside, *Trans. Am. Soc. Artif. Organs*, 29: 501–505.
114. Pfeiffer, E.F. and W. Kerner. 1983. Comparison of the ability of glucose-controlled and preprogrammed insulin delivery systems to achieve normoglycemia, in *Artificial Systems for Insulin Delivery*, P. Brunetti et al., Eds., Raven Press, New York, 573–583.
115. Rabinovitch, B., W.F. March, and R.L. Adams. 1982. Noninvasive glucose monitoring of the aqueous humor of the eye. I. Measurement of very small optical rotations, *Diabetes Care*, 5(3): 254–258.
116. Reasbeck, M.P. 1982. Modelling and Control of Postoperative Pain, Ph.D. dissertation, University of Oxford.
117. Reibnegger, G., D. Fuchs, et al. 1987. Theoretical implications of cellular immune reactions against helper lymphocytes infected by an immune system retrovirus, *Proc. Natl. Acad. Sci. U.S.A.*, 84: 7270–7274.
118. Reid, J.A. and G.N.C. Kenny. 1987. Evaluation of closed-loop control of arterial pressure after cardiopulmonary bypass, *Br. J. Anaesth.*, 59: 247–255.
119. Shah, S.A. 1990. An Adaptive, IPFM/SDC Controller for Regulating Post-Operative Blood Pressure, M.S. dissertation, University of Connecticut, Storrs.
120. Shah, S., E.A. Woodruff, and R.B. Northrop. 1988. An adaptive IPFM/SDC controller for the closed-loop sodium nitroprusside regulation of post-operative blood pressure, in *Proc. 10th Annu. IEEE/EMBS Conf.*, New Orleans, November 4 to 7, 0519–0520.
121. Sheppard, L.C. 1979. Computer-controlled infusion of vasoactive drugs in post cardiac surgical patients, in *Proc. IEEE Conf. Eng. in Medicine & Biology*, Denver, 280–284.
122. Sheppard, L.C. 1981. Computer control of the infusion of vasoactive drugs, *Ann. Biomed. Eng.*, 8: 431–444.
123. Sheppard, L.C. and B. McA. Sayers. 1977. Dynamic analysis of the blood pressure response to hypotensive agents, studied in postoperative cardiac surgical patients, *Comput. Biomed. Res.*, 10: 237–246.
124. Sherwin, R.S. et al. 1974. A model of the kinetics of insulin in man, *J. Clin. Invest.*, 53: 1481–1492.

125. Shimauchi, T., N. Kugai, N. Nagata, and O. Takatani. 1988. Microcomputer-aided insulin dose determination in intensified conventional insulin therapy, *IEEE Trans. Biomed. Eng.*, 35(2): 161–171.
126. Slate, J.B. 1980. Model-Based Design of a Controller for Infusing Sodium Nitroprusside During Postsurgical Hypertension, Ph.D. dissertation, University of Wisconsin.
127a. Slate et al. 1979. *IEEE EMBS Proc.*, Session 13.
127. Slate, J.B. and B.S. Sheppard. 1982. Automatic control of blood pressure by drug infusion, *IEE Proc.*, 129(9)A: 639–644.
128. Smith, O.J.M. 1959. A controller to overcome dead time, *ISA J.*, 6: 28–33.
129. Smolen, V., R. Barile, and D. Carr. 1979. Design and operation of a system for automatic feedback-controlled administration of drugs, *Med. Dev. Diagn. Ind.*, 1: 55.
130. Stark, L. 1968. *Neurological Control Systems*, Plenum Press, New York.
131. Stoelting, R.K. 1979. Attenuation of blood pressure response to laryngoscopy and tracheal intubation with sodium nitroprusside, *Anesth. Analg.*, 58(2): 116–119.
132. Talbot, S.A. and U. Gessner. 1973. *Systems Physiology*, John Wiley & Sons, New York.
133. Tamborlane, W.V. et al. 1979. Reduction to normal of plasma glucose in juvenile diabetes by subcutaneous administration of insulin with a portable insulin pump, *New Engl. J. Med.*, 300(11): 573–578.
134. Tinker, J.H. and J.D. Michenfelder. 1976. Sodium nitroprusside: pharmacology, toxicology and therapeutics, *Anesthesiology*, 45(3): 340–354.
135. Tomović, R. 1966. *Introduction to Nonlinear Automatic Control Systems*, Wiley, London.
136. Tompkins, W.J. and J.G. Webster. 1981. *Design of Microcomputer-Based Medical Instrumentation*, Section 3.2, Prentice-Hall, Englewood Cliffs, NJ.
137. Tourville, J. 1975. Sodium nitroprusside, *Drug Intell. Clin. Pharm.*, 9: 361–364.
138. Truxal, J.G. 1955. *Automatic Feedback Control System Synthesis*, McGraw-Hill, New York.
139. Tuffs, P.S. and D.W. Clarke. 1985. Self-tuning control of offset: a unified approach, *IEE Proc.*, 132(3): Part D, 100–110.
140. Voit, E.O. 1989. New nonlinear methodologies for modeling molecular and cellular systems, in *Modelling and Control in Physiological Systems*, C. Cobelli. and L. Mariani, Eds., IFAC/Pergamon Press, Oxford, 217–228.
141. West, J.B., Ed. 1985. *Best & Taylor's Physiological Basis of Medical Practice*, 11th ed., Williams & Wilkins, Baltimore.
142. Wise, D.L. 1989. *Applied Biosensors*, Butterworths, Boston.
143. Woodruff, E.A. 1989. The Closed-Loop Regulation of the Blood Glucose Concentration in Type 1 Diabetics When Sparse Glucose Measurements Are Available, Ph.D. dissertation, University of Connecticut, Storrs.
144. Woodruff, E.A. and R.B. Northrop. 1987. Closed-loop regulation of a physiological parameter by an IPFM/SDC (integral pulse frequency modulated/Smith delay compensator) controller, *IEEE Trans. Biomed. Eng.*, 34(8): 595–602.
145. Woodruff, E.A., S. Gulaya, and R.B. Northrop. 1988. The closed-loop regulation of blood glucose in diabetics, in *Proc. 14th Annu. Northeast Bioeng. Conf.*, J.R. LaCourse, Ed., IEEE Press, New York, 54–57.
146. Woodruff, E.A., S. Shah, and R.B. Northrop. 1988. A discussion of the performance of the Smith delay compensator for systems with appreciable time delay, in *Proc. 10th Annu. IEEE/EMBS Conf.*, New Orleans, November 4 to 7, 0505–0506.

147. Yates, F.E., D.J. Marsh, et al. 1973. Modelling metabolic systems and the attendant data handling problem, in *Regulation and Control in Physiological Systems,* A.S. Iberall and A.C. Guyton, Eds., Instrument Society of America, Pittsburgh, 464–468.
148. Zaharko, D.S., R.L. Dedrick, et al. 1971. Methotrexate tissue distribution: prediction by a mathematical model, *J. Natl. Cancer Inst.,* 46: 775–784.

WEB SITES

Note: These references are ephemera; that is, they exist until someone updates or removes them from the host server. They have short-term archival value.

Leptin

149. http://server.stedwards.edu/contrib/Chemistry/CHEM43/Leptin/REGCNTRL.HTML
150. http://www.rndsystems.com/search/mfs/02/cb/cbsu96/cbsu96a2.html
151. http://arbl.cvmbs.colostate.edu/hbooks/pathphys/digestion/pregastric//ob_gene.html

Immune System

152. http://www.savba.sk/logos/books/scientific/node21.html (*1.3.4 Macrophages and Monocytes.*)
153. http://www.merck.com/!!vl1VU25Jrvl1XH2ZoM/pubs/mmanual/html/mkiidjdc.htm (*The Merck Manual,* Section 2-18, Biology of the Immune System.)
154. http://bioscience.org/1996/v1/d/cohn1/htmls/5.htm (Cohn, M. and R.E. Langman. 1996. The immune system: A look from a distance.)
155. http://panther.acp.edu/web/studproj/antibody/isotypes.htm
156. http://edcenter.med.cornell.edu/CUMC_PathNotes/Immunopathology/Immuno_01.html (Mellors, R.C. 1998. Immunopathology: normal immune system.)
157. http://www.bio.cam.ac.uk/dept/path/teaching/ittmac1/lectures/Lec10/lec10.html (Part 1B Pathology: Lecture 10, *The Complement System,* 1998.)
158. http://ulmo.stud.slu.se:8001/~b6stiast/saker/cb2/stuff/102.htm (On MHC molecules.)
159. http://www.ultranet.com/%7Ejkimball/BiologyPages/H/HLA.html (On MHC molecules and antigen presentation.)
160. http://gcrc1.pci.upmc.edu/~davis/IVV/Mar96/ifna.html (*In Vivo Veritas,* Cytokine of the month: interferon alpha.)
161. http://www.ifmss.org.uk/publics/msmgmt/nov96/reingold.htm (Reingold, S.C. 1996. *The Future with Interferons.*)
162. http://www.critpath.org/aric/library/art006.htm (Duke, R.C. and J.D. Young. 1998. Cell suicide in health and disease.)
163. http://cobra.cabm.rutgers.edu/~white/review.html (White, E. 1998. Overview on apoptosis.)
164. http://www.oncolink.upenn.edu/speciality/chemo/support/hemat1a.html (Yee, G.C. and D.L. Stanley. 1994. Highlights on antineoplastic drugs [specifically, IL1, IL3, IL6, PIXY321 & IL11].)

165. http://www2.igh.cnrs.fr/bioscience/1997/v2/d/feghali1/htmls/list.htm (On cytokines, 1997.)

HIV AND AIDS

166. http://www.critpath.org/aric/library/img002.htm (Color image of innards of HIV virion.)
167. http://www.critpath.org/aric/library/art004.htm (Harden, V. and P. D'Souza. 1996. Chemokines and HIV second receptors: a short history of a recent breakthrough, *Nature Med.,* 2(12): 1293–1300.)
168. http://www.savba.sk/logos/books/scientific/node33.html (Slavkovsky, P. 1995. 1.4.7.1 Chemokines.)
169. http://histo.cryst.bbk.ac.uk/WWWFiles/cd_table.html (Table of CD antigens.)
170. http://cicr.pa.uq.edu.au/lect2.htm (McMillan, N. MY350 Lectures 12: HIV: a lesson in complexity.)
171. http://www.path.ox.ac.uk/wsj/hiv_rece.htm (James, W. 1995. The receptor for HIV.)
172. http://www.cdc.gov/ncidod/EID/vol3no3/smith.htm (McNicholl, J.M., D.K. Smith, et al. 1997. Host genes and HIV: the role of the chemokine receptor gene CCR5 and its allele [Δ32 CCR5].)

Appendix: Discussion of Simulation Languages for Physiological/Phamacokinetic and Chemical Kinetic Systems

1. INTRODUCTION

The earliest practical way to accurately simulate and predict the behavior of large, nonlinear dynamic systems was by *analog computer,* using op amps to sum, subtract, integrate, and filter analog voltages representing physical quantities such as chemical concentrations or the roll, yaw, and pitch angles of aircraft, ships, missiles, etc. Nonlinear operations were also done with the assistance of op amps. These included multiplication (and squaring) by quarter-square multipliers, division, square rooting, and general nonlinear transfer functions generated using biased diodes. Analog computers had their heyday from the late 1940s through the mid- 1970s. They were tedious to use because of the need to time and amplitude scale the voltages to remain within the linear operating regions and bandwidths of the op amps, multipliers, etc. One advantage of analog computation was that the output voltages could be viewed on a multitrace strip chart recorder or an oscilloscope, as they were computed. Accuracy seldom exceeded equivalent 8-bit.

Through the development of large mainframe digital computers in the 1960s and onward, there was a trend to solve simulation problems through digital computation. The efficiency of digital simulation, even though output graphics were crude by today's standards, eventually led to the demise of analog computing. IBM Corp. developed a FORTRAN-based, application-oriented, simulation language called Digital Simulation Language-90 (DSL-90) for its 7090 mainframe digital computers. When the IBM System-360 mainframe was developed, DSL-90 was modified to a version called S/360 CSMP (Continuous System Modeling Program). CSMP was then upgraded to have better graphics and other enhancements and was called CSMP III. In the mid-1980s, California Scientific Software adapted the batch-process, mainframe CSMP III to the IBM PC and its clones as Micro-CSMP©. The author first used Micro-CSMP, which in the era of 9-pin dot-matrix printers still had primitive graphics output, for modeling studies of physiological control systems.

Various other modeling programs written as PC applications began to appear in the late 1980s and early 1990s. These included Simnon™ and Protosim™. Most recently, Matlab®'s Simulink® has found wide acceptance among control engineers. The latter three applications are considered below.

2. SIMNON™

Simnon was developed at the Department of Automatic Control at the Lund Institute of Technology, Sweden in the late 1980s. The program in its early versions (V1.0 to V3.2) was written to run under DOS on PCs. The author found in 1988 that Simnon V2.0 was particularly well suited for simulation of compartmental pharmacokinetic systems, chemical kinetic systems, and physiological regulators and control systems because its input modality is in the form of algebraic ordinary differential equations which arise naturally from the analysis and modeling of these three classes of systems. Simnon is well suited to solve sets of stiff ordinary differential equations. The user has the choice of one of four integration algorithms: Runge–Kutta–Fehlberg 2nd/3rd, Runge–Kutta–Fehlberg 4th/5th, Euler, and Dormand–Prince 4th/5th. Solutions of sets of nonlinear ordinary differential equations by Simnon can be displayed in the time domain or parametrically as phase-plane plots. Simnon has quality graphics output on CRTs and laser printers. The author currently runs DOS Simnon V3.2 on a Pentium PC with the Windows NT® 4.0 operating system. Simnon V3.2 has color graphics which can be printed out as such with a suitable color printer. Simnon V3.2 costs about $750, and the student version costs $95. The Windows version of Simnon, Simnon/PCW 2.0, costs about $1045, and the student version costs $95. Simnon/PCW 2.0 handles up to 300 states and can handle up to 50 time delays and store up to 100 variables. With Simnon, it is possible to simulate a large, continuous, nonlinear dynamic system and at the same time simulate a discrete controller for that system.

A new windows version, Simnon/PCW 3.0, was available in late 1998 from SSPA Maritime Consulting (P.O. Box 24001, S-400 22 Göteborg, Sweden, e-mail simnon@sspa.se). Its price is ECU 99 (about US $122) from SSPA. Simnon/PCW 3.0 has many enhancements and new features. The CD includes a complete instruction manual for the program.

3. SIMULINK®

Simulink runs with Matlab; both are products of The MathWorks (Natick, MA). (As of November 1998, Simulink was at V2.2; Matlab and Simulink run on PCs with Windows 95, 98, or NT). Simulink is an icon-driven, dynamic simulation package that allows the user to represent a nonlinear dynamic process as a block diagram. As the block diagram is built, the user has to specify numerical values for the parameters in the blocks and, of course, the interconnections between them. Before the simulation is run, the user specifies the integration routine to be used, the step size, and start and stop times. Selectable integration routines include R-K 23, R-K 45, Euler, Gear, Adams, and linsim (plain vanilla for purely linear state systems). Gear is recommended for stiff, nonlinear systems. The classroom version of Simulink, V2.0, apparently does not offer integrator options; presumably it uses linsim. Because Simulink runs in the Matlab "shell," it can make use of all of the many features of Matlab and its various toolboxes.

4. PROTOSIM™

ProtoSim is a product of Systems Engineering Associates, Inc. (Pittsfield, MA). Early ProtoSim V1.11 runs on PCs under DOS. ProtoSim also uses a block diagram input format similar to Simulink. In addition to the customary linear signal operations (e.g., integration, summing, subtraction, delays, multiplication by a constant, filtering, etc.), there is a rich library of nonlinear operations including trig functions, quantization, 0-crossing detection, various algebraic functions, multiplication, division, etc. Its output graphics are relatively poor in V1.11, compared with Simulink and Simnon. Like Simulink, ProtoSim is best suited to simulate small- to medium-sized nonlinear control systems, such as autopilots. The author does not recommend it for modeling compartmental pharmacokinetic or chemical kinetic systems.

5. SUMMARY

Simulation of physiological/biochemical systems and compartmental pharmacokinetic/physiological control systems places special requirements on the simulation application used because of the frequent need to handle stiff, nonlinear ordinary differential equations. In the author's experience, Simnon and Simulink are the best simulation programs for the applications above. Both programs are easy to use and have excellent output graphics.

Index

A

Active transport, 224
Adaptive controllers, 304
Adenyl cyclase, 218
Adrenocorticotropin (ACTH), 27
AIDS, 394
Aldosterone (ALDO), 233
Anesthesia, 127
Angiogenesis, 386
Antibodies, 360, Fig. 10.3
Antidiuretic hormone (ADH), 229
Anti-idiotypic antibodies, 380
Antigen presentation, 368
Apoptosis, 358, 381
Aqueous humor, 22
Arachidonic acid, 377
Artificial beta cell, 273
Artificial endocrine pancreas, 273
Aspirin, 380
Autacoids, 353

B

B-Cells, 360
Biostator, 257
Blood volume regulation, 232 (Figure 6.8)
Bolus injection pump, 279
Bowman's capsule, 222

C

Calcitonin (CATN), 239
Calmodulin, 218
Cancer, 385
Catenary compartmental model, 130
Characteristic impedance
    (of transmission line), 34
Chelated Ca^{++}, 237 (Figure 6.12)
Chemokines, 398
Cholecalciferol (CC), 239
Circle root locus, 64
Cluster of differentiation (CD) antigens, 424
Colloid osmotic pressure, 224
Compartment, 128
Compensation filter
    lag, 57
    lag/lead, 57
    lead, 57
Complement system, 362
Complexity, 1
Cyclic AMP, 218
Cytotoxic T-cells, 358

D

Delay, systems with, 107
Describing functions, 90, 154
D-Glucose, 248
Diabetes mellitus
    Type 1, 272
    Type 2, 272
Diapedesis, 377
Diffusion
    carrier-mediated, 38
    ligand-gated, 38
    voltage-gated, 39
Drug infusion controller, 156
Duffing's equation, 14
Dynorphins, 332

E

Eclosion (of HIV), 397
Eicosanoids, 377
Endorphins, 332
Enkephalins, 328, 332
Epitope, 358

F

Fas, Fas ligand, 382
Fc binding site, 362
Fentanyl, 331
Fick's law (diffusion), 40
Fick's constant, 40
Filtration fraction (kidney), 223
Final value theorem in z, 175
Flugge-Lotz controller, 202
Fluid capacitance, 32
Fluid inertance, 34
Fluid resistance, 32

G

Glomerular filtration rate (GFR), 223
Glucagon, 248, 251
Gluconeogenesis, 251
Glucose, 248
Glucose phosphatase, 251, 251
Glucose sensors, 273
Glucose tolerance test, 261
Glucose utilization, 254
Glycogen, 249
Granulocytes, 355

H

Hapten, 368
Hill function, 12, 255, 260
Homeostasis, 2
Hormones, 217
Hormone receptor, 217
Human immune deficiency virus (HIV), 394
Hydroxyapatite, 241
Hypercalcemia, 237
Hypocalcemia, 237
Hysteresis, 157 (Table 3.3, Figure 5.4)

I

Implicit summing point, 6, 219
Inflammation, 326, 365
Inulin, 227
Instantaneous frequency, 172
Insulin, 249
Insulin-sensitive cells, 247, 245

Integrase, 399
Interferons, 376
Interleukins, 372
Intermodulation distortion, 11
Interstitial fluid water volume (IFWV), 231
Intraocular pressure, 23
Intravenous glucose tolerance test
 (IVGTT), 261, 264
IPFM bolus controller, 170, 279

J

Jump resonance, 13
Juxtaglomerular complex, 224

K

Kidneys
 transport maxima, 225
 tubules, 224

L

Laminar flow, 31
Leptin, 249, 253
Limit cycle, 90, 187
Linear system, 8
Loop of Henle (kidney), 225

M

Macrophages, 356
Macula densa (kidney), 224
Major histocompatibility complex (MHC)
 Class I, 358 (Figure 10.1)
 Class II, 356, 358 (Figure 10.2)
Mammilary compartmental model, 130
Mass-action kinetics, 41
Mast cells, 365
Mean arterial pressure, 295
Megakaryocytes, 355
Membrane attack complex, 367
Memory B-cells, 362
Michaelis-Menton reaction, 42, 139, 341
Model-reference adaptive controller
 architecture (MRAC), 304
Model reference controller architecture
 (MRCA), 113

Models
   descriptive, 3
   explanatory, 3
   gray-box, 4
   isomorphic, 4
   predictive, 3
Modified z-transform, 174
Monocytes, 356
Multiple-model adaptive controller (MMAC), 304, 309

N

Natural killer (NK) cells, 363
Nephron, 222
Net hepatic glucose balance (NHGB), 256
Neuropeptide Y (NPY), 253
Non-insulin sensitive cells, 255
Nonlinear decoupling, 162
Nonlinear system, 11
Nyquist stability criterion, 77

O

ON/OFF controllers, 154
Oncogenesis, 385
Opiates, 330
Opioids, 330
Opsonization, 362
Osmolarity, 227
Osmoreceptor neurons, 229
Osmotic pressure, 228
Osteoblasts, 241
Osteoclasts, 241
Osmotic pressure, 228
Osteoblasts, 241
Osteoclasts, 241
Osteocytic membrane system, 241
Osteoporosis, 237

P

Pain
   neurological, 326
   psychological, 336
Pang, 337
Parameter-switching controller, 193
Parametric control, 15
Parathyroid hormone (PTH), 239
Patient-controlled analgesia (PCA), 325
Perforin, 358
Phase-plane
   portrait, 184
   trajectory, 185
Phosphorylase, 251
Physiology, 2
PI controller, 59
PID controller, 59
Platelets, 366
Platelet activating factor (PAF), 366
Pleiotropy, 354, 372
Popov's stability criterion, 102
Proportional infusion, 153
Prostaglandins, 377
Pseudo-random binary noise, 141, 298

R

Radial damping (RADD) controller, 195
RANTES, 399
Rate saturation, 12
Receptor proteins, 217
Reductionism, 1
Regulator, 2
Relaxation pulse frequency modulation (RPFM), 343
Renal fraction, 223
Retrovirus, 394
Reverse transcriptase, 399
Reynolds number, 32
Root locus
   angle criterion, 62, 67
   magnitude criterion, 62, 67
   plotting rules, 62
Routh-Hurwitz stability test, 74

S

Self-tuning regulators, 304
Semi-permeable membrane, 228
Separated stochastic controller, 339
Smith delay compensator (SDC), 113, 300
Sodium nitroprusside (SNP), 297
Soluble CD4, 399
Somatostatin, 253
Spontaneous regression of tumor, 386
Subharmonic generation, 12
Superposition, 8

Suppressor T-cells (Ts), 359
Switching boundary, 194
Switching controller, 193, 202
Switching law, 202

T

T-lymphocytes
  cytotoxic T-cells (CTL), 357
  helper T-cells (Th), 357
  suppressor T-cells (Ts), 359
Total water volume (TWV), 231
Trajectory, 185
Transport maximum (kidney), 225
Tubular load (kidney), 225
Tumor necrosis factor (TNF), 376
Type 0 system, 56
Type 1 system, 56

U

Uridine diphosphate glucose (UDG), 249

V

Vasopressin, 229
Viscosity, 32

W

Washout curve, 140, 191

Z

z,m-transforms, 174
Zona glomerulosa (adrenal cortex), 234

DATE DUE

DUE DATE SUBJECT TO CHANGE
IF A RECALL IS REQUESTED

OCT - 8 2003	
Rtnd-TO	JUN 1 9 2003